Human Paleontology and Prehistory

Vertebrate Paleobiology and Paleoanthropology Series

Edited by

Eric Delson
Vertebrate Paleontology, American Museum of Natural History
New York, NY 10024, USA
delson@amnh.org

Eric J. Sargis
Anthropology, Yale University
New Haven, CT 06520, USA
eric.sargis@yale.edu

Focal topics for volumes in the series will include systematic paleontology of all vertebrates (from agnathans to humans), phylogeny reconstruction, functional morphology, Paleolithic archaeology, taphonomy, geochronology, historical biogeography, and biostratigraphy. Other fields (e.g., paleoclimatology, paleoecology, ancient DNA, total organismal community structure) may be considered if the volume theme emphasizes paleobiology (or archaeology). Fields such as modeling of physical processes, genetic methodology, nonvertebrates or neontology are out of our scope.

Volumes in the series may either be monographic treatments (including unpublished but fully revised dissertations) or edited collections, especially those focusing on problem-oriented issues, with multidisciplinary coverage where possible.

More information about this series at http://www.springer.com/series/6978

Human Paleontology and Prehistory

Contributions in Honor of Yoel Rak

Edited by

Assaf Marom

Department of Anatomy and Anthropology, Sackler Faculty of Medicine, Tel-Aviv University, Tel-Aviv, Israel

Erella Hovers

Institute of Archaeology, The Hebrew University of Jerusalem, Jerusalem, Israel

 Springer

Editors
Assaf Marom
Department of Anatomy and Anthropology
Sackler Faculty of Medicine
Tel-Aviv University
Tel-Aviv
Israel

Erella Hovers
Institute of Archaeology
The Hebrew University of Jerusalem
Jerusalem
Israel

ISSN 1877-9077   ISSN 1877-9085   (electronic)
Vertebrate Paleobiology and Paleoanthropology Series
ISBN 978-3-319-83553-2   ISBN 978-3-319-46646-0   (eBook)
DOI 10.1007/978-3-319-46646-0

Cover Illustration: Amud Cave during the 1992 excavation season. (*Photo Erella Hovers*)

Printed on acid-free paper

This Springer imprint is published by Springer Nature
The registered company is Springer International Publishing AG
The registered company address is: Gewerbestrasse 11, 6330 Cham, Switzerland

Dedicated to Yoel Rak on the occasion of his 70th birthday, from his many friends and colleagues

Yoel Rak and Skull 5 from Dmanisi (*Photo* Avishag Ginzburg)

Foreword

On a Personal Note

I am grateful to Erella Hovers and Assaf Marom for the invitation to jot down these thoughts and to Erella for playing back many shared memories of our years of friendship with Yoel.

I first met Yoel Rak in 1978, shortly after he began his graduate studies in Clark Howell's lab at UC Berkeley. At that time I was a graduate student at Kent State University and research assistant to Don Johanson in the Cleveland Museum of Natural History's physical anthropology lab. Clark called me one day to ask if I could host a student who was going to be visiting the lab to study the Hadar fossils of the newly minted species *Australopithecus afarensis*, which were then on loan for research from the Ethiopian government. "He's Israeli and doesn't speak English all that well," Clark explained in his inimitable matter-of-fact way, "but he's interested in the skull, so you two will get along fine." And so we did, and thus was born a lasting professional partnership and deep friendship that I am proud to celebrate on the occasion of Yoel's retirement from Tel Aviv University.

Yoel's dissertation was on the facial structure of *Australopithecus* – he was (and is) fascinated by the unusual anatomy of the robust australopiths. His research (published as *The Australopithecine Face*, Academic Press, 1983) displayed a connoisseur's appreciation of morphology that has become his trademark: meticulous attention to detail and an unsurpassed ability to describe it beautifully in prose and synthesize it in visually arresting graphics. (More than 30 years later, I still insist that my graduate students read Yoel's book as a hedge against the rush to 3-D digitizing, which, as Yoel would say, "misses the morphology between the measurement points!") His work revealed previously unappreciated distinctions between southern and eastern African "robust" australopiths and detected in the face of *Australopithecus africanus* unique morphological ties to *Australopithecus robustus* (*cognoscenti* will recall the buzz around the "anterior pillar"), which supported the still-fresh Johanson-White phylogenetic proposal.

Spurred by the discovery of the Kebara skeleton in 1982, Yoel turned his sharp eye to the Neanderthals, in whose unique skull and pelvic morphology he found compelling evidence for a deep-rooted phylogenetic separation from modern humans. (As the physical anthropologists on the Kebara project divided up the new skeleton for study, Yoel was offered the pelvis, at that point the only complete and undistorted Neanderthal specimen. "After everybody got the part they wanted, they gave me the *tuchus*," Yoel used to say, "but it was the weirdest, most interesting part of the skeleton.") Fascinated by the mechanical workings of morphology, Yoel always seeks functional explanations for apparently anomalous anatomy on the wayward branches of our family tree and then weaves these into sharply etched evolutionary scenarios. Yoel's collaboration with Bill Hylander on the biomechanics of the Neandertal mandible is one recent outcome of this perspective.

Yoel and I have worked together in the field since the early 1990s, in both Israel (Amud Cave) and Ethiopia (Hadar). I think one of his proudest professional moments was his discovery of the long-anticipated complete skull of Lucy's species, *A. afarensis*, at Hadar in 1992. Up to that time, most of our knowledge about the *A. afarensis* skull had come from fragments found in the 1970s at the A.L. 333 locality; recovering an intact skull was one of our top priorities during renewed field work in the 1990s.

I vividly recall Yoel's excitement at the moment of discovery. We had been out on survey one afternoon when he came across two small fragments of a hominin subadult's occipital bone in a gully at the base of an outcrop. Suspecting that additional pieces might be lying on the adjacent hillside, he climbed up to find skull fragments clumped in a recently cut rivulet. (These, it turned out, were from a second individual, fully adult, that became known as the "first" adult skull of *A. afarensis*, A.L. 444-2.) I was sieving at a hominin locality nearby when I heard excited shouts from several of our team's local Afar fossil collectors—always a signal of an important discovery. I headed toward the uproar, and rushing around an outcrop, ran head-on into Yoel, who was coming the other way. "I can't believe it, I found a f—ing skull," he cried (his colloquial English had improved by then). When we reached the locality, climbing the outcrop, we saw pieces of the maxilla and cranial base poking out of the gully infill, with other vague shapes just visible beneath the colluvial crust. It took weeks of hard work to extract all of the pieces of the skull from the sediment, and, after months of cleaning and reconstruction in the National Museum in Addis Ababa, we had assembled the skull of a huge adult male of *A. afarensis*.

Yoel holds strong, sometimes controversial, opinions about the fossil record, but he is humble about his knowledge and revels in new discovery. Our study of the *A. afarensis* skull, published in a 2004 monograph (with Don Johanson), was the result of an intense period of collaboration. We argued and we cajoled, but not a day passed without my learning something new. Joining Yoel at a table full of fossils is an amazing learning experience, and I can honestly say that, still, after working together all these years, I am stunned by his observational skills and insights. The shape of the *foramen rotundum*? Well, of course. The times we've shared with the fossils are the most cherished of my career.

The practicalities of fieldwork were a challenge and great fun for Yoel. Always thinking out of the box, he would come up with novel ideas on setting up the field camp more efficiently. His original design for our field showers remains the Hadar camp standard. And his mechanized pulley system for transporting field gear up the near-vertical, 30-m-high wall of the Nahal Amud made our excavations at Amud Cave feasible.

One of his more memorable, though short-lived, innovations dealt with survey under the hot Afar sun and how to make it, well, less hot. The incongruous sight of a large red and white beach umbrella moving fitfully across the barren landscape is etched in the memories of those of us who were fortunate enough to witness it (Erella, despite being brought to tears from laughter, managed to capture the moment in the accompanying photo).

As anyone who spent any time with Yoel knows, he is an entertaining story-teller. For him, great stories reside in the minutiae of everyday life, which makes him a great dinner (and breakfast, and lunch…) companion. Meals with Yoel are always memorable occasions because of the humor and warmth he spreads around the table. It helps that he cherishes good food (seafood and Asian cuisine being favorites) and wine (Zinfandel, especially), and he is himself an accomplished cook who loves to feed friends and family. I still recall with a chuckle a dinner many years ago in a wonderful Chinese restaurant in Cleveland. Yoel was particularly taken with the Mu Shu Pork and asked the proprietor for the recipe, assuring her, in thickly accented English, that, being from Israel, he would not divulge it to a competitor. "Sorry, Missouri is not far enough," she replied, parrying his appeal to geographical isolation.

The eldest son of Polish Holocaust survivors (he was born in 1946, in a displaced-persons camp in Germany), Yoel bears a deeply humanistic philosophy. He is skeptical of all forms of "belief" and is harshly critical of political institutions. But he is an unalloyed optimist when it comes to the potential of individuals to improve the lives of those around them through learning and teaching (he holds the *Talmud* in high esteem, despite being a self-avowed atheist)—which doubtlessly will extend far into his retirement years. While those of us fortunate enough to be counted among his colleagues and friends will continue to enjoy the benefits of these years, it is already clear that his wife Ricka, his children Ariel, Benjamin and Carmi, and (so far) four grandchildren will be the main beneficiaries of his time away from the university. To this, I say *"L'Chaim*, Yoel!"

William H. Kimbel
Arizona State University

Preface

"A boring story"

For several decades now, medical students at the Sackler School of Medicine at Tel-Aviv University have been meeting Professor Yoel Rak primarily as their anatomy teacher in their first year. Still equipped with his drawing chalk and classical anatomical charts, and only seldom with PowerPoint presentations, Rak's method of anatomy gained much popularity over the years, whether he was using a model of an upside-down table to illustrate the basic framework of the sphenoid bone, or leading the students through the complexities of the peritoneal ligaments. As a professor of human anatomy, he always insisted that all bodily systems and organs should be equally presented in the anatomy course curriculum, not only the ones that have unequivocal clinical significance. His students' comments at the end of the course repeat themselves every year in stating that even the dullest parts of the material became fascinating topics, and that what they remember most of all are the evolutionary correlations Yoel frequently adds to his lectures. In doing so, he reminds his students that human anatomical structure is but one example of vertebrate anatomy and constantly encourages them to think beyond the definitions of their anatomy text-book. Naturally, it follows that during the long years he had served as Chair of the Department of Anatomy and Anthropology, he also insisted that anatomy should be taught to medical students mainly through dissection of the human cadaver. In his own words to first year students, "the efforts a student makes in the strenuous and often frustrating process of searching for an anatomical structure are at the heart of studying anatomy, far more than the mere act of identifying it" (AM, pers. obs.). According to Rak, there exists a reliable correlation between a student's knowledge of anatomy and how stained her laboratory coat is at the end of a dissection.

Along these lines, the physician *sensu stricto* may be thought of as a highly trained technician, who must memorize protocols for diagnosis and treatment. The physician *sensu lato* is an educated and well-informed professional, equipped with thorough knowledge of all the basic sciences, anatomy included (Marom and Tarrasch 2015). Yoel's pedagogical principles, well expressed in both his lectures and active participation in curricular committees, most certainly adhere to the latter. His approach to anatomical studies has made it feasible for scores of medical students to study, but also to love studying anatomy, echoing some beloved phrases from Anton Chekhov's "A Boring Story", narrated by no other than an anatomy professor: "And we proceed in the following order: in front walks Nikolai with the slides or atlases, I come after him, and after me, his head humbly lowered, strides the cart horse; or else, if necessary, a cadaver is carried in first, after the cadaver walks Nikolai, and so on. At my appearance, the students rise, then sit down, and the murmur of the sea suddenly grows still. Calm ensues.[1]"

---

[1] A Boring Story: From the Notebook of an Old Man. In *Selected Stories of Anton Chekhov* (R. Pevear & Larissa Volokhonsky, translators) Modern Library, New York. 2000.

A Matter of Character

Scholars of today's academic world are highly specialized: whether their research interests are the three dimensional architecture of a trans-membrane channel or the effect of a drug on gene expression, they keep refining their line of work. In the face of a rapidly increasing rate of knowledge expansion, this tendency may very well be an unavoidable outcome, and arguably has many advantages. However, one disadvantage that must not be overlooked is the price the scientific world pays for this high level of specialization: the increasing dearth of scholars with broad-spectrum knowledge and the ensuing ability of integration within and between scientific disciplines. One study even demonstrates how scientists view broad projects as riskier and less important than deeper projects (Bateman and Hess 2015). To many students, beginning scholars and colleagues, Yoel Rak has always represented a rare example to the contrary. His broad knowledge of human anatomy and deep acquaintance with the fossil record, but also his vast knowledge of the animal world, geology, geography, and archaeology, has allowed him to explore almost any part of the skeleton: from ear ossicles and the osseous labyrinth to the pelvic girdle; from the finest details of dental anatomy to the facial masks of the robust australopiths and the Neanderthals. To put it in Hennigian terms (Cracraft and Eldredge 1979), connecting the dots of Rak's professional accomplishments reveals the central motif of his search for characters, character states, and their positions along the morphoclines of each trait. In Rak's view, every character counts (and is legitimate to use) in the process of reconstructing phylogenies. In the debate between the anagenetic and cladogenetic views of evolution, even the *adaptive values* conferred by anatomical structures – which are outcomes of one's understanding of said anatomy – were recruited by Rak *as characters* of no less significance than the anatomical traits themselves (Marom 2013). [The squamosal suture of *Australopithecus boisei* (Rak 1978) or the Neanderthal face (Rak 1986), are cases in point.] Rak's appreciation of genetic evidence in the form of DNA sequences – as additional albeit discrete characters, that are by no means superior to anatomical traits – is probably the best example for this argument. In other words, his loyalty to the scientific "game rules" and his natural curiosity are the two main driving forces behind the process of his scientific thought and academic achievements. This is in fact what enabled him to present mandibular evidence in support of the premise that *Homo neanderthalensis* does not play a role in modern human ancestry (Rak 2002). Along the same lines, he later pointed to gorilla-like anatomy in *Australopithecus afarensis*, suggesting an evolutionary link between *Au. afarensis* and the robust australopiths (Rak et al. 2007). In both cases this was done elegantly by employing the rules of cladistic analysis and the logic of parsimony. Rak's interest in characters and their status along the primitive-derived axis should be viewed here as stemming from the high potential he ascribes to them in falsifying (or corroborating) a suggested phylogeny.

What's in the Book?

In early 2013, we embarked on the project of editing this volume, thinking that the occasion of Yoel's 70th birthday would be an excellent opportunity to celebrate his numerous achievements in science (and wish him many more), as well as to the many friendships he has struck throughout the years. This volume presents a collection of original papers contributed by many of Yoel's friends and colleagues from all over the globe, many of whom have collaborated with their students, thus keeping the flame burning, so to speak. The papers in this volume touch upon diverse ways of thinking about human evolution. Many of these approaches are among the topics that Yoel has been studying during his productive career.

The papers fall roughly into three broad categories: Reflections on some of the broad theoretical questions of evolution, and especially about human evolution; the early hominins, with special emphasis on *Australopithecus afarensis* and *Paranthropus*; and the Neanderthals, that contentious group of our closest extinct relatives. Within and across these categories, nearly every paper addresses combinations of methodological, analytical and theoretical questions that are pertinent to the whole human evolutionary time span.

Three chapters are concerned with the history of research, changing ideas about human evolution and the mechanisms that drive it. Wool (2017) reviews influential evolutionary ideas, and their particular applications to the evolution of humans to the effect that human evolution was once deemed unique compared to other species. He reiterates [as does Newman (2017) in his historical review of Darwin's and Wallace's personal and scientific interactions] Wallace's understanding of the unique evolution of humans due to "glorious qualities which raise us so immeasurably above our fellow animals" (Wallace 1891:190). It was Wallace's opinion that this process could not have occurred by blind natural selection. Some sixty years later, Dobzhansky (1962:199) promoted "intelligence, ability to use linguistic symbols, and culture which man has developed" as the leading forces in "a whole new evolutionary pattern...which is human rather than animal". Notably, in this later view the novel evolutionary pattern was formed by the combined effects of nature and nurture rather than by the invisible power that had been Wallace's driving force. Dobzhansky (1955, 1962) raised the point that cultural evolution is faster and more efficient than biological evolution, as the pathways of information transfer can take many forms within ever-broadening audiences (due to technologies of information transmission such as writing, printing, and later the radio, television, and the internet).

Tattersall's (2017) contribution takes off from this very question when he asks why human evolution was so fast. The empirical archaeological record demonstrates that human evolution started off slowly and accelerated only over the last 100,000 years. (Notably, many archaeologists may place this acceleration at somewhat earlier or later dates.). Indeed, Holloway (2017) reiterates the anatomical evidence from the *Australopithecus* endocasts that speaks to the slow pace of brain evolution in early hominins. Tattersall suggests that around 100,000 years ago the human brain switched to a different, more efficient processing algorithm, based on a connection-sensitive brain rather than sheer brain tissue volume. While the evolutionary mechanisms underlying this acceleration in human evolutionary tempo are not exclusive to this lineage and do not require that special evolutionary mechanisms be invoked in explanation, material culture must be incorporated as part of the evolutionary process in order to explain the change of pace. The operation of genes and culture in tandem, combined with the demographic structure and spatial spread of human populations, were the forces enabling the implementation and spread of novel cultural ideas and practices that we refer to as inventions. The resulting ratchet effect propelled the evolution of the genus *Homo*, specifically in *Homo sapiens*, inevitably leading to increasing levels of complexity and symbolic information processing.

Finally, Wool (2017) draws attention to a different aspect – both exciting and somber – of humans' rapidly increasing ability for cultural evolution. Galton's nineteenth century eugenics aspirations to direct human evolution have become that much more feasible with the great advances made since the mid-twentieth century in understanding the cogs and wheels of hereditary mechanisms. It is now that much (some would say, too much) easier to actively shape the direction of many biological, cognitive and social aspects of future human evolution.

A number of papers in the volume explore paradigmatic ideas about the ways in which modern research approaches key elements of the evolutionary process. Holloway's (2017) chapter is a cautionary tale about methodology and analytical tools. He reflects on state-of the-art techniques of studying early hominin enodcasts and on the contribution of such techniques to our knowledge, emphasizing the limitations of paleoneurology and the ever-growing necessity for comparative data regarding the human and ape neuroanatomical patterns. He concludes that, when applied to the early members of the hominin lineage, novel techniques of segmentation and reconstruction do not necessarily add new information or resolve old controversies, albeit they may be more informative with regard to later hominin species such as the Neanderthals. Within the framework of their study of the robust and hyper-robust hominins of the Early Pleistocene, Wood and Schroer (2017) question the ability of hard tissue morphology to recover phylogenetic relationship, and raise doubts about the assumption of non-independence of some of the traits often used in cladistics analysis.

Several papers raise questions about the epistemological implications of studies of eco-logical constraints, dietary preferences/restrictions, food-getting behaviors and life histories of early hominins. Harrison (2017) discusses Laetoli, where a number of faunal proxies lead to contrasting understanding of the habitats populated by *Paranthropus boisei* versus earlier ones in which *Au. afarensis* fossils were found. Harrison uses this case study to ask whether there is a direct cause-and-effect relationship between the appearance of a new hominin species and ecological change in a region of mosaic environments. The Laetoli analysis seems to demonstrate that such is not the case, although Harrison does not rule out the possibility that when studied over larger geographic regions, environmental changes may show such correspondence.

Cartmill and Brown (2017) consider in Chap. 6 one of the defining traits of the hominin lineage, assessing empirically hypotheses that link the emergence of hominin bipedality with feeding behavior. Their study of the anatomy of the gerenuk, an African gazelle that feeds bipedally, indicates that its post-cranial anatomy shows only few of the apomorphies expected to occur given the animal's feeding habits. Importantly, they do not recognize the expected diagnostic morphologies of bipedality in early hominins (*Ardipithecus* or *Australopithecus*). To them, this result argues (albeit not conclusively, in the absence of decisive fossil material) against the bipedal feeding hypothesis of hominin origins.

A number of papers in this volume contribute to the burgeoning paleoanthropological literature dedicated to mandibular morphology and dentition of the robust and hyper-robust australopiths, a group that has been a special favorite of Yoel's given their highly specialized cranial and dental morphologies. These papers ask how these traits inform us about feeding and dietary practices of robust and hyper-robust australopiths. Perhaps surprisingly, their results are not clear-cut but rather more speculative than one might expect given the current paradigm.

Wood and Schroer's contribution (2017) discusses the taxonomy and phylogeny of the megadont (*Paranthropus robustus*) and hyper-megadont hominins (*P. boisei* and geologically older *P. aethiopicus*), known from southern Africa (ca. 2.0–1.0 Ma) and eastern Africa (ca. 2.6–1.3 Ma), respectively. They accept that *P. boisei* and *P. robustus* are two separate species sharing a number of derived characters, based on the features of the mandibular bone and dentition of these hominins, yet they are concerned that these characters are not truly inde-pendent. Wood and Schroer favor homoplasy of these features in the eastern and southern robust hominins as a valid possibility because various studies of the dentition of the two species did not yield results that would be expected under the hypothesis of monophyly.

The papers by Hylander (2017); Daegling and Grine (2017); and Glowacka et al. (2017) speak to this very point when discussing dietary habits, ontogenetic processes and biome-chanical forces that affect some or all the components of the masticatory system of early hominins. The implication is that such components cannot be treated as independent traits in cladistic analyses.

Current views on the diet of *P. boisei* are discordant, due to the differing inferences from tooth wear analyses and stable isotope data (from which absence of hard objects and inclusion of sedges and grasses in the diet were inferred) on the one hand, and biomechanical models and comparative studies of dental enamel (indicating specialization for hard-object feeding) on the other. In Chap. 9, Daegling and Grine (2017) ask whether the jaws of *P. boisei* can be interpreted as functionally consistent with a herbivorous diet with only an insignificant component of hard foodstuffs. They find that a "massive" mandible may represent a structural solution to either forceful biting and chewing, persistent and prolonged mastication, or both. In the case of *P. boisei*, it is plausible that the bulk processing of low quality fibrous foods was the target of natural selection in this lineage.

Hylander (2017) focuses on the vertical shortening of canines, a characteristic that makes an early appearance in hominins, and evaluates the premise that canine reduction has an impact on the mechanical efficiency of the masticatory apparatus, in particular due to its relation with the measure of gape. Based on a large catarrhine database, Hylander attempts to understand the links among gape, jaw length and canine overlap. Interestingly, reduced sexual dimorphism is associated with relatively smaller (in humans) and larger (in hylobatids) gapes, and there are considerable differences in the amount of gape relative to jaw length on the interspecific level. Hylander concludes that a major benefit for canine height reduction in early hominins was functionally linked to increased mechanical efficiency of the jaws, and he demonstrates that gape can be predicted by the independent variables of projected jaw length and canine overlap. While he hypothesizes that the driving force for canine height reduction is gape reduction, he remains cautious about the ultimate reason for this change, because this largely depends on the identification of the earliest hominins. Explanatory scenarios would differ if this role is assigned to *Australopithecus anamensis/afarensis* (in which case a hypothesis of dietary shifts is advocated), or to *Ardipithecus ramidus* [(in which case canine reduction was due to a combination of dietary shifts with social factors (e.g., mating patterns) and tool use (presumably invisible to the archaeological eye)].

Glowacka et al. (2017) focus on the importance of dynamic ontogenetic development of the mandibular corpus in *A. afarensis*, using an expanded sample from Hadar to investigate this question. They find that the pattern of mandibular corpus growth in *A. afarensis* is neither exactly human-like nor chimpanzee-like. In chimpanzees, slow canine formation and their late emergence in relation to the other permanent teeth affect the size and shape of the anterior corpus throughout most of mandibular ontogeny, and the final adult size is reached later in the individual's life. Yet differences between chimps, humans and *A. afarensis* are more subtle for later dental emergence stages, because the growth of teeth does not appear to influence corpus morphology throughout all of mandibular ontogeny or in all parts of the mandibular corpus. Glowacka et al. (2017) propose a mechanism to explain such ontogenetically changing relationship. Although they do not dismiss hypotheses suggesting that differences in feeding behavior may have led to differences in symphyseal form [thus converging to Hylander's (2017) discussion, although they invoke a different mechanism], they emphasize the need to consider the effect of mandibular ontogenetic growth in addition to the often-used biomechanics of the adult form.

Human evolutionary research within the time frame of the Middle and Late Pleistocene has undergone a revolution with the introduction of paleogenetic research on the Neanderthals [Pääbo (2014) is a personal account of this scientific revolution, which includes a list of many of the influential studies in the field], the identification of new hominin groups (Denisovans), and following the actual genetic, demographic, and geographic histories of hominins during the Late and even the Middle Pleistocene. This body of research has revealed a complex and rich record of dispersals, interbreeding and bottlenecks that has been hard to decipher from the fossil record alone. In this newly emerging research-scape, it is nearly too easy to forget that a fundamental requirement for these advanced analyses is the actual presence of fossils. Importantly, researchers have been able to retrieve ancient DNA from only a small fraction of the available fossils, even when dealing with relatively late extinct hominins such as the

Neanderthals or early moderns. Geographically, such fossils are constrained to Europe; in key regions such as Africa or the Levant fossils (and archaeology) are currently the only way to directly address questions about any aspects of human evolution. A number of chapters in this volume refer to the recent body of research on Neanderthal genetics [Bailey et al. (2017); Caspari et al. (2017)]. Still, most of the contributions in the group of papers that deal with the Middle and Late Pleistocene records focus on the actual fossils, using various methodologies and analytical tools to address questions of phylogeny, activity patterns and ecological adaptations. Some of these studies [Collard and Cross, (2017); Pearson and Sparacello (2017); Weinstein-Evron and Zaidner (2017); Frayer (2017)] raise – explicitly or implicitly – questions about the type of relationship that should be expected between fossil anatomy and the archaeological behavioral record.

Rightmire's (2017) null hypothesis states that the Middle Pleistocene Eurasian (Arago, Petralona, Sima de los Huesos, Zuttiyeh, Mauer) and African (Kabwe, Bodo, Elandsfontein) lineages represent paleodemes of one species, *Homo heidelbergensis*. He uses the relatively complete Bröken Hill (Kabwe 1) cranium from Zambia and Petralona cranium from Greece to test this hypothesis. Rightmire concludes from his detailed review of the anatomy, as well as of independent studies of scaling and of geometric morphometric properties of facial features, that there is little basis for distinguishing these mid-Pleistocene individuals. Additionally, neither Petralona nor Arago can be linked more closely to later European populations (i.e., Neanderthals) than can the African mid-Pleistocene hominins. Since the null hypothesis cannot be rejected, he suggests that Petralona and Kabwe be viewed as representatives of paleodemes of a single, widely spread evolutionary lineage, which split relatively late in the Middle Pleistocene. Such a conclusion is in broad agreement with the favored interpretation of an mtDNA analysis of the ~400,000 year-old Sima de los Huesos fossils (Meyer et al. 2014).

Collard and Cross (2017) revisit a consensual notion about the relationship between thermoregulation, body shape and body size in *Homo erectus* and Neanderthals, two groups believed to represent warm and cold adaptations, respectively. Their analysis is novel in that it looks at the effects of body segment differences in surface area, skin temperature, and rate of movement, in addition to the typically discussed whole-body thermoregulation. They complied data from published material on a sample consisting of Holocene modern humans, Pleistocene *H. sapiens*, Eurasian and African *H. erectus*, and Neanderthals, and used a series of equations to model the more complex parameters in their study. Admittedly their model simplifies past conditions, as the task of estimating the thermal responses of extinct, culture-using hominins has the potential to be extremely complex, especially given the less than ideal resolution of environmental backgrounds. With this caveat in mind, Collard and Cross (2017) find that whole-body and whole-limb heat loss estimates were consistent with the consensual notions for *H. erectus* and Neanderthals, and with the notion that there are thermoregulation-related differences in body size and shape within *H. erectus* and *H. neanderthalensis*. However, differences between the proximal and distal limb segments did not follow any particular trend. Thus, the immediate conclusion from their research is that the current consensus requires some modification: while the basic idea that thermoregulation influenced the evolution of body size and shape in *H. erectus* and *H. neanderthalensis* seems to hold, differences in limb segment size may not be linked to thermoregulation. Collard and Cross point out some future studies that may be useful for checking on "assumption errors" in their current study and for explanations of these unexpected results. Interestingly, they imply that such explanations may derive from cultural as well as biological factors.

Pearson and Sparacello (2017) also direct their attention to post-cranial remains. Against a background of competing hypotheses about the levels of robusticity of long bones in Southeast Asian Neanderthals and the early moderns of the Skhul-Qafzeh group, this contribution presents an evaluation of size-adjusted strength of the limbs in the two populations. The comparison with a suite of other groups of Pleistocene fossil hominins and populations of recent humans that differ in lifeways, geographic origin, and ecogeographic adaptations results in a mosaic pattern. In the lower limb, European and Southwest Asian Neanderthals resemble a diverse array of modern agriculturalists and intensive foragers that are generally active but not highly mobile over long distances. While disparities in the indices of Neanderthal humeri and radii may be a species-level characteristic of this group, the (well-documented) flattening of the radius mid-shaft is known from other populations sampled for this study and may well develop from intensive physical activity or activities. Of the repertoire of activities associated with Neanderthals on the basis of use-wear studies of lithics, the authors suggest that scraping rather than spear-thrusting could account for the patterns in the upper limb bones. In contrast, the people from Skhul and Qafzeh are quite distinct from recent samples but bear a degree of resemblance to Khoesan and Zulu males and females, Kebaran foragers (Ohalo 2), and also Amud 1. The bone shape and size-adjusted strength indices suggest that each of these groups had patterns of physical activity that did not place high or frequent mechanical demands on their upper limb. Pearson and Sparacello (2017) conclude that Neanderthals and the Skhul-Qafzeh humans seem to have had highly different lifeways, and that early moderns rather than Neanderthals may have faced an uncommon set of mechanical demands on their limbs. Having said that, Pearson and Sparacello (2017) caution that these differences need not necessarily reflect fixed species-level differences, given the considerable variation that exists among ethnographically documented foragers. They also remind us that archaeologists have long seen the different cultural remains of the Levantine Middle Paleolithic as a continuum of responses to environmental and ecological conditions rather than a dichotomy in behavior between the two Levantine Middle Paleolithic populations.

Weinstein-Evron and Zaidner (2017) report on the Middle Paleolithic site of Misliya Cave in Israel. Although no diagnostic human remains have been reported from this site, the authors link it to the debates about the origins of modern humans. The Middle Pleistocene chronology of the two cultural units in the site – the Lower Paleolithic Acheulo-Yabrudian and the early Middle Paleolithic (EMP) – provides a context for delineating the cultural developments that may attest to an important behavioral shift. Major collapses of the cave mask the actual boundary between the two cultural units, yet a robust TL chronology of the sequence places the boundary between them at around 250,000 years ago, in general accordance with the chronology of Tabun, Hayonim and Qesem Caves. Weinstein-Evron and Zaidner (2017) suggest that the marked technological break between the two cultural complexes could have been associated with the arrival of a new population in the Levant. They concur with previous accounts (e.g., Hovers and Belfer-Cohen 2013) that in terms of its lithic technology, toolkit composition and potentially also the settlement patterns in the EMP, this cultural phase differs from both the earlier Acheulo-Yabrudian and the later Late Middle Paleolithic (LMP), but whether the shift to the LMP is linked to the arrival of another dispersing population remains a moot point. Regardless of demographic changes (if there were any) the authors argue that in the majority of behavioral characteristics the EMP hominins of Misliya Cave were similar to their late Mousterian counterparts, even if their biological identity still eludes us. This archaeological case study raises again familiar questions (Lieberman and Bar-Yosef 2005; Hovers 2006; Hovers and Belfer-Cohen 2006) about the links between cultural and biological evolution and whether we should expect close correlations between these trajectories.

Tabun Cave is a site where such questions are most pressing, given its long and controversial Middle Paleolithic chronology and the suggestion that the two hominin specimens found within it – Tabun C1 and Tabun C2 – represent two different groups [e.g., Rak (1998) even though they had been found in – presumably – the same stratigraphic unit. In fact, Garrod (Garrod et al. 1937) herself mentions doubts about the stratigraphic origin of the Tabun C1 skeleton]. Other researchers assign this specimen to either the Neanderthals or modern humans. Harvati and Lopez (2017) tackle this problem through a 3-D geometric morphometric analysis of the Tabun C2 mandible in a comparative sample of 26 mandibles of Middle and Late Pleistocene *H. heidelbergensis*, Neanderthals, and *H. sapiens*. Despite the greater analytical rigor of this analysis, the results remain inconclusive in showing that the overall mandibular shape cannot be easily accommodated either within the Neanderthal or the early modern human range of variation. In their analysis, Tabun C2 does not group with the Neanderthals; it differs from Upper Paleolithic modern human specimens; and its relationship to early anatomical moderns of Skhul and Qafzeh is unclear. Thus its affinities with any of the Levantine Middle Paleolithic hominins remain unresolved. Harvati and Lopez (2017) found, to their surprise, that Tabun C2 fell closest to the much older Middle Pleistocene European specimens in their sample, presenting a mosaic of traits, possibly because the large size of the specimen contributes to its archaic morphology. Finally, the authors also consider the scenario that Tabun C2 represents a hybrid between Neanderthals and early modern humans, but they are concerned that their methods are inadequate to assess this possibility. Interestingly, hybridization in the late Middle–early Late Pleistocene has come to the forefront with the recent publication of ancient DNA analyses suggesting early interbreeding in the Levant and gene flow *from* early moderns to the (eastern) Neanderthals prior to the Late Pleistocene interbreeding showing a different direction of gene flow (Kuhlwilm et al. 2016). Given the complex statistics that led to this interpretation and the complexity of the suggested interbreeding processes on the Eurasian scene throughout this time frame, the applicability of this scenario to the Tabun C2 (or any other) specimen should be validated by future paleoanthropological and paleogenetic data.

Bailey et al. (2017) examine in their contribution whether a pattern of dental trait frequencies can be used to statistically distinguish *H. sapiens* from Neanderthals. Reviewing the recent literature on hybridization, they suggest that if Neanderthals and *H. sapiens* did interbreed extensively, one might expect to find morphological evidence of such admixture in their dentition, although no specific model is offered for how this might occur. They hypothesized that if an identifiable modern human dental pattern emerged early in our lineage, then in their analysis the earliest *H. sapiens* should classify predominantly as *H. sapiens*. If, on the other hand, the earliest *H. sapiens* are characterized by a primitive dental pattern, then their classification should be ambiguous. A second prediction is that if there had been a significant admixture event in Western Asia (as suggested by some paleogenetic studies), then a higher percentage of *H. sapiens* in Western Asia would be misclassified as *H. neanderthalensis* in comparison to Africa. Their results suggest that in most cases the predominance of primitive features, rather than derived Neanderthal traits, drove the classification. Bailey et al. (2017) find a strong modern signal at two of the earliest *H. sapiens* sites (Qafzeh and Skhul), which suggests that dental modernity appeared early in our lineage. They posit that this also argues against significant admixture between Neanderthals and *H. sapiens* in this region. However, marked heterogeneity in their African sample (independent of geographic distance) suggests that Late Pleistocene Africans were not a dentally homogeneous group, such that some populations appear to have retained higher frequencies of primitive characteristics than others. In the face of a moderate frequency of African material classifying as Neanderthal, Bailey et al. (2017) conclude their contribution with some methodological reflections regarding their method's ability to test Neanderthal – *H. sapiens* admixture, especially in the absence of detailed modeling of how the dental traits may track population history and/or gene flow.

Frayer's (2017) short paper focuses on the cognitive abilities of Neanderthals, specifically those associated with speech. He reviews the osteological evidence for sound-producing abilities as reflected by the hyoid bone, the first fossil of which was discovered in Kebara Cave in the 1980s, since then augmented by additional hyoid finds. These show how the Neanderthal hyoid differed from that of earlier *A. afarensis*, with a morphology closer to that of modern apes (implying similar, limited sound-producing abilities), and how similar it was to the modern configuration. Frayer's review (2017) of studies of the auditory anatomy, base of the skull, brain lateralization and handedness, as well as the presence of the gene FOXP2 in its modern form in Neanderthals, suggests to him that in all these the Neanderthals had the modern configuration and therefore cannot be denied the ability that defines modern humans – the ability of language.

Been et al. (2017) used a 3-D model from CT scans of the Kebara 2 Neanderthal partial skeleton to provide, for the first time, a complete 3-D virtual reconstruction of the spine of an extinct hominin. This reconstruction demonstrates that the upright posture of Kebara 2 was slightly different from that of the average modern human. When compared to modern humans, the spine of Kebara 2 exhibits a combination of a vertical sacrum and a small lumbar lordosis together with a nearly average thoracic kyphosis. As a result, the spinopelvic alignment of this specimen was different from that of modern humans, suggesting locomotor and weight-bearing differences between the two groups. Neanderthals might have been better adapted to carry heavy loads and, potentially, to engage in generally more rigorous upper body activities (as is also discussed briefly by Pearson and Sparacello 2017). On the other hand, it suggests that Neanderthals potentially had a shorter stride length and slower walking speed on a flat terrain in comparison with modern humans. If validated, these observations may have important implications for understanding the organization of activities of Neanderthals and their energetic costs while moving across various types of terrain.

In the last chapter, Caspari et al. (2017) ponder the place of the Neanderthals in human evolution, looking at the evidence from the perspective of three topics. They consider body form (focusing on the pelvis); population structure (paleodemography), and breeding behavior (as seen from the genetic evidence). Their extensive review and testing of the evidence lead them to conclude that in body form, demography and population structure, Neanderthals were unlike modern humans in the Upper Paleolithic or later. In some cases (the pelvis shape, paleodemographic curves), they seem to indeed reflect the ancestral condition. Demographic factors in particular (adult survivorship, the ratio of older to younger individuals) negatively affected the resilience of Neanderthal groups to stochastic fluctuations in size and also their densities on the landscape and their ability to formulate extensive and lasting social and economic networks. The increased survivorship and longevity in the Upper Paleolithic eventually led to social pressures that Caspari et al. (2017) associate with extensive trade networks and more complex systems of cooperation and competition between groups. In that sense, "modern human behavior" (a problematic term by many archaeological accounts) is a response to demographic pressures. The Neanderthal archaeological record shows glimpses of this behavior, but it is less frequent and less sophisticated than in the Upper Paleolithic, a reflection of their archaic life history pattern. Caspari et al. (2017) tie these inferences with their long-standing view, now arguably (e.g., Holliday et al. 2014) bolstered by ancient DNA studies demonstrating hybridization between Neanderthals and moderns during the Middle and Upper Paleolithic, that phylogenetically Neanderthals do not constitute a different taxon. They represent another way of being human.

Bones, indeed, can tell a lot, if coaxed in the right way. The papers in this volume provide diverse perspectives on what it means to be human and how our present is an outcome of our evolutionary past. The papers differ in their interests, questions, methodological approaches and analytical tools and provide quite a number of take-home messages. One insight that stands out is that the discourse between 'hard core' human paleoanthropology and the many other disciplines that seek to understand the social and biological evolution of humans yields the most interesting results. We hope that this volume helps to promote such interdisciplinary work in the future.

Last but certainly not least, we thank all the 33 authors who contributed to this volume. Many colleagues and friends of Yoel agreed kindly to act as reviewers of the papers presented here; their efforts were essential to the final product in front of you. We are grateful to three reviewers who chose to remain anonymous, and to Oren Ackerman, Zerasenay Alemseged, Berhane Asfaw, Alon Barash, Anna Barney, Anna Belfer-Cohen, Miriam Belmaker, Michael Berthaume, René Bobé, Emiliano Bruner, Adeline le Cabec, Michael Chazan, Michelle Drapeau, Raphael Falk, John Fleagle, Sarah Friedline, Dan Graur, Joel Irish, Eva Jablonka, William Kimbel, Zacharay Kofran, Kornelius Kupczik, Marta Lahr, Emanuel Marx, Marie-Helène Moncel, Olga Panagiotopoulou, Smadar Peleg, Kaye Reed, Chris Robinson, Chris Ruff, Michael Ruse, Dan Schmitt, Liza Shapiro, Gonen Sharon, Jay T Stock, Chris

Stringer, Anne-Marie Tillier, Scott Williams and Bernard Wood. Avishag Ginzburg provided invaluable help in organizing the graphic materials and their preparation for publication. Special thanks are due to the series editors, and especially Eric Delson, for their guidance, diligent quality control, and patience throughout the lengthy process of bringing this volume to publication.

Erella Hovers
Assaf Marom

References

Bailey, S. E., Weaver, T. D., & Hublin, J.-J. (2017). The dentition of the earliest modern humans: How 'modern' are they? In A. Marom & E. Hovers (Eds.), *Human paleontology and prehistory. Contributions in honor of Yoel Rak* (pp. 215–232). Cham: Springer.

Bateman, T. S., & Hess, A. M. (2015). Different personal propensities among scientists relate to deeper vs. broader knowledge contributions. *Proceedings of the National Academy of Sciences USA, 112,* 3653–3658.

Been, E., Gómez-Olivencia, A., Kramer, P. A., & Barash, A. (2017). 3D reconstruction of the spinal posture of the Kebara 2 Neanderthal. In A. Marom & E. Hovers (Eds.), *Human paleontology and prehistory. Contributions in honor of Yoel Rak. Contributions in Honor of Yoel Rak* (pp. 239–251). Cham: Springer.

Cartmill, M., & Brown, K. (2017). Posture, locomotion and bipedality: The case of the gerenuk (*Litocranius walleri*). In A. Marom & E. Hovers (Eds.), *Human paleontology and prehistory. Contributions in honor of Yoel Rak* (pp. 53–70). Cham: Springer.

Caspari, R., Rosenberg, K. R., & Wolpoff, M. H. (2017). Brother or other? The place of Neanderthals in human evolution. In A. Marom & E. Hovers (Eds.), *Human paleontology and prehistory. Contributions in honor of Yoel Rak* (pp. 253–271). Cham: Springer.

Collard, M., & Cross, A. (2017). Thermoregulation in *Homo erectus* and Neanderthals: A reassessment using a segmented model. In A. Marom & E. Hovers (Eds.), *Human paleontology and prehistory. Contributions in honor of Yoel Rak* (pp. 161–174). Cham: Springer.

Cracraft, J., & Eldredge, N. (1979). *Phylogenetic Patterns and the evolutionary process.* New York: Columbia University Press.

Daegling, D. J., & Grine, F. E. (2017). Feeding behavior and diet in *Paranthropus boisei*: The limits of functional inferences from the mandible. In A. Marom & E. Hovers (Eds.), *Human paleontology and prehistory. Contributions in honor of Yoel Rak* (pp. 109–125). Cham: Springer.

Dobzhansky, T. (1955). *Evolution, genetics, and man.* New York: J. Wiley & Sons.

Dobzhansky, T. (1962). *Mankind evolving.* New Haven: Yale University Press.

Frayer, D. W. (2017). Talking hyoids and talking Neanderthals. In A. Marom & E. Hovers (Eds.), *Human paleontology and prehistory. Contributions in honor of Yoel Rak* (pp. 233–237). Cham: Springer.

Garrod, D. A. E., Bate, D. M. A., McKown, T. D., & Keith, A. (1937). *The Stone Age of Mt. Carmel* vol. I. Oxford: Clarendon Press.

Glowacka, H., Kimbel, W. H., & Johanson, D. C. (2017) Aspects of mandibular ontogeny in *Australopithecus afarensis*. In A. Marom & E. Hovers (Eds.), *Human paleontology and prehistory. Contributions in honor of Yoel Rak* (pp. 127–144). Cham: Springer.

Harrison, T. (2017). The paleoecology of the Upper Ndolanya Beds, Laetoli, Tanzania, and its implications for hominin evolution. In A. Marom & E. Hovers (Eds.), *Human paleontology and prehistory. Contributions in honor of Yoel Rak* (pp. 31–44). Cham: Springer.

Harvati, K., & Lopez, E. N. (2017). A 3-D look at the Tabun C2 jaw. In A. Marom & E. Hovers (Eds.), *Human paleontology and prehistory. Contributions in honor of Yoel Rak* (pp. 203–213). Cham: Springer.

Haylander, W. L. (2017). Canine height and jaw gape in catarrhines with reference to canine reduction in early hominins. In A. Marom & E. Hovers (Eds.), *Human paleontology and prehistory. Contributions in honor of Yoel Rak* (pp. 71–93). Cham: Springer.

Holliday, T. W., Gautney, J. R., & Friedl, L. (2014). Right for the wrong reasons: Reflections on modern human origins in the post-Neanderthal genome era. *Current Anthropology, 55,* 696–724.

Holloway, R. L. (2017). The australopithecine brain: Controversies perpetual. In A. Marom & E. Hovers (Eds.), *Human paleontology and prehistory. Contributions in honor of Yoel Rak* (pp. 45–52). Cham: Springer.

Hovers. E. (2006). Neandertals and Modern Humans in the Middle Paleolithic of the Levant: What kind of interaction? In N. Conard (Ed.), *When Neandertals and moderns met* (pp. 65–86). Tübingen: Kerns Verlag.

Hovers E., & Belfer-Cohen A. (2006). "Now you see it, now you don't" – modern human behavior in the Middle Paleolithic. In E. Hovers & S. L. Kuhn (Eds.), *Transitions before the transition: Evolution and stability in the Middle Paleolithic and Middle Stone Age.* (pp. 295–304). New York: Springer.

Hovers, E., & Belfer-Cohen, A. (2013). On variability and complexity: Lessons from the Levantine Middle Paleolithic Record. *Current Anthropology 54*, S337–S357.

Kuhlwilm, M., Gronau, I., Hubis, M. J., de Filippo, C., Prado-Martinez, J., Kircher, M., et al. (2016). Ancient gene flow from early modern humans into Eastern Neanderthals. *Nature, 530*, 429–433.

Lieberman, D. E., & Bar-Yosef, O. (2005). Apples and oranges: Morphological versus behavioral transitions in the Pleistocene. In D. E. Lieberman, R. J. Smith & J. Kelley (Eds.), *Interpreting the past: Essays on human, primate, and mammal evolution in honor of David Pilbeam* (pp. 275–296). Boston: Brill Academic Publishers.

Marom, A. (2013). Mechanical implications of the unique Neandertal facial skeleton. Ph.D. Dissertation, Tel Aviv University

Marom, A., & Tarrasch, R. (2015). On behalf of tradition: An analysis of medical student and physician beliefs on how anatomy should be taught. *Clinical Anatomy 28*, 980–984.

Meyer, M., Fu, Q., Aximu-Petri, A., Glocke, I., Nickel, B., Arsuaga, J.-L. et al. (2014). A mitochondrial genome sequence of a hominin from Sima de los Huesos. *Nature, 505*, 403–406.

Neuman, J. (2017). Wallace's controversy with Darwin on man's mental evolution, on the position of the natives in human evolution and his anticipation of cultural evolution, as distinct from biological evolution. In A. Marom & E. Hovers (Eds.), *Human paleontology and prehistory. Contributions in honor of Yoel Rak* (pp. 11–20). Cham: Springer.

Pääbo, S. (2014). *Neanderthal man: In search of lost genomes*. New York: Basic Books.

Pearson, O. M. (2013). Hominin evolution in the Middle-Late Pleistocene: Fossils, adaptive scenarios, and alternatives. *Current Anthropology 54*, S221–S233.

Pearson, O. M., & Sparacello, V. S. (2017). Behavioral differences between Near Eastern Neanderthals and the early modern humans from Skhul and Qafzeh: An assessment based on comparative samples of Holocene humans. In A. Marom & E. Hovers (Eds.), *Human paleontology and prehistory. Contributions in honor of Yoel Rak* (pp. 175–186). Cham: Springer.

Rak, Y. (1978). The functional significance of the squamosal suture in *Australopithecus boisei*. *American Journal of Physical Anthropology 49*, 71–78.

Rak, Y. (1986). The Neanderthal: A new look at an old face. *Journal of Human Evolution, 15*, 151–164.

Rak, Y. (1998). Does any Mousterian cave present evidence of two hominid species? In T. Akazawa, K. Aoki & O. Bar-Yosef (Eds.), *Neanderthals and modern humans in western Asia* (pp. 353–366). New York: Plenum Press.

Rak, Y., Ginzburg, A., & Geffen, E. (2002). Does *Homo neanderthalensis* play a role in modern human ancestry? The mandibular evidence. *American Journal of Physical Anthropology 119*, 199–204.

Rak, Y., Ginzburg, A., & Geffen, E. (2007). Gorilla-like anatomy on *Australopithecus afarensis* mandibles suggests *Au. afarensis* link to robust australopiths. *Proceedings of the National Academy of Sciences USA, 104*, 6568–6572.

Rightmire, G. P. (2017). Middle Pleistocene *Homo* crania from Broken Hill and Petralona: Morphology, metric comparisons, and evolutionary relationship. In A. Marom & E. Hovers (Eds.), *Human paleontology and prehistory. Contributions in honor of Yoel Rak* (pp. 145–150). Cham: Springer.

Tattersall, I. (2017). Why was human evolution so rapid? In A. Marom & E. Hovers (Eds.), *Human paleontology and prehistory. Contributions in honor of Yoel Rak* (pp. 1–9). Cham: Springer.

Wallace, A. R. (1891). *Natural selection and tropical nature*. London: Macmillan.

Weinstein-Evron, M., & Zaidner, Y. (2017). The Acheulo-Yabrudian – Early Middle Paleolithic sequence of Misliya Cave, Mount Carmel, Israel. In A. Marom & E. Hovers (Eds.), *Human paleontology and prehistory. Contributions in honor of Yoel Rak* (pp. 187–201). Cham: Springer.

Wood, B., & Shcroer, K. (2017). *Paranthropus*: Where do things stand? In A. Marom & E. Hovers (Eds.), *Human paleontology and prehistory. Contributions in honor of Yoel Rak* (pp. 95–107). Cham: Springer.

Wool, D. (2017). Man's place in past and future Evolution: A historical survey of remarkable ideas. In A. Marom & E. Hovers (Eds.), *Human paleontology and prehistory. Contributions in honor of Yoel Rak* (pp. 21–29). Cham: Springer.

Major Publications of Yoel Rak

Books

Rak, Y. (1983). *The australopithecine face*. New York: Academic Press.
Kimbel, W. H., Rak, Y., & Johanson, D. C. (2004). *The skull of* Australopithecus afarensis. London: Oxford University Press.

Selected Articles and Reviews

Rak, Y., & Howell, F. C. (1978). Cranium of a juvenile *Australopithecus boisei* from the lower Omo basin, Ethiopia. *American Journal of Physical Anthropology, 48*, 345–366.
Rak, Y., (1978). The functional significance of the squamosal suture in *Australopithecus boisei*. *American Journal of Physical Anthropology, 49*, 71–78.
Rak, Y., & Clarke, R. (1979). Ear ossicle of *Australopithecus robustus*. *Nature, 279*, 62–63.
Rak, Y., & Clarke, R. (1979). Aspects of the middle and external ear of early South African hominids. *American Journal of Physical Anthropology, 51*, 471–474.
Rak, Y. (1981). The morphology and architecture of the Australopithecine face. Ph.D. dissertation, University of California, Berkeley.
Cronin, J. E., Boaz, N., Stringer, C. B., & Rak, Y. (1981). Tempo and mode in hominid evolution. *Nature, 292*, 113–122.
Kimbel, W. H., & Rak, Y. (1985). Functional morphology of the asterionic region in extant hominoids and fossil hominids. *American Journal of Physical Anthropology, 66*, 31–54.
Rak, Y. (1985). Australopithecine taxonomy and phylogeny in light of facial morphology. *American Journal of Physical Anthropology, 66*, 281–287.
Arensburg, B., Bar-Yosef, O., Chech, M., Goldberg, P., Laville, H., Meignen, L., et al. (1985). Une sepulture néandertalienne dans la grotte de Kebara (Israel). *Comptes Rendus de l'Academie des Sciences Paris, 6*, 227–230.
Rak, Y. (1985). Sexual dimorphism, ontogeny and the beginning of differentiation of the robust Australopithecine clade. In P. V. Tobias (Ed.), *Past, present and future of hominid evolution* (pp. 233–237). New York: Alan R. Liss.
Bar-Yosef, O., Vandermeersch, B., Arensburg, B., Goldberg, P., Laville, H., Meignen, L., et al. (1986). New data on the origin of modern man in the Levant. *Current Anthropology, 27*, 63–64.
Rak, Y. (1986). The Neanderthal: A new look at an old face. *Journal of Human Evolution, 15*, 151–164.
Rak, Y., & Arensburg, B. (1987). Kebara 2 Neanderthal pelvis: First look at a complete inlet. *American Journal of Physical Anthropology, 73*, 227–231.
Valladas, H., Joron, J., Valladas, G., Arensburg, B., Bar-Yosef, O., Belfer-Cohen, A., et al. (1987). Thermoluminescence dates for the Neanderthal burial in Kebara cave, Mt. Carmel, Israel. *Nature, 330*, 159–160.
Rak, Y. (1988). On variation in the masticatory system of *A. boisei*. In F. Grine (Ed.), *The evolutionary history of the robust Australopithecines* (pp. 193–199). New York: Aldine.
Tillier, A., Arensburg, B., Rak, Y., & Vandermeersch, B. (1988). Les sepultures Néanderthaliennes du Proche-Orient: Etat de la question. *Paléorient, 14*, 130–136.
Bar-Yosef, O., Laville, H., Meignen, L., Tillier, A., Vandermeersch, B., Arensburg, B., et al. (1988). La Sepulture Neandertalienne de Kebara. In M. Otte (Ed.), *L'Homme de Neandertal* (pp. 17–24). Liège: ERAUL.

Arensburg, B., Tillier, A.-M., Vandermeersch, B., Duday, H., Scheperts, L. A., & Rak, Y. (1989). A Middle Paleolithic human hyoid bone. *Nature, 338*, 758–760.

Meignen, L., Vandermeersch, B., Tillie, A.-M., Laville, H., Arensburg, B., Bar-Yosef, O., et al. (1989). Néanderthaliens et hommes modernes au proche-orient: chronologie et comportements culturels. *Bulletin de la Société Préhistorique Française, 10*, 354–362.

Rak, Y. (1990). On the differences between two pelvises of Mousterian context from the Qafzeh and Kebara Caves, Israel. *American Journal of Physical Anthropology, 81*, 323–332.

Rak, Y. (1991). Lucy's pelvic anatomy: Its role in bipedal gait. *Journal of Human Evolution, 20*, 283–290.

Hovers, E., Rak, Y., & Kimbel, W. H. (1991). Amud Cave: 1991 season. *Journal of the Israel Prehistoric Society, 24*, 152–157.

Tillier, A.-M., Arensburg, B., Rak, Y., & Vandermeersch, B. (1991). L'apport de Kebara à la palethnologie funeraire des Néanderthaliens du Proche-Orient. In O. Bar-Yosef & B. Vandermeersch (Eds.), *Le squelette Mousterien de Kebara 2* (pp. 87–92). Paris: CNRS.

Rak. Y. (1991). The Kebara pelvis. In O. Bar-Yosef & B. Vandermeersch (Eds.), *Le squelette Mousterien de Kebara 2* (pp. 143–152). Paris: CNRS.

Bar-Yosef, O., Vandermeersch, B., Arensburg, B., Belfer-Cohen, A., Goldberg, P., Laville, H., et al. (1992). The excavations in Kebara Cave, Mt. Carmel. *Current Anthropology, 33*, 497–550.

Kimbel, W. H., & Rak, Y. (1993). The importance of species taxa in paleoanthropology and an argument for the phylogenetic concept of the species category. In W. H. Kimbel & L. B. Martin (Eds.), *Species, species concept and primate evolution* (pp. 461–484). New York: Plenum.

Rak, Y. (1993). Morphological variation in *Homo neanderthalensis* and *Homo sapiens* in the Levant: A biological model. In W. H. Kimbel & L. B. Martin (Eds.), *Species, species concept and primate evolution* (pp. 523–536). New York: Plenum.

Kimbel, W. H., Johanson, D. C., & Rak, Y. (1994). The first skull and other discoveries of *Australopithecus afarensis* at Hadar, Ethiopia. *Nature, 368*, 449–451.

Rak, Y., Kimbel, W. H., & Hovers, E. (1994). A Neandertal infant from Amud Cave, Israel. *Journal of Human Evolution, 26*, 313–324.

Rak, Y. (1994). The middle ear of *Australopithecus robustus*: Does it bear evidence of a specialized masticatory system? In R. S. Corruccini & R. L. Ciochon (Eds), *Integrative paths to the past: Paleoanthropological advances in honor of F. Clark Howell* (pp. 223–227). Englewood Cliffs: Prentice Hall.

Rak, Y., Kimbel, W. H., & Johanson, D. C. (1996). The crescent of foramina of *Australopithecus afarensis*. *American Journal of Physical Anthropology, 101*, 93–99.

Kimbel, W. H., Walter, R. C., Johanson, D .C., Reed, K. E., Aronson, J. L., Assefa, Z., et al. (1996). Late Pliocene *Homo* and Oldowan tools from the Hadar Formation (Kada Hadar Member), Ethiopia. *Journal of Human Evolution, 31*, 549–561.

Kimbel, W. H., Johanson, D. C., & Rak, Y. (1996). Systematic assessment of a maxilla of *Homo* from Hadar, Ethiopia. *American Journal of Physical Anthropology, 103*, 235–262.

Rak, Y. (1998). Does any Mousterian cave present evidence of two hominid species? In T. Akazawa, K. Aoki & O. Bar-Yosef (Eds.), *Neandertals and modern humans in Western Asia* (pp. 353–366). New York: Plenum Press.

Vallada, H., Mercier, N., Froget, L., Hovers, E., Joron, J. -L., Kimbel, et al. (1999). TL Dates for the Neanderthal site of the Amud Cave, Israel. *Journal of Archeological Science, 26*, 259–268.

Rak, Y., Ginzburg, A., & Geffen, E. (2002). Does *Homo neanderthalensis* play a role in modern human ancestry? The mandibular evidence. *American Journal of Physical Anthropology, 119*, 199–204.

Drapeau, M. S. N., Ward, C. V., Kimbel, W. H., Johanson, D. C., & Rak, Y. (2005). Associated cranial and forelimb remains attributed to *Australopithecus afarensis* from Hadar, Ethiopia. *Journal of Human Evolution, 48*, 593–642.

Kimbel, W. H., Lockwood, C. A., Ward, C. V., Leakey, M. G., Rak, Y., & Johanson, C. (2006). Was *Australopithecus anamensis* ancestral to *A. afarensis*? A case of anagenesis in the hominin fossil record. *Journal of Human Evolution, 5*, 134–152.

Rak, Y., Ginzburg, A., & Geffen, E. (2007). Gorilla-like anatomy on *Australopithecus afarensis* mandibles suggests *Au. afarensis* link to robust australopiths. *Proceeding of the National Academy of Sciences USA, 104*, 6568–6572

Rak, Y., & Hylander, W. (2008). What else is the tall mandibular ramus of the robust australopiths good for? In C. J. Vinyard, M. J. Ravosa & C. E. Wall (Eds.), *Primate craniofacial function and biology* (pp. 431–442). New York: Springer.

Quam, R., & Rak, Y. (2008). Auditory ossicles from southwest Asian Mousterian sites. *Journal of Human Evolution, 54*, 414-433.

Kimbel, W. H., & Rak, Y. (2010). The cranial base of *Australopithecus afarensis*: New insights from the female skull. *Philosophical Transactions of the Royal Society B, 365*, 3365–3376.

Lordkipanidze, D., Ponce de Leon, M. S., Margvelashvili, A., Rak, Y., Rightmire, G. P., Vekua, A., et al. (2013). A complete skull from Dmanisi, Georgia, and the evolutionary biology of early *Homo*. *Science, 342*, 326–331.

Kimbel. W. H., Suwa, G., Asfaw, B., Rak, Y., & White, T. D. (2014). *Ardipithecus ramidus* and the evolution of the human cranial base. *Proceeding of the National Academy of Sciences USA, 111*, 948–953.

Contents

1   Why Was Human Evolution So Rapid?   1
    Ian Tattersall

2   Wallace's Controversy with Darwin on Man's Mental Evolution,
    on the Position of the Natives in Human Evolution, and His Anticipation
    of Cultural Evolution, as Distinct from Biological Evolution   11
    Joseph Neumann

3   Man's Place in Past and Future Evolution: A Historical Survey
    of Remarkable Ideas   21
    David Wool

4   The Paleoecology of the Upper Ndolanya Beds, Laetoli, Tanzania,
    and Its Implications for Hominin Evolution   31
    Terry Harrison

5   The Australopithecine Brain: Controversies Perpetual.   45
    Ralph L. Holloway

6   Posture, Locomotion and Bipedality: The Case of the Gerenuk
    (*Litocranius walleri*)   53
    Matt Cartmill and Kaye Brown

7   Canine Height and Jaw Gape in Catarrhines with Reference
    to Canine Reduction in Early Hominins   71
    William L. Hylander

8   *Paranthropus*: Where Do Things Stand?   95
    Bernard Wood and Kes Schroer

9   Feeding Behavior and Diet in *Paranthropus boisei*: The Limits
    of Functional Inference from the Mandible.   109
    David J. Daegling and Frederick E. Grine

10 Aspects of Mandibular Ontogeny in *Australopithecus afarensis*   127
    Halszka Glowacka, William H. Kimbel and Donald C. Johanson

11 Middle Pleistocene *Homo* Crania from Broken Hill and Petralona:
    Morphology, Metric Comparisons, and Evolutionary Relationships   145
    G. Philip Rightmire

12 Thermoregulation in *Homo erectus* and the Neanderthals:
    A Reassessment Using a Segmented Model   161
    Mark Collard and Alan Cross

13 Behavioral Differences Between Near Eastern Neanderthals and the Early Modern Humans from Skhul and Qafzeh: An Assessment Based on Comparative Samples of Holocene Humans 175
Osbjorn M. Pearson and Vitale S. Sparacello

14 The Acheulo-Yabrudian – Early Middle Paleolithic Sequence of Misliya Cave, Mount Carmel, Israel 187
Mina Weinstein-Evron and Yossi Zaidner

15 A 3-D Look at the Tabun C2 Jaw 203
Katerina Harvati and Elisabeth Nicholson Lopez

16 The Dentition of the Earliest Modern Humans: How 'Modern' Are They? 215
Shara E. Bailey, Timothy D. Weaver and Jean-Jacques Hublin

17 Talking Hyoids and Talking Neanderthals 233
David W. Frayer

18 3D Reconstruction of Spinal Posture of the Kebara 2 Neanderthal 239
Ella Been, Asier Gómez-Olivencia, Patricia A. Kramer and Alon Barash

19 Brother or Other: The Place of Neanderthals in Human Evolution 253
Rachel Caspari, Karen R. Rosenberg and Milford H. Wolpoff

Index 273

Contributors

Shara E. Bailey Department of Anthropology, New York University, New York, NY, USA

Alon Barash Faculty of Medicine in the Galilee, Bar-Ilan University, Zefat, Israel

Ella Been Faculty of Health Professions, Physical Therapy Department, Ono Academic College, Kiryat Ono, Israel; Sackler Faculty of Medicine, Department of Anatomy and Anthropology, Tel Aviv University, Tel Aviv, Israel

Kaye Brown Department of Anthropology, Boston University, Boston, MA, USA

Matt Cartmill Department of Anthropology, Boston University, Boston, MA, USA

Rachel Caspari Department of Sociology, Anthropology and Social Work, Central Michigan University, Mount Pleasant, MI, USA

Mark Collard Human Evolutionary Studies Program and Department of Archaeology, Simon Fraser University, Burnaby, BC, Canada; Department of Archaeology, University of Aberdeen St. Mary's Building, Aberdeen, UK

Alan Cross Human Evolutionary Studies Program and Department of Archaeology, Simon Fraser University, Burnaby, British Columbia, Canada

David J. Daegling Department of Anthropology, University of Florida, Gainesville, FL, USA

David W. Frayer Department of Anthropology, University of Kansas, Lawrence, KS, USA

Halszka Glowacka Institute of Human Origins, School of Human Evolution and Social Change, Arizona State University, Tempe, AZ, USA

Asier Gómez-Olivencia IKERBASQUE. Basque Foundation for Science & Facultad de Ciencia y Tecnología, Department o de Estratigrafía y Paleontología, Euskal Herriko Unibertsitatea, UPV-EHU, Bilbao, Spain; Département de Préhistoire, Muséum National d'Histoire Naturelle, Musée de l'Homme, Paris, France; Centro UCM-ISCIII de Investigación sobre Evolución y Comportamiento Humanos, Madrid, Spain

Frederick E. Grine Departments of Anthropology and of Anatomical Sciences, Stony Brook University, Stony Brook, NY, USA

Terry Harrison Department of Anthropology, Center for the Study of Human Origins, New York University, New York, NY, USA

Katerina Harvati Paleoanthropology, Senckenberg Center for Human Evolution and Paleoenvironments, Department of Geosciences, Eberhard Karls Universität Tübingen, Tübingen, Germany

Ralph L. Holloway Department of Anthropology, Columbia University, New York, NY, USA

Jean-Jacques Hublin Department of Human Evolution, Max Planck Institute for Evolutionary Anthropology, Leipzig, Germany

William L. Hylander Department of Evolutionary Anthropology, Duke University, Durham, NC, USA

Donald C. Johanson Institute of Human Origins, School of Human Evolution and Social Change, Arizona State University, Tempe, AZ, USA

William H. Kimbel Institute of Human Origins, School of Human Evolution and Social Change, Arizona State University, Tempe, AZ, USA

Patricia A. Kramer Departments of Anthropology and of Orthopaedics and Sports Medicine, University of Washington, Seattle, WA, USA

Elisabeth Nicholson Lopez Department of Basic Science and Craniofacial Biology, New York University College of Dentistry, New York, USA

Joseph Neumann Department of Molecular Biology and Ecology of Plants, Tel-Aviv University, Tel-Aviv, Israel; Department of Philosophy, Tel-Aviv University, Tel-Aviv, Israel

Osbjorn M. Pearson Department of Anthropology, University of New Mexico, Albuquerque, NM, USA

G. Philip Rightmire Department of Human Evolutionary Biology, Harvard University, Peabody Museum, Cambridge, USA

Karen R. Rosenberg Department of Anthropology, University of Delaware, Newark, DE, USA

Kes Schroer Neukom Institute Fellow in Computational Sciences and Department of Anthropology, Dartmouth College, Hanover, USA

Vitale S. Sparacello Department of Anthropology, University of New Mexico, Albuquerque, NM, USA

Ian Tattersall Division of Anthropology, American Museum of Natural History, New York, NY, USA

Timothy D. Weaver Department of Anthropology, University of California, Davis, CA, USA

Mina Weinstein-Evron Zinman Institute of Archaeology, University of Haifa, Haifa, Israel

Milford H. Wolpoff Department of Anthropology, University of Michigan, Ann Arbor, MI, USA

Bernard Wood CASHP, George Washington University, Washington, USA

David Wool Department of Zoology, Tel Aviv University, Tel Aviv, Israel

Yossi Zaidner Zinman Institute of Archaeology, University of Haifa, Haifa, Israel; Institute of Archaeology, The Hebrew University of Jerusalem, Jerusalem, Israel

Chapter 1
Why Was Human Evolution So Rapid?

Ian Tattersall

Abstract Nowhere in the entire fossil record of life do we find more dramatically accelerated accumulation of evolutionary novelty than we do in the genus *Homo*. Quite simply, and by whatever criteria you measure it, our species *Homo sapiens* is more different from its own precursors of two million years ago than is any other species living in the world today. What might account for this unusually rapid rate of evolution? A major influence was almost certainly material culture, though not in the gene-culture co-evolutionary context envisaged by the evolutionary psychologists. Rather, material culture enhances the ability of hominid populations to disperse at times when conditions are favorable for expansion, while incompletely insulating the resulting enlarged populations from environmental stress when circumstances deteriorate. In other words, by facilitating expansion beyond normal physiological limits in good times, culture makes populations more vulnerable to fragmentation in bad ones. Over the course of the Pleistocene, short-term but large-scale local environmental changes became increasingly frequent over large tracts of the Old World, further amplifying the stress-and-response cycle. Since the fixation probabilities of evolutionary novelties of all kinds (as well as of local extinctions) are promoted by population fragmentation and consequent small effective population sizes, we see in the synergy between environmental effects and material culture a sort of ratchet effect which would have acted to leverage rates of accumulating change. This interaction explains the extraordinarily fast tempo of evolution within the genus *Homo* by invoking perfectly routine evolutionary processes; and it eliminates any need for special pleading in the hominid case, at least in terms of mechanism. Apparent recent diminution in human brain size may result from greater algorithmic efficiency.

Keywords Evolutionary rates • Tachytely • Hominids • Hominins • Material culture • Morphological change • Rapid evolution

There are many extraordinary things about our species *Homo sapiens*. The most obvious of these reside in our unique symbolic cognitive style, and in the physical correlates of our unusual form of striding bipedal locomotion. Much has been written about conspicuous features such as these, and about how they may have evolved. But there is something else about our species and its precursors that is equally striking, but that has somehow contrived to escape as much attention as it merits: namely, the rapidity with which the human lineage has evolved. By virtually any measure, *Homo sapiens* is more different from its own ancestors of only two million years ago, both in its morphology and in the way it processes information, than is any other contemporary mammal species.

The genus *Homo* has been in existence as a morphologically coherent entity for less than two million years (Myr) (Wood and Collard 1999; Collard and Wood 2015; Schwartz and Tattersall 2005; Tattersall and Schwartz 2009). Material culture, as inferred from the deliberate manufacture of stone tools, has been a property of at least some hominid lineages for a little longer: the earliest clear evidence for it goes back as far as about 2.5 Ma (Semaw et al. 1997). This also happens to be the age of the earliest claimed "early *Homo*" fossils (e.g., Schrenk et al. 1993; Kimbel et al. 1997), as well as of an inferred *Kenyanthropus* lineage (Leakey et al. 2001); but whether or not the fossils concerned are appropriately allocated, the current best guess is that stone tool fabrication was introduced into the hominid behavioral repertoire by archaically-proportioned australopiths (de Heinzelin et al. 1999). The earliest stone tool makers were thus terrestrially upright bipeds; but they were relatively small-bodied, and had archaic limb proportions and a host of morphological features, especially of the forelimbs and

I. Tattersall (✉)
Division of Anthropology, American Museum of Natural History, New York, NY 10024, USA
e-mail: iant@amnh.org

© Springer International Publishing AG 2017
Assaf Marom and Erella Hovers (eds.), *Human Paleontology and Prehistory*,
Vertebrate Paleobiology and Paleoanthropology, DOI: 10.1007/978-3-319-46646-0_1

upper body, that attest to a partially arboreal way of life (Susman et al. 1984). Additionally, their skulls were constructed much as in today's great apes. They had large, protruding faces, hafted in front of tiny neurocrania that had contained brains no larger than one would expect of an ape of similar body mass. In all these features they contrasted dramatically with the tall, slender, long-legged *Homo sapiens*, which exhibits a large, balloon-like braincase with a tiny face retracted beneath its front (see Fig. 1.1).

Despite various "advanced" features reported in the newly described 2.0 Myr-old *Australopithecus sediba* (Pickering et al. 2011), the exact evolutionary roots of the genus *Homo* remain obscure. But on present evidence there is little doubt that it is from a form possessing the general morphological features of an australopith, and that lived at some time between about 2.5 and 2.0 Ma, that *Homo sapiens* ultimately descended. This represents a remarkable transformation that was accomplished very fast. It was not, of course, linear. Rather, as Fig. 1.2 shows, it was achieved in the context of vigorous evolutionary experimentation. Since the very beginning, numerous hominid species have apparently been pitchforked out on to the ecological stage, to succeed – or, more likely, to fail – over a period of intensely unstable climatic and environmental conditions.

A pattern of diversity of the kind represented in the figure is typical of successful mammalian families; but in morphological as well as in behavioral terms this particular transformation was distinctive by virtue of being both vast in scale, and exceptionally fast in time. To put it in perspective, two million years is approximately the amount of time that has elapsed since the divergence of the two species of *Pan*, *P. troglodytes* and *P. paniscus* (Stone et al. 2010). And while there are certainly noticeable differences in both behavior and morphology between these two species, they vanish alongside those separating an australopith from a modern human. To take another example, Fig. 1.3 compares the crania of two other hominoid genera, each one, like *Pan* and *Homo*, the other's closest living relative. On the right is a gibbon, *Hylobates*. On the left is a siamang, *Symphalangus*. What makes the comparison of these two morphologically similar genera particularly instructive is that, almost exactly as in the case of *Pan* and *Homo* (Stone et al. 2010), the best molecular estimate is that these two hominoids last shared an ancestor some seven million years ago, plus or minus a million years or so (Matsudaira and Ishida 2010).

The hylobatid case is a rather routine illustration of what G. G. Simpson (e.g., 1944, 1953) called "horotely," namely evolution at "normal" rates. The morphological differences between gibbons and siamangs appear to be pretty much what one would expect for closely related creatures with a divergence time in this general range (see reviews of primate evolutionary patterns in Hartwig 2002). The human/chimpanzee case, on the other hand, is very different. Nobody would dispute that modern humans are much more unlike

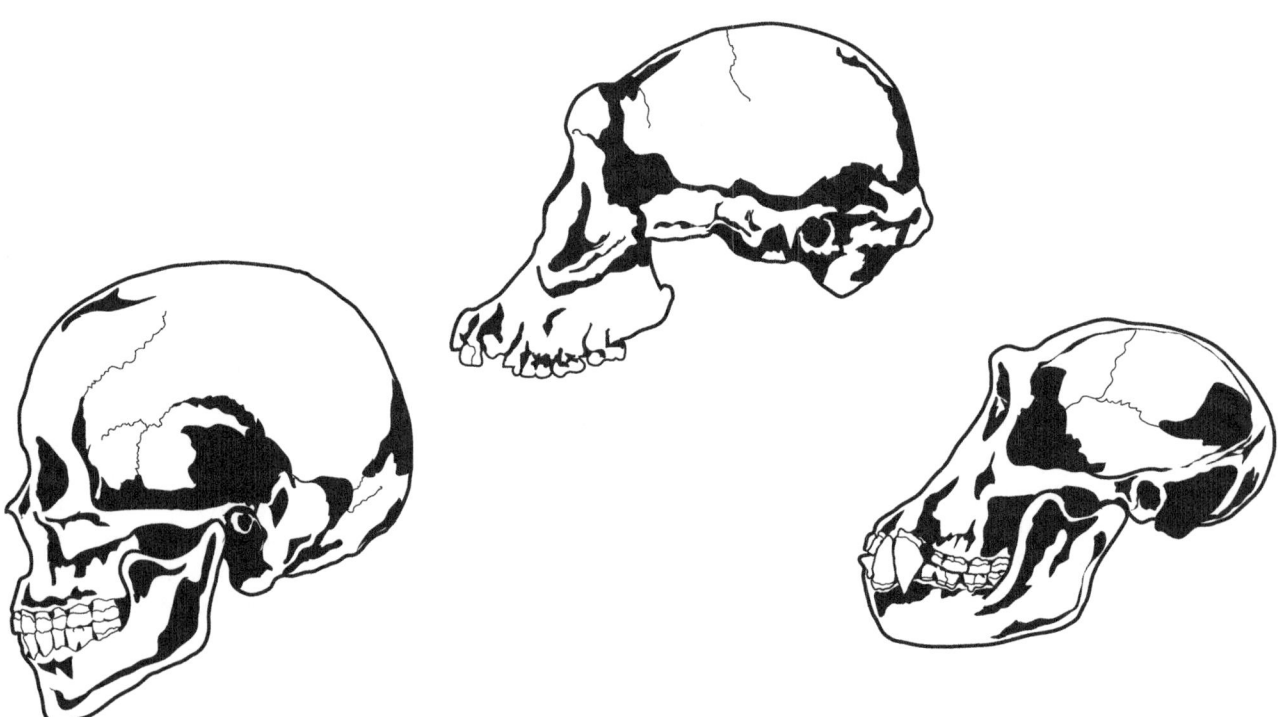

Fig. 1.1 Lateral views of the crania of: (*left*) a modern human, *Homo sapiens*; (*center*) *Australopithecus afarensis*; (*right*) a modern chimpanzee, *Pan troglodytes*. Drawing by Jennifer Steffey

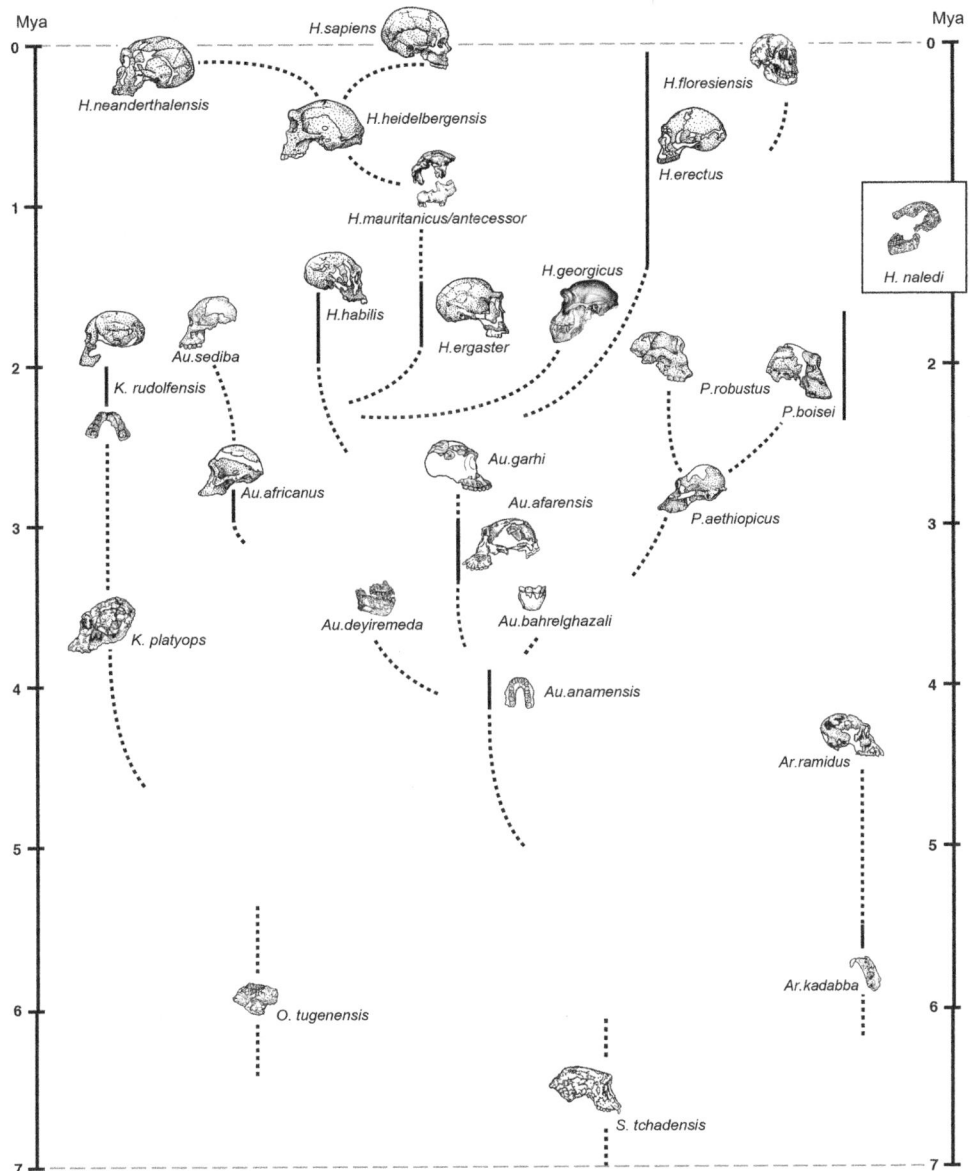

Fig. 1.2 Highly tentative phylogeny of the hominid family, showing the diversity of species currently known within the group, and indicating some possible lines of descent. Multiple hominid lineages have typically existed in parallel. Artwork by Jennifer Steffey

the common ancestor than modern chimpanzees are; and the human lineage thus seems to provide us with an example – and an extreme one – of what Simpson called "tachytely," the very fast accumulation of evolutionary change. Simpson believed that tachytelic episodes are often implicated in the origin of higher taxa, and by extension are responsible for many of the "systematic deficiencies" of the fossil record (Simpson 1953). He also observed that tachytely was to be expected when populations "are shifting from one major adaptive zone to another, and especially when a threshold is crossed" (1953: 334). This certainly appears significant when we contrast the hominoid cases just discussed. For it is certainly true that, while the brachiating siamang and gibbon

lineages remain restricted to the ancestral closed tropical forests, human precursors crossed a major adaptive/habitat threshold on at least two occasions over the last seven million years or so.

The first time was when archaically-proportioned hominids committed themselves to an at least part-time terrestrial bipedal existence and a generalist diet (Sponheimer and Lee-Thorpe 2007), even as they retained a suite of climbing adaptations. The second was when early members of the genus *Homo* more or less entirely emancipated themselves from the trees, by acquiring basically modern body form and today's familiar striding locomotion. In one sense, then, the hominids conformed to Simpson's expectations by

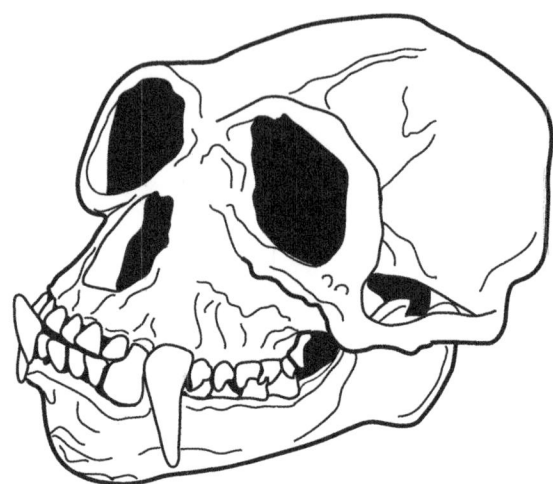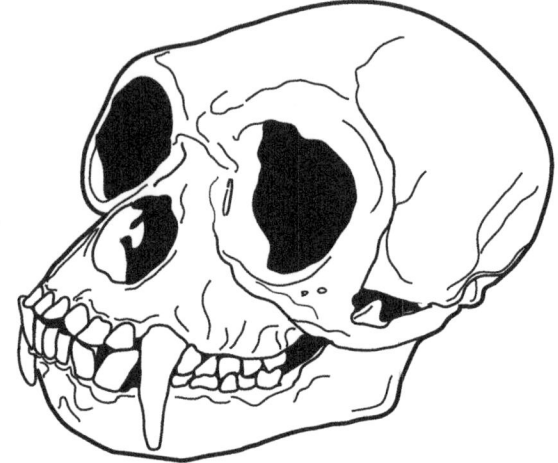

Fig. 1.3 Three-quarter views of two hylobatid crania. *Left*: siamang, (*Symphalangus syndactylus*). *Right*: gibbon (*Hylobates lar*). Drawn by Jennifer Steffey

undergoing rapid major morphological transformations in concert with major adaptive shifts. But in other respects – including having left an excellent fossil record of their transformation – hominids have departed dramatically from his predicted pattern. According to Simpson (1953: 333), "Evolution at exceptionally high rates cannot long endure. A tachytelic line must soon become horotelic, bradytelic [slow-evolving], or extinct." Yet, particularly since the birth of the genus *Homo*, when the last adaptive zone shift was achieved, high rates of both behavioral and morphological change have been remarkably consistent themes in hominid evolution. The most famous example of a consistent long-term hominid trend is, of course, the startling increase in brain size within multiple lineages of the genus *Homo* over the span of the Pleistocene (see data in Holloway et al. 2004).

Some two million years ago, hominids had brains that were, both absolutely and relatively, about the size of those of the already highly-encephalized apes. A million years later, the average hominid brain was twice as big. And today, after the lapse of another million years, it has doubled in size again. This observed increase in mean hominid brain sizes may well have been due to the success of larger-brained forms in inter-species competition for ecological space, rather than to the reproductive success of larger-brained individuals within in a gradually-modifying single lineage (Tattersall 2008). But whatever the case, this apparently steady trend represents a marked departure from the kind of tachytely that Simpson had in mind when he was seeking mechanisms for the origination of higher taxa. Clearly, the definitive abandonment of hominid dependence on trees has to count as one of the most radical shifts in adaptive zone ever made by any vertebrate, ever since the very first tetrapod heaved itself out of the water and on to terra firma. But once hominids had

made their new ecological commitment – which eventually expressed itself in the occupation of an altogether remarkably wide range of open habitats – there must have been other factors at work to maintain both their persistently high rate of brain size increase and the associated morphological changes.

For reasons that are not entirely clear to me, morphologists have tended to avoid this issue, leaving the field clear for speculation by evolutionary psychologists. The reason for these scientists' intense interest is partly, of course, that the cognitive peculiarities of our species *Homo sapiens* are at least as striking as our physical ones. But perhaps more importantly, it is because if you are looking for a satisfyingly reductionist feedback scenario to explain the startling increases in hominid brain size – and, by extension, in cognitive complexity – over the span of the Pleistocene, nothing fits the bill better than a mutually reinforcing link between genes and culture. This link has been energetically promoted by evolutionary psychologists ever since Charles Lumsden and Ed Wilson published their book *Genes, Mind and Culture: The Coevolutionary Process* in 1981. There, with much mathematical folderol, these authors elaborated a notion of "gene-culture coevolution" in which, as they summarized it the following year:

> "culture is shaped by biological imperatives while biological traits are simultaneously altered by genetic evolution in response to cultural history" (Lumsden and Wilson 1982: 1).

Having made this sweeping general pronouncement, Lumsden and Wilson proceeded to apply its principles specifically to human cognition:

> "genetic and cultural evolution are inseverable, and ... the human mind has tended to evolve so as to bias individuals toward certain patterns of cognition and choice rather than others" (Lumsden and Wilson 1982: 1).

A mini-industry had been founded.

Three decades on, the bandwagon continues to roll with undiminished vigor, led now by Peter Richerson and Rob Boyd. In their tellingly-titled book *Not by Genes Alone: How Culture Transformed Human Evolution*, Richerson and Boyd (2005) make a marvelous theoretical case for taking a hardened neodarwinian approach to understanding how humans acquired their extraordinary cognitive powers. From their viewpoint, evolutionary change sums out simply to the steady operation of natural selection on generation after generation of individuals, with a strong positive feedback between cultural and physical innovation. As a sweeping explanation for hominid uniqueness this is an appealing idea; after all, to smart members of a smart species, it seems intuitively obvious that being smarter is a Good Thing, and that even being a tiny bit smarter than your neighbor would be a significant advantage in the race for reproductive success. In fact, from a purely neodarwinian perspective it's hard to imagine how being smarter, or possessing any other excellent heritable quality, would not virtually *oblige* you to reproduce more successfully. In which case, as a result of the inherent feedback between genes and culture, human precursors were virtually condemned to become progressively more complex and intelligent, and by extension to have been predisposed to rapid and continuous evolutionary change.

But though it may provide reductively compelling examples in particular instances, maybe this isn't actually the whole story. For one thing, large brains are metabolically expensive as well as presumptively advantageous for reasons we cannot at present specify in any detail. For another, being smarter doesn't necessarily make you more reproductively attractive, or fleeter of foot, or keener of eye, or stronger, or more aggressive, or socially more adroit. Or any of the other things that, in a random and complicated world, might help to make you more fortunate – or simply less unfortunate – both reproductively and in simply staying alive. The bottom line here is that, in the end, it has to be the whole organism – an astonishingly complex and integrated genetic entity, with a limited number of genes doing a huge amount of work – that, for whatever reasons, succeeds or fails in the evolutionary stakes. Individually, none of the particular traits into which our orderly minds would like to dissect the whole organism can be singled out by natural selection for favor or disfavor – unless it has an unusually powerful effect on reproductive success not just at any particular point in time but consistently enough, and over a long enough period, to make a biologically meaningful difference. For most of the characteristics that paleoanthropologists are able to observe or to infer, this is rather improbable, especially in a world that was as unpredictable and constantly fluctuating as the one in which our Pleistocene precursors lived.

Still, these objections are as theoretical as the original argument; and to their great credit Richerson and Boyd readily concede how crucial empirical observation is in this context. In their words, "the world is *so* complex that without sound empirical data the theorists are blind" (2005: 257). The relevant empirical data must necessarily come from the archaeological record, which is our only source of reasonably direct proxies for ancient hominid behaviors. And, blurry and incomplete as the Paleolithic record may be as an archive of the undoubtedly rich and complex behavioral repertoires and social lives of earlier hominids, the picture it yields is not the pattern of gradual improvement that the neodarwinian feedback model of cognitive refinement predicts. Instead, we find just the opposite: major innovations tended to occur relatively suddenly, interspersed with immensely long periods during which nothing much occurred beyond the occasional refinement.

Thus, the first deliberately manufactured stone tools show up rather abruptly in the record at about 2.5 Ma (Semaw et al. 1997). There is then a wait of a million years, with one single outlier (Lepre et al. 2011), before a substantially new kind of tool is introduced, in the form of the Acheulean handaxe. And while handaxes became generally slimmer and more elegant over time, it was another million years before a new concept in stone tool making – core preparation – began to appear. To cut a long story short, what we are *not* finding here is a smooth increase in technological complexity and refinement over the past 2.5 million years. The spirit of questing and innovation we are so familiar with today simply was not expressed in the material expressions of our precursors until very recently indeed. What is more, in the period prior to the appearance of the new spirit, technological innovation (as opposed to refinement) was both sporadic and rare. What this simple observation clearly reveals, is that our modern cognitive style hardly serves as a reliable model for the ways in which our precursors dealt with information. Intellectually, they were not merely less gifted versions of us: they were doing business in entirely different ways.

This shows up in dramatic behavioral contrasts. While we modern *Homo sapiens* tend to invent new kinds of tools for new purposes, earlier hominids evidently responded to (sometimes rapidly) changing environmental circumstances by repurposing old tools. This is not to deny that those hominids were skilled, resourceful and intelligent. But it does suggest that their cognitive style was not ours. Indeed, perhaps the most telling of all of the innovations which begin to pile up toward the end of the Pleistocene was not the presumed bodily ornamentation, or the engraved symbols, or the cave art, or any of the other many striking individual expressions of the modern symbolic cognitive style that show up in that time range. Rather, it was a profound shift in the tempo of change itself. Technological innovation became the norm, rather than the exception. This implies a relatively abrupt, qualitative change in mental information processing, rather than simply an incremental improvement on what was there before (Tattersall 2008, 2012). What is more, the example of the non-symbolic and comparatively plesiomorphic but nonetheless large-brained Neanderthals demonstrates that neither our cranial configuration, nor our unusual cognitive status, can be interpreted as merely a passive consequence of our large brain size. We thus cannot view either of these modern human features as merely an extrapolation of long-running established trends.

All of this suggests that the apparent long-term feedback between culture and morphology in human evolution is an artifact of evolutionary model, rather than something we can hypothesize from empirical evidence. Indeed, even the culture/biology link seems tenuous, at least as proposed. But if the high average rate of morphological and cognitive change among hominids was not driven by the acquisition of modern body form; and if it was not driven in a linear way by a feedback between incremental cognitive/cultural improvements and reproductive success, then what *was* the factor that drove the extraordinary tachytely in Pleistocene *Homo*?

Perhaps oddly in light of what I have just said, in answering this key question I am nonetheless going to implicate culture, which has certainly been omnipresent as a central and basically unique fact of hominid life throughout the tenure of the genus *Homo*. Culture is, of course, a famously slippery concept, and the word means very different things to different people. By the narrowest definition, culture may not even be unique to humans (Mercader et al. 2007); and the issue is undoubtedly complicated by the fact that the incredibly complex behavioral expressions we see today in *Homo sapiens* are a reflection of our unique and recently acquired cognitive mode, rather than linear extrapolations of simpler behavioral forms that may have preceded them. Here I shall use "culture" in its narrow material sense, as reflected by the tangible products of technology, and the behaviors directly associated with manufacturing and using those products. I am not concerned with any wider social or cognitive implications.

This restricted definition has one singular advantage. For everyone can agree that, by substantially extending the phenotype, material culture constituted a major element in our precursors' ability to respond to the external and adventitious climatic changes that regularly assailed them over the course of the Pleistocene (e.g., van Andel and Davies 2003). And it is in the context of those external changes, rather than as an expression of any intrinsic dynamic, that the effects of material culture would have made themselves felt among Pleistocene *Homo*. Here's why. It has been clearly understood for many years that both small effective sizes and physical isolation are essential for the fixation of genetic novelty in populations of complex mammals like primates (Eldredge and Cracraft 1980). In large and continuous populations there is simply too much genetic inertia for either chance or selection to drive the incorporation of heritable novelties while, in contrast, within small ones the incorporation of such novelties – whether for chance or for selective reasons – seems to be routine. And, as it happens, the conditions in which Pleistocene members of the genus *Homo* evolved were hugely propitious for the fixation of genetic changes. These ancient hominids were almost certainly widely but thinly spread across the landscape, in small groups that probably belonged to relatively isolated population clusters. At the same time they were amazingly mobile, and it appears that demographic pressures toward expansion *within* those sparse populations were probably fairly intense, as we can fairly infer from the extremely rapid rate of spread of early *Homo* species. For example, hard on the heels of the first appearance of *Homo* in Africa, hominids widely considered to be of our genus had already ventured as far afield as Dmanisi in the Caucasus (e.g., Gabunia et al. 2000), even though the cool temperate environment there was very unlike any of the habitats the hominids' predecessors had ever had to cope with in their home continent (Messager et al. 2011).

As an immediate consequence of their first known movement out of Africa, hominids were thus already occupying a range of environments far broader than any documented for even the most eurytopic of primates today. Almost certainly, this penetration of new ecological zones was made possible by some form of cultural accommodation to local conditions. Indeed, much as Phillip Tobias (1995) observed in another context entirely, it seems likely that even at this early point cultural *accommodation* had become more important than biological *adaptation* as a factor governing hominid history. Exactly what the factor was that facilitated the Dmanisi hominids' penetration of the dry temperate zone must remain conjectural, since in terms of preserved technology there is no conceptual difference between the stone tools produced at Dmanisi and those that had already been produced in Africa for hundreds of thousands of years (e.g., Gabounia et al. 2002). But it seems reasonable to hazard that it was cultural accommodation that made it possible for the Dmanisi hominids to flourish in unfamiliar environmental conditions. And even were this not the case, it is evident that in later times it was material cultural innovations such as clothing, fire use, and shelter construction that eventually made possible later range expansions by *Homo* populations into yet more difficult environments to the north and west of the Caucasus.

Yet, while to some extent they almost certainly insulated hominids from the direct effects of biological selection, at least prior to the modern era technology and material culture had their functional limits. Of course, there can be little doubt that even simple material cultures would have had the potential to buffer hominid populations from some effects of the environment, and to allow its more efficient exploitation.

In favorable times cultural practices would certainly have facilitated geographic expansion of hominid populations into new environments. Equally likely, they would have made this occupation possible at higher population densities than would otherwise have been the case. But, in an age of dramatic climatic swings, material culture would not always have sufficed to maintain those larger populations in marginal zones when conditions became less propitious. At times of climatic deterioration, such as the onset of drought or extreme cold, hominid populations would have had to abandon difficult territories, becoming locally extinct where technological compensation failed. Even where culture may have allowed the survival of reduced and isolated population remnants, the new demographic and geographical circumstances would have enhanced the probabilities of biological divergence through drift alone, though it is not possible to preclude some biological adaptation to the new conditions.

From this perspective, one may consider material culture to be a factor that confers enhanced survival in isolation, and thus to be a potential initial trigger for both diversification and possible speciation. But by allowing generous range expansions in good times, material culture would also have made the "artificially" enlarged hominid populations more vulnerable to fragmentation in unfavorable conditions for which it could not completely compensate. Over the span of the climatically unsettled Pleistocene, multiply repeated sequences of such events would frequently have created the ideal conditions, in numerous and widely scattered hominid subpopulations, for the fixation of genetic novelties. Hence the tachytely we observe among Pleistocene hominids, expressed in parallel accelerated rates of accumulation of such novelties in multiple hominid lineages.

Of course, the basic dynamic involved here is a normal and fairly elementary feature of the evolving world. But, among hominids, cultural accommodation to unpredictably varying conditions would have created a sort of ratchet effect for evolutionary innovation. In good times, populations would have expanded into areas that lay climatically beyond their purely physiological limits. But when environmental conditions deteriorated beyond what prevailing material culture could cope with, those populations would have been fractured into small, genetically unstable units. If they contrived to avoid extinction these would, in turn, have been reunited when climatic amelioration occurred.

If speciation had intervened during the period of isolation, the result would have been competition among the newly reunited populations and the eventual elimination of some of them, potentially leading to the "trends" we discern in the fossil record. In the absence of speciation the entirely different phenomenon of reintegration would have occurred; but biologically it would have been equally significant, allowing the incorporation into the expanded population of genetic novelties that could never have become fixed if the earlier fragmentation had not happened. A further possibility is that cultural accommodation would have served to keep genetic novelties alive in populations that would otherwise have gone extinct, and have taken those novelties with them into oblivion.

Still, whatever the exact mechanism at work in any particular case, the possession by hominids of material culture in a fluctuating world would have had a profound effect on the evolutionary pattern we see in retrospect, reflected in the fossil record. In this perspective it is externally-mediated effects of this kind, rather than any internal dynamic, which place culture as such a powerful putative facilitator and accelerator of hominid evolutionary change, on both the physical and cognitive levels.

In a world of perpetual climatic and environmental oscillation, both the limits and the upside potential of technology provide us with plausible starting-points from which to examine the extraordinarily rapid rate of accumulation of morphological novelty in hominid populations over the last two million years. In its dual roles as facilitator of geographic expansion in good times, and as incomplete insulator in bad ones, material culture certainly seems more plausible as an explanatory agent for hominid tachytely than any amount of feedback between cognitive prowess – or any morphological factor – and individual reproductive success.

Finally, the most powerful metaphor for rapid hominid change over the Pleistocene is the remarkable rate of brain expansion in this group. Yet it is notable that, within the single surviving species *Homo sapiens*, the last 20 kyr or so have seen a trend toward brain size *reduction* (see Hawks 2011 and references therein). For example, Holloway et al. (2004) cite a mean brain size for a worldwide sample of recent humans of 1,330 ml. This contrasts with a mean of 1499 ml for a sample of 29 Late Pleistocene *Homo sapiens* calculated from Appendix 1 of the same source: a figure some 12.7% greater than the contemporary one. Various explanations have been put forward for this phenomenon, which cannot be explained by commensurately shrinking body size (Hawks 2011). Attempts have been made, for example, to associate endocranial volumes with climate, or more narrowly with prevailing temperatures (e.g., Beals et al. 1984). However, Bailey and Geary (2009) reject such climatic hypotheses in favor of a "dumbing-down" notion, whereby membership in increasingly complex societies placed decreasing intellectual demands on the individual. In essence, these authors argue that more elaborate social safety nets substituted for raw brain power. Wrangham (2011) blames "self-domestication" for the diminution of the human brain:body size ratio (brain sizes are typically reduced some 10–15% in domestic forms compared to their wild counterparts), while Hawks (2011) more vaguely associates smaller endocranial volumes with "higher fitness" resulting from unspecified causes.

Yet, one obvious explanatory possibility for brain diminution has been overlooked. The earliest anatomically modern *Homo sapiens*, known from eastern Africa in the period following 200 ka, had large brains that appear to have functioned much as the Neanderthals' equally large brains did (Tattersall 2012). The fateful shift to the symbolic information processing mode already referred to appears to have happened significantly later, in the period following about 100 ka. Once this shift had occurred, the metabolically expensive human brain found itself working on a new and different processing algorithm: one that was less dependent on the sheer volume of brain tissue than on the specific nature of the operations and connections within it. Quite simply, a more efficient algorithm may have permitted a reduction in the quantity of energy-hungry brain tissue, while simultaneously making possible a qualitative leap in processing power.

Acknowledgments My gratitude goes to Clive and Geraldine Finlayson, and to Darren Fa, for inviting me to the splendid Calpe'12 Conference on "The Human Niche: Ecology, Behavior and Culture," for which the thoughts in this essay were originally gathered. And I equally warmly thank Assaf Marom and Erella Hovers for enabling me to express them in appreciation of our great friend and colleague Yoel Rak. The perceptive comments of two anonymous reviewers improved the manuscript, and Jennifer Steffey kindly prepared the illustrations.

References

Bailey, D. H., & Geary, D. C. (2009). Hominid brain evolution: Climatic, ecological, and social competition models. *Human Nature, 20,* 67–79.

Beals, K. L., Smith, C. L., & Dodd, S. M. (1984). Brain size, cranial morphology, climate and time machines. *Current Anthropology, 25,* 301–330.

Collard, M., & Wood, B. (2015). Defining the genus *Homo.* In W. Henke & I. Tattersall (Eds.), *Handbook of paleoanthropology* (2nd ed., pp. 1575–1610). Berlin: Springer.

de Heinzelin, J., Clark, J. D., White, T., Hart, W., Renne, P., WoldeGabriel, G., et al. (1999). Environment and behavior of 2.5 million-year-old Bouri hominids. *Science, 284,* 625–629.

Eldredge, N., & Cracraft, J. L. (1980). *Phylogenetic patterns and the evolutionary process.* New York: Columbia University Press.

Gabunia, L., Vekua, A., Lordkipanidze, D., Swisher, C. C., Ferring, R., Justus, A., et al. (2000). Earliest Pleistocene hominid cranial remains from Dmanisi, Republic of Georgia: Taxonomy, geological setting and age. *Science, 288,* 1019–1025.

Gabounia, L., de Lumley, M.-A., Vekua, A., Lordkipanidze, D., & de Lumley, H. (2002). Découverte d'un nouvel hominidé à Dmanissi (Transcaucasie, Géorgie). *Comptes Rendus Palevol, 1,* 243–253.

Hartwig, W. (Ed.). (2002). *The primate fossil record.* Cambridge, UK: Cambridge University Press.

Hawks, J. (2011). Selection for smaller brains in Holocene human evolution. arXiv:1102.5604v1.

Holloway, R. L., Broadfield, D. C., & Yuan, M. S. (2004). *The human fossil record, vol. 3: Hominid endocasts, the paleoneurological evidence.* New York: Wiley-Liss.

Kimbel, W. H., Johanson, D. C., & Rak, Y. (1997). Systematic assessment of a maxilla of *Homo* from Hadar, Ethiopia. *American Journal of Physical Anthropology, 103,* 236–262.

Leakey, M. G., Spoor, F., Brown, F. H., Gathogo, P. N., Leakey, L. N., & McDougall, I. (2001). New hominin genus from eastern Africa shows diverse middle Pliocene lineages. *Nature, 410,* 433–440.

Lepre, C. J., Roche, H., Kent, D. V., Harmand, S., Quinn, R. L., Brugal, J.-P., et al. (2011). An earlier origin for the Acheulian. *Nature, 477,* 82–85.

Lumsden, C., & Wilson, E. (1981). *Genes, mind and culture: The coevolutionary process.* Cambridge, MA: Harvard University Press.

Lumsden, C., & Wilson, E. (1982). Précis of *Genes, mind and culture. Behavioral and Brain Sciences, 5,* 1–7.

Matsudaira, K., & Ishida, T. (2010). Phylogenetic relationships and divergence dates of the whole mitochondrial genome sequences among three gibbon genera. *Molecular Phylogenetics and Evolution, 55,* 454–459.

Mercader, J., Barton, H., Gillespie J., Harris, J., Kuhn, S., Tyler, R., et al. (2007). 4,300-year-old chimpanzee sites and the origins of percussive stone technology. *Proceedings of the National Academy of Science, USA, 104,* 3043–3048.

Messager, E., Lebreton, V., Marquez, L., Russo-Ermoli, E., Orain, R., Renault-Miskovsky, J., et al. (2011). Palaeoenvironments of early hominins in temperate and Mediterranean Eurasia: New palaeobotanical data from Palaeolithic key-sites and synchronous natural sequences. *Quaternary Science Reviews, 30,* 1439–1447.

Pickering, R., Dirks, P. G. M., Jinnah, Z., de Ruiter, D., Churchill, S. E., Herries, A. I. R., et al. (2011). *Australopithecus sediba* at 1.977 Ma and implications for the origins of the genus *Homo. Science, 333,* 1421–1422.

Richerson, P. J., & Boyd, R. (2005). *Not by genes alone: How culture transformed human evolution.* Chicago: University of Chicago Press.

Schrenk, F., Bromage, T., Betzler, C., Ring, U., & Juwayeyi, Y. (1993). Oldest *Homo* and Pliocene biogeography of the Malawi Rift. *Nature, 365,* 833–836.

Schwartz, J. H., & Tattersall, I. (2005). *The human fossil record, vol. 3: Genera* Australopithecus, Paranthropus, Orrorin, *and overview.* New York: Wiley-Liss.

Semaw, S., Renne, P., Harris, J. W. K., Feibel, C. S., Bernor, R. L., Fesseha, N., et al. (1997). 2.5 million-year-old stone tools from Gona. *Ethiopia. Nature, 385,* 333–336.

Simpson, G. G. (1944). *Tempo and mode in evolution.* New York: Columbia University Press.

Simpson, G. G. (1953). *The major features of evolution.* New York: Columbia University Press.

Sponheimer, M., & Lee-Thorp, J. (2007). Hominin paleodiets: The contribution of stable isotopes. In W. Henke & I. Tattersall (Eds.), *Handbook of Paleoanthropology* (Vol. 1, pp. 554–585). New York: Springer.

Susman, R., Stern, J. T., & Jungers, W. L. (1984). Arboreality and bipedality in the Hadar hominids. *Folia Primatologica, 43,* 113–156.

Stone, A. C., Battistuzzi, F. U., Kubatko, L. S., Perry, G. H., Jr., Trudeau, E., Lin, H., et al. (2010). More reliable estimates of divergence times in *Pan* using complete mtDNA sequences and accounting for population structure. *Philosophical Transactions of the Royal Society B, 365,* 3277–3288.

Tattersall, I. (2008). An evolutionary framework for the acquisition of symbolic cognition by *Homo sapiens. Comparative Cognition and Behavior Reviews, 3,* 99–114.

Tattersall, I. (2012). *Masters of the planet: The search for our human origins*. New York: Palgrave Macmillan.

Tattersall, I., & Schwartz, J. H. (2009). Evolution of the genus *Homo*. *Annual Reviews of Earth and Planetary Sciences, 37*, 67–92.

Tobias, P. V. (1995). *The communication of the dead: Earliest vestiges of the origin of articulate language*. Amsterdam: Kroon Lectures, Stichting Nederlands Museum voor Anthropologie en Praehisoire.

Van Andel, T. H., & Davies, W. (Eds.). (2003). *Neanderthals and modern humans in the European landscape during the last glaciation*. Cambridge, UK: McDonald Institute for Archaeological Research.

Wood, B., & Collard, M. (1999). The human genus. *Science, 284*, 65–71.

Wrangham, R. (2011). Chimpanzees, bonobos and the self-domestication hypothesis. *American Journal of Primatology, 73*, 33–45.

Chapter 2
Wallace's Controversy with Darwin on Man's Mental Evolution, on the Position of the Natives in Human Evolution, and His Anticipation of Cultural Evolution, as Distinct from Biological Evolution

Joseph Neumann

Abstract Darwin argued that man, including his mental faculties, developed from his sub-human ancestors by natural selection, sexual selection, and the use and disuse of organs (in the Lamarckian mode). He rejected any non-natural involvement in this process, and described a large number of behavioral and mental properties, including language, which can be found in rudimentary form in some animals. However, he assumed this, prior of the discovery of the crucial differences between the instinctive and specific calls of animals, and the symbolic language of humans. His major conclusion was that although the gap in the mental properties between humans and their closest relatives is enormous, it is quantitative rather than qualitative. With regard to the different human races, Darwin suggested that they differ in their inherited mental properties, but belong to a single species. In contrast to Darwin, Wallace did not regard modern human "primitives" as candidates that could fill the gap between humans and apes. He envisioned two steps in human evolution: first, the development of upright posture and freeing of the hands, brought about by natural selection, and then a second step that involved mainly the evolution of the brain and the mind. Wallace subsequently argued that some of the higher human mental abilities (mathematics, art, or the use of abstract concepts) were not the result of natural selection, since they are beyond utility. He claimed that these properties developed as a result of the action of a "higher intelligence", which guides human intelligence and morality, and the whole evolutionary process, purposefully. There is some disagreement as to whether Wallace's belief in the action of a "higher intelligence", and his descent from Darwin on this issue, were the result of his support of spiritualism or was based on purely scientific arguments. Darwin, on his part, forcefully rejected Wallace's support of the involvement of non-natural causes in evolution of human mental faculties and provided arguments that they were the result of the same mechanisms that acted in the formation of the body, and generally in species evolution. Later, S. J. Gould pointed out that the rapid rate of the development of several mental functions, which Wallace had regarded as an indication of a lack of role in the struggle of life are actually the result of cultural evolution. Both Darwin and Wallace did not pay sufficient attention to the large diversity in human mentality, and the rare and unique existence of individuals with outstanding achievements ("geniuses"). The latter's unusual and unique creativity in various artistic, philosophical and related activities apparently developed intrinsically, from some "inner resources", unrelated to the Darwinian "struggle for life".

Keywords Darwin • Guiding intelligence • History of science • "Struggle for life" • Wallace

Introduction

It is known that the publication of the "Origin of Species" by Darwin in 1859 was provoked by a short assay by Alfred Russel Wallace, who outlined a similar theory, and sent it to Darwin for review and publication. In the "Origin", Darwin devoted just a single sentence to man: "Light will be thrown on the origin of man and his history". The detailed discussion of human evolution had to wait till 1871, with the publication of "The Descent of Man", which was too, in a sense, a response to Wallace, who at about this time had abandoned natural selection as a cause for the formation of human higher mental faculties, and replaced it by the action of a "higher intelligence".

It should be noted that in "The Descent", Darwin extended his theory to man, without having the benefit of the

J. Neumann (✉)
Department of Molecular Biology and Ecology of Plants, Tel-Aviv University, 69978 Tel-Aviv, Israel
e-mail: jnoy@post.tau.ac.il

J. Neumann
Department of Philosophy, Tel-Aviv University, 69978 Tel-Aviv, Israel

Assaf Marom and Erella Hovers (eds.), *Human Paleontology and Prehistory*, Vertebrate Paleobiology and Paleoanthropology, DOI: 10.1007/978-3-319-46646-0_2

evidence of a single subhuman fossil. His arguments in the "Descent" were based on his own observations, on the scientific and popular publications of others, and occasionally, on some anecdotes.

Darwin's thesis was opposed to the widely accepted view of his time. According to Darwin, "many authors insisted, that man is divided by an insuperable barrier from all the lower animals in his mental faculties. … man alone is capable of progressive improvement; that he alone makes use of tools or fire, domesticated other animals or possesses property; that no animal has the power of abstraction, or of forming general concepts, is self-conscious and comprehends himself; that no animal employs language; that man alone has a sense of beauty, is liable to caprice, has feeling of gratitude, mystery etc.; believes in God or is endowed with a conscience" (Darwin 2009, p. 70).

In opposition to this view, Darwin believed that man descended from an ancestral form, common to man and the anthropoid apes, by the same mechanisms that were active in the evolution of other species, namely, natural selection (based on the laws of variation and heredity), sexual selection, the inherited effects of use and disuse, (in the Lamarckian mode), and "correlated variation".[1]

Darwin insisted that both human body and mental faculties, including intellectual, moral and spiritual capacities, have been derived from their rudiments in the lower animals, through the above mentioned mechanisms. He presented many observations, showing that the rudiments of most, if not all mental and moral faculties of man are present in some animals. Thus, certain animals exhibit distinct acts of reasoning, curiosity, imitation, attention, wonder and memory; some of their behaviors may be interpreted as displays of kindness toward their fellows; some exhibit pride, contempt, shame, suspicion, pleasure, pain, happiness, misery and fear, as well as courage and timidity; some exhibit behavior that suggests the power to deceive; many animals exhibit maternal affection; grief; attention; jealousy; some adopt youngsters, even from other species; and the love of the dog (a domesticated beast) for his master is well known.[2]

As for the origin of the mental powers Darwin wrote: "In what manner the mental powers developed in the lower organisms, is as hopeless an inquiry as how life originated" (Darwin 2009: 61).

Turning to the development of intellect, Darwin endorsed the premise that the size of the brain is closely correlated with the development of the intellectual faculty. This he thought is supported by the "skulls of savage and civilized races, of ancient and modern people and by the comparison of the whole vertebrate series" (Darwin 2009: 52).

One important feature separating humans from other animals is language. According to Darwin, language also developed in the process of evolution; it depended on, and was enhanced by sociality. Darwin compared the similarity of the formation of the different languages, with the formation of the species, indicating that the former developed also through a gradual process (Darwin 2010: 33).[3]

Darwin assumed that the human "vocal organs" became adapted through the inherited effect of use for the utterances of articulate language. He stressed the similarity between human language and the calls made by certain animals, suggesting that the two may have developed by comparable mechanisms. Some animals, indeed, utter different sounds, to their fellows or their young, each which a different message. However, he wrote this, before the discovery of the crucial difference between the instinctive calls of animals, and human symbolic language.[4]

All in all, Darwin's major conclusion was that the difference in mental abilities between man and the higher

---

[1]Darwin noted that since an organism is an integrated whole, an adaptive change in one part of the organism, may entail non-adaptive changes in other parts (Darwin 2009: 44).

[2]Note that here Darwin drew conclusions about the existence of feelings and emotions, like fear, anger and pleasure, which are subjective, from the observation of behavior – an objective property. Still, it should be mentioned that Darwin did speculate about the relation between the brain and the mind – "The brain, for example, might secrete thoughts as the liver secreted bile" (quoted by Richards 2005: 169).

[3]Modern support for the evolutionary origin of language was discussed in Pinker (1994). Pinker regards language as an ability unique to humans, formed during evolution, in order to solve the specific problem of communication among social hunter-gatherers. He compared language to other species' adaptations, such as spiders' web-weaving or beavers' dam-building behavior, designating all three "instincts".

[4]Unlike human language, which is based on a large vocabulary, that can still be enlarged, animals possess a limited number of sounds, each one directed to a specific aim. Animals are unable to increase the number of their sounds, or transform their emotional cries into sounds with different meanings. Human language, on the other hand, is composed of symbols (Cassirer 1944), with a wide range of meanings, including the capacity to refer to past and future events. A symbol is not an element of reality, like mass or energy; it is a sign that a humans refer to an entity, by arbitrary convention.

According to the philosopher Karl Popper (1972), "Human languages share with animal languages the two lower functions: (1) self-expression and (2) signaling. Animal language is symptomatic of the state of the organism; whereas the signaling or release function can cause a response in another organism".

On the other hand, human languages have in addition, many other functions. And the two most important according to Popper (1972) are: the descriptive function and the argumentative function. "It is to the development of these higher functions that we owe our human reason. They are also a condition for acquiring knowledge".

One should add, that humans use language for many other functions, like asking questions, giving promises or giving orders; it is also a prerequisite for the development of a complex human culture (see below).

Finally, today we know that the sounds of animals depend on the activity of an evolutionary older part of the brain, the "limbic system", whereas human language is based on the activity of the neo-cortex.

animals, although immense, is one of degree and not of kind; it is quantitative and not qualitative.

Darwin on Human Races[5]

Darwin's opinion on human races was equivocal. It has been argued that Darwin was not a racist. He actively opposed the mistreatment of other races and opposed slavery. During his voyage on the 'Beagle' he described the Fuegians as a starving, dirty, ill clad, and war like people, who would kill and eat their elderly women before they devour their hunting dogs. On the other hand he wrote: "The Fuegians rank among the lowest barbarians, [but]…the three natives on board H.M.S. 'Beagle', who have lived some years in England … resembled us in disposition and in most our mental faculties" (Darwin 2009: 60).

Darwin claimed that until paleontological evidence of human origin were discovered, the best case for human evolution could be made by assuming that the most primitive human groups could be shown to be behaviorally as little different as possible from the great apes.

Belonging to the cultural milieu of the mid-19th century Victorian England, Darwin believed in a racial gradation tracing back to the ape. The less culturally advanced people were regarded as living fossils, both culturally and physically, without a clear differentiation between the two.

In the "Descent" (quoted by Eiseley 1961: 288) Darwin "implied marked differences in the inherited mental faculties between the members of the different existing races, postulating that in the lowest savages many of these faculties are very little advanced from the condition in which they appear in the higher animals, and some are very inferior in comparison to those that appear in the civilized races".

In addition, Darwin, like many thinkers of his time, argued that the cultures had changed from the simple to the complex, by gradually, developing from an original type that was perhaps less different, from that of the great apes, than it was from the most advanced modern societies. He assumed that all civilized nations were once barbarous, which he supported by observation, of the low conditions, customs, beliefs, language etc. in the societies of the natives of his day.

All in all, according to Darwin, the western nations of Europe immeasurably surpassed their former savage progenitors and stand now at the summit of civilization; still he maintained, that all human races descended from a single ancestral population, thus believing in monogenism as against polygenism, according to which the different races, represent different lineages of origin.

Alfred Russel Wallace

Wallace was a naturalist who spent a considerable time among the tribal societies in South America and South-East Asia under conditions where his existence depended on their help. Observing their life extensively, he concluded that these people, as far as their behavior and habits are concerned, were indeed retarded in comparison to the Europeans, but basically they are neither intellectually nor morally inferior to them; and with proper training, could rapidly reach their level. Unlike Darwin, Wallace did not explain human races as representing successive stages of evolution leading up to the Europeans; and maintained that there were no essential differences between civilized and savage men. Further breaking from Darwin, he did not regard "the modern primitives as almost filling the gap between man and ape" (Eiseley 1961, p. 305). Wallace rejected Darwin's conclusion that the mental faculties of the savages are very little advanced from their conditions in the higher animals, and that they are much inferior in comparison to those possessed by the civilized races. In his description of the natives, Wallace betrays scarcely a trace of the superiority so common in nineteenth-century European scientific circles (Eiseley 1961).

With regard to human evolution, Wallace accepted Darwin's basic conclusion that human's bodily structure descended from an ancestral form, common to man and the anthropoid apes, by natural selection. However, in a paper published in 1864 (quoted in Darwin 2009: 107), he presented a novel idea, according to which the rise of the human brain had altogether altered the nature of the evolutionary process (Eiseley 1961). Wallace maintained that human evolution took place in two stages: the first was indeed a product of natural selection and resulted in the physical changes of the body, culminating in the bipedal posture and the freeing of the hands, as implements to carry out the dictates of the brain; however, in a second stage whose postulation constituted Wallace's original contribution to the evolution of man (Eiseley 1961) nature had at last produced an organism that was not confined to any narrow category of existence, but rather was potentially capable of endless inventions (by which Wallace alluded to cultural evolution, see below), a being whose mind was of vastly greater importance than his bodily structure – "a true culture-producing brain" (Eiseley 1961: 318).

Wallace pointed out that the bodily differences between man and the great apes were small, but the gap in mental and cranial characters was vast. He surmised that the evolution of

---

[5]In modern times, some anthropologists (e.g., Alland 1973) have claimed that the term "race" should be restricted to sociological analyses, since according to this view, it is not a valid taxonomic unit in biology.

human cranial size was a very long process, perhaps lasting as long as ten million years (Eiseley 1961, p. 307).[6]

Wallace's "Apostasy"

Several years after publishing the paper about the two phases of human evolution, Wallace made a radical change in his attitude to the development of mind (sometimes dubbed as "apostasy"). In a paper published in 1869, Wallace came to the conclusion that "natural selection and its purely utilitarian approach to life could not account for many aspects and capacities of the human brain" (quoted in Eiseley 1961, p. 310). "We must therefore admit, that man's large brain could never have been solely developed by any of those laws of evolution, whose essence is that they lead to a degree of organization exactly proportionate to the wants of each species never beyond those wants" (Shanahan 2004, p. 252). "There had come into existence, (Wallace emphasized), a being in whom mind was of vastly greater importance than bodily structure"; this view, "neither requires us to depreciate the intellectual chasm which separates man from the apes, nor refuses the full recognition of the striking resemblances to them, which exists in other parts of his structure" (quoted in Eiseley 1961: 308). Furthermore, "Natural selection...could have endowed the savage with a brain a little superior to that of an ape, whereas he actually possesses one but very little inferior to that of the average member of our learned societies" (Wallace's quoted in Eiseley 1961: 311).

Commenting on this statement, Loren Eiesley (1961: 311) wrote: "Today when careful distinctions are made between natural genetic endowment and cultural inheritance, such a remark does not sound particularly iconoclastic. In Wallace's time, however, it was a direct challenge to western ethnocentrism and the whole conception of the natives as a living fossil".

Wallace pointed out that "among the lowest savages with the least copious vocabularies, the capacity of uttering a variety of distinct articulate sounds, and of applying them to an almost infinite amount of modulation and inflection, is not in any way inferior to that of the higher races. Thus, the problem posed by human evolution was the failure of natural selection to explain the enlarged human brain (event of the savages), compared to that of the apes, (since as far as we know, the brains of savages are neither smaller nor more poorly organized than our own),[7] as well as the organ of speech. *An instrument has been developed in advance of the needs of its possessor"* (my emphasis); Wallace quoted in Eiseley (1961: 311).

Wallace reminded us that Darwin maintained in the "Origin" that "natural selection tends only to make each organic being as perfect as or slightly more perfect than, the other inhabitants of the same country with which it has to struggle for existence; "Natural selection will not produce absolute perfection". Thus, Wallace concluded that natural selection and its purely utilitarian approach cannot account for many aspects and capacities of the human brain.

Though, like all his contemporaries, Wallace did not doubt the superiority of the European culture, he believed that all human groups had *innately* equal intellectual capacities.

With regard to the role of natural selection in the development of human mental evolution, Darwin did not concur. "Man in the rudest state in which he now exists it the most dominant animal that has ever appeared on this earth... He manifestly owes this superiority to his intellectual faculties, to his social habits, which laid him to aid and defend his fellows, and to his corporeal structure, ...through his power of intellect, articulate language has been evolved... He has invented and is able to use various weapons, tools traps etc., by which he defends himself... He has made canoes for fishing or for crossing to neighboring fertile islands. He discovered the art of making fire... These several inventions, by which man in the rudest state has become so pre-eminent are the direct result of the development of his power of observation, memory, curiosity, imagination and reason. I cannot therefore understand how it is that Mr. Wallace maintains that natural selection could only have endowed the savage with a brain a little superior to that of ape" (Darwin 2009: 48).

The intellectual and moral faculties of man are variable and probably heritable, "therefore if they were formerly of high importance to primeval man and to his ape-like

---

[6]Since Darwin's and Wallace's time, a number of highly important "proto-human" fossils were discovered. Some of these could be arranged (in hindsight!) as a series of "missing links" leading to modern humans. Based on these discoveries, it is indeed by now agreed, that human bipedal posture and the freeing of the hands preceded the large end very fast rate (on an "evolutionary time scale") expansion of the brain.

Unlike Wallace's supposition, that this process took perhaps 10 million years, there is now substantial evidence that the brain increased over the last 2 million years from about 500 cc (a size only slightly over that of non-human primates) to about almost 1400 cc. This fast rate of change was probably not the result of ecological change, but of fierce social competition (e.g., Foley 1995).

The social competition was expressed by Richard Dawkins as the dictate "to be smart and outsmart the other", a type of competition which led to an "arms race", i.e., a process of "evolutionary interactions, within a species or between two species, in which each player becomes adapted as a result of interaction with the other player" (Sterelny 2007: 199).

[7]Thomas Henry Huxley responded to Wallace's challenge by pointing out, that the life of primitive people actually required extraordinary mental feats. "The intellectual labor of a good hunter or warrior considerably exceeds that of an ordinary Englishman" (Shanahan 2004: 253).

progenitors, they would have been perfected or advanced through natural selection" (Darwin 2009: 107). Thus, Darwin concluded that both the intellectual and moral faculties have been increased by natural selection.

Darwin speculated that in the civilized society, perhaps those of superior intellect tend to rear a greater number of children hence producing "some tendency to an increase in both number and standard of the intellectually able". He claimed that those individuals who were the most sagacious, who invented and used the best weapons, would rear the greatest number of offspring. In the same vain, the tribes that included the greatest numbers of such men would increase in number and supplant other tribes.

Moreover, Darwin claimed that since there are gradations in mental capacity between a savage and a Newton or a Shakespeare,[8] gradual changes are possible between civilized people and brutes, and between the latter and some primeval man (Darwin 2009: 60).

Wallace's descent from Darwin, concerning the alleged insufficiency of natural selection in the formation of various mental faculties in man, was supported by several observations and arguments. Regarding the mathematical faculty, Wallace claimed that in the lower races, this faculty is either absent or quite unexercised, if at all present. Bushmen are unable to count beyond two; and many Australians tribes can count only to six, whereas people in civilized races can count up to hundred thousand. Moreover, the development of the mathematical faculty in its broad sense depended on the introduction (in the sixteenth century), of the decimal notation, after which, it developed very rapidly and widely, particularly in the last three centuries. This fast development, Wallace argued, could not be the result of natural selection, since it did not serve as a means in the struggle for life, neither between individuals nor between tribes or nations (Wallace 1889: 277).

The musical faculty resembles the mathematical. Among the savages, music as we understand it, hardly existed; no elements of harmony, or other essential features of modern music were present, and little progress took place, until the fifteenth century. From that point on, however, the musical faculty advanced rapidly and in curious tandem with the advance of mathematics, with great musical geniuses appearing suddenly among different nations, at about the same time (Wallace 1889: 280).

Again, like the mathematical faculty, Wallace argues, this fast development is unrelated to the struggle of life, and he continues, "It seems to have arisen as a *result* of social and intellectual advancement" (Wallace 1889: 280).

Alluding to the metaphysical faculty, which enables us to form abstract concepts remote from any practical applications, such as the concept of cause, the nature and qualities of matter, the existence of the will and the existence of the conscience, Wallace states that they appear suddenly, and develop very rapidly. They are unique to humans and are not derived from animals.

Considering the development of the mathematical faculty, Wallace claimed: "we are limited to two possible theories": either the natives did not possess this faculty, or else they possessed it, but had neither the means nor the incentive for its exercise. In the former case, we have to ask by what means had this faculty appeared, and rapidly developed in the civilized races, reaching the level of a Newton, a La Place or a Gauss;[9] what motive power caused this development? (Wallace 1889: 278). What advantage has this extremely fast development of the mathematical faculty for the individual possessor in the struggle for life, in the struggle of tribe with tribe, of race with race?; if it had no such advantage, it could not have developed by natural selection.

As an alternative explanation, Wallace considered the possibility of the existence of the above mentioned properties in a latent form, which became activated under particular circumstances, very much later; he claims that this option, posed even a greater difficulty. Any property formed by natural selection must have some advantage at the time and place of its formation; no property can be formed by this mechanism for future use; no creature can be improved beyond the necessary existence.

In addition, anticipating Darwin's response, Wallace argued that "to prove continuity and the progressive development, of the intellectual (and moral) faculties leading from animals to man, is not the same as proving that these faculties have been developed by natural selection". In Wallace words, "Because man's physical structure has been developed from an animal form by natural selection, it does not

---

[8]Newton and Shakespeare are regarded as "geniuses", a quality defined by Rubens as "evincing of exceptional range of vision, and exceptional technique for conveying that vision". All the epithets used here imply that genius is extremely rare (Rubens 2012: 78–85).

More important and relevant to Darwin's conclusion, in regarding Newton or Shakespeare as indicating "degrees" of human mental evolution, is the fact he is referring to their phenotype (and not their genotype, concepts unknown to Darwin, and other biologists at his time), and therefore irrelevant to evolution.

[9]See footnote 10. In addition, it should be noted, that both Darwin and Wallace did not address the problem of the existence of the enormous mental differences among men. In a book published about 60 years after the Wallace-Darwin dispute, the anthropologist Alexander Alland J. wrote: "Acceptance of the problem of [the mental] differences [should be searched] in historical, rather than genetic terms ... [in] importance of contact between people as stimulant to creative thinking. It is an exchange of ideas, not of genes [that matters] ... The accomplishments of Greek philosophers and scientist, Elizabethan writers, Flemish painters, German musicians, are understandable not in terms of biological changes that occurred antecedent to their periods of intense activity, but in light of peculiar conjunctions of outlooks and juxtapositions of contrasting world views" (Alland 1973: 167).

necessarily follow that his mental nature, even though developed *pari passu* (side by side) with it, has been developed by the same causes only" (Wallace 1889: 277).[10]

In addition, Wallace pointed out that the fast development of the mental faculties in the fields of music, mathematics or metaphysics is confined to a very small segment of the population, claiming, that "natural selection cannot work on extreme variations that crop up in only a tiny proportion of the population … Natural selection cannot work on extreme variation…" (Wallace 1889: 280).

This statement calls for some qualifications. To the extent that "geniuses" have some inborn (today we shall call it genetic) components, (a possibility that was supported at the time of Darwin by Francis Galton), it should be pointed out (in hindsight, and again based on our present knowledge) that the problem is not their rarity, but the question whether these outstanding people had any advantage, as far as differential reproduction is concerned, which in some famous individuals, like Kant, Newton or Schubert, who were childless, they evidently had not.

In summary, Wallace's major conclusion was that man's higher mental abilities, his intellectual (and moral) faculties have not been developed by natural selection, but were formed by some other "influence" for a special purpose (resembling man, who can direct and select in the process of artificial selection of plants or animals, certain properties); they "point to the existence in man of something which has not been derived from his animal progenitors – something which we may best refer to as being under spiritual essence … we may perceive that the love of truth, the delight of beauty, the passion for justice ….are the working within us of a higher nature which has not been developed by means of the struggle for material existence" (Wallace 1889: 282); it also explains the enormous influence of ideas and beliefs over man's action and his whole life. It is pertinent to mention (as Wallace does not) that this capacity seems to be a mixed blessing![11]

Furthermore, Wallace claims that "the nobler qualities of justice, mercy and humanity…have been steadily increasing in the world" (Wallace 1892: 284). The statement reflects perhaps the rather myopic view of a nineteenth century Victorian thinker, but becomes very questionable in the 21st century!

Against the expected argument, this belief in a "higher intelligence" introduces a new cause in the continuous process of evolution. Wallace reminds us that the three new powers had been introduced (in the development of the organic world), which caused a breach of continuity: the change from the inorganic to the organic (introducing vitality), the introduction of sensation or consciousness into the animal kingdom, and the third one, discussed above. The latter "raises [man] furthest above the brutes and opens up possibilities of almost indefinite advancement". In this phase Wallace includes "the constancy of the martyr, the unselfishness of the philanthropist, the devotion of the patriot…the love for beauty and more" (Wallace 1889: 282).

"These three distinct stages of progress, from the inorganic world of matter and motion up to man, point clearly to an unseen universe – to a world of spirit, to which the world of matter is altogether subordinate" (Wallace 1889: 283). The existence of a spiritual world would also remove the sense of despair about the ultimate fate of the universe (referring to the "heat death", as a result of the second law of thermodynamics). In Wallace's words, "we who accept the existence of a spiritual world, can look upon the universe as a grand consistent whole, adapted in all its parts to the development of spiritual beings capable of indefinite life and perfectibility… To us the whole purpose, the only *raison d'être* (reason for existence) of the world … was the development of the human spirit in association of the human body" (Wallace 1889: 284).

It is known that Wallace turned to spiritualism,[12] believing (among other supernatural phenomena) that departed souls can communicate through mediums with humans still living on Earth (Wallace 1892). He attended séances, and claimed to obtain messages from dead friends.

In addition, Wallace was known to be a reformer and a socialist who was passionately concerned with struggles for justice and well-being for humanity – values that were inconsistent, in his view, with a materialistic philosophy (Wallace 1892).

There is some disagreement as to whether Wallace's turn to spiritualism affected his position regarding his dispute with Darwin. According to Cartwright (2001: 17): "What seems to have prompted Wallace's apostasy from the cause of

---

[10]Today, such a separation between the body (or the brain) and the mental systems, as is implied by Wallace's description, would be rejected by most philosophers and neuroscientist. For example, the philosopher John Searle wrote: "We know that human and some animal brains are conscious. Those living systems with certain sorts of nervous systems are the only systems in the world that we know for a fact are conscious" (Searle 1997: 170).

[11]The psychologist Charles Rycroft wrote: "As both religious and political history show, men who in their private life may be kind and tolerant are prepared to kill, persecute and engage in heresy-hunting at the behest of abstract nouns, whether these be God, Liberty, Equality, Fraternity, the Fatherland or the Party." (Rycorft 1985: 293). Note also that here once again, Wallace disregards the extreme diversity among men with respect to the above mentioned properties.

[12]Spiritualism is the name applied to a belief in a series of abnormal phenomena, including the possibility to communicate with the dead, through mediums. Spiritualists claim that their beliefs are founded on evidence and proven beyond any reasonable doubt. In addition spiritualism is based on the belief that the whole material universe exists for the purpose of spiritual development, and that death is simply a transition from material existence to spirit life.

naturalism was his conversion, around 1866 to spiritualism. Like many of his British contemporaries, including Francis Galton … and some Americans, like William James".

Kottler (1974), in a detailed and closely argued paper, also claimed that Wallace's belief in spiritualism was a major cause of his departure from Darwin. On the other hand, Harman (2004), in reviewing Michael Shermer's "In Darwin's Shadow: The Life and Science of Alfred Russell Wallace", argued that according to the latter, "[Wallace's] spiritualism did not influence his science or his teleological evolutionary worldview… He simply assumed that a guiding intelligence was a more likely inference from reality than the reductionist view, ascribing the mystery of mind to the properties of matter" (Shermer 2002; Harman 2004: 470–473).[13]

How did Darwin react to Wallace's "apostasy"? He concurred that humans indeed have a powerful ability to adapt to new life conditions by inventing weapons, tools, clothes and dwellings, and making fire. They aid their fellow men in many ways, and anticipate future events; even in remote periods humans practiced some form of division of labor. However, contrary to Wallace, Darwin claimed that since the intellectual and moral faculties of man are variable and probably heritable, "therefore if they were formerly of high importance to primeval man and to his ape-like progenitors, they would have been perfected or advanced through natural selection" (Darwin 2009: 107).

As for the introduction of a "higher intelligence", Darwin was no less than dismayed by Wallace's "heresy" and his response is by now notorious: "I hope you have not murdered too completely your own and my child" (quoted in Eiseley 1961: 313). He was worried that his co-discoverer of evolution had lost his nerve when it came to consider the case of humans. Darwin vehemently opposed Wallace's conclusion about the involvement of some "higher intelligence" in the formation of human intellectual and moral faculties; "he could never endure miraculous additions at any one stage of ascent" (Eiseley 1961: 313). "Darwin's aim [in the "Descent of Man] was to elaborate a thoroughly naturalistic account of human characteristics physical and mental" (Shanahan 2004: 254).

With regard to Wallace's belief in evolutionary progress it is fitting to quote Howard (1982: 77): "Perfection and progress were abstractions which had no place in Darwin's pragmatic and relativistic scheme… "perfection" in biological organization could be defined only in relation to the environment in which an animal or plant live". However,

Darwin's attitude to the idea of progress in evolution of species, and the evolution of man is in dispute.

According to Shanahan, who summarized Darwin's idea of progress in the "Descent", "Darwin's evolutionary progress is both a well-grounded theoretical prediction derived from the theory of natural selection, and an established empirical fact confirmed by geological evidence" (Shanahan 2004: 192). A contrary view is presented by Foley (1995), and it is worthwhile to quote in length from his book. "Along with the growth of knowledge of animal behavior has come a greater understanding of the diversity of human life, and to some extent to which humans could be said to be above the swamp of animal brutishness. The camps of Dachau and Belsen, the millions killed in religious wars, and the almost boundless capacity of humans to do damage to each other at national and personal levels, in the twentieth century, rather dented human self esteem." (Foley 1995: 39).

According to S.J. Gould, who studied extensively the question of progress in evolution, "…the overarching aim of his book *Full House* is to present the general argument for denying that progress defined the history of life or even exists as a general trend at all" (Shanahan 2004: 207).

Gould's Criticism of Wallace

The prominent paleontologist Stephen Jay Gould contested Wallace's conclusion that the development of man's mental faculties depended on the action of a "higher intelligence". To begin with, he pointed out that unlike Darwin, who repeatedly emphasized that "natural selection has been the chief, but not the only agent of change" (during evolution), Wallace (according to Gould) was a "pan-selectionist", believing that each and every property of the organism was the result of natural selection leading to an improved adaptation.

It is known that Darwin added "sexual selection" to the principle of natural selection – the competition between males for females, for reproduction (independent of the availability of any resources) and "female choice", where the female selects the more agreeable partner.[14] Wallace rejected sexual selection, (particularly "female choice" where there was an element of "volition"). Darwin assigned a rather important role to "sexual selection" in the formation of the different human races.[15]

---

[13]It is of some interest to note that the distinguished American Philosopher, Thomas Nagel, has recently published a book – "Mind and Cosmos", (2012), in which he claimed that Neo-Darwinism is probably unable to explain the formation of life and the appearance of mind; he proffered to believe in the existence of some hitherto unknown, teleological laws acting in evolution.

---

[14]"The whole case for sexual selection is in fact an enormous appendage to Darwin's book, *The Descent of Man and Selection in Relation to Sex (1870),*" quoted in Howard (1982: 55).

[15]"In the *Descent of Man*, sexual competition and sexual choice were invoked to explain some of the physical attributes of man that did not seem to contribute directly to the general biological advantage. The general lack of body hair compared with man's ape-like relatives and its

Against Wallace's conclusion, that higher human mental properties could not have been developed by natural selection, Gould argued that natural selection could build an organ 'for' a specific 'purpose', but this 'purpose' need not fully specify its capacity.

"Our large brains may have originated "for" some set of necessary skills, such as gathering food, socializing, or whatever; but these skills do not exhaust the limits of what such a complex machine can do. Fortunately for us, those limits include among other things an ability to read and to write, and for some creative people to compose poems and symphonies" (Gould 1980: 57). In other words, "historical origin and current function are different properties of biological traits" (Gould 1988: 122).

As a variation on the same idea, it is enlightening to consider Tennant's comment that "the human mind once having attained in the course of evolution to ideation, social intercourse and language, is in a position to develop spontaneously, no longer controlled by mechanical selection (which is but rejection) but by his own interest and intrinsic potencies. From intelligence and emotional sensibility, that are biological useful, it may proceed to disinterested science, to pure mathematics, having no relation to the needs of life, to art, morality and religion", and he adds, probably hinting to Wallace's 'higher intelligence', "without requiring any unexpected intervention" (quoted in Eiseley 1961: 322).

Furthermore, in reference to the Cro-Magnon people, who lived about 40,000 years ago, Gould wrote that it is known that they produced marvelous paintings in their caves. He asserted that these men had a brain that was not smaller (perhaps even greater) than ours, and all that we have accomplished since then is the product not of biological evolution but of cultural evolution (Gould 1980; and see below).

In addition Gould wrote, again referring to the brain: "here side consequences may overwhelm the original purposes ... consider for example our knowledge of personal mortality. Nothing in our large brain ... has proved more frightening and of weighty import. Surely no one would argue that our brains increased in order to teach us this unpleasant truth." (Gould 1988: 122).

It may be of interest to point out that Darwin preceded Gould in suggesting a similar idea (albeit with some hesitation), writing: "If it could be proved that certain high mental powers, such as the formation of general concepts, self-consciousness, etc. were absolutely peculiar to man, which seems extremely doubtful, it is not improbable that these qualities are merely the incidental results of other highly-advanced intellectual faculties; and these again

mainly the result of the continuous use of a perfect language" (Darwin 2009: 106). Related to this sort of explanation is also Darwin's concept of correlated change (see footnote 1).

Cultural Evolution Versus Biological Evolution

As mentioned earlier, Wallace came close to realizing that in man there occur two distinct processes: biological evolution and cultural evolution. According to the anthropologist Loren Eiseley: "Wallace's contribution to anthropology...[was] the recognition that man had transferred to his tools and mechanical devices the specialized evolution which so totally involves the plants and animals..." (Eiseley 1961: 313).

The concept of cultural evolution preceded the Darwinian theory of evolution, or both were seen as aspects of a single process, for example by Herbert Spencer.[16] The distinction between these two processes depended on the discovery of the hereditary units of biological evolution by Mendel (latter dubbed genes), or rather their "re-discovery" in 1900, independently by three different biologists.

Man originated from his progenitors, like all other species, by the slow process of biological evolution. At some point in the past, based on his developed cognitive abilities and his sociality (which was crucial for a weak organism who lacked devices for self-defense), a new process was superadded to the biological evolution – cultural evolution.[17] Instead of passively adapting to the environment, man began to change the environment actively and consciously according to his needs. He used various natural implements as tools, invented tools, made clothes and dwellings, exploited various sources of energy and much more.[18]

Some aspects of culture (like the use of simple tools) are found in certain groups of animals, but only in humans is cultural change cumulative, resulting in a very wide gap

---

(Footnote 15 continued)
different distribution in males and females, Darwin attributed to sexual preference" (Howard 1982, p. 69).

[16]See, for example "Social Darwinism in American Thought". R. Hofstadter. Beacon Press, Boston (1944).

[17]According to Medawar 1981, "cultural evolution is not a very good description of this process, because it could be taken to connote evolution of culture, instead of evolution mediated through culture", thus he prefers "exogenetic" or "exosomatic" evolution. Separating these two aspect seems to be rather important; they can be lucidly exemplified for example by "The Great Transition" from nomadic life to permanent settlement that took place same 15,000 years ago. This transition produced a profoundly altered social environment: among other changes, society became more hierarchical with all the consequences.

Julian Huxley (1955: 17) preferred the term "psycho-social evolution".

[18]According to the anthropologist Edward Tylor (1924), culture is "that complex whole which includes knowledge, belief, art, morals, law, custom and any other capabilities and habits acquired by man as a member of society".

between the modest beginnings of culture among animals and human culture (Neumann 2013).[19]

One major difference between biological and cultural evolution is their rate of change.[20] The latter is several orders of magnitude faster compared to the former. Biological evolution depends on the rare appearance of "useful" chance mutations, and their proliferation in the population, through an increase in the relative rate of the reproduction of individuals in whom they reside. Thus, the minimum time for the transmission of a novel change is one generation. Cultural innovations, on the other hand may be transmitted very quickly, whether by imitation, learning,[21] indoctrination and most importantly through man's symbolic language (a major event in human history).

The transmission of a favorable genetic mutation can take place only "vertically", from parents to children. In cultural change the transmission can be "vertical", in both directions (from parents to children and vice versa) and most important, "horizontally", from one individual to another, in the population. New discoveries by some individuals (sometimes even by a single individual!) can quickly spread to the entire society and indeed across the world.

Thus, the fast rate of cultural evolution is a pertinent answer to Wallace's claim, who pointed to the very fast speed of some of the cultural innovation, mentioned above, in the last centuries in the arts, music, or mathematics.

As a matter of fact, both Darwin and Wallace provided many examples of man's behavior and action, such as hunting and fishing, using weapons and many other activities, without being aware that these processes are part of culture and not biological traits.

Addendum

Wallace versus Darwin: On the Relation of Consciousness[22] to the Brain

Wallace quoted with approval John Tyndall's remarks in 1868: "…the passage from the physics of the brain to the corresponding facts of consciousness is unthinkable. Granted that a definite thought and a definite molecular action in the brain occur simultaneously, we do not possess the intellectual organ, nor apparently any rudiment of the organ, which would enable us to pass by a process of reasoning from the one phenomenon to the other…"

This quotation was aimed to oppose the materialistic position of Thomas Henry Huxley, who reduced the thinking process to the molecular level. Huxley wrote: "Consciousness is a function of nervous matter, when that nervous matter has attained a certain degree of organization, just as we know the other actions, to which the nervous system ministers, such as reflex action and the like…" (Slotten 2004: 283).

Wallace surmised that Huxley's theory "was not only untestable but inconsistent with accurate conceptions of molecular physics". He continued by describing the almost infinite complexity of molecular combination, which enables us to comprehend the possibility of vegetative life. "But this increasing complexity, even if carried out, could not have the slightest tendency to originate consciousness in such molecules or groups of molecules…or to produce a self-conscious existence". And Wallace concluded: there was no escaping from the dilemma: "Either all matter was conscious, or consciousness was something distinct from matter" (Slotten 2004: 283).

Furthermore, Slotten (2004: 284), wrote "that after accusing Huxley of using words "to which we can attach no clear conception", Wallace made statements equally abstruse. Matter was force and nothing but force…He identified two types of force: the first was "primary force", which included gravitation, cohesion, heat and electricity. The second was what he called will-force, which he defined as a power that directed the action of the forces stored up in the body…The origin of the will-force could be traced not to something inside, but to something outside humans – the will of higher intelligences or of one Supreme Intelligence".

According to Slotten, Wallace's response to the critics of the above statements (regarding the existence of the Higher Intelligence etc.) was to conclude the *Homo sapiens* differed in kind from other animals (Slotten 2004: 286).

Darwin on Consciousness

Gould (1977) refers to Darwin's ideas on consciousness, as described in the so-called "M" and "N" notebooks, written in 1838 and 1839. He claims that these sketches indicate that "Darwin supported materialism – the postulate that matter is

---

[19]This does not mean, that humans are independent of the action of genes. According to Ernest Gellner, "humans are still subject to genetic control, but "Humans are the way they are, because their genes do no determine their behaviour, but rather permit great variation and flexibility" (quoted in Foley 1995: 197).

[20]Another major difference is the fact that biological evolution is irreversible, whereas cultural change is reversible.

[21]Learning involves the capacity to respond to stimuli with appropriate behavior (it is an example of phenotypic plasticity). In man this capacity has been highly developed, including the capacity to learn a language and a culture.

[22]A common sense definition of consciousness is given by Searle: 'consciousness' refers to those state of sentience or awareness

---

(Footnote 22 continued)

that typically began when we wake from a dreamless sleep and continue through the day, until we fall asleep again, die, go into a come or otherwise become 'unconscious' (Searle 2002: 21).

the stuff of all existence and that all mental and spiritual phenomena are its by-products. … mind – however complex and powerful is simply a product of the brain".

It is noteworthy that in his commentary on the "M" and "N" notebooks, Gruber labeled materialism as "at that time more outrageous than evolution" (quoted by Gould).

One should add that the relation of consciousness to the brain, is a major controversial issue in philosophy, psychology, neurophysiology and related areas, dubbed in its modern version as (part of) the "Mind-Body" problem (see for example, Searle 2004).

Perhaps it should also be mentioned that according to some philosophers, not only it is an unsolved problem, but it is unsolvable! (e.g., McGinn 1989).

References

Alland, A. Jr. (1973). *Human diversity*. Garden City, New York: Anchor Books

Cartwright, J. (2001). *Evolution and human behavior*. Cambridge, MA: The MIT Press.

Cassirer, E. (1944). *An essay on man*. New Haven: Yale University Press.

Darwin, C. (2009). *The descent of man and selection in relation to sex* (2nd ed. First Published in 1874). A Digireads.com Book.

Darwin, C. (2010). *The descent of man*. Abridged with an Introduction by Ghiselin M. Minnesota. New York: Dover Publications

Eiseley, L. (1961). *Darwin's century*. New York: Anchor Books.

Foley, R. (1995). *Humans before humanity*. Oxford: Blackwell Publishers.

Gould, S. J. (1977). *Ever since Darwin: Reflections on natural history*. London: W. W. Norton & Company.

Gould, S. J. (1980). *The panda's thumb*. New York: W.W. Norton & Company.

Gould, S. J. (1988). *An urchin in the storm*. London: Collins Harvill.

Harman, O. S. (2004). The evolution of a naturalist. *American Scientist, 92*, 470–473.

Hofstadter, R. (1944). *Social Darwinism in American thought*. Boston: Beacon Press.

Howard, J. (1982). *Darwin*. New York: Hill and Wang.

Huxley, J. (1955). *Evolution in action*. New York: Penguin Books.

Kottler, M. J. (1974). Alfred Russell Wallace, the origin of man and spiritualism. *Isis, 65*, 145–192.

McGinn, C. (1989). Can we solve the mind-body problem? *Mind, 98*, 349–356.

Medawar, P. B. (1981). Stretch genes. Genes, mind and culture: The co evolutionary process, by C.L. Lumsden and E.O. Wilson. *NYB, 28*, 45–48.

Nagel, T. (2012). *Mind and cosmos: Why the materialist neo-Darwinian conception of nature is almost certainly false*. Oxford: Oxford University Press.

Neumann, J. (2013). Biology and culture. *Journal of Life Sciences, 7*, 322–333.

Pinker, S. (1994). *The language instinct*. New York: Allen Lane The Penguin Press.

Popper, K. R. (1972). *Objective knowledge—an evolutionary approach*. Oxford: Oxford University Press.

Richards, R. J. (2005). Darwin's metaphysics of mind. In V. Hosle & C. Illies (Eds.), *Darwinism and philosophy* (pp. 166–180). Notre Dame Indiana: University of Notre Dame Press.

Rubens, T. (2012). *Genius and changes in social context in politics and neo-Darwinism and other essays*. Exeter UK: Societas, Imprint Academic.

Searle, J. R. (1997). The mystery of consciousness. *The New York Review of Books*.

Searle, J. R. (2002). *Consciousness and language*. Cambridge: Cambridge University.

Searle, J. R. (2004). *Mind – a brief introduction*. Oxford: Oxford University Press.

Slotten, R. A. (2004). *The heretic in Darwin's court*. New York: Columbia University Press.

Shanahan, T. (2004). *The evolution of Darwinism*. Cambridge: Cambridge University Press.

Shermer, M. (2002). *In Darwin's shadow*. Oxford: Oxford University Press.

Sterelny, K. (2007). *Dawkins vs. Gould*. Cambridge: Icon Books Ltd.

Tylor, E. B. (1924). *Primitive culture. Researches into the development of mythology, philosophy, art and custom* (7th ed.). New York: Brentano's

Rycroft, C. (1985). *Psychoanalysis and beyond*. Chicago: The University of Chicago Press.

Wallace, A. R. (1892, 1980). 'Spiritualism'. Excerpt from 'spiritualism' chambers encyclopedia. In N. G. Coley & M. D. Hall (Eds.), *Darwin to Einstein. Primary sources on science and belief* (pp. 645–648). London: Longman in Association with The Open University Press.

Wallace, A. R. (1889, 1980). 'Darwinism applied to Man'. Excerpt from 'Darwinism Applied to Man' In N. G. Coley & M. D. Hall (Eds.), *Darwin to Einstein. Primary sources on science and belief* (pp. 274–299). London: Longman in Association with The Open University Press.

Chapter 3
Man's Place in Past and Future Evolution: A Historical Survey of Remarkable Ideas

David Wool

Abstract Most evolutionary biologists, after Lamarck and Darwin, were concerned with evolutionary processes in the natural world with no special mention of Man – taking him as just another animal. Some dealt with the human species only, and were concerned with human descent from the apes – or, like Wallace and Dobzhansky, with the unique abilities of humans to protect themselves from nature. Still others, like Galton and the Eugenicists, were interested in controlling or improving the future qualities of the human population. That human activities, as a dominant species, affect the natural environment was already noted by Lyell in the early 19th century, but the effect of mankind on the rest of the biological world became of public concern only recently. The implications of human activities for the future evolution – and fate – of our entire planet, seems to be of only limited academic concern, and Man [=mankind, the "international community"] is either uninterested or unable to do anything positive about it.

Keywords Apes • Darwin • Eugenics • Genetics • Heredity • Intelligence quotient • Lamarck • Nature and nurture • Wallace

The Origin

The year 1809 marks two important events. In February of this year, Charles Darwin was born. Fifty years later he framed a new theory and suggested a new mechanism for the evolution of the biological world, the human species included, and produced a profound effect on humanity (Darwin 1898a). In August of the same year, the French biologist

D. Wool (✉)
Department of Zoology, Tel Aviv University,
69978 Tel Aviv, Israel
e-mail: dwool@post.tau.ac.il

Jean Baptiste Lamarck – aged 65 – published his book "Zoological philosophy" (Lamarck 1984) in which he suggested that the biological world evolved from simple to more complex organisms, and thus became the first evolutionist (Graur et al. 2009).

Lamarck was an outstanding biologist. Apart from establishing the taxonomy of invertebrates, he published many books on a variety of subjects. True, almost all of his original ideas were either ignored, or ridiculed, or rejected as false – as his colleague and bitter adversary, Georges Cuvier, did not hesitate to point out in his eulogy after Lamarck's death (Lamarck 1984: 434–435). Even the mechanism he envisioned as the driving force of evolution – the inheritance of acquired characters ["use and disuse"] – was totally rejected already at the end of the 19th century (Weismann 1891 I:85). But among his "scientific rubbish" (Graur et al. 2009), he had some original ideas which eventually were accepted by all. One of them was the common descent of all organisms from simple "monads". Another was his vision of the evolution of the human species from the apes:

> If some race of quadrumanous animals, especially one of the most perfect of them, were to lose…the habit of climbing trees and grasping the branches with its feet – and if individuals of this race were forced, for a series of generations, to use their feet only for walking…- furthermore, if the individuals of which I speak were impelled by the desire to command a large and distant view, and hence endeavored to stand upright, and continually adopted that habit from generation to generation, there is no doubt that their feet would acquire a shape suitable to supporting them in an erect attitude (Lamarck 1984 [1809]: 170).

Lamarck goes on to argue that other anatomical changes – as the shape of the skull – would follow as a consequence of the upright posture: the throat and tongue would be used to utter sounds for communication and will develop into language – "the marvelous faculty of speaking" would give that race an advantage over all other animals. All this could happen "If man were distinguished from animals only by his organization, and if his origin were not different from theirs".[1]

Fifty years later, Darwin assimilated the idea of evolution from simple to complex structure in the "Origin of Species", but denied any influence of Lamarck's on his theory. In a letter to his friend Joseph Hooker – often cited but misquoted – Darwin writes,

> Heaven forefend me (sic) from Lamarck's nonsense of 'a tendency to progression', adaptation from the slow willing of animals – but the conclusions I am led to are not widely different from his, though the means of change are wholly so[2] (Darwin to Hooker, 11.1.1844. Darwin, F. 1887, II: 23).

The "means of change" Darwin suggested is, of course, natural selection – although he did not entirely abandon the Lamarckian "use and disuse". In his book "The Descent of Man" (Darwin 1874 [1871]), 12 years after the publication of "the Origin of Species", Darwin first stated clearly that man and the apes had a common ancestor. In his autobiography he explained that in 1859, he worried that a clear statement may hamper the public response to "The Origin of Species" (Barlow 1958: 130).

Man and the Apes

In an article in 1863, entitled "Man's Place in Nature", Darwin's friend and ally, Thomas Henry Huxley, published a detailed comparison of human and ape skeletons, illustrating their close similarity to support their common origin (Huxley 1900 [1863]). Even the anatomist Richard Owen, Darwin's great scientific opponent, had to admit the skeletal similarity of man and gorilla:

> I cannot shut my eyes to the significance of that all-pervading similitude of structure, every tooth, every bone strictly homologous – which makes the determination of the difference between *Homo* and *Pithecus* the anatomist's difficulty (Owen 1857, quoted by Huxley 1900 [1863], footnote on p. 153).

All evolutionary biologists agreed on the common origin of man and the apes, but there was no fossil evidence to support this origin. The "pygmy", the first African primate presented to the Royal Society, was described in 1699 by Eduard Tyson as an intermediate between an ape and man, but when its skeleton was examined by Huxley in 1863, it was identified as a young Chimpanzee (Gould 1985).

A fragment of a human skull was discovered in 1829 in the Engis cave in Belgium, together with bones of extinct mammals. Another fossil fragment was discovered in the Neander valley in Germany in 1856, and is listed as the "type" specimen of the Neanderthal humans. These discoveries raised great interest among biologists – in an article entitled "On some fossil remains of man", Huxley described in detail the fossil skulls, their dimensions, and the opinions of different contemporary anatomists over more than 60 pages (Huxley 1900). The verdict was that "the Neanderthal cranium has most extraordinary characters…it belonged to one of the wild races of Northern Europe" (Huxley 1900: 168). However,

> In no sense can the Neanderthal bones be regarded as the remains of a human being Intermediate between man and the apes (Huxley 1900: 205).[3]

In 1876, the German biologist Ernst Haeckel – Darwin's greatest supporter in Germany – felt that the gap between Man and his predecessors in the paleontological record must be filled. He invented a creature which fitted his perception of how this "missing link" must have looked like, and hired an artist to draw it. It resembled Darwin's suggestion:

> We thus learn that man is descended from a hairy, tailed quadruped, probably arboreal in its habits, and an inhabitant of the Old World (Darwin 1952 [1871]: 911).

Haeckel added that the creature had longer arms and crooked, shorter legs than modern humans, "possibly similar to the black Africans". Sharing with Cuvier and Huxley the idea that the main difference between man and ape is the ability to speak, Haeckel gave the hypothetical animal a scientific name – *Pithecanthropus alatus* – ["speechless monkey-man"]. Intrigued by the need to find this "missing link", the Dutch physician Eugene Dubois searched for and discovered some fossil human remains in Java, Indonesia in 1891. These creatures are referred to as *Pithecanthropus erectus*.

The first major finds of African hominids were collected many years later. In 1925, R.A. Dart described a hominid skull ["the Taung baby"] from South Africa, aged 2–3 MY, and classified as *Australopithecus africanus*. Early finds by Mary Leakey in 1974 near the Olduvai Gorge in Tanzania were dated to be 3–4 million years old. Abundant skeletal remains were found in the Afar region in Ethiopia. The most famous of these finds is of course "Lucy", with about 40% of

---

[1]Lamarck's vision is remarkable since no fossils intermediate between man and monkey were known in 1809. Also, he did not hesitate to suggest that Man and the apes shared a common origin!

[2]While the first part of the sentence is often quoted, the second is rarely if ever mentioned.

---

[3]Many more Neanderthal fossils were since discovered in Europe (and in Israel. Rak and Arensburg 1987). They are considered a parallel species, not in the line of descent of *Homo sapiens*. Interest in the Neanderthals was greatly renewed in the late 20th century, when molecular studies discovered some mtDNA sequence similarity between Neanderthal and *Homo* genomes (Krings et al. 1997). The overlap in the ranges of the two species raised the possibility – still debated – of some interbreeding. A complete Neanderthal mtDNA sequence [taken from a single specimen] indicated that it is outside the range of modern human variation (Clark 2008).

the skeleton recovered (Johanson and White 1979). All these early hominids were bipedal, although their cranial capacity was no larger than that of apes. Clearly, the upright posture preceded the increase in brain size.[4] These early African hominids were assigned to the species *Australopithecus afarensis*. Haeckel's "Pithecanthropus alalus" remained an evolutionary curiosity.

Brain and Intellect

The volume of the brain – the cranial capacity – was considered by many as one, perhaps the most important, measure of intellect, in which mankind differed greatly from the apes. The American physician, Samuel Morton, measured the cranial capacity of several hundred human skulls to "prove scientifically" that Europeans are intellectually superior to all other human races. It was an unpleasant surprise for Darwin and Huxley when in 1869, Alfred Russel Wallace – co-discoverer of the theory of natural selection, who supported Darwin and evolution all his life – made an exception of man (Wallace 1869). Wallace accepted the common origin of mankind from the apes, but insisted that humans are unique in being able to protect themselves from the forces of natural selection. In particular, the intellectual and moral characteristics of humans, those "glorious qualities which raise us so immeasurably above our fellow animals" [Wallace 1891] – could not have evolved by natural selection and the "survival of the fittest". Wallace was familiar with Morton's work, extended the data to include brains of native "savages", and noticed that the savage brains were no smaller than those of civilized men, but were much larger than the brains of primates [gorilla and orang-utan] although their body weights are similar to humans (Wallace 1891: 190).

> This being the case, we cannot fail to be struck with the apparent anomaly that many of the lowest savages should have as much brains as average Europeans. This idea is suggestive of a surplusage [sic] of power – of an instrument beyond the needs of its possessor (Wallace 1891: 190).

Wallace claimed that an invisible power planned and directed human evolution. Even some of man's physical faculties, like the versatile uses of the modern human hand, afford proof that "there are other and higher existences than ourselves, from which those qualities may have been derived" (Wallace 1891: 190).

The anatomist Richard Owen insisted that despite the structural similarity, Man cannot be "a modified ape": if brain-size differences are not clear proof, he claimed, there was a great structural difference in the brains of the two species. He did not change his mind even when Huxley proved, in a careful investigation in 1860, that no such difference existed – although Huxley did not rule out that a subtle difference did once exist: "I by no means believe that it was any original difference of cerebral quality, or quantity, which caused the divergence between the human and the pithecoid stirpes, which has ended in the present enormous gulf between them. And believing as I do, with Cuvier, that the possession of articulate speech is the grand distinctive character of man … I find it very easy to comprehend that some equally inconspicuous structural difference may have been the primary cause" [of the difference]. (Huxley 1900 [1863], footnote on p. 142–3).

Huxley rejected the claims that a common origin with the apes is a disgrace for Man,

> Is it indeed true that the Poet, or the Philosopher, or the Artist, whose genius is the glory of his age, is degraded from his high estate by the undoubted historical probability, not to say certainty, that he is the direct descendant of some naked and bestial savage, whose intelligence was just sufficient to make him a little more cunning than the fox, and by so much more dangerous that the tiger? (Huxley 1900 [1863]: 153–154).

Heredity and Human Evolution

A major problem for 19th century evolutionists was "the great mystery" of heredity. Darwin's theory of natural selection required hereditary variation in populations, but there was little direct evidence for [phenotypic] variation in natural populations. Darwin wrote in 1844: "natural populations vary very little", but added that heritable variation must be there, since there was no lack of evidence that artificial selection in animals and plants is very effective [Darwin accumulated the available data in his book on Domestication (Darwin 1898b [1865]). In particular, there was little evidence for heritable variation in human populations, apart from some aberrant individual cases as albinism, 6-digited hands, and "porcupine skin", which were exhibited to the public as curiosities. The pattern of transmission of the sex-limited disease, Hemophilia, was described in detail but not understood. "The laws governing inheritance are for the most part unknown" (Darwin, in the Origin). Darwin substituted his own "rules" to explain the phenomena of inheritance – in particular his hypothesis of "pangenesis" (Darwin 1898b [1865], ch. 7) which was disproved experimentally by Darwin's cousin, Francis Galton, shortly afterwards.

Galton decided to collect data on inheritance in human populations and build a data base for studies on heredity. He

---

[4]As the early hominid fossils – such as "Lucy" – proved, upright stature – and not the increase in brain size, as many evolutionary biologists assumed for years – was the first step in the human line (Rak 1991). Lamarck had it right!

was first interested in the inheritance of special talents, the carriers of which are rare in the population – referred to as "geniuses" ("one in a million, or one in ten million"). Galton searched history books and social records in Europe in the 17th–18th centuries, and listed about 400 persons whom their contemporaries, and later generations, considered outstanding – arguing that reputation is an indication of quality. Among these "geniuses" were famous Chief Justices, army commanders, artists, musicians and scientists. He then searched their pedigrees for the occurrence of outstanding people among their descendants. In his book "Hereditary Genius" (Galton 1962 [1869]), Galton showed that the frequency of outstanding people in the lineages of "geniuses" was higher than their frequency in the general public – thus supporting his conclusion that the characters of a "genius" were heritable.

Galton then turned to collect data on the inheritance of ordinary human characters – such as weight, stature, color of eyes, intellectual and artistic tendencies, and occurrence of diseases – and did so by a technique which is common today but unheard-of at the time – a public survey. He published notices in newspapers, promising payment to anyone who will provide data on these variables, of all his family members for three generations: the magnitude of the reward depending on the quality of the data. Thirty-seven families responded (and are listed in Galton's book "Natural Inheritance" (1889)). Galton – one of the founders of statistics as a science – declared that his aim was

> to show that a large part is always played by chance in the course of hereditary transmission, and to establish the importance of an intelligent use of the laws of chance and the statistical methods based on them, in expressing the conditions under which heredity acts (Galton 1889: 171).

His analysis yielded interesting insights, the common conclusion being that all these characters are at least in part heritable. Yet from his analysis of pedigrees of "geniuses", as well as his breeding of dogs, Galton realized that breeding from exceptional parents does not ensure that all their offspring carry their exceptional characters. He attributed this to a "Law of Regression to the Mean".

> Now a man is not only the product of his father, but of all his past ancestry – and… the mean of that ancestry is probably not far from that of the general population. In the tenth generation, a man has 1024 tenth-generation grandparents. He is eventually a product of a population of this size… It is the heavy weight of this mediocre ancestry which causes the son of an exceptional father to regress towards the general population mean; it is the balance of this sturdy commonplaceness which enables the son of a degenerate father to escape the whole burden of the parental ill. (Pearson 1900: 456).

His friend and colleague, the statistician Karl Pearson, regarded "Galton's Law" as perhaps equal in importance to Newton's law of gravity (cited in Moore 1986).

Eugenics – Controlling Human Evolution

Galton reasoned that if all characters, and in particular the outstanding characters of "geniuses", are heritable, it should be possible to improve the quality of future human generations by selection. Galton thus became the founder of the Eugenics movement, whose goal was to direct the evolution of mankind.

> Eugenics is the new science, concerned with the study of agencies under social control, that may improve the racial qualities of future generations, either physically or mentally (Pearson 1949 [1892]).

Concern with the deteriorating quality of the human race was expressed as early as 1816. The physician Sir William Lawrence complained that unlike the breeding of domestic animals, the principles of selective breeding are not employed in human populations. "All the native deformities of mind and body are handed down to posterity and tend to degrade the race". The strongest illustration of this fact, he argued, will be found in the state of many Royal Houses of Europe, who are confined by custom and prejudice to intermarriages with each other (cited in Wells 1972: 327).

Galton suggested that the frequency of "geniuses" – or highly-endowed people – in the population should be increased by assortative mating, thereby affecting the evolution of the human race and improving its quality.

> Looked at from the social standpoint, we see how exceptional families, by careful marriages, can within even a few generations obtain an exceptional stock, and how directly this suggests assortative mating as a moral duty for the highly endowed. (Pearson 1900).

Galton suggested that better-endowed young people – especially women – should be encouraged to marry early with similarly-endowed partners and produce more children. Encouragement should be by providing such couples with cheap housing and by generous grants (Galton 1901). He added that the expense is economically justified: if such exceptional people could be detected as children, "procurable by money and reared as Englishmen, it would be a cheap bargain for the nation to buy them at the rate of many hundred or some thousands of pounds per head" – in view of their value for British science and economy. However,

> The idea is smiled at as most interesting in itself, and possibly worth of academic discussion, but absolutely out of the question as a practical problem. (Galton 1901)

The discovery in 1900 of the old (1865) paper by Gregor Mendel, which provided the key to understanding of inheritance, became a landmark in evolution. Mendelian inheritance of discrete characters was first considered an alternative theory, in opposition to Darwinian evolutionary theory: Galton and Pearson remained dedicated to the selection of continuous characters and blending inheritance! But later, the science of Genetics gave a tremendous boost to the theory of Eugenics. Enthusiasm led to the belief that all human characters – not only physical but also mental and intellectual traits, alcoholism, crime, feeble-mindedness and disease – are inherited as simple, Mendelian unit characters. University courses and "chairs" for the study of Eugenics were established, and research was generously funded by rich private foundations in America (see Cravens 1978).

The subsequent two or three decades many Americans may wish to forget. In 1917, the "intelligence quotient" (IQ) was first measured on a large scale in recruits to the US Army, illustrating that recruits coming from poor neighborhoods – and evidently mostly black – received much lower grades than white recruits. Science combined with racial hatred and social prejudices led to the notion that the poor inhabitants of the slums in the cities, where crime and disease prevailed, are in that state only because they carry undesirable genes: active steps should be taken to curb the reproduction of these "degenerate" people. Legislation advocating confinement in closed institutions and enforced sterilization was passed in several States and put into practice. Karl Pearson contributed theoretical support for this kind of practical eugenics.

> On the other hand, the exceptionally degenerate isolated in the slums of our modern cities can easily produce permanent stock also; a stock which no change of environment will permanently elevate, and which nothing but mixture with better blood will improve. But … we do not want to eliminate bad stock by watering it down with good, but by placing it under conditions where it is relatively or absolutely infertile. (Pearson 1900).
>
> It is a false view of human solidarity, a weak humanism, which regrets that a capable and stalwart race of white men, should replace a dark-skinned tribe which can neither utilize his land for the full benefit of mankind, nor contribute its quota to the common stock of human knowledge (Pearson 1949 [1892]).[5]

After the Second World War – and the horrors of the Holocaust – the eugenics movement faded away. One of the leading geneticists and evolutionary biologists of the 20th century, Theodosius Dobzhansky, rejected the ideas of the "Social Darwinians" and eugenicists of improving the human race by selection. Their basic belief that all human traits are controlled by genes only – was flawed: they ignored the effect of the environment.

A person is what he is because of his nature and his nurture. His genes are his nature. His upbringing is his nurture. The same is true of mankind as a whole. (Dobzhansky 1962: 24).

Biological evolution does not transmit cultural, or for that matter, physical traits ready-made: what it does is determine the response of the developing organism to the environment in which the development takes place…We inherit genes, nor genotypes, of our parents, and we transmit our genes, not our genotypes, to our children. A caste originally recruited from persons of high ability will contain some less-able individuals in the following generations. (Dobzhansky 1962: 21)

Science and Human Evolution

The advancement of science in the 20th and 21st centuries brings us more and more closely to the control of human evolution – for better or for worse. As early as 1923, the evolutionist John B. Sanderson Haldane was fascinated by the progress of science and the prospect of improvement of human life:

> Bad as our urban conditions are, there is not a slum in the country which has a third of the infantile mortality – of the Royal family in the Middle Ages!

In a utopian essay entitled "Daedalus",[6] he drew up his vision of Britain when science is applied to everyday life: food will be "cheap as sawdust" when sugar and starch will be produced in factories from elementary materials like coal and atmospheric nitrogen – or by bacteria breaking down cellulose – and agriculture will become a luxury, flower beds replacing the slaughterhouses near the cities. Electric energy will be produce by a network of windmills, and the electricity used to electrolyze water – the resulting oxygen and hydrogen will be stored and used to produce power for industry and transportation to replace coal and oil. Biologically, Haldane had a vision:

> Now that the technique is fully developed, we can take an ovary from a woman and keep it growing in a suitable fluid for as long as twenty years, producing a fresh ovum each month, of which 90 per cent can be fertilized, and the embryos grown successfully for nine months, then brought out into the air (Haldane 1923).

Haldane foresaw great opposition to the idea of in-vitro fertilization of humans ["the biological invention tends to begin as a perversion and end as a ritual"] but unlike the

---

[5]It is not difficult to realize that these lines of thought were the same as the racist ideology of Hitler's followers in Germany in the 1930s – and its disastrous application in the Holocaust. Thus eugenics, which began as a scientific theory directing human evolution for a better future, turned – in the wrong hands – to a means of destruction of humanity.

[6]Daedalus symbolizes applied science. In Greek mythology, Daedalus was a sculptor who carved the statues of the gods. He also designed and built the wings which enabled him and his son Icarus to escape from the minotaurs. Icarus flew too close to the sun, the wax in his wings melted, and he fell into the sea and drowned.

other parts of Haldane's vision, in-vitro fertilization and selection of desirable genotypes is no longer a dream.[7]

Mankind in Nature: A Special Case?

Darwin's theory of natural selection had a profound effect on social philosophers. Late in the 19th century, Karl Pearson quoted the German biologist Ernst Haeckel:

> The theory of selection teaches us that in human life, exactly as in animal and plant life, at each place and time only a small privileged minority can continue to flourish. The great mass must starve and more or less prematurely perish in misery (Haeckel 1776, cited by Pearson 1949 [1892]).

At the start of the 20th century, this approach permeated the social thinking [millionaires considered themselves products of selection because they out-competed adversaries in business]. Karl Pearson described what he called "socialism" [not the ordinary meaning of the term].

> The struggle for existence involves not only competition between individuals, but also between societies and between nations. We have always to remember that, hidden beneath diplomacy, commerce and adventure there is a struggle between modern nations – which is none less real if it does not take the form of open warfare.
>
> Every society is interested in developing its resources. Each society strives to educate, train and organize its members [for its own interests], because only in this way the society can survive in the struggle for existence. Socialism is a direct result of the principle of evolution. (Pearson 1900: 368).

The history of the past century – including the devastating effects of two World Wars, the rise and fall of empires, and re-partitioning of the surface of the globe between nations – illustrate that the picture called by Pearson "socialism" is still valid.

Late in his life, Thomas Henry Huxley resented the argument that natural selection is a built-in factor in human society. In his last lecture (1893), entitled "Evolution and Ethics", he argued that in human populations the struggle for existence is in reality "the struggle for the means of enjoyment". Huxley discussed the need of human societies to replace the "Law of the Cosmos" – where competition and the struggle for existence lead to "the survival of the fittest" – by the Law of Ethics, which strives to make as many people fit as possible. If a part of wild nature is intended to become a beautiful garden, the gardener must curb the spread of each individual plant, prevent competition, and supply the necessities to all species within the garden walls. Similarly in human populations, Huxley argued, the law of ethics requires that individuals restrain their personal desires for means of enjoyment for the common good (Huxley 1989 [1894]).[8]

Two of Dobzhansky's books (1955, 1962) are dedicated to human evolution. Dobzhansky emphasized the unique features of man, making him distinct from the primates:

> Biologists have been so pre-occupied with proving that man is a product of organic evolution, that they have scarcely noticed that man is an extra-ordinary and unique product of this evolution. The leading forces of human evolution are intelligence, ability to use linguistic symbols, and culture which man has developed (Dobzhansky 1955: 320)
>
> The question may again be raised as to whether upright stance, tools, constant sexual receptivity of females, symbolic language, monogamous family, change in food habits, or relaxation of male aggressiveness came first… What we are dealing with is the emergence of a whole new evolutionary pattern, a transition to a novel way of life which is human rather than animal (Dobzhansky 1962: 199)

The unique feature of human evolution, Dobzhansky emphasized, is cultural evolution. Cultural evolution is faster and more efficient than biological evolution: it is not limited to transferring information from parents to offspring, but proceeds by learning, spreading written information and technical and scientific knowledge – today also by radio, television and the internet – across generations and across national and geographical barriers.

Man and the Rest of the World: Ecology and Evolution

Human effects on the rest of the biological world were not an issue for evolutionary biologists, from Darwin to the "Modern Synthesis" (Huxley 1942). They discussed the evolution of animals apart from man, and when interested in the evolution of the human species, the rest of the world was apparently of no concern. An exception was the geologist, Charles Lyell.

As early as 1830–1832, in his monumental book "Principles of Geology",[9] Charles Lyell described in detail the negative effects of human population on the environment. He reported that the felling of forests caused extreme land

---

[7]In-vitro fertilization and maintenance of human embryos – even deep-freezing them for long-term storage – is today, only ninety years after "Daedalus", common practice in modern hospitals. The road is open for selection of embryos – [true, now only for medical reasons – but who knows?] – directly affecting the future evolution of mankind. Aldous Huxley's and George Orwell's fictional new worlds may yet become a reality.

[8]Past experience shows that the "law of ethics" may prevail – in part – within human populations, but rarely if ever between populations and nations.

[9]Charles Darwin took Lyell's book with him on the Beagle, and the book had a major influence on forming his ideas about evolution. Lyell was instrumental in helping Darwin in his career, but did not accept the theory of evolution by natural selection until late in his life.

erosion in the parts of the USA which he visited, and that in Europe, swamps and bogs replaced the forests historically removed by man – among other reasons, because they provided shelter for wolves and outlaws. The expanding European population into newly-acquired lands necessarily had detrimental consequences for the local flora and fauna:

> When a powerful European colony lands on the shores of Australia, and…imports a multitude of plants and large animals from the opposite extremity of the earth, and begins rapidly to extirpate many of the indigenous species, a mightier revolution is effected in a brief period than the first entrance of a savage tribe, or their continued occupation of the country for many centuries, can possibly be imagined to have produced. (Lyell 1853: 150).

Lyell was familiar with the changes in the fossil fauna in adjacent geological formations. He was aware that the world was subject to the action of the forces of erosion, changing the face of the earth as well as climate at different localities with time. He suggested that the changes in the fauna followed the changes of the environment.

At the time of writing "Principles of Geology", Lyell believed in Creation, but he introduced his concept of unlimited time – into the creation process (Lyell 1853: 582). Surprisingly, his belief in the creation and immutability of species, brought him to modern concepts of ecology – years before the term Ecology was first used (by Haeckel) and the science of ecology was born (Wool 2001). Since species were immutable, Lyell argued, the changes in the composition of the fauna – as observed in the geological record – must have been the result of replacement: when the sea receded from the land, marine animals were replaced by terrestrial ones, and vice versa when land was inundated. Lyell was careful not to assume that new species were created, only that they migrated from some other locality. Thus the ecological forces of migration and colonization, rather than transmutation and evolution, explained the paleontological pattern (Wool 2001).

As for the changes caused by mankind, Lyell did not find them outstanding: any species expanding its range does so at the expense of other species that formerly occupied the area. Lyell recognized that human activity constantly limits the diversity of the world's flora and fauna, replacing wild species by domesticated ones that can be used for human needs.

> It may perhaps be said that Man has, in some degree, compensated for the appropriation to himself of so much food, by artificially improving the natural productiveness of soils, by irrigation, manure, and a judicious intermixture of mineral ingredients conveyed from different localities. But it admits of reasonable doubt whether, upon the whole, we fertilize or impoverish the lands which we occupy. This assertion may seem startling to many; because they are so much in the habit of regarding the sterility or productiveness of land in relation to the wants of man himself, and not as regards the organic world generally. (Lyell 1953: 681).

But since Man was created to rule the earth, and was endowed with the ability to accomplish this mission, it was only natural that his effect on the environment would be noticeable.

Man and Future Biological Evolution

One of the thinkers who alerted the public to the effect of humans on evolution – in particular on the future of humanity itself – was the biologist and humanist Julian Huxley. Huxley was an active member of the British Eugenics Society (and served as its president for some years). He was rather alarmed by the difficulties of providing food for the expanding human population, and advocated limiting its growth rate – with a Eugenics flavor:

> With this, the population problem has entered on a new phase: It is no longer primarily a race between population and food production, but between death-control and birth control. If nothing is done to control this flood of people, mankind will drown in its own increase…the world economy will burst at the seams, and mankind will become a planetary cancer. (Huxley 1957a, b: 16).
>
> It is the quality of people, not merely quantity, is what we must aim at, and therefore a concerted effort is required to prevent the present flood of population increase from wrecking all our hopes for a better world. (Huxley 1957a, b: 16).

More than 50 years ago, in a short contribution entitled *Transhumanism*, Julian Huxley had this message for humanity:

> It is as if Man has been suddenly appointed managing director of the biggest business of all, the business of evolution – appointed without being asked if he wanted it, and without proper warning and preparation. What is more, he cannot refuse the job.
>
> Whether he wants to or not, whether he is conscious of what he is doing or not, he is in fact determining the future of evolution on this earth. That is his inescapable destiny, and the sooner he realizes it and starts believing in it, the better for all concerned (Huxley 1957a, b: 13).

Will Mankind Rise to the Challenge?

Unlike prophecies and visions, predictions can only be based on previous experience and information.

In the years after Lyell, and until recently, the negative effects of man on the biological world were only occasionally mentioned – if at all – in an evolutionary context. Only towards the end of the 20th century did the diminishing of natural species diversity, species extinction and destruction of natural habitat – as noted by Lyell 180 years ago – become a public interest. The conflict between preserving the natural environment and the needs of providing food for the increasing human population – in particular the destruction of the global oxygen-generating vegetation of the Amazon basin in Brazil to make room for agriculture – is widely discussed today. Air pollution and human effects on climate [global warming and the "greenhouse effect"] have raised scientific and public concern and even international conferences were dedicated to the problem. Unfortunately, so far with very limited practical results. The diminishing biological diversity is of limited, mostly academic concern.

Science has made giant steps forward since Haldane's vision in Daedalus. Although his vision of cheap artificial production of starch and sugar is still a dream, molecular and genetic advances – like genetic engineering of plants – opened new pathways for increasing food production. The medical profession benefits greatly from the progress in molecular genetics and the use of the known human genome sequence to prepare new drugs against genetic diseases. These advances hold great hopes for the future of mankind.

However, Julian Huxley's challenge may never be met. "Man" as a whole, the hypothetical collective entity which Huxley deemed "managing director" responsible for the "business" of evolution, is represented by the "international community" and its international institutions – an aggregate of nations with conflicting political, economic, and other interests. The needs and interests of the rich, industrialized countries are in constant conflict with the under-developed (and highly populated) ones. For example, the need for new sources of energy for the expanding industry led to the use of corn – staple food for millions of people and livestock in under-developed countries – for production of "green fuel" usable by the better-endowed part of mankind who possess cars. The medical industry invests huge sums of money in search for cures for cancer, but not for malaria – a devastating source of human mortality in the Third World.

The "international community" does get together to help different countries in cases of large-scale local, humanitarian disasters, such as famine, earthquakes, floods and tsunamis, but political and economic interests make active international cooperation on ecological problems very difficult. The international institutions either do not realize their responsibility, or lack the ability to take action for the welfare of the planet or the direction of biological evolution, mankind included.

Acknowledgments I thank my colleagues, Dan Gerling and Dan Graur, for their comments on an earlier draft of the manuscript. The comments of the two anonymous reviewers, who pointed out inaccuracies and requested useful additions to the text, are greatly appreciated.

References

Barlow, N. (Ed.). (1958). *Autobiography of Charles Darwin, 1809–1882*. London: Collins.

Cravens, H. (1978). *The triumph of evolution*. Philadelphia: The University of Pennsylvania press.

Clark, A. G. (2008). Genome sequences from extinct relatives. *Cell, 134*, 388–389.

Darwin, C. (1874 [1871]). *The descent of man, and selection in relation to sex* (2nd ed.). New York: Hurst & Co.

Darwin, F. (Ed.). (1887). *Life and letters of Charles Darwin* (Vols. I–III, 3rd ed.). London: Murray.

Darwin, C. (1898a). *The origin of species by means of natural selection, or the preservation of favoured races in the struggle for life* (6th ed.). London: Murray.

Darwin, C. (1898b). *The variation of plants and animals under domestication* (Vols. I, II). New York: Appleton.

Darwin, C. (1952). *The origin of species and the descent of man*. New York: The Modern Library, Random House.

Dobzhansky, T. (1955). *Evolution, genetics, and man*. New York: Wiley.

Dobzhansky, T. (1962). *Mankind evolving*. New Haven: Yale University Press.

Galton, F. (1889). *Natural inheritance*. London: Richard Clay.

Galton, F. (1901). The possible improvement of the human breed under the existing conditions of law and sentiment. *Nature, 64*, 659–665.

Galton, F. (1962 [1869]). *Hereditary genius*. New York: World Publishing

Gould, S. J. (1985). To show an ape. In S. J. Gould (Ed.), *The flamingo smile* (pp. 263–280). London: Norton.

Graur, D., Gouy, M., & Wool, D. (2009). In retrospect: Lamarck's treatise at 200. *Nature, 460*, 688–689.

Haldane, J. B. S. (1923). Daedalus, or science and the future. Paper read at the University of Cambridge.

Huxley, J. (1942). *Evolution, the modern synthesis*. London: George Allen & Unwin.

Huxley, J. (1957a). Transhumanism. In J. Huxley (Ed.), *New bottles for new wine* (pp. 13–17). New York: Harper.

Huxley, J. (1957b). Man's place and role in nature. In J. Huxley (Ed.), *New bottles for new wine* (pp. 41–59). New York: Harper.

Huxley, T. H. (1900 [1863]). *Man's place in nature, and other essays*. London: Macmillan.

Huxley, T. H. (1989 [1894]). Evolution and ethics. In J. Paradis & G. C. Williams (Eds.), *Evolution and ethics: T.H. Huxley's evolution and ethics with new essays on its Victorian and sociobiological context*. Princeton: Princeton University Press.

Johanson, D. C., & White, T. D. (1979). A systematic study of African hominids. *Science, 203*, 321–330.

Krings, M., Stone, A., Schmitz, R. W., Krainitzki, H., Stoneking, M., & Pääbo S. (1997). Neanderthal DNA sequences and the origin of modern humans. *Cell, 90*, 19–30.

Lamarck, J. B. (1984 [1809]). *Zoological philosophy*. Chicago: University of Chicago press

Lyell, C. (1853). *Principles of geology* (9th ed.). London: Murray.

Moore, J. A. (1986). Science as a way of knowing. III. Genetics. *American Zoologist, 26*, 583–747.

Pearson, K. (1900). *The grammar of science* (2nd ed.). London: Adam and Charles Black.

Pearson, K. (1949 [1892]). *The grammar of science*. New York: Everyman.

Rak, Y. (1991). Lucy's pelvic anatomy: Its role in bipedal gait. *Journal of Human Evolution, 20*, 283–290.

Rak, Y., & Arensburg, B. (1987). Kebara 2 Neanderthal pelvis: First look at a complete inlet. *American Journal of Physical Anthropology, 73*, 227–231.

Wallace, A. R. (1869). Sir Charles Lyell on geological climates and the origin of species. *Quaternary Review, 1869*, 362–364.

Wallace, A. R. (1891). *Natural selection and tropical nature*. London: Macmillan.

Wells, K. D. (1972). Lawrence, W. [1787–1867]. A study of pre-Darwinian ideas on heredity and variation. *Journal of Historical Biology, 4*, 307–317

Weismann, A. (1891). *Essays on heredity and kindered biological problems*. Oxford: Clarendon Press.

Wool, D. (2001). Charles Lyell – "the father of geology" – as a fore-runner of modern Ecology. *Oikos, 94*, 385–391.

Chapter 4
The Paleoecology of the Upper Ndolanya Beds, Laetoli, Tanzania, and Its Implications for Hominin Evolution

Terry Harrison

Abstract Evidence from the Pliocene hominin site of Laetoli in northern Tanzania demonstrates that there was a taxonomic turnover of the mammalian fauna between the Upper Laetolil Beds (3.6–3.85 Ma) and the Upper Ndolanya Beds (2.66 Ma). *Paranthropus aethiopicus* was one of the novel species that appeared locally as part of the restructured fauna. This turnover coincides with a major climatic shift at ∼2.8–2.5 Ma, which had an important impact on the local environment and the composition of the faunal community. Investigation of the paleoecology of the Upper Ndolanya Beds provides critical evidence about how the vegetation and fauna at Laetoli, including the hominins, responded to these environmental changes. The preponderance of alcelaphin bovids and the reduced frequency of browsing ungulates, in conjunction with evidence from ecomorphology, mesowear and stable isotopes, indicate that the Upper Ndolanya Beds sample drier habitats with a greater proportion of grasslands compared with the earlier Upper Laetolil Beds. However, paleoecological inferences based on ostrich eggshells, rodents, and terrestrial gastropods present a more complicated picture, indicating instead that Upper Ndolanya habitats were more mesic and dominated by dense woodlands. Such confounding results can be reconciled as a consequence of the differential impact of climatic and environmental change on a global, regional and local scale.

Keywords Climate change • Fauna • Environment • *Paranthropus* • Pliocene

Introduction

The Pliocene site of Laetoli in northern Tanzania is well known for the fossil remains of *Australopithecus afarensis* and associated trails of hominin footprints from the Upper Laetolil Beds (ULB) dating to 3.6–3.85 Ma (Fig. 4.1) (Leakey 1987a, b; Harrison 2011a). In addition, *Paranthropus aethiopicus* has been recovered from the younger Upper Ndolanya Beds (UNB) at 2.66 Ma (Harrison 2011a). A major focus of recent research at Laetoli has been to reconstruct the paleoecology of the hominins using evidence from a wide spectrum of different sources, including modern-day ecosystems, sedimentology, paleobotany, stable isotopes, mesowear, ecomorphology, faunal studies and community structure analyses (Kovarovic et al. 2002; Su 2005, 2011; Su and Harrison 2007, 2008; Kingston and Harrison 2007; Kovarovic and Andrews 2007, 2011; Musiba et al. 2007; Andrews and Bamford 2008; Peters et al. 2008; Andrews et al. 2011; Bamford 2011a, b; Bishop 2011; Bishop et al. 2011; Ditchfield and Harrison 2011; Gentry 2011; Harrison 2005, 2011b, c, d, e; Hernesniemi et al. 2011; Kaiser 2011; Kingston 2011; Reed 2011; Reed and Denys 2011; Rossouw and Scott 2011; Tattersfield 2011). Study of the paleoecology provides important contextual evidence that is critical for interpreting hominin habitat preferences, ecology and paleobiology (see Su and Harrison 2008). It is obviously not possible to make deductions about key events that shaped human evolution from the narrow vantage point of individual paleontological sites, but detailed studies of sites such as Laetoli do provide small-scale temporal and spatial snapshots of past ecosystems that can be used to assemble a regional and continent-wide paleoenvironmental montage. The latter can then be used to test macroevolutionary models about hominin speciation, diversification, and extinction. The utility of such models is, however, entirely contingent upon the detail and precision of the paleoecological interpretations of the individual sites. In this regard, Laetoli offers an informative case study.

T. Harrison (✉)
Department of Anthropology, Center for the Study of Human Origins, New York University, New York, NY 10003, USA
e-mail: terry.harrison@nyu.edu

© Springer International Publishing AG 2017
Assaf Marom and Erella Hovers (eds.), *Human Paleontology and Prehistory*,
Vertebrate Paleobiology and Paleoanthropology, DOI: 10.1007/978-3-319-46646-0_4

Conflicting interpretations of the paleoecology at Laetoli can be formulated using different lines of evidence, and possible explanations for these confounding results have implications for understanding the impact of climatic and environmental changes on hominin evolution at the local and regional scales.

Study of the time-successive faunas at Laetoli demonstrates that there was a taxonomic turnover of the

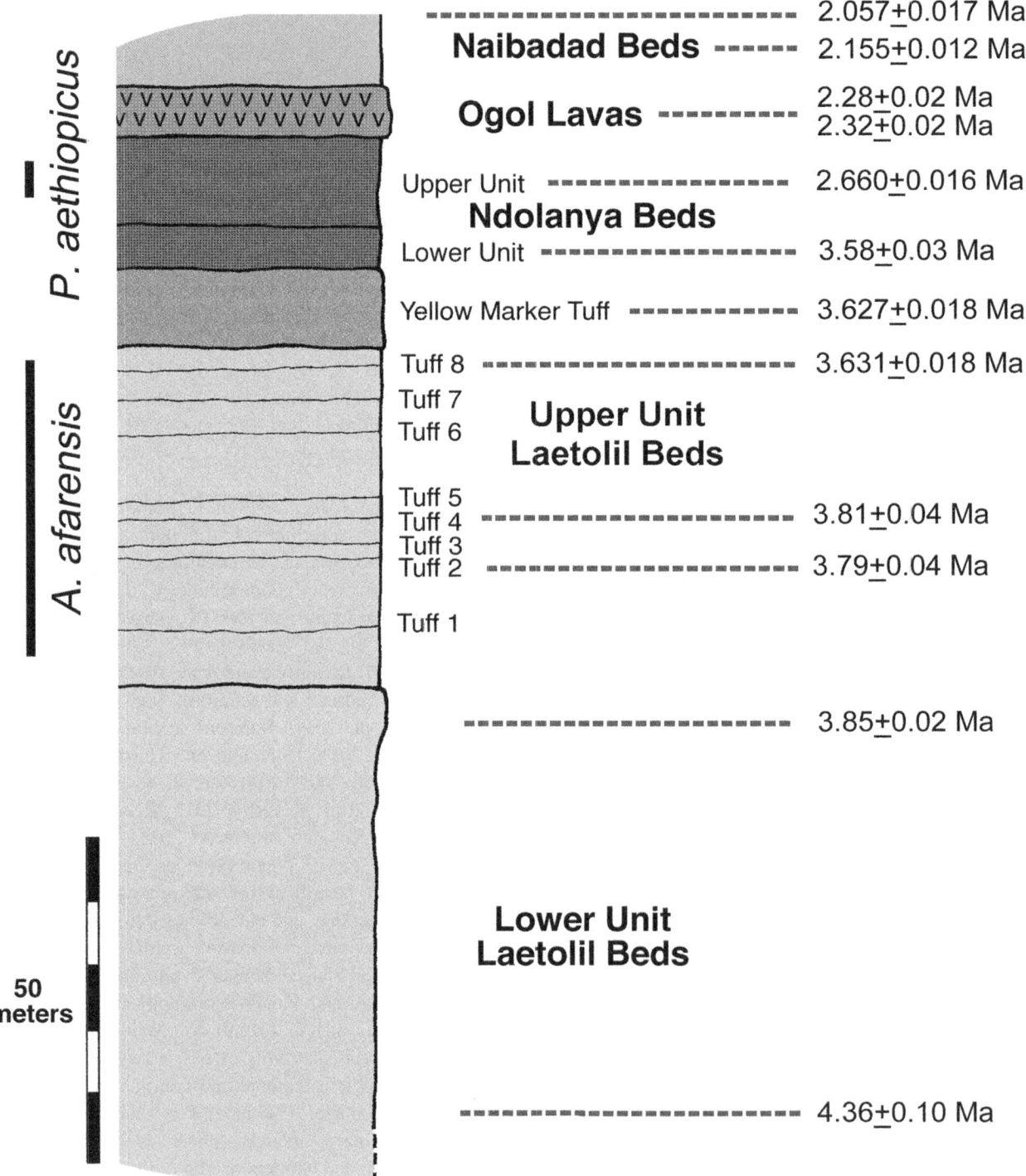

Fig. 4.1 Stratigraphic column and radiometric dating of the lower part of the sequence at Laetoli (adapted from Harrison 2011a). The temporal distribution of the Pliocene hominins is shown (left). Data from Hay (1987); Drake and Curtis (1987); Ndessokia (1990); Manega (1993); Mollel et al. (2011); Deino (2011); Harrison (2011a)

mammalian fauna between the ULB and the UNB. *Paranthropus aethiopicus* was one of the new species that appeared locally as part of the restructured faunal community. This turnover coincides with a major climatic shift in eastern Africa at ∼2.8–2.5 Ma. Climate change at this time has been associated with increased intensification of northern hemisphere glacial cycles and greater aridity in eastern Africa (Bobe and Behrensmeyer 2004; Bonnefille et al. 2004; deMenocal 2004, 2011; Feakins et al. 2005; Sepulchre et al. 2006; Feakins and deMenocal 2010; Bonnefille 2010). However, recent work on lake-levels in East Africa has shown that climate change at 2.6 Ma coincides with 400 kyr eccentricity maxima, which resulted in greater climate variability and relatively high moisture levels (Deino et al. 2006; Kingston et al. 2007; Trauth et al. 2005, 2007, 2009, 2010; Maslin and Trauth 2009).

A number of lines of evidence indicate that the UNB fauna is associated with a shift from a woodland-bushland-grassland mosaic in the ULB to habitats that were somewhat drier, with a greater proportion of grasslands (Kovarovic et al. 2002; Kovarovic 2004; Kingston and Harrison 2007; Kovarovic and Andrews 2007, 2011; Gentry 2011; Harrison 2011b, c; Hernesniemi et al. 2011; Kaiser 2011; Kingston 2011; Rossouw and Scott 2011; Su 2011; Barboni 2014). Given the length of the hiatus between the ULB and the UNB (almost 1 myrs), it is possible that the change in the ecology was the result of multiple shifts over an extended period of time. Such an ecological transition between the ULB and UNB would be fully consistent with expectations of increased aridity as a result of global climate change during the mid-Pliocene. However, alternative lines of evidence point to the UNB being more comparable to the ULB, with habitats that continued to be dominated by woodland mosaics. This latter scenario fits better with an inferred period of increased moisture availability, rather than increased aridity.

Accurate reconstruction of the paleoecology of the UNB and a better understanding of the nature of the ecological changes at Laetoli during the Pliocene are important because the UNB samples a key period in human evolution that witnessed the local extinction of *Australopithecus* and the origin and divergence of *Paranthropus*. The ecological changes that took place at Laetoli and at other localities in eastern Africa during this time period potentially provide valuable clues to understanding what environmental factors may have contributed to these major evolutionary events. This chapter aims to critically examine the evidence available to reconstruct the paleoecology of the UNB, and to offer a possible explanation for how contradictory lines of evidence might be reconciled. It provides a more nuanced and synthetic approach to understanding possible paleoecological change at Laetoli. The findings also have implications for contemporary debates about the relationship between climatic change and cladogenesis among Pliocene African hominins (Potts 1998, 2013; deMenocal 2004, 2011; Bobe and Behrensmeyer 2004; Maslin and Trauth 2009; Reed and Russack 2009; Bobe and Leakey 2009; Trauth et al. 2010; Harrison 2011a; Macho 2014).

Paleoecology of the Upper Ndolanya Beds

The consensus view, based on multiple lines of evidence, is that the ULB was dominated by a mosaic of closed woodland, open woodland, shrubland and grassland, with riverine woodlands and forests along ephemeral watercourses (see Harrison 2011c). It was certainly more densely wooded and more mesic than the modern-day Laetoli ecosystem (Andrews and Bamford 2008; Andrews et al. 2011). Changes in the vertebrate fauna between the ULB and the UNB provide clear evidence of a shift in the ecology, but the precise nature of what those changes mean in terms of the overall structure of the habitat is less evident.

The greater preponderance of bovids (especially alcelaphins) and the reduced frequency of large browsing herbivores in terms of number of specimens, in conjunction with evidence derived from ungulate ecomorphology, mesowear and stable isotopes, suggests that the UNB samples drier habitats with a greater predominance of grasslands compared with the ULB (Kovarovic et al. 2002; Kovarovic 2004; Kovarovic and Andrews 2007, 2011; Hernesniemi et al. 2011; Kaiser 2011; Bishop et al. 2011; Gentry 2011; Su 2011). However, stable isotope data from ostrich eggshells (Kingston and Harrison 2007; Kingston 2011) and the community structure of the rodents (Reed and Denys 2011; Denys 2011) and terrestrial gastropods (Peters et al. 2008; Tattersfield 2011) indicate that the picture is much more complicated, and that the UNB was a relatively mesic habitat dominated by woodlands.

The evidence supporting an ecological shift in the UNB towards drier habitats dominated by open woodlands and grasslands comes from a number of independent avenues of investigation. The taxonomic and paleobiological composition of the large mammal fauna provides one such important line of evidence. Of the 24 large mammal taxa from the UNB identified to the species level (including those identified as cf. and aff.), 62.5% also occur in the younger ULB (Table 4.1). Only 9 large mammal species make their first appearance at Laetoli in the UNB (i.e., *Paranthropus aethiopicus, Eurygnathohippus cornelianus, Ceratotherium simum, Metridiochoerus andrewsi, Giraffa pygmaea, Parmularius altidens, Parmularius parvicornis, Megalotragus kattwinkeli/isaaci* and *Antidorcas recki*). Using a combination of stable isotopes, dental mesowear and ecomorphology (Kingston 2011; Kaiser 2011; Bishop 2011), all of the new

Table 4.1 List of the mammalian taxa from the Upper Laetolil Beds (ULB) and Upper Ndolanya Beds (UNB) (after Harrison 2011c)

Order	Family	Genus and species	ULB	UNB
Macroscelidea	Macroscelididae	*Rhynchocyon pliocaenicus*	X	
Tubulidentata	Orycteropodidae	*Orycteropus* sp.	X	
Proboscidea	Deinotheriidae	*Deinotherium bozasi*	X	?
		Anancus ultimus	X	
	Stegodontidae	*Stegodon* sp. cf. *Stegodon kaisensis*	X	
		Loxodonta exoptata	X	X
Primates	Galagidae	*Laetolia sadimanensis*	X	
	Cercopithecidae	*Parapapio ado*	X	X
		Papionini indet.	X	
		cf. *Rhinocolobu*s sp.	X	X
		Cercopithecoides sp.	X	
	Hominidae	*Australopithecus afarensis*	X	
		Paranthropus aethiopicus		X
Rodentia	Sciuridae	*Paraxerus meini*	X	X
		Xerus sp.	X	
		Xerus janenschi	X	X
	Cricetidae	*Gerbilliscus satimani*	X	
		Gerbilliscus winkleri		X
		Gerbilliscus cf. *inclusus*	X	
		Dendromus sp.	X	
		Steatomys sp.	X	
		Saccostomus major	X	cf.
		Saccostomus sp.		X
	Muridae	*Aethomys* sp.	X	
		Thallomys laetolilensis	X	X
		Mastomys cinereus	X	
		Mus sp.	X	
	Thryonomyidae	*Thryonomys wesselmani*		X
	Bathyergidae	*Heterocephalus quenstedti*	X	
	Hystricidae	*Hystrix leakeyi*	X	
		Hystrix makapanensis	X	X
		Xenohystrix crassidens	X	
	Pedetidae	*Pedetes laetoliensis*	X	
		Pedetes sp.		X
Lagomorpha	Leporidae	*Serengetilagus praecapensis*	X	X
Soricimorpha	Soricidae	?*Crocidura* sp.		X
Carnivora	Canidae	?*Nyctereutes barryi*	X	
		cf. *Canis* sp. A	X	
		cf. *Canis* sp. B	X	
		aff. *Otocyon* sp.	X	
	Mustelidae	*Propoecilogale bolti*	X	X
		Mellivora sp.	X	
		Mustelidae indet.	X	
	Viverridae	*Viverra leakeyi*	X	
		Genetta sp.	X	
		aff. Viverridae	X	
	Herpestidae	*Herpestes palaeoserengetensis*	X	
		Herpestes ichneumon	X	
		Galerella sp.	X	
		Helogale palaeogracilis	X	X
		Mungos dietrichi	X	X
		Mungos sp. nov.	X	

(continued)

Table 4.1 (continued)

Order	Family	Genus and species	ULB	UNB
	Hyaenidae	*Crocuta dietrichi*	X	X
		Parahyaena howelli	X	
		Ikelohyaena cf. *I. abronia*	X	?
		Lycyaenops cf. *L. silberbergi*	X	
		?*Pachycrocuta* sp.	X	
	Felidae	*Dinofelis petteri*	X	X
		Homotherium sp.	X	X
		Panthera sp. aff. *P. leo*	X	
		Panthera sp. cf. *P. pardus*	X	X
		Acinonyx sp.	X	
		Caracal sp. or *Leptailurus* sp.	X	X
		Felis sp.	X	X
Perissodactyla	Equidae	*Eurygnathohippus* aff. *hasumense*	X	
		Eurgnathohippus aff. *cornelianus*		X
	Chalicotheriidae	*Ancylotherium hennigi*	X	
	Rhinocerotidae	*Ceratotherium efficax*	X	X
		Ceratotherium cf. *simum*		X
		Ceratotherium sp.		X
		Diceros sp.	X	
Artiodactyla	Suidae	*Notochoerus euilus*	X	
		Notochoerus jaegeri	X	
		Nyanzachoerus kanamensis	X	
		Potamochoerus afarensis	X	
		Kolpochoerus heseloni	X	X
		Metridiochoerus andrewsi		X
	Giraffidae	*Giraffa stillei*	X	aff.
		Giraffa jumae	aff.	
		Giraffa pygmaea		aff.
		Sivatherium maurusium	X	aff.
	Camelidae	*Camelus* sp.		X
	Bovidae	*Tragelaphus* sp.	X	
		Tragelaphus sp. cf. *T. buxtoni*		X
		Simatherium kohllarseni	X	
		Brabovus nanincisus	X	
		Bovini sp. indet.	X	X
		Cephalophini sp.	X	?
		Hippotragus sp.	X	
		Hippotragus sp. aff. *cookei*?		X
		Oryx deturi	X	
		Oryx sp.		X
		Parmularius pandatus	X	
		Parmularius altidens		X
		Parmularius parvicornis		X
		Alcelaphini, larger sp. indet.	X	
		Alcelaphini, small sp.	?	
		Megalotragus kattwinkeli or *M. isaaci*		X
		?*Connochaetes* sp.		X
		Reduncini sp. indet.	X	X
		Madoqua avifluminis	X	X
		?*Raphicerus* sp.	X	X
		Aepyceros dietrichi	X	
		Aepyceros sp.		X
		"*Gazella*" *kohllarseni*	X	
		Gazella janenschi	X	X
		Gazella granti	?	?
		Gazella sp.		X
		Antidorcas recki		X

ungulate taxa, with the exception of *Giraffa pygmaea*, can be deduced to be mixed feeders (*Ceratotherium simum, Metridiochoerus andrewsi, Parmularius parvicornis, Antidorcas recki*) or grazers (*Eurygnathohippus cornelianus, Megalotragus kattwinkeli/isaaci, Parmularius altidens*). At the same time, many of the large browsing mammals in the ULB, such as *Anancus ultimus, Deinotherium bozasi, Ancylotherium hennigi, Diceros* sp., *Giraffa jumae, Simatherium kohllarseni* and *Brabovus nanincisus*, are no longer present in the UNB. Consequently, the UNB witnessed a significant shift in its large herbivore dietary guild to one with a greater emphasis on taxa that included a significant proportion of grasses in their diets. In the ULB, only 41% of ungulate species are grazers or mixed feeders, whereas the proportion increases to 59% in the UNB. A further indicator of the decline in large browsing mammals in the UNB is provided by the reduction in the number of giraffids. Giraffids comprise only 4.6% of the ruminant specimens in the UNB, compared with 15.7% in the ULB (Harrison 2011b, c; Robinson 2011).

Differences in the taxonomic composition of the bovid fauna provide further support for an ecological difference between the UNB and ULB. The UNB has a much higher proportion of alcelaphin and antilopin bovids (77.4% of bovid specimens), which are predominantly mixed feeders and specialist grazers (Gagnon and Chew 2000), compared with the ULB (only 50.1%) (Table 4.2). The small gazelle in the ULB, *Gazella janenschi*, continues into the UNB, but is replaced as the dominant antilopin by the medium-sized and more hypsodont *Antidorcas recki* (Gentry 2011). Similarly, the dominant alcelaphin in the ULB, the medium-sized *Parmularius pandatus*, is replaced by a greater diversity of alcelaphins in the UNB, ranging in size from the small *Parmularius parvicornis* to the large *Megalotragus* sp., with most species having more hypsodont molars.

Stable carbon isotope data (Kingston and Harrison 2007; Kingston 2011) confirms a shift towards a greater emphasis on C_4 diets among equids and alcelaphin bovids in the UNB compared with the ULB (Fig. 4.2). The mean $\delta^{13}C_{enamel}$ for UNB alcelaphins is 0.1‰, which is significantly higher than the −2.4‰ in the ULB (Student's t-test, p = 0.01). The hipparionine equid *Eurygnathohippus* exhibits a similar trend, although less pronounced, with a shift in mean $\delta^{13}C_{enamel}$ from −1.0‰ to 0.2‰ from the ULB to the UNB (Student's t-test, p = 0.05). None of the other mammals demonstrate a significant difference in carbon isotopic signatures between the ULB and UNB.

Analyses of ungulate dental mesowear provide complementary results (Kaiser 2011). Mesowear scores are an indication of the degree of abrasiveness of the diet of large herbivores and they provide a guide to overall dietary behavior. The mesowear scores for the ULB are much lower than for the UNB, which Kaiser (2011) interprets as a shift to a predominance of grazing species in the UNB (57%) compared with that in the ULB (6%). However, few species are sampled from the UNB (n = 7) and the overall mesowear score is heavily influenced by the high scores for alcelaphin bovids and equids, which fall within the specialist grazer end of the spectrum.

Kovarovic et al. (2002) and Andrews (2006), using an ecological diversity approach (including all mammals, except bats), conclude that the UNB is predominantly a semi-arid bushland-grassland that was distinctly drier and more open than the ULB. This is due to the high proportion of terrestrial taxa and grazing herbivores, and the low incidence of frugivores. Ecomorphological studies of bovid postcranials (Kovarovic and Andrews 2007; Bishop et al. 2011) indicate that, although woodland and forest habitats continued to be present at Laetoli during the UNB, the majority of UNB bovids had a preference for open or lightly vegetated habitats (65.4% in the UNB versus 23.3% in the ULB).

Finally, paleobotanical remains are scarce in the UNB (no pollen or macrobotanical remains are known), but phytoliths have been recovered (Rossouw and Scott 2011). Although the abundance of phytolith in grasses can lead to an over-estimation of the extent of grassland habitats in

Table 4.2 Relative proportions of bovid tribes in the Upper Laetolil and Upper Ndolanya Beds

Taxon (Tribe)	Upper Laetolil Beds		Upper Ndolanya Beds	
	NISP[a]	%	NISP[a]	%
Alcelaphini	561	28.1	171	49.6
Antilopini	440	22.0	96	27.8
Bovini	14	0.7	8	2.3
Cephalophini	12	0.6	2	0.6
Hippotragini	332	16.6	6	1.7
Neotragini	629	31.5	41	11.9
Reduncini	2	0.1	0	0.0
Tragelaphini	7	0.4	21	6.1
Total	**1997**	**100.0**	**345**	**100.0**

Data from Gentry and Su (2011)

[a]NISP, number of individual specimens

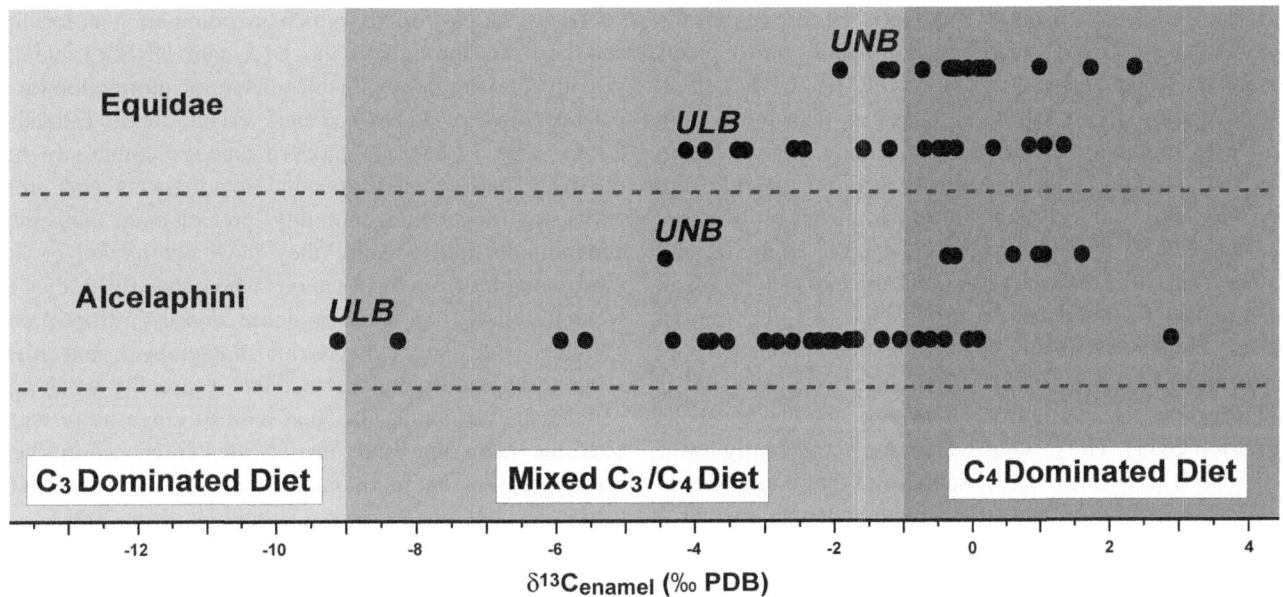

Fig. 4.2 Stable carbon isotope analysis of the dental enamel of hipparionine equids and alcelaphin bovids from the Upper Laetolil Beds (*ULB*) and the Upper Ndolanya Beds (*UNB*). The data points represent values for individual specimens. The dietary categories (in different shades of grey) correspond to browser (left), mixed feeders (middle), and grazers (right), respectively. Note that the UNB values are more strongly skewed towards the C_4 dominated end of the spectrum. (data from Kingston 2011)

paleoenvironmental reconstructions, a critical interpretation of the evidence indicates that grasses were ubiquitous in the ULB and UNB, but were never the dominant vegetation cover. Instead, the phytoliths indicate that the vegetation was heterogeneous throughout the sequence, and included a combination of C_3 and C_4 grasses. Even so, the phytoliths do indicate that there was a relatively higher frequency of C_4 grasses in the UNB compared with the upper part of the ULB. This is supported by stable carbon isotope analyses of soil carbonates that indicate a shift from woodland in the ULB to grassy woodland or grassy bushland habitats that were dominated by C_4 grasses in the UNB (Cerling 1992).

In summary, the combined evidence from multiple proxies indicates that there was a shift (or probably multiple shifts) in the paleoecology at Laetoli during the depositional hiatus between the ULB and UNB. Data from stable isotopes, mesowear, ecomorphology, phytoliths and the mammalian community structure all provide support for the conclusion that the paleoecology of the UNB was somewhat drier with a greater proportion of grassland than in the ULB.

However, it is important to emphasize that this evidence does not indicate that woodland gave way to grassland. Rather, the ULB and UNB both represent a spectrum of woodland-bushland-grassland habitats, in which the ULB is inferred to be at the mesic and more wooded end of that range, while the UNB is inferred to be slightly more arid with a somewhat greater coverage of grasses. As noted by Kovarovic and Andrews (2007), the types of habitats in the area did not change between the ULB and UNB, only the

relative proportions of vegetation types. This relationship is reflected in the marked continuity in the large mammal faunas between the ULB and UNB, with 62.5% of UNB species also occurring in the ULB, despite the substantial temporal gap. Another important point to note is that the evidence in support of a significant ecological change in the UNB is driven to a large extent by taxonomic and paleobiological changes in the bovids and equids. The UNB witnessed the arrival of just a few new species of antilopin and alcelaphin bovids, as well as a replacement species of hipparionine equid, many of which were more specialized for grazing (based on hypsodonty, mesowear and stable isotopes), and presumably better adapted postcranially for increased cursoriality in open country settings (based on ecomorphology) than their earlier counterparts in the ULB. The significance of these observations will be made apparent in the concluding discussion.

As noted above, other lines of evidence run counter to the interpretation that the UNB was characterized by drier and more open habitats compared with the ULB. Ostrich eggshells are ubiquitous throughout the sequence at Laetoli (Harrison and Msuya 2005), and can be attributed to two time-successive species – *Struthio kakesiensis* in the Lower Laetolil Beds and lower part of the ULB and the extant *Struthio camelus* in the upper part of the ULB and UNB (Harrison and Msuya 2005). Studies of the carbon and oxygen isotopes from the ostrich eggshells provide evidence that contradicts the conclusion that the UNB samples drier and more open habitats than the preceding ULB.

The $\delta^{13}C_{OES}$ demonstrates that ostriches throughout the Laetoli sequence were foraging predominantly on C_3 plants (Kingston 2011) (Table 4.3). However, the UNB ostrich eggshells show more depleted ^{13}C values than those from the ULB, implying that habitats were likely more mesic in the UNB. In addition, the eggshells of *Struthio camelus* from the UNB are significantly thicker (14% thicker on average) than those from the upper part of the ULB (Harrison and Msuya 2005). It is known that extant ostriches with access to better quality food and those living in areas of higher rainfall produce eggs with relatively thicker shells (Sauer 1968; Harrison and Msuya 2005). These inferences are further supported by studies of oxygen isotopes in the eggshells (Kingston 2011). The $\delta^{18}O_{OES}$ values are significantly lower in the UNB than in the ULB, suggesting that conditions in the UNB were cooler and more humid than in the ULB (Kingston 2011) (Table 4.3).

The rodent fauna from the Upper Ndolanya Beds is dominated by the ground squirrel, *Xerus janenschi* (58.8% of the rodent fauna) and the gerbil, *Gerbilliscus winkleri* (20.0%) (Denys 2011) (see Table 4.1). Modern-day *Xerus* occurs in semi-arid open woodland, wooded grassland and subdesert habitats (Kingdon 1997; Waterman 2013), while *Gerbilliscus* has broad habitat tolerances, ranging from forest edge mosaics to woodlands and grasslands (Kingdon 1997; Campbell et al. 2011; Reed 2011; Granjon and Dempster 2013). In addition to the greater frequency of *Xerus* in the UNB compared with the ULB (where *Xerus* represents less than 1% of the rodent fauna), the absence of *Heterocephalus*, the rarity of *Pedetes* and the appearance of *Thryonomys* in the UNB represent important differences (Reed and Denys 2011). Today, the naked mole-rat, *Heterocephalus,* and the spring hare, *Pedetes*, have minimal geographical overlap, but they both prefer dry grassland and open woodlands with firm, well-drained soils. The occurrence of *Thryonomys*, the cane rat, in the UNB (comprising 11.3% of the rodent fauna) is indicative of habitats with dense grass cover and reliable precipitation, such as open woodlands, wooded grasslands, reed beds and swamps (Kingdon 1997; Happold 2013). Overall, the rodent fauna from the UNB implies open woodlands and sparsely wooded grasslands that were more mesic than those in the ULB (Reed and Denys 2011).

Terrestrial gastropods, which are common at all localities and horizons throughout the ULB and UNB (Table 4.4), provide an extremely valuable source of information on the paleoecology of Laetoli (Peters et al. 2008; Tattersfield 2011). This is because modern analogs commonly have relatively narrow environmental requirements and preferences (i.e., vegetation, humidity, precipitation, temperature and altitude) and because they have not moved or been transported far from the locations where they lived, died and were fossilized. As a consequence, fossil gastropod communities are likely to provide fine-grained and highly accurate indicators of local habitats, especially when compared with vertebrate taxa that tend to range more widely over the landscape and have a greater chance of being transported (as entire or partial carcasses) by mammalian carnivores and avian raptors (Su and Harrison 2008).

The terrestrial snail community changes during the course of the ULB sequence (Peters et al. 2008; Tattersfield 2011) (Table 4.5; Fig. 4.3). Below Tuff 7, *Subulona, Kenyaella,* and *Achatina* are the dominant taxa, which indicate the presence of woodland and forest habitats. Above Tuff 7, *Gittenedouardia* and *Trochonanina* indicate a less mesic period, with a predominance of woodland and wooded grassland. However, the rare occurrence of *Halolimnohelix* and *Subulona* suggest that forest and dense woodland habitats continued to persist. The gastropod fauna confirms that the ULB was more mesic than present-day Laetoli, and indicates that the ULB ecosystem was heavily vegetated with extensive woodland and forest habitats.

The common genera of terrestrial snails represented in the UNB fauna, especially *Kenyaella* and *Subuliniscus*, are restricted today to forest and closed woodland habitats with relatively high levels of precipitation (Table 4.5). This is inconsistent with inferences based on the large mammal fauna that the UNB was a relatively dry open woodland-grassland mosaic. A possible explanation for these contradictory results could be that the fossil gastropods are being sampled from heavily vegetated microhabitats that are patchily distributed across the local landscape, and are not representative of the wider ecosystem. However, if this were the case, one would expect to find marked heterogeneity in the gastropod communities between different localities, reflecting both the dominant vegetation type and the mosaic of different

Table 4.3 Comparison of carbon and oxygen isotopes from ostrich (*Struthio camelus*) eggshells from the Upper Laetolil Beds and Upper Ndolanya Beds

	Upper Laetolil Beds				Upper Ndolanya Beds				
	n	Mean ‰	SE	Range ‰	n	Mean ‰	SE	Range ‰	p
$\delta^{13}C_{OES}$	45	−7.8	0.267	−4.1 to −12.1	12	−9.6	0.570	−6.1 to −11.7	0.003
$\delta^{18}O_{OES}$	45	3.7	0.328	−1.8 to 9.1	12	1.7	0.570	−2.2 to 4.7	0.007

Data from Kingston (2011)

Table 4.4 List of gastropod taxa from the Upper Laetoli Beds (ULB) and Upper Ndolanya Beds (UNB) (after Tattersfield 2011; Harrison 2011c) (see Fig. 4.3)

Family	Genus and species	ULB	UNB
Cerastidae	*Gittenedouardia laetoliensis*	X	
Subulinidae	*Subulona pseudinvoluta*	X	
	Pseudoglessula (Kempioconcha) aff. *gibbonsi*	X	
	Kenyaella leakeyi	X	
	Kenyaella harrisoni		X
	Subuliniscus sp. A		X
Streptaxidae	*Streptostele (Raffraya)* aff. *horei*	X	X
	Streptostele sp. A	X	
	Gulella sp. A	X	
Achatinidae	*Burtoa nilotica*	X	
	Limicolaria martensiana	X	
	Achatina (Lissachatina) indet.	X	
Urocyclidae	*Trochonanina* sp. B	X	X
	Urocyclinae sp. A	X	X
	Urocyclinae sp. B	X	X
	Urocyclinae sp. C	X	X
	Urocyclinae sp. D	X	X
	Urocyclinae sp. E	X	X
	Urocyclinae sp. F	X	X
Halolimnohelicidae	*Halolimnohelix rowsoni*	X	

Table 4.5 Stratigraphic distribution and inferred habitat preferences of fossil terrestrial gastropods from the Upper Laetoli Beds and Upper Ndolanya Beds

Stratigraphic unit	Horizon	#1 Ranked taxon	#2 Ranked taxon	#3 Ranked taxon	Paleoecological inference
Upper Ndolanya Beds		*Kenyaella* (72%)	*Subuliniscus* (16%)	*Streptostele* (4%)	Closed woodland and forest Rainfall: 760–1500 mm
Upper Laetolil Beds	Above Tuff 7	*Gittenedouardia* (45%)	*Trochonanina* (30%)	*Subulona* (10%)	Woodland and wooded grassland; forest patches Rainfall: 500–1270 mm
	Between Tuffs 5 & 7	*Subulona* (50%)	*Achatina* (23%)	*Trochonanina & Burtoa* (both 6%)	Woodland and forest Rainfall: 760–1500 mm
	Between Tuffs 3 & 5	*Subulona* (44%)	*Kenyaella* (41%)	*Achatina* (10%)	Closed woodland and forest Rainfall: 760–1500 mm
	Below Tuff 3	*Kenyaella* (77%)	*Achatina* (6%)	*Pseudoglessula* (6%)	Woodland and forest Rainfall: 700–1200 mm

See Fig. 4.1 for reference to stratigraphic units, horizons and radiometric dating
Data on Laetoli gastropods from Harrison (unpublished) and Tattersfield (2011)
Data on habitat preferences and rainfall from Verdcourt (1963, 1987), Pickford (1995, 2004, 2009) and Tattersfield (2011)

microhabitats. This is not the pattern observed. The same gastropod community occurs uniformly at all of the UNB localities, implying that woodland-forest habitats were widespread rather than patchily distributed. Additional support for this inference comes from urocyclid slugs, which are particularly sensitive to humidity and precipitation (Fig. 4.3). Slugs do occur in the Laetoli area today, but they are active only during or immediately following the rainy season and they are ecologically restricted to densely vegetated areas where leaf-litter and fallen tree trunks offer suitable habitats for estivation during the dry season. The ubiquitous

occurrence of fossil slugs at UNB localities (they comprise 88.8% of all fossil gastropods recovered from the UNB) offers incontrovertible evidence that woodland habitats were present and relatively widespread.

The conflicting paleoecological evidence presented above for the UNB is not easily reconciled. The evidence derived primarily from the large mammal fauna suggests that the faunal turnover between the ULB and UNB was associated with increased aridity and a change in the composition of the woodland-shrubland-grassland mosaic in favor of a greater representation of grassland. In contrast, the stable isotope

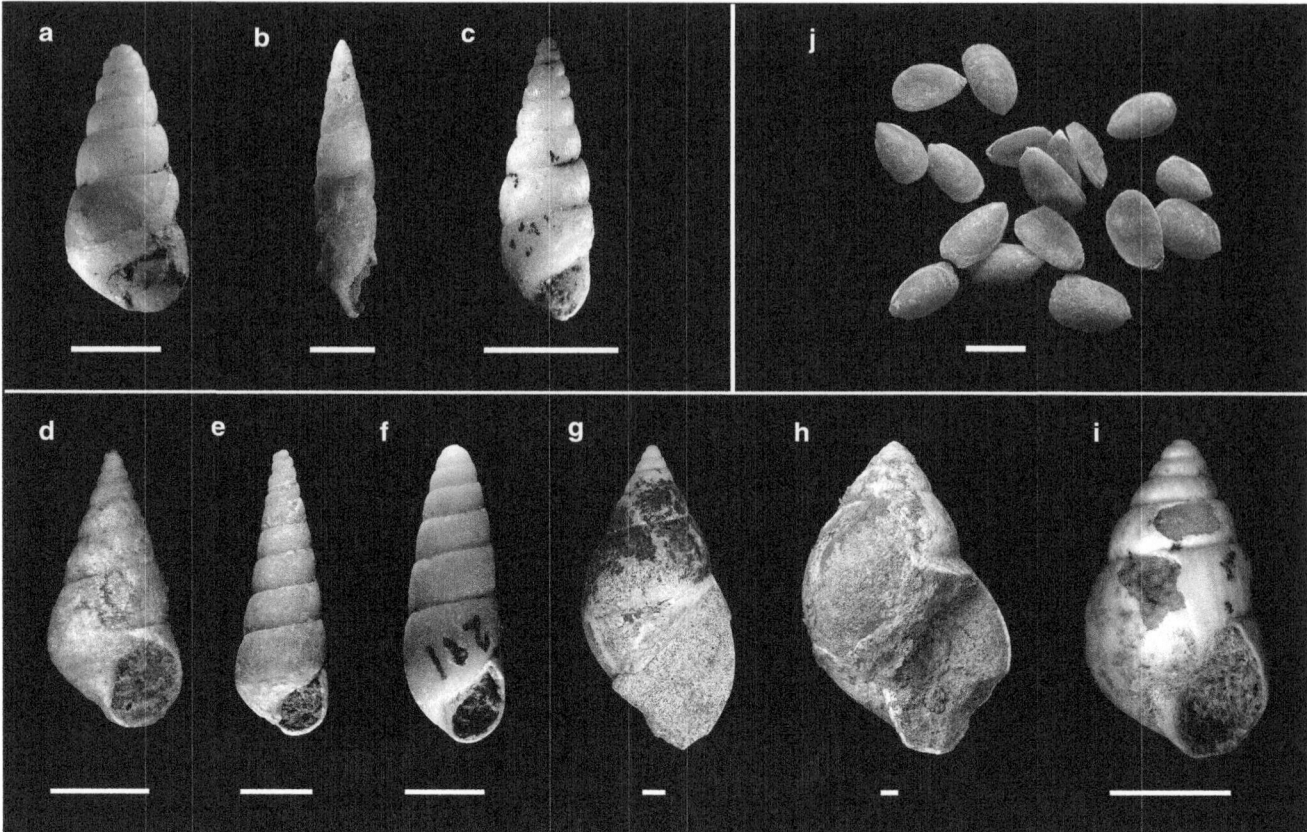

Fig. 4.3 Representative shells of the most common fossil terrestrial gastropods at Laetoli. Upper Ndolanya Beds: (*a*) *Kenyaella harrisoni*; (*b*) *Subuliniscus* sp. A; (*c*) *Streptostele* aff. *horei*. Upper Laetolil Beds: (*d*) *Gittenedouardia laetoliensis*; (*e*) *Subulona pseudinvoluta*; (*f*) *Kenyaella leakeyi*; (*g*) *Achatina* (*Lissachatina*) indet.; (*h*) *Burtoa nilotica*; (*i*) *Pseudoglessula* aff. *gibbonsi*. Urocyclid slugs: (*j*) assorted shells. a–j at approximately the same shell height. Scale bars = 5 mm. Images a–i courtesy of P. Tattersfield

data from ostrich eggshells, the rodent fauna, and the composition of the gastropod communities, indicate that the UNB was relatively more mesic in comparison to the ULB, and that woodlands persisted as the dominant habitat type.

Discussion and Conclusions

These contradictory findings appear, at first glance, to be difficult to reconcile. However, a more critical assessment of the nature of the evidence, along with a more nuanced appreciation of the significance of spatial scale and evolutionary processes, may offer the possibility to develop a unified and coherent paleoecological model that is consistent with all available lines of evidence. The solution to the problem may have implications for how one perceives the relationship between paleoecological reconstruction of fossil sites and hominin evolution.

There is near-universal agreement that the ULB and UNB represent a mosaic of woodland, shrubland and grassland. What is less certain is the relative proportion of grasslands

that were represented in these mosaic habitats, and whether or not there were significant shifts in the paleoecology between the ULB and UNB. The consensus view, based mainly on evidence derived from the large mammal fauna, is that the ULB ecosystem was composed predominantly of closed and open woodlands with large tracts of grassland, while the UNB was more arid with a higher proportion of grassland. However, contradictory evidence implies that the UNB was relatively mesic and that woodlands continued to be the dominant habitat type.

Two key questions need to be answered to settle the impasse. Which of the two alternative scenarios is most likely given the nature of the evidence? If one scenario is preferred over the other, how can the contradictory evidence be reconciled? First, it is important to highlight that the different lines of evidence offer insights into the paleoecology of Laetoli on different spatial scales. For example, ostrich eggshells, terrestrial snails and micromammals have a limited capacity for dispersal and/or transportation, and modern gastropods and rodents are often characterized by relatively narrow habitat preferences. As a consequence, these sources of evidence tend to reflect fine-grained ecological differences over relatively small spatial scales, and they are likely to provide a high-precision and reliable indicator of local environmental conditions. Large mammals, on the other hand, tend to range more widely across the landscape, with a greater capacity to traverse and occupy a broad range of different habitats, including those that are marginal or lie outside what would be considered their preferred habitats. As such, paleoecological interpretations based on large mammals tend to be more coarse-grained and applicable over larger spatial scales. The graininess of the environment is clearly dependent on the size of the organism. A small patch of dense vegetation represents a complex ecosystem for a small gastropod, whereas to an elephant the same patch is likely an inconsequential component of a much larger ecosystem. If one accepts the validity of this general premise, then it follows that the evidence derived from ostrich eggshells, rodents and gastropods should be given the greatest weight and has the potential to provide the most accurate reading of the local ecology. In this case, the most likely interpretation of the paleoecology is that the UNB was dominated by woodland and was not substantially different, at least in the general composition of the major vegetation types, from the ULB. This is not to imply that there was no discernable difference in the ecology between the ULB and UNB, but the differences may have been far subtler than has been proposed previously based on analyses of the large mammal fauna.

If this is the preferred scenario, then one has to account for the contradictory evidence that indicates a profound ecological shift in the UNB. It is important to reiterate two points made earlier: (1) many of the species of large mammals in the ULB continued unchanged in the UNB; (2) the main difference in the dietary and locomotor profiles of the large mammals is principally a consequence of the reduced diversity of the browsing ungulates and the appearance of new species of equids and bovids that were more specialized for cursoriality and grazing. However, the temptation is to presume a close correspondence between changes in the composition of the fauna at Laetoli and the local ecology, but this does not take into account the broader paleoenvironmental and evolutionary changes that were taking place across eastern Africa at this time. As noted above, the UNB coincides with a major global climatic shift associated with increased intensification of northern hemisphere glaciation, and this likely led to greater aridity and the expansion of grasslands in eastern Africa (deMenocal 2004, 2011; Bonnefille 2010; Barboni 2014). However, the impact of global climate change was modulated and amplified to a greater or lesser degree by synchronous regional and local influences on the environment (Bobe and Behrensmeyer 2004; Feakins et al. 2005; Trauth et al. 2005, 2009, 2010; Kingston 2007; Kingston et al. 2007; Bobe and Leakey 2009; Bailey et al. 2011; Levin et al. 2011; Barboni 2014). For example, the

complex interplay between regional climatic variability, tectonic activity and lake formation, produced a diversity of ecological settings regionally that presumably represented important loci for speciation and endemism (Potts 1998, 2013; Trauth et al. 2005, 2009, 2010; Maslin and Christensen 2007; Bailey et al. 2011; Macho 2014). It is important, therefore, to place Laetoli in a broader regional context when attempting to interpret the ecological implications of faunal change.

It could be argued that changes in the composition of the large mammal fauna in the UNB has been influenced more by biotic responses to environmental change at a regional level than it has at the local level. In other words, climate change and the accompanying expansion of grasslands in eastern Africa at ~ 2.8–2.5 Ma was associated with extinctions and speciation events in mammalian lineages across the region in response to local environmental changes and varied selection pressures. Successful new species capable of extending their geographical ranges beyond the confines of their original centers of endemism were potentially able to occupy new areas with somewhat different environments and become new constituent members of previously established local faunal communities. Of the changes that took place in the UNB fauna, it is the appearance of a new species of *Eurygnathohippus* and of several new alcelaphins and antilopins that had the greatest impact on the stable isotope, mesowear and ecomorphology results. These taxa, many of which were more specialized for exploiting more open country environments than their ULB counterparts, had presumably originated elsewhere in eastern Africa in response to regional climatic and environmental changes. These more sophisticated specialists were better able to take advantage of the availability of the grasslands that existed within the woodland-grassland mosaic at Laetoli. The best interpretation of the faunal evidence suggests that the overall ecology remained broadly similar between the ULB and UNB (although it is likely that there was a slight increase in aridity and the proportion of grassland in the UNB), but the composition of the large mammal fauna changed with the arrival of more advanced and specialized herbivores that were better adapted for exploiting the grassland component of the Laetoli ecosystem. Such a model implies that changes in the community structure of local faunas may not necessarily be indicative of significant changes in local ecosystems, but potentially reflect speciation and evolutionary events operating on a broader regional scale.

Testing these ideas and understanding their relationship to hominin evolution will require much more data from many more sites in eastern Africa, but the conclusions presented here do allow us to question whether the appearance of *Paranthropus* in the UNB was directly related to ecological change at Laetoli. The evidence suggests that *Paranthropus* occupied habitats at Laetoli that were not substantially different from those of *Australopithecus afarensis* earlier in time. We can conclude from this case study that there is a not a simple correspondence between the local appearance of new hominin species and changes in the immediate ecology. Attempts to model the causal factors driving hominin speciation events using paleoenvironmental evidence obtained at the local level are highly unlikely to lead to meaningful interpretations and conclusions. One has to look to broader regional environmental changes to understand the nature of the underlying factors that led to the extinction of *A. afarensis* and the origin of the *Paranthropus* lineage.

Acknowledgements I am very pleased to have this opportunity to acknowledge the remarkable influence that Yoel Rak's pioneering and seminal research has had on the study of human evolution. Several generations of students and scholars of paleoanthropology, myself included, have been profoundly influenced by his landmark contributions to hominin morphology and paleobiology. I extend my warmest appreciation, gratitude and congratulations to Yoel for his many outstanding and important scientific contributions and for being such a wonderful and supportive colleague. I thank Assaf Marom and Erella Hovers for the invitation to contribute to this honorary volume. I would also like to acknowledge the following colleagues for helping to shape the research and ideas presented here (although none should be held responsible for the final outcome): Peter Andrews, Christiane Denys, Mikael Fortelius, Alan Gentry, John Kingston, Thomas Kaiser, Kris Kovarovic, Amandus Kwekason, Chris Robinson, and William Sanders. Denise Su deserves special thanks for her valuable discussions about many of the themes presented in this paper. The manuscript benefited from the wisdom and critical comments of three anonymous reviewers. Research in Tanzania was granted by the Tanzania Commission for Science and Technology, the Department of Antiquities, and the National Museum of Tanzania. I am grateful to the following institutions and their staff for access to the fossil and skeletal collections in their care: National Museum of Tanzania, National Museum of Kenya, American Museum of Natural History and The Natural History Museum, London. The National Geographic Society, the Leakey Foundation and NSF (Grants BCS-0309513, BCS-0216683 and BSC 1350023) provided funding.

References

Andrews, P. (2006). Taphonomic effects of faunal impoverishment and faunal mixing. *Palaeogeography, Palaeoclimatology, Palaeoecology, 241*, 572–589.

Andrews, P., & Bamford, M. (2008). Past and present ecology of Laetoli, Tanzania. *Journal of Human Evolution, 54*, 78–98.

Andrews, P., Bamford, M., Njau, E.-F., & Leliyo, G. (2011). The ecology and biogeography of the Endulen-Laetoli area in northern Tanzania. In T. Harrison (Ed.), *Paleontology and geology of Laetoli: Human evolution in context. Vol. 1: Geology, geochronology, paleoecology, and paleoenvironment* (pp. 167–200). Dordrecht: Springer.

Bailey, G. N., Reynolds, S. C., & King, G. C. P. (2011). Landscapes of human evolution: Models and methods of tectonic geomorphology and the reconstruction of hominin landscapes. *Journal of Human Evolution, 60*, 257–280.

Bamford, M. (2011a). Fossil leaves, fruits and seeds. In T. Harrison (Ed.), *Paleontology and geology of Laetoli: Human evolution in context. Vol. 1: Geology, geochronology, paleoecology, and paleoenvironment* (pp. 217–233). Dordrecht: Springer.

Bamford, M. (2011b). Fossil woods. In T. Harrison (Ed.), *Paleontology and geology of Laetoli: Human evolution in context. Vol. 1: Geology, geochronology, paleoecology, and paleoenvironment* (pp. 235–252). Dordrecht: Springer.

Barboni, D. (2014). Vegetation of northern Tanzania during the Plio-Pleistocene: A synthesis of the paleobotanical evidences from Laetoli, Olduvai, and Peninj hominin sites. *Quaternary International, 322–323*, 264–276.

Bishop, L. C. (2011). Suidae. In T. Harrison (Ed.), *Paleontology and geology of Laetoli: Human evolution in context. Vol. 2: Fossil hominins and the associated fauna* (pp. 327–337). Dordrecht: Springer.

Bishop, L. C., Plummer, T. W., Hertel, F., & Kovarovic, K. (2011). Paleoenvironments of Laetoli, Tanzania as determined by antelope habitat preferences. In T. Harrison (Ed.), *Paleontology and geology of Laetoli: Human evolution in context. Vol. 1: Geology, geochronology, paleoecology, and paleoenvironment* (pp. 355–366). Dordrecht: Springer.

Bobe, R., & Behrensmeyer, A. K. (2004). The expansion of grassland ecosystems in Africa in relation to mammalian evolution and the origin of the genus *Homo*. *Palaeogeography, Palaeoclimatology, Palaeoecology, 207*, 399–420.

Bobe, R., & Leakey, M. G. (2009). Ecology of Plio-Pleistocene mammals in the Omo-Turkana basin and the emergence of Homo. In F. E. Grine, J. G. Fleagle, & R. E. Leakey (Eds.), *The first humans: Origins of the genus* Homo (pp. 173–184). Dordrecht: Springer.

Bonnefille, R. (2010). Cenozoic vegetation, climate changes and hominid evolution in tropical Africa. *Global and Planetary Change, 72*, 390–411.

Bonnefille, R., Potts, R., Chalié, F., Jolly, D., & Peyron, O. (2004). High-resolution vegetation and climate change associated with Pliocene *Australopithecus afarensis*. *Proceedings of the National Academy of Sciences USA, 101*, 12125–12129.

Campbell, T. L., Lewis, P. J., & Williams, J. K. (2011). Analysis of the modern distribution of South African *Gerbilliscus* (Rodentia: Gerbillinae) with implications for Plio-Pleistocene palaeoenvironmental reconstruction. *South African Journal of Science, 107*, Art. #497.

Cerling, T. E. (1992). Development of grasslands and savannas in East Africa during the Neogene. *Palaeogeography, Palaeoclimatology, Palaeoecology, 97*, 241–247.

Deino, A. (2011). ^{40}Ar/^{39}Ar dating of Laetoli, Tanzania. In T. Harrison (Ed.), *Paleontology and geology of Laetoli: Human evolution in context. Vol. 1: Geology, geochronology, paleoecology, and paleoenvironment* (pp. 77–97). Dordrecht: Springer.

Deino, A. L., Kingston, J. D., Glen, J. M., Edgar, R. K., & Hill, A. (2006). Precessional forcing of lacustrine sedimentation in the late Cenozoic Chemeron basin, central Kenya rift, and calibration of the Gauss/Matuyama boundary. *Earth and Planetary Science Letters, 247*, 41–60.

deMenocal, P. B. (2004). African climate change and faunal evolution during the Plio-Pleistocene. *Earth and Planetary Science Letters, 220*, 3–24.

deMenocal, P. B. (2011). Climate and human evolution. *Science, 331*, 540–542.

Denys, C. (2011). Rodents. In T. Harrison (Ed.), *Paleontology and geology of Laetoli: Human evolution in context. Vol. 2: Fossil hominins and the associated fauna* (pp. 15–53). Dordrecht: Springer.

Ditchfield, P., & Harrison, T. (2011). Sedimentology, lithostratigraphy and depositional history of the Laetoli area. In T. Harrison (Ed.), *Paleontology and geology of Laetoli: Human evolution in context. Vol. 1: Geology, geochronology, paleoecology, and paleoenvironment* (pp. 47–76). Dordrecht: Springer.

Drake, R., & Curtis, G. H. (1987). K-Ar geochronology of the Laetoli fossil localities. In M. D. Leakey & J. M. Harris (Eds.), *Laetoli: A Pliocene site in northern Tanzania* (pp. 48–52). Oxford: Clarendon Press.

Feakins, S. J., & deMenocal, P. B. (2010). Global and African regional climate during the Cenozoic. In L. Werdelin & W. J. Sanders (Eds.), *Cenozoic mammals of Africa* (pp. 45–55). Berkeley: University of California Press.

Feakins, S. J., deMenocal, P. B., & Eglinton, T. J. (2005). Biomarker records of late Neogene changes in northeast African vegetation. *Geology, 33*, 977–980.

Gagnon, M., & Chew, A. E. (2000). Dietary preferences in extant African Bovidae. *Journal of Mammalogy, 81*, 490–511.

Gentry, A. W. (2011). Bovidae. In T. Harrison (Ed.), *Paleontology and geology of Laetoli: Human evolution in context. Vol. 2: Fossil hominins and the associated fauna* (pp. 363–465). Dordrecht: Springer.

Gentry, A. W., & Su, D. F. (2011). Bovidae, Appendix. In T. Harrison (Ed.), *Paleontology and geology of Laetoli: Human evolution in context. Vol. 2: Fossil hominins and the associated fauna* (pp. 413–465). Dordrecht: Springer.

Granjon, L., & Dempster, E. R. (2013). Genus *Gerbilliscus* gerbils. In D. Happold. (Ed.), *Mammals of Africa. Volume III: Rodents, hares and rabbits* (pp. 268–270). London: Bloomsbury.

Happold, D. C. D. (2013). Genus *Thryonomys* cane rats. In D. Happold. (Ed.), *Mammals of Africa. Volume III: Rodents, hares and rabbits* (pp. 685–688). London: Bloomsbury.

Harrison, T. (2005). Fossil bird eggs from Laetoli, Tanzania: Their taxonomic and paleoecological implications. *Journal of African Earth Sciences, 41*, 289–302.

Harrison, T. (2011a). Hominins from the Upper Laetolil and Upper Ndolanya Beds, Laetoli. In T. Harrison (Ed.), *Paleontology and geology of Laetoli: Human evolution in context. Vol. 2: Fossil hominins and the associated fauna* (pp. 141–188). Dordrecht: Springer.

Harrison, T. (2011b). Laetoli revisited: Renewed paleontological and geological investigations at localities on the Eyasi Plateau in northern Tanzania. In T. Harrison (Ed.), *Paleontology and geology of Laetoli: Human evolution in context. Vol. 1: Geology, geochronology, paleoecology, and paleoenvironment* (pp. 1–15). Dordrecht: Springer.

Harrison, T. (2011c). Introduction: The Laetoli hominins and associated fauna. In T. Harrison (Ed.), *Paleontology and geology of Laetoli: Human evolution in context. Vol. 2: Fossil hominins and the associated fauna* (pp. 1–14). Dordrecht: Springer.

Harrison, T. (2011d). Cercopithecids (Cercopithecidae, Primates). In T. Harrison (Ed.), *Paleontology and geology of Laetoli: Human evolution in context. Vol. 2: Fossil hominins and the associated fauna* (pp. 83–139). Dordrecht: Springer.

Harrison, T. (2011e). Coprolites: Taphonomic and paleoecological implications. In T. Harrison (Ed.), *Paleontology and geology of Laetoli: Human evolution in context. Vol. 1: Geology, geochronology, paleoecology, and paleoenvironment* (pp. 279–292). Dordrecht: Springer.

Harrison, T., & Msuya, C. P. (2005). Fossil struthionid eggshells from Laetoli, Tanzania: Their taxonomic and biostratigraphic significance. *Journal of African Earth Sciences, 41*, 303–315.

Hernesniemi, E., Giaourtsakis, I. X., Evans, A. R., & Fortelius, M. (2011). Rhinoceroses. In T. Harrison (Ed.), *Paleontology and geology of Laetoli: Human evolution in context. Vol. 2: Fossil hominins and the associated fauna* (pp. 275–294). Dordrecht: Springer.

Hay, R. L. (1987). Geology of the Laetoli area. In M. D. Leakey & J. M. Harris (Eds.), *Laetoli: A Pliocene site in northern Tanzania* (pp. 23–47). Oxford: Clarendon.

Kaiser, T. M. (2011). Feeding ecology and niche partitioning of the Laetoli ungulate faunas. In T. Harrison (Ed.), *Paleontology and geology of Laetoli, Tanzania: Human evolution in context. Vol. 1: Geology, geochronology, paleoecology and paleoenvironment* (pp. 329–354). Dordrecht: Springer.

Kingdon, J. (1997). *The Kingdon field guide to African mammals*. San Diego: Academic Press.

Kingston, J. D. (2007). Shifting adaptive landscapes: Progress and challenges in reconstructing early hominid environments. *Yearbook of Physical Anthropology, 50*, 20–58.

Kingston, J. (2011). Stable isotopic analyses of Laetoli fossil herbivores. In T. Harrison (Ed.), *Paleontology and geology of Laetoli, Tanzania: Human evolution in context. Vol. 1: Geology, geochronology, paleoecology and paleoenvironment* (pp. 293–328). Dordrecht: Springer.

Kingston, J. D., & Harrison, T. (2007). Isotopic dietary reconstructions of Pliocene herbivores at Laetoli: Implications for early hominin paleoecology. *Palaeogeography, Palaeoclimatology, Palaeoecology, 243*, 272–306.

Kingston, J. D., Edgar, A. L., Deino, R. K., & Hill, A. (2007). Astronomically forced climate change in the Kenyan Rift Valley 2.7–2.55 Ma: Implications for the evolution of early hominin ecosystems. *Journal of Human Evolution, 53*, 487–503.

Kovarovic, K. (2004). Bovids as palaeoenvironmental indicators. An ecomorphological analysis of bovid postcranial remains from Laetoli, Tanzania. Ph.D. Dissertation, University of London.

Kovarovic, K., & Andrews, P. (2007). Bovid postcranial ecomorphological survey of the Laetoli paleoenvironment. *Journal of Human Evolution, 52*, 663–680.

Kovarovic, K., & Andrews, P. (2011). Environmental change within the Laetoli fossiliferous sequence: Vegetation catenas and bovid ecomorphology. In T. Harrison (Ed.), *Paleontology and geology of Laetoli, Tanzania: Human evolution in context. Vol. 1: Geology, geochronology, paleoecology and paleoenvironment* (pp. 367–380). Dordrecht: Springer.

Kovarovic, K., Andrews, P., & Aiello, L. (2002). An ecological diversity analysis of the Upper Ndolanya Beds, Laetoli, Tanzania. *Journal of Human Evolution, 43*, 395–418.

Leakey, M. D. (1987a). The Laetoli hominid remains. In M. D. Leakey & J. M. Harris (Eds.), *Laetoli: A Pliocene site in northern Tanzania* (pp. 108–117). Oxford: Clarendon Press.

Leakey, M. D. (1987b). The hominid footprints: Introduction. In M. D. Leakey & J. M. Harris (Eds.), *Laetoli: A Pliocene site in northern Tanzania* (pp. 490–496). Oxford: Clarendon Press.

Levin, N., Brown, F. H., Behrensmeyer, A. K., Bobe, R., & Cerling, T. E. (2011). Paleosol carbonates from the Omo Group: Isotopic records of local and regional environmental change in East Africa. *Palaeogeography, Palaeoclimatology, Palaeoecology, 307*, 75–89.

Macho, G. A. (2014). An ecological and behavioral approach to hominin evolution during the Pliocene. *Quaternary Science Reviews, 96*, 23–31.

Manega, P. (1993). Geochronology, geochemistry and isotopic study of the Plio-Pleistocene hominid sites and the Ngorongora volcanic highlands in northern Tanzania. Ph.D. Dissertation, University of Colorado at Boulder, Boulder.

Maslin, M. A., & Christensen, B. (2007). Tectonics, orbital forcing, global climate change, and human evolution in Africa. *Journal of Human Evolution, 53*, 443–464.

Maslin, M. A., & Trauth, M. H. (2009). Plio-Pleistocene East African pulsed climate variability and its influence on early human evolution. In F. E. Grine, J. G. Fleagle, & R. E. Leakey (Eds.), *The first humans: Origin and early evolution of the genus* Homo (pp. 151–158). Dordrecht: Springer.

Mollel, G. F., Swisher, C. C., III, Feigenson, M. D., & Carr, J. D. (2011). Petrology, geochemistry and age of Satiman, Lemagurut and Oldeani: Sources of the volcanic deposits of the Laetoli area. In T. Harrison (Ed.), *Paleontology and geology of Laetoli, Tanzania: Human evolution in context. Vol. 1: Geology, geochronology, paleoecology and paleoenvironment* (pp. 99–120). Dordrecht: Springer.

Musiba, V., Magori, C., Stoller, M., Stein, T., Branting, S., & Vogt, M. (2007). Taphonomy and paleoecological context of the Upper Laetolil Beds (Localities 8 and 9), Laetoli in northern Tanzania. In R. Bobe, Z. Alemseged, & A. K. Behrensmeyer (Eds.), *Hominin environments in the East African Pliocene: An assessment of the faunal evidence* (pp. 257–278). Dordrecht: Springer.

Ndessokia, P. N. S. (1990). The mammalian fauna and archaeology of the Ndolanya and Olpiro Beds, Laetoli, Tanzania. Ph.D. Dissertation, University of California, Berkeley.

Peters, C. R., Blumenschine, R. J., Hay, R. L., Livingstone, D. A., Marean, C. W., Harrison, T., et al. (2008). Paleoecology of the Serengeti-Mara ecosystem. In A. R. E. Sinclair, C. Packer, S. A. R. Mduma, & J. M. Fryxell (Eds.), *Serengeti III: Human impacts on ecosystem dynamics* (pp. 47–94). Chicago: University of Chicago Press.

Pickford, M. (1995). Fossil land snails of East Africa and their palaeoecological significance. *Journal of African Earth Sciences, 20*, 167–226.

Pickford, M. (2004). Palaeoenvironments of early Miocene hominoid-bearing deposits at Napak, Uganda, based on terrestrial molluscs. *Annales de Paleontologie, 90*, 1–12.

Pickford, M. (2009). Land snails from the early Miocene Legetet Formation, Koru, Kenya. *Geo-Pal Kenya, 2*, 1–88.

Potts, R. (1998). Environmental hypotheses of hominin evolution. *Yearbook of Physical Anthropology, 41*, 93–136.

Potts, R. (2013). Hominin evolution in settings of strong environmental variability. *Quaternary Science Reviews, 73*, 1–13.

Reed, D. (2011). Serengeti micromammal communities and the paleoecology of Laetoli, Tanzania. In T. Harrison (Ed.), *Paleontology and geology of Laetoli: Human evolution in context. Vol. 1: Geology, geochronology, paleoecology, and paleoenvironment* (pp. 253–263). Dordrecht: Springer.

Reed, D., & Denys, C. (2011). The taphonomy and paleoenvironmental implications of the Laetoli micromammals. In T. Harrison (Ed.), *Paleontology and geology of Laetoli: Human evolution in context. Vol. 1: Geology, geochronology, paleoecology, and paleoenvironment* (pp. 265–278). Dordrecht: Springer.

Reed, K., & Russack, S. M. (2009). Tracking ecological change in relation to the emergence of Homo near the Plio-Pleistocene boundary. In F. E. Grine, J. G. Fleagle, & R. E. Leakey (Eds.), *The first humans: Origin and early evolution of the genus* Homo (pp. 159–171). Dordrecht: Springer.

Robinson, C. (2011). Giraffidae. In T. Harrison (Ed.), *Paleontology and geology of Laetoli: Human evolution in context. Vol. 2: Fossil hominins and the associated fauna* (pp. 339–362). Dordrecht: Springer.

Rossouw, L., & Scott, L. (2011). Phytoliths and pollen, the microscopic plant remains in Pliocene volcanic sediments around Laetoli,

Tanzania. In T. Harrison (Ed.), *Paleontology and geology of Laetoli, Tanzania: Human evolution in context. Vol. 1: Geology, geochronology, paleoecology and paleoenvironment* (pp. 201–215). Dordrecht: Springer.

Sauer, E. G. F. (1968). Calculations of struthious egg sizes from measurements of shell fragments and their correlation with phylogenetic aspects. *Cimbebasia Series A, 1*, 27–55.

Sepulchre, P., Ramstein, G., Fluteau, F., Schuster, M., Tiercelin, J.-J., & Brunet, M. (2006). Tectonic uplift and eastern African aridification. *Science, 313*, 1419–1423.

Su, D. (2005). The paleoecology of Laetoli, Tanzania: Evidence from the mammalian fauna of the Upper Laetolil Beds. Ph.D. Dissertation, New York University.

Su, D. F. (2011). Large mammal evidence for the paleoenvironment of the Upper Laetolil and Upper Ndolanya Beds of Laetoli, Tanzania. In T. Harrison (Ed.), *Paleontology and geology of Laetoli: Human evolution in context. Vol. 1: Geology, geochronology, paleoecology, and paleoenvironment* (pp. 381–392). Dordrecht: Springer.

Su, D. F., & Harrison, T. (2007). The paleoecology of the Upper Laetolil Beds at Laetoli: A reconsideration of the large mammal evidence. In R. Bobe, Z. Alemseged, & A. K. Behrensmeyer (Eds.), *Hominin environments in the East African Pliocene: An assessment of the faunal evidence* (pp. 279–313). Dordrecht: Springer.

Su, D. F., & Harrison, T. (2008). Ecological implications of the relative rarity of fossil hominins at Laetoli. *Journal of Human Evolution, 55*, 672–681.

Tattersfield, P. (2011). Terrestrial Mollusca. In T. Harrison (Ed.), *Paleontology and geology of Laetoli: Human evolution in context. Vol. 2: Fossil hominins and the associated fauna* (pp. 567–587). Dordrecht: Springer.

Trauth, M. H., Maslin, M. A., Deino, A., & Strecker, M. R. (2005). Late Cenozoic moisture history of East Africa. *Science, 309*, 2051–2053.

Trauth, M. H., Maslin, M. A., Deino, A. L., Strecker, M. R., Bergner, A. G. N., & Dühnforth, M. (2007). High- and low-latitude forcing of Plio-Pleistocene East African climate and human evolution. *Journal of Human Evolution, 53*, 475–486.

Trauth, M. H., Larrasoaña, J. C., & Mudelsee, M. (2009). Trends, rhythms and events in Plio-Pleistocene African climate. *Quaternary Science Reviews, 28*, 399–411.

Trauth, M. H., Maslin, M. A., Deino, A. L., Junginger, A., Lesoloyia, M., Odada, E. O., et al. (2010). Human evolution in a variable environment: The amplifier lakes of eastern Africa. *Quaternary Science Reviews, 29*, 2981–2988.

Verdcourt, B. (1963). The Miocene non-marine mollusca of Rusinga Island, Lake Victoria and other localities in Kenya. *Palaeontographica, 121*(A), 1–37.

Verdcourt, B. (1987). Mollusca from the Laetolil and Upper Ndolanya beds. In M. D. Leakey & J. M. Harris (Eds.), *Laetoli: A Pliocene site in northern Tanzania* (pp. 438–450). Oxford: Clarendon Press.

Waterman, J. M. (2013). Genus *Xerus* ground squirrels. In D. Happold (Ed.), *Mammals of Africa. Volume III: Rodents, hares and rabbits* (pp. 93–100). London: Bloomsbury.

Chapter 5
The Australopithecine Brain: Controversies Perpetual

Ralph L. Holloway

Abstract While paleoneurology has undergone major changes relevant to hominid evolution, largely through newer computer-driven segmentation techniques using CT, laser, MRI, and other imaging technologies, so-called state-of-the-art techniques still require expert understanding of underlying endocranial morphology. The australopithecine endocranial remains, whether from natural endocasts such as Taung, Sts60, SK1585, or those made from rubber, silicone-based reagents, such as AL 444-2, or CT scans (MH1), still occasion major differences of interpretation and thus controversy, and the controversy initiated by Dart in 1925 for the Taung specimen is still alive and well. The newer non-invasive techniques have much to offer human paleontologists regarding the evolution of the brain as long as basic anatomical realities are appreciated.

Keywords *Australopithecus afarensis* • *Asutralopithecus africanus* • Brain reorganization • Cortex • Endocast • Paleoneurology • Prefrontal cortex

It has been my honor, privilege, and pleasure to have worked with Prof. Yoel Rak on aspects of hominin evolution, and in particular with one of the most difficult endocasts I've ever encountered, AL 444-2 (perhaps matched by Stw505). I have often discussed the agony and ecstasy of paleoneurology. Without the help of my former student, Dr. Michael Yuan, the 444-2 endocast project (Fig. 5.1) would have been more agony than ecstasy. I bring up AL 444 first, because the work we did was prior to availability of CT scans,

whether medical or micro, and of software packages that could manipulate fragments as is now beautifully shown in many recent publications (Neubauer et al. 2012; Weber et al. 2012; Zollikofer and Ponce de León 2013; Spoor et al. 2015). Back then the work we did was based on adding plasticine to missing regions, cutting apart plaster casts, re-aligning various sections, and trying at least 3 levels of reconstruction to gain some appreciation of possible maximum and minimum values for the endocranial volume of that hominin. The quality of internal table of bone made convolutional details almost impossible to interpret, which is one of those "agonies" inherent in paleoneurological research. The beautiful illustrations on pp. 50–54 of Holloway et al. (2004a) and p. 124 of Kimbel et al. (2004) are a small part of the occasional "ecstasy".

It would be a useful and hopefully an educational experience to have the CT scans of the AL 444-2 cranium, and see what the expertise of the current crop of segmentation artists such as Gunz et al. (2009), Neubauer et al. (2012), Weber et al. (2012) would find. I'd even be willing to give it a shot although I doubt I have the level of expertise needed for this difficult specimen, and some of those below.

Could such modern techniques improve upon the beautiful images of AL 162-28, AL 288-1 (about which more later), AL 333-45 with its missing frontal and temporal poles portions, or the enigmatic, but exquisite basal detail of the infant/child AL 133-105 with temporal lobes fully hominin and non-pongid? These were beautifully illustrated by my friend and colleague, John Gurche in Holloway et al. (2004a) (Fig. 5.4). Surely, these deserve the attention and expertise that is potentially available by segmenting techniques and morphometric rendering. Are these fragments worth the effort? Indeed they are, because aside from the Taung specimen, these are among the best evidence for

R.L. Holloway (✉)
Department of Anthropology, Columbia University, New York, NY 10027, USA
e-mail: rlh2@columbia.edu

Assaf Marom and Erella Hovers (eds.), *Human Paleontology and Prehistory*, Vertebrate Paleobiology and Paleoanthropology, DOI: 10.1007/978-3-319-46646-0_5

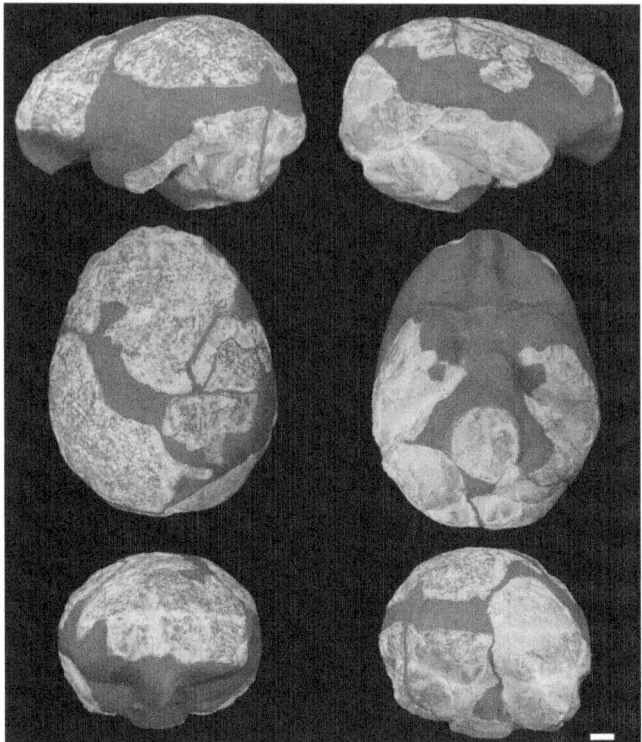

Fig. 5.1 Dr. Michael Yuan's reconstruction of the Hadar *Australopithecus afarensis* AL 444 endocast, which yielded an ECV of 545 ml

some kind of brain reorganization prior to brain enlargement in the genus *Australopithecus,* however one wishes to assign a species designation.

Ongoing Controversies

For all the discussion of AL 288-1 (Kimbel et al. 1982; Johanson and Blake 2006), as well as all the human evolution texts too numerous to mention, the possibility that *A. afarensis* had important brain reorganizational features in addition to a brain size within modern great ape limits, seems to have been ignored. Yes, I know, paleoneurology is "controversial", but I find it surprising that basic neuroanatomical details which might show some cerebral convolutional details are not considered in scenarios of early hominin evolution but simply ignored, as if brain size were the only neural variable under selection. When cast in such a limited scenario of size increases only, the rich fabric of human brain evolutionary changes that formed the genera *Australopithecus* and *Homo* becomes deprived of comparative neuroscience and selection

events. Both the occipital fragment of AL 288-1 (Fig. 5.2) and AL 162-28 provide tantalizing evidence for reorganization of the occipital and parietal lobes, in which the relative amount of primary visual striate cortex (Holloway et al. 2001, 2003, 2004a), Brodmann's area 17 was relatively reduced in *A. afarensis* compared to the African apes. Admittedly, finding a *clear-cut* lunate sulcus (LS) on the occipital on AL 228-1 is difficult and probably controversial; the lateral and dorsal contours are suggestive of a reduced occipital lobe as would be defined by the lunate sulcus.

AL 162-28 (Fig. 5.3), however, is a different matter. Here, there is good evidence for a possible intraparietal sulcus (IP) between superior and inferior parietal lobules which posteriorly abuts against the remnant of the lambdoid suture. The distance from the occipital pole (clearly visible) to the posterior end of the IP is 15 mm. The estimated cranial capacity for this specimen is between 385 and 400 ml, which is within the range for modern chimpanzees, where the distance from occipital pole to the lunate sulcus is on average 35 mm in distance. That is 5 S.D.'s larger than AL 162-28 (Holloway 1983; Holloway and Kimbel 1986). What

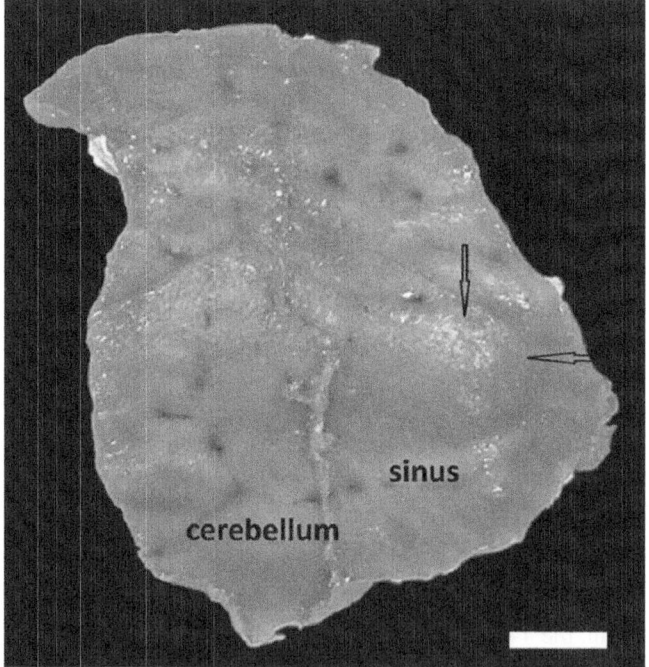

Fig. 5.2 Occipital and parietal fragments of AL 288-1. The arrows point to the dorsal and lateral edges of the suggested lunate sulcus bounding Brodmann's area 17, or primary visual striate cortex. If accurate, the posterior position indicates reorganization involving reduction of primary visual cortex and relative expansion of parietal association cortex

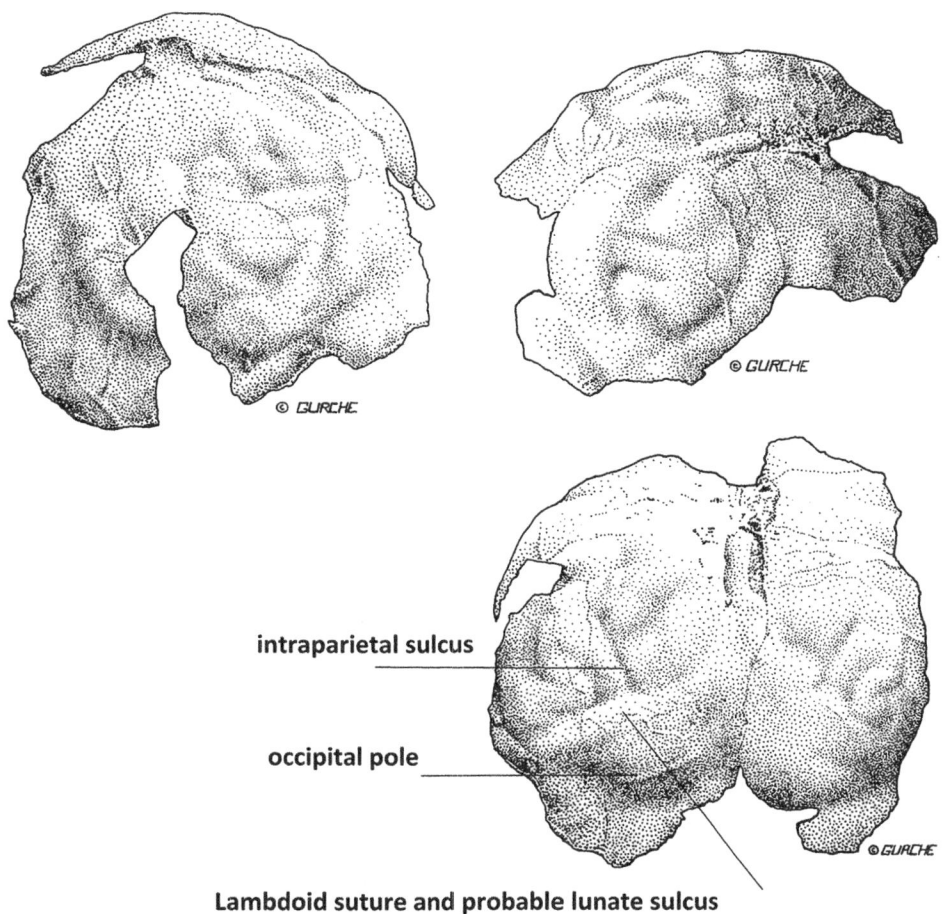

Lambdoid suture and probable lunate sulcus

Fig. 5.3 John Gurche's rendition of the parietal and occipital portions of the AL 162-28 endocast. The intraparietal sulcus (IP) always abuts the lunate sulcus in great apes, and while the lambdoid suture and probable lunate sulcus appear to coincide in these illustrations, the distance from the most posterior aspect of IP to occipital pole is 15 mm, while in chimpanzees of similar ECV (ca. 385 ml), that distance is usually around 30 mm. (as described in Holloway and Kimbel (1986))

more could one ask for to make a good case that the brain of *A. afarensis* probably differed in important ways from the chimpanzee? The only way it was possible for AL 162-28 to be considered as proof of a pongid status was for Falk (1985a) to orient the fragment so that the cerebellar lobes protruded more posteriorly than the occipital lobes, which left bregma so anterior and shifted downward that this *A. afarensis* had to be walking on its forehead! Of course, this was a time of considerable disagreement between me (Holloway 1984, 1985, 1988, 1991, 2014; Holloway and Broadfield 2012) and Falk (1985b) over the question of brain reorganization of the Taung *A. africanus* specimen.

Figure 5.4 showcases John Gurche's artistry in illustrating the detailed basal portion of the A.L. 133-105 infant endocast.

Some Recent Issues

Most recently, Falk et al. (2012) have tried to show that the Taung child had a metopic suture from glabella to bregma, and this allowed postnatal widening of the prefrontal cortex, suggesting some reorganization being possible after going through the "pelvic dilemma". Unfortunately, micro-CT scans of Taung demonstrate that the only metopism visible was limited to the glabellar region (Fig. 5.5), and that contrary to their claims of metopism being common in early *Homo*, all published account suggests the metopic closes early, and leaves a remnant at glabella only (Holloway et al. 2014). Additionally, one can find bonobos and chimpanzees with broader prefrontals than Taung even with smaller brain volumes. If neural reorganization did occur in *A. africanus*

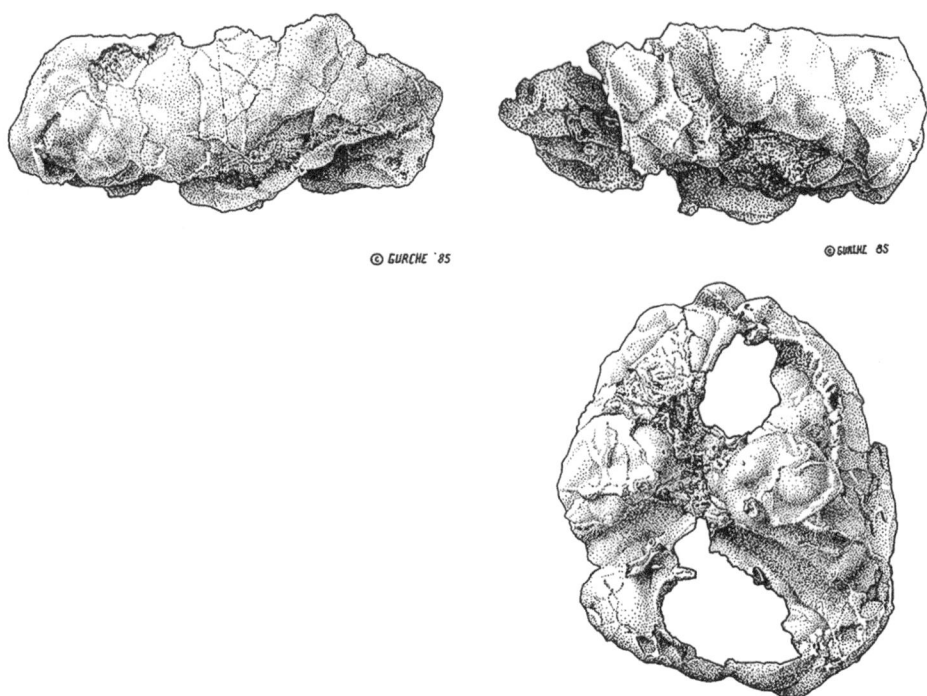

Fig. 5.4 The left, right, and basal views of AL 333-105, an infant *Australopithecus afarensis*. Such sulcal details are extremely rare

or some taxon before, this will need to be shown by actual cerebral convolutional details either in the occipital/parietal region, or the prefrontal regions. Thus far, the best case is for the possible, and probable, posteriorly oriented lunate sulcus is on the Taung child, just as Dart declared back in 1925 (but see Stw 505 below). The question of prefrontal reorganization is not clearly answered by the MH1 *A. sediba* specimen which Carlson et al. (2011) have so beautifully illustrated in their article (see below). Both Broadfield et al. (2015) and Hurst et al. (2015) have suggested more study is needed to rule out distortion, although the new prefrontal portion of the Taung child, extracted using micro CT scan as described in Holloway et al. (2014), does show some slight asymmetries of the frontal poles.

Of course, Stw 505 (Holloway et al. 2004b), an *A. africanus* specimen shows a clear-cut lunate in a relatively posterior position. Or at least I thought that was the case up until Falk's (2014) paper in Frontiers in Human Neuroscience. Here she has proposed that the crescentic-shaped furrow facing (concave) medially, and in a relatively posterior and inferior position, is possibly a lateral calcarine sulcus (LC) and not a lunate sulcus (LS) as we described (Holloway 2004b). Here, the neuroanatomy is so incorrectly examined, both in actual ape brains and endocasts that it is

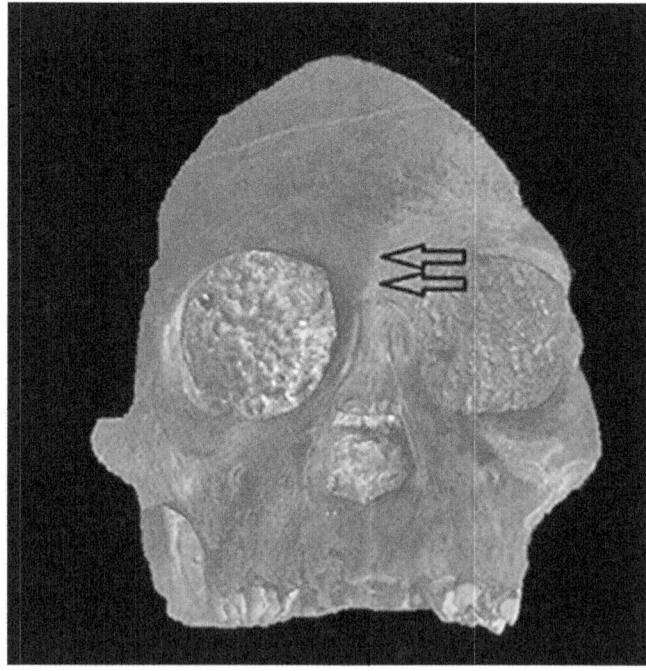

Fig. 5.5 Micro-CT Scan of the Taung cranium in frontal view showing the metopic suture limited to the nasion region. (Modified after Holloway et al. 2014)

necessary to criticize her paper. Figures 5.6 and 5.7 provide views showing both Falk's and my interpretation of the Stw 505 occipital portion. Notice that in her rendition, the LC doesn't even follow the same contour as the true LS. Examining the occipital lobes of some 54 ape hemispheres from 31 brains, mostly chimpanzee, not a single case provides an example where the curvature of the inferior part of the LC matches that of the LS! (see Fig. 5.8). What is even more problematic is that the LC never appears on any of the 200+ ape endocast occipital lobes that I have examined (see Fig. 5.9 for examples), and the LC does not appear in *Homo* brains or on endocasts on the lateral surface! (Holloway et al. 2015). Holloway et al. (2004a, Part 6) present an extended discussion of possible cognitive consequences of such a reorganizational change and a speculative evolutionary scenario (Table 5.1; and see Holloway et al. 2015).

Perhaps this quote from a recent paper on the lateral occipitotemporal cortex (LOTC) will illustrate why the matter of an anterior or posterior placement of the lunate sulcus in early hominins is of such importance (Table 5.1):

"LOTC…encodes many related dimensions of action. These include representations of: simple and complex patterns of motion; the appearance, uses, and characteristic motions of manipulative artifacts, such as tools; the shape of human bodies and body parts as well as their movements, and verbal material referring to actions symbolically." (Lingnau and Downing 2015).

Of course, evolutionary speculative scenarios are what we do as paleoanthropologists, and none of us is exempt from this charge. But it is surprising how newer non-invasive techniques are considered improvements over older techniques, although they still require a good understanding of

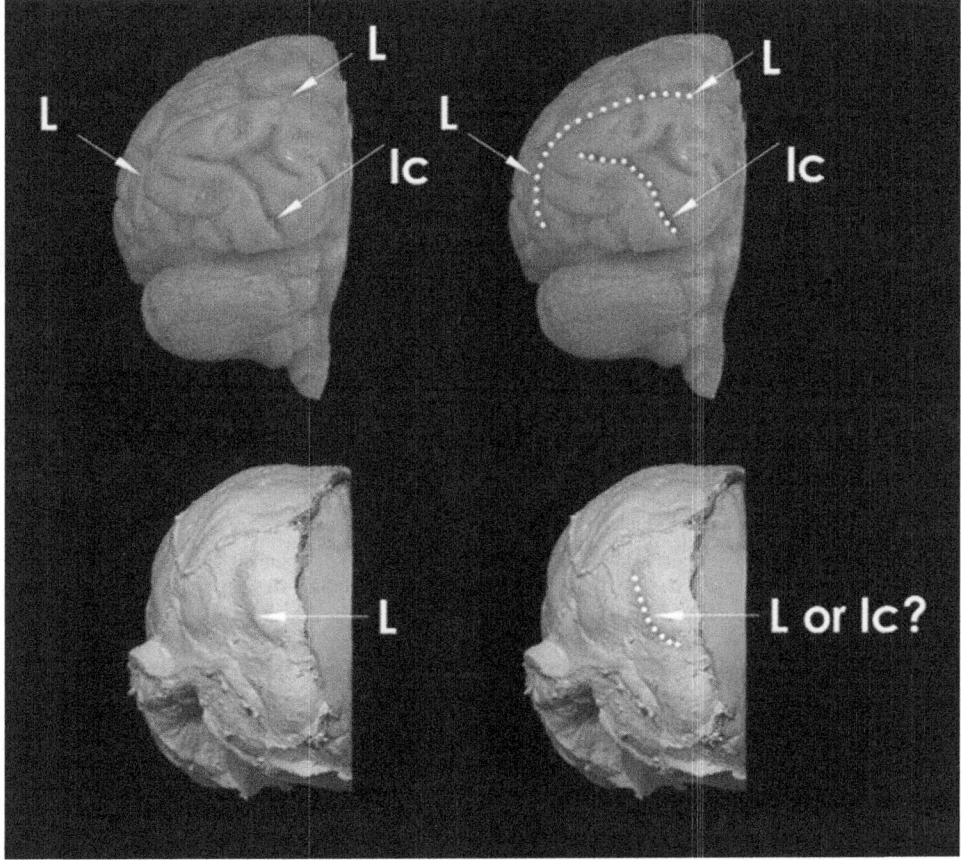

Fig. 5.6 Comparison of Stw 505 endocast (below) and chimpanzee (above). Falk suggested that the lunate sulcus (L) described by Holloway et al. (2004b) is probably a lateral calcarine (lc) sulcus. Notice how the curvature of the chimpanzee lc is totally in the opposite direction of the *L* lunate sulcus. Holloway et al. (2015) show that (**a**) no hominins, including modern *Homo*, show a lateral calcarine sulcus on the lateral brain surface; (**b**) no ape endocasts ever show a lc on the endocast surface; (**c**) there is no sulcal morphology suggesting a lunate sulcus anterior to what is depicted as L in this illustration; (**d**) None of the 30+ brain occipital lobes of chimpanzee, gorilla, or orang known to me show a lateral calcarine sulcus curving like a lunate sulcus

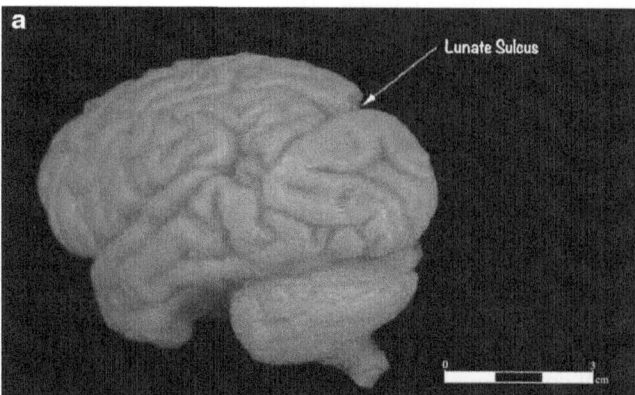

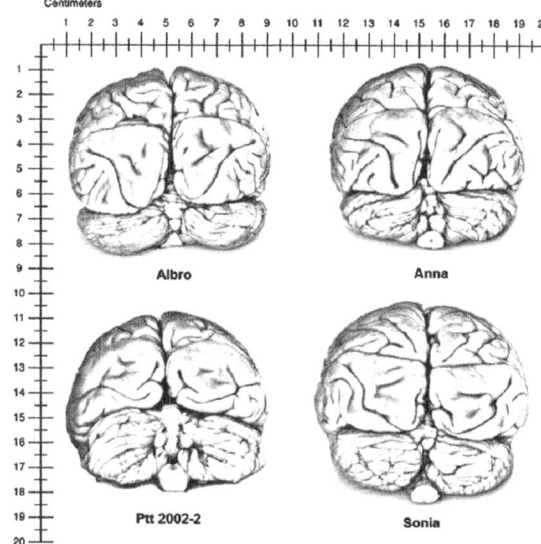

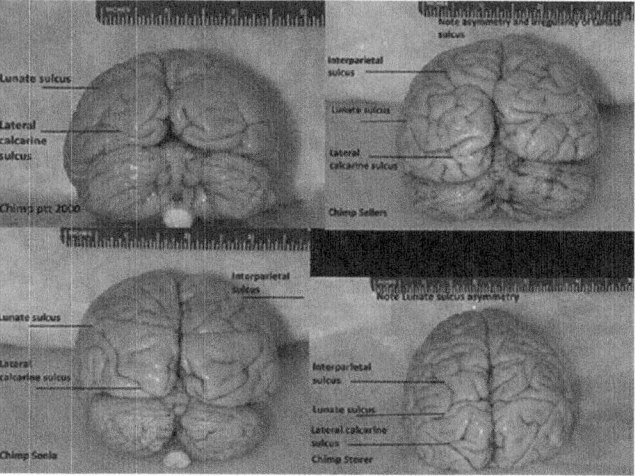

Fig. 5.7 An oblique view of a chimpanzee brain cast (**a**) and the Stw 505 endocast (**b**) showing the position of the lunate sulcus relative to the midline. It should be noted that if this were indeed a lateral calcarine sulcus, one would expect to find the lunate sulcus and intraparietal sulcus well anterior to it. This illustration makes the additional point that the intraparietal sulcus does not always imprint on the parietal endocast surface

Fig. 5.8 Above Four chimpanzee brains in occipital view, as drawn by Shawn Hurst. Note that the lateral calcarine sulcus does not follow the curvature of the lunate sulcus; Below occipital views of actual chimpanzee brains showing details of sulci and gyri

neuroanatomical detail, as the Falk/Holloway controversies above suggest.

Maybe Yoel could enter the fray regarding Falk's claim that the difference between the pointed prefrontals of robust australopithecines and gracile ones suggests brain reorganization in the latter. Here, one wonders how such very robust dentition and facial structures might modify prefrontal breadth, but not function, in robust australopithecine growth (e.g., Rak 1983).

A beautifully illustrated paper by Carlson et al. (2011), using micro-CT scans on the newly discovered MH1 *A. sediba* specimen has tentatively suggested that the prefrontal region did show some reorganizational change toward a *Homo* pattern that was different from what has been seen thus far in *A. africanus* endocasts. However, as Hurst et al. (2015) have suggested, the MH1 *A. sediba* prefrontal region requires some reconstruction given the obvious distortion of the frontal bec and temporal poles. Alas, the posterior portion is missing, so nothing can be said regarding the likely position of the lunate sulcus. A tentative reconstruction as shown in Fig. 5.10 yielded a volume of 442 ml. Notice that the conjoint occurrence of a new specimen and better techniques contributes to this fascinating possibility of early frontal lobe reorganization as Falk (2014) suggests.

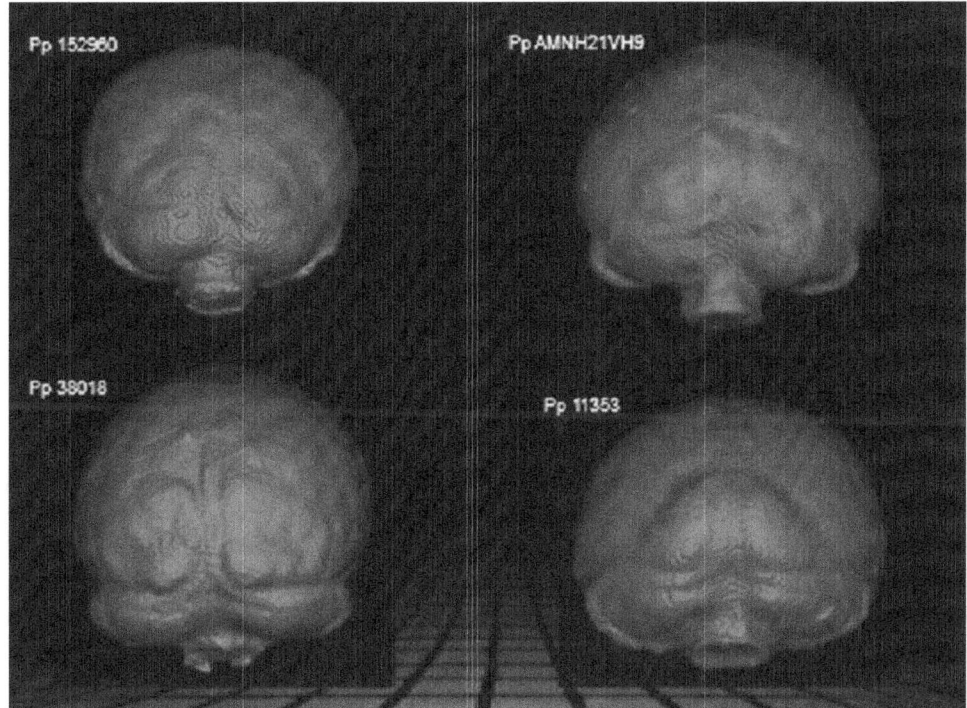

Fig. 5.9 Occipital view of four bonobo endocasts. There is no sign of any lateral calcarine sulci commencing from the occipital pole. The lateral calcarine sulcus does not appear on any of the 200+ ape endocasts examined

Table 5.1 This chart shows reorganizational changes to the hominin brain based on endocast morphology. Adapted from Holloway (2015)

Reorganizational brain changes	Taxon
(1) Reduction of primary visual striate cortex, area 17, and a relative increase in posterior parietal and temporal cortex, Brodmann areas 37, 39, 40, as well as 5 and 7	*Australopithecus afarensis, Australopithecus africanus*
(2) Reorganization of frontal lobe (3rd inferior frontal convolution, Broca's areas 44, 45, 47)	*Homo rudolfensis; early Homo*
(3) Cerebral asymmetries, left-occipital right frontal petalias	Australopithecines and early *Homo*
(4) Refinements in cortical organization to a modern *Homo sapiens* pattern	*Homo erectus* to present

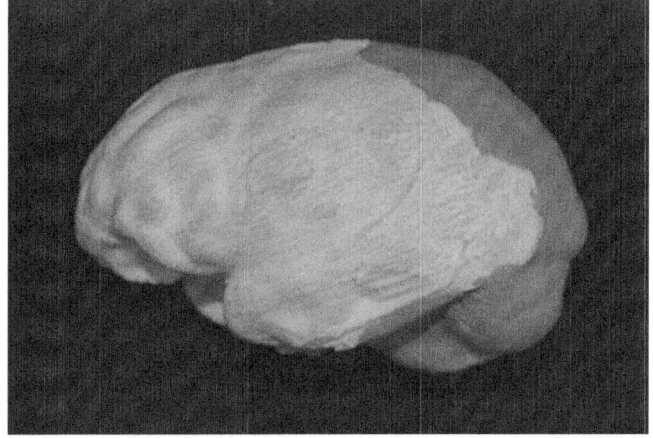

Fig. 5.10 Left lateral view of my reconstruction of the MH1 *A. sediba* partial endocast, using plasticine to sculpt the missing posterior portions, based on the Taung and ST60 endocasts

Conclusion

I have tried to limit my discussions to the australopithecines that Yoel has worked on, and where he has made such significant contributions to our knowledge of early hominin anatomy and function. Were I to take on the Neandertals and recent techniques and speculations (particularly regarding the occipital lobes), it would be amiss. Allow me to simply say that modern techniques are all well and fine, but let us not lose sight of anatomical realities when using these techniques.

Acknowledgments I am grateful to the editors of this volume honoring Professor Yoel Rak for inviting me to participate. Many colleagues helped me with the illustrations: Drs. Michael Yuan, Doug Broadfield, Jill Shapiro, Tom Schoenemann, and Mr. Shawn Hurst. I also thank Dr. Chet Sherwood and his neuroscience group at George Washington University for the illustrations of chimpanzee brains used here.

References

Broadfield, D. C., Carlson, K. J., Chapelle, K. E. J., Hurst, S. D., & Holloway, R. L. (2015). New micro-CT scan of the imbedded prefrontal of the Taung endocast. *American Journal of Physical Anthropology, 156* (Suppl. 60), 93.

Carlson, K. J., Stout, D., Jashashvili, T., de Ruiter, Darryl J., Tafforeau, P., Carlson, K., et al. (2011). The endocast of MH1, *Australopithecus sediba. Science, 333,* 1402–1407.

Falk, D. (1985a). Hadar AL 162-28 endocast as evidence that brain enlargement preceded cortical organization in hominid evolution. *Nature, 313,* 45–47.

Falk, D. (1985b). Apples, oranges, and the lunate sulcus. *American Journal of Physical Anthropology, 67,* 313–315.

Falk, D. (2014). Interpreting sulci on hominin endocasts: Old hypotheses and new findings. *Frontiers in Human Neuroscience, 8. Art., 134,* 1–11.

Falk, D., Zollikofer, C. P. E., Morimoto, N., & Ponce de León, M. (2012). Metopic suture of Taung (*Australopithecus africanus*) and its implications for hominin brain evolution. *Proceedings of the National Academy of Sciences, 109,* 8467–8480.

Gunz, P., Mitteroecker, P., Neubauer, S., Weber, G. W., & Bookstein, F. L. (2009). Principles for the virtual reconstruction of hominin crania. *Journal of Human Evolution, 57,* 48–62.

Holloway, R. L. (1983). Cerebral brain endocast patten of *Australopithecus afarensis* hominid. *Nature, 303,* 420–422.

Holloway, R. L. (1984). The Taung endocast and the lunate sulcus: A rejection of the hypothesis of its anterior position. *American Journal of Physical Anthropology, 64,* 285–287.

Holloway, R. L. (1985). The past, present, and future significance of the lunate sulcus in early hominid evolution. In P. V. Tobias (Ed.), *Hominid evolution: Past, present, and future* (pp. 47–62). New York: A.R. Liss Inc.

Holloway, R. L. (1988). Some additional morphological and metrical observation on *Pan* brain casts and their relevance to the Taung endocast. *American Journal of Physical Anthropology, 77,* 27–33.

Holloway, R. L. (1991). On Falk's 1989 accusations regarding Holloway's study of the Taung endocast: A reply. *American Journal of Physical Anthropology, 84,* 87–88.

Holloway, R. L. (2014). Paleoneurology, resurgent! In E. Bruner (Ed.), *Human paleoneurology* (pp. 1–10). Dordrecht: Springer.

Holloway, R. L. (2015). The evolution of the hominid brain. In Tattersall & W. Henke (Eds.), *Handbook of paleoanthropology,* (2nd ed., pp. 1961–1987). Berlin: Springer.

Holloway, R. L., & Broadfield, D. (2012). Letter to the editor: Response to Falk and Clarke regarding Taung midline. *American Journal of Physical Anthropology, 143,* 326.

Holloway, R. L., & Kimbel, W. H. (1986). Endocast morphology of Hadar hominid AL 162-28. *Nature, 321,* 536.

Holloway, R. L., Broadfield, D. C., & Yuan, M. S. (2001). Revisiting Australopithecine visual striate cortex: Newer data from chimpanzee and human brains suggest it could have been reduced during Australopithecine times. In D. Falk & K. R. Gibbon (Eds.), *Evolutionary anatomy of the primate cerebral cortex* (pp. 177–186). Cambridge: Cambridge University Press.

Holloway, R. L., Broadfield, D. C., & Yuan, M. S. (2003). Morphology and histology of chimpanzee primary visual striate cortex indicate that brain reorganization predated brain expansion in early hominid evolution. *Anatomical Record, 273A,* 594–602.

Holloway, R. L., Yuan, M. S., & Broadfield, D. C. (2004a). *Brain endocasts: Paleoneurological evidence.* New York: Wiley.

Holloway, R. L., Clarke, R. J., & Tobias, P. V. (2004b). Posterior lunate sulcus in *Australopithecus africanus*: Was Dart right? *Comptes Rendue Palevol, 3,* 287–293.

Holloway, R. L., Broadfield, D., & Carlson, K. (2014). New high-resolution CT data of the Taung partial cranium and endocast and their bearing on metopism and hominin brain evolution. *Proceedings of the National Academy of Sciences, 111,* 13022–13027

Holloway, R. L., Hurst, S. D., Broadfield, D. C., & Schoenemann, P. T. (2015). The new and old in hominid brain evolution: Why paleoneurology needs the lunate sulcus. *American Journal of Physical Anthropology, 156* (Suppl 60), 167–168.

Hurst, S. D., Holloway, R. L., Schoenemann, T., Broadfield, D. C., & Hunt, K. D. (2015). The new and old in hominin brain evolution, part II: Why paleoneurology needs a chimpanzee brain atlas. *American Journal of Physical Anthropology, 156* (Suppl 60), 173–174.

Johanson, D., & Blake, E. (2006). *From Lucy to language.* New York: Simon and Schuster.

Kimbel, W. H., Johanson, D. C., & Coppens, Y. (1982). Pleistocene hominid cranial remains from the Hadar Formation, Ethiopia. *American Journal of Physical Anthropology, 57,* 453–499.

Kimbel, W. H., Johanson, D. C., & Rak, Y. (2004). *The skull of Australopithecus afarensis.* Oxford: Oxford University Press.

Lingnau, A., & Downing, P. E. (2015). The lateral occipitotemporal cortex in action. *Trends in Cognitive Sciences, 19,* 268–277.

Neubauer, S., Gunz, P., Weber, G. W., & Hublin, J.-J. (2012). Endocranial volume of *Australopithecus africanus*: New CT-based estimates and the effects of missing data and small sample size. *Journal of Human Evolution, 62,* 498–510.

Rak, Y. (1983). *The Australopithecine face.* New York: Academic Press.

Spoor, F., Gunz, P., Neubauer, S., Stelzer, S., Scott, N., Kwekason, A., et al. (2015). Reconstructed *Homo habilis* type OH7 suggests deep-rooted species diversity in early *Homo. Nature, 519,* 83–86.

Weber, G. W., Gunz, P., Neubauer, S., Mitteroecker, P., & Bookstein, F. L. (2012). Digital South African fossils: Morphological studies using reference-based reconstruction and electronic preparation. In S. C. Reynolds & A. Gallagher (Eds.), *African genesis: Perspectives on Hominin evolution* (pp. 298–316). New York: Cambridge University Press.

Zollikofer, C. P. E., & Ponce de León, M. (2013). Pandora's growing box: Inferring the evolution and development of hominin brains from endocasts. *Evolutionary Anthropology, 22,* 20–33.

Chapter 6
Posture, Locomotion and Bipedality: The Case of the Gerenuk (*Litocranius walleri*)

Matt Cartmill and Kaye Brown

Abstract Most explanations for the origin of hominin bipedality cannot be comparatively tested, because there are no other striding bipeds among mammals. However, there are other mammals that stand bipedally for long periods of time. One such is the gerenuk (*Litocranius walleri*), an African gazelle that browses while standing bipedally, with extended hips and knees and a marked lumbar lordosis. Despite these behavioral resemblances to humans, Richter's (1970) extensive comparative study of gerenuk anatomy found only one skeletal apomorphy specifically related to bipedality – namely, a reduction in the lumbar spinous processes, which permits that lumbar lordosis. Our data show that gerenuks lack two other features – an expanded cranial sector of the acetabular semilunar surface, and "wedging" of the lumbar vertebral bodies – that we had expected from their bipedal positional behavior. We infer that even prolonged and extensive postural bipedality results in little or no postcranial remodeling, unless selection favoring the maintenance of efficient quadrupedal locomotion is relaxed. This conclusion undercuts theories, such as Hunt's (1994) "postural feeding hypothesis," that portray early hominin postcranial apomorphies as having originated as adaptations to bipedal feeding postures rather than to bipedal locomotion.

Keywords Bipedality • Locomotion • Posture • *Litocranius* • Bovidae

M. Cartmill (✉) · K. Brown
Department of Anthropology, Boston University,
232 Bay State Rd, Boston, MA 02215, USA
e-mail: cartmill@bu.edu

K. Brown
e-mail: kaybrown@bu.edu

Introduction

In their terrestrial locomotion, hominins and birds are unique among living animals in being obligate bipeds. However, many other terrestrial animals occasionally stand up or move around on their hind legs alone; and some form of this so-called facultative bipedality must have been an initial stage in the evolution of obligate bipedality.

In extant animals, facultative bipedality serves diverse functions. For some mammals (e.g., bears and gorillas), standing up on their hind legs serves to make them look larger and more dangerous as part of a threat display. Many mammals (e.g., ground squirrels, meerkats) have bipedal vigilance postures, which allow them to see over nearby obstacles and detect possible dangers at longer distances. Bipedality is used in high-speed locomotion in various animals – including some insects, lizards, kangaroos, and rodents – that usually move more slowly on all fours or all sixes (Full and Tu 1991; Alexander 2004). Gibbons, whose long, limber arms are not well adapted to bearing loads under compression, often stand on their hindlimbs alone when running on top of branches or on the ground. However, when moving more slowly on the ground, even gibbons frequently drop to all fours (Vereecke et al. 2006).

For many other mammals, facultative bipedality is a subsistence strategy. Goats and many other browsing artiodactyls often stand on their hind legs while feeding, especially when they are on the ground reaching up for food items in low branches of trees. Chimpanzees and other primates also sometimes feed while standing bipedally, in the trees as well as on the ground (Stanford 2002). Hunt (1994, 1996) found that among Mahale Forest chimpanzees, over 80% of all bipedalism occurred during feeding.

From this datum, Hunt (1994) went on to suggest that bipedal feeding postures had played a crucial role in human evolution; that "bipedalism evolved more as a terrestrial feeding posture than as a walking adaptation" in early hominins; and that "... a bipedal postural feeding adaptation

Assaf Marom and Erella Hovers (eds.), *Human Paleontology and Prehistory*,
Vertebrate Paleobiology and Paleoanthropology, DOI: 10.1007/978-3-319-46646-0_6

may have been a preadaptation for the fully realized loco-motor bipedalism apparent in *Homo erectus*." Variations of this idea, that bipedal postural feeding was a precursor to bipedal locomotion, have been proposed by others (Tuttle 1975; Köhler and Moyà-Solà 1997; Thorpe et al. 2007).

Is there in principle any way of testing the notion that habitual bipedal feeding preceded habitual bipedal locomotion in human evolution? We can test this idea against the fossil record only if we can find some way of distinguishing positional or postural adaptations to bipedality from adaptations to bipedal locomotion – say, in such early hominins as *Sahelanthropus* or *Ardipithecus*. But to do this, we need first to ask, "Are there are in fact any morphological adaptations correlated with habitual bipedal feeding postures *per se*?" To answer that question, we would like to compare species of mammals that feed bipedally, but do not walk bipedally, with closely related species that do neither. Unfortunately, no primates fill the bill, because (as far as we know) all primates are facultative postural bipeds and often stand on their hind legs while feeding.

One group that provides the necessary comparisons is the family Bovidae. All bovids habitually walk and run on all fours; but goats (Goetsch et al. 2010) and other browsing bovids often assume bipedal postures in feeding on overhead branches, whereas sheep, cows, and many other grazing bovids rarely rise to their hind legs. Bipedal browsing appears to be generally more common in smaller bovids, where it is facilitated by their smaller size and by the relatively greater concentration of muscle mass in the hindquarters (Grand 1997). However, perhaps the most frequently bipedal bovid is a medium-sized (30–45 kg) gazelle: the gerenuk, *Litocranius walleri* (Fig. 6.1a), which inhabits arid areas in and around the Horn of Africa. An exclusive browser, the gerenuk feeds mainly on the leaves of trees exceeding 1 m in height (Elliot 1897; Lydekker 1908, pp. 273–278; Leuthold 1978), and "habitually rises on its hindlegs to reach a zone over 2 m high" (Kingdon 2004). In captivity, baby gerenuks begin trying to stand on their hind legs two weeks after birth, and start feeding in this position some two weeks later (Leuthold and Leuthold 1973).

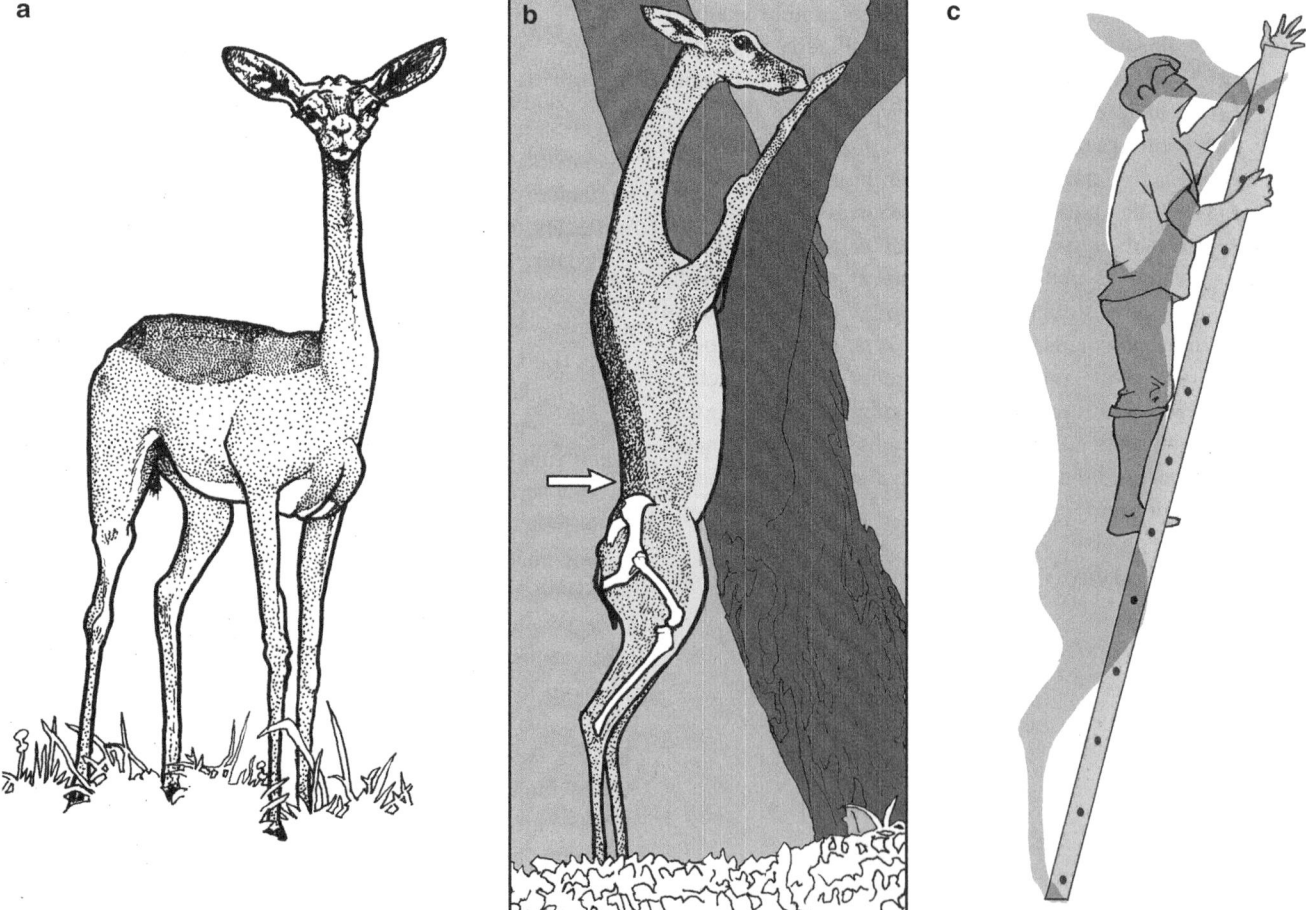

Fig. 6.1 The gerenuk (*Litocranius walleri*). **a** standing quadrupedally. **b** feeding on overhead leaves in a characteristic bipedal posture, showing the lumbar lordosis (white arrow). Diagrammatic outlines of the pelvis, femur, and tibia are superimposed to show the joint angulations. **c** The statics of a feeding gerenuk resemble those of a human standing on a ladder. Drawings of the animals are traced from photographs taken in the Miami (Florida) Metrozoo

Gerenuks habitually forage in an upright posture, standing on their elongated hind legs with knees and hips extended (Fig. 6.1b), deploying their elongated forelimbs in the branches to steady themselves and reaching up with their strikingly elongated necks to browse in the branches overhead. The resulting configuration is structurally and functionally similar to that of a man standing on a ladder (Fig. 6.1c).

Gerenuks display anatomical as well as behavioral specializations for bipedal browsing, the most obvious being the striking elongation of the neck. When standing bipedally, the gerenuk exhibits a visible lumbar lordosis (Fig. 6.1b), which helps to balance the center of mass over the hind feet in bipedal postures. Richter (1970: 460) concluded that when the hindlimb is maximally extended in a bipedal gerenuk, the body is raised into a more vertical position than in other ungulates. Noting that this might be regarded as an evolutionary convergence with humans, he accordingly undertook a comparison of the musculo-skeletal anatomy of *Litocranius* with that of other antelopes to determine whether they exhibited morphological adaptations for bipedal standing. Richter found two differences that he thought could be interpreted in those terms:

1. *Gaps between lumbar spinous processes.* The tips of the spinous processes of a gerenuk's lumbar vertebrae are less flared, and therefore more widely spaced, than those of related gazelles (Fig. 6.2). Richter interpreted this difference as an adaptation to permit lumbar lordosis in the gerenuk during bipedal standing.

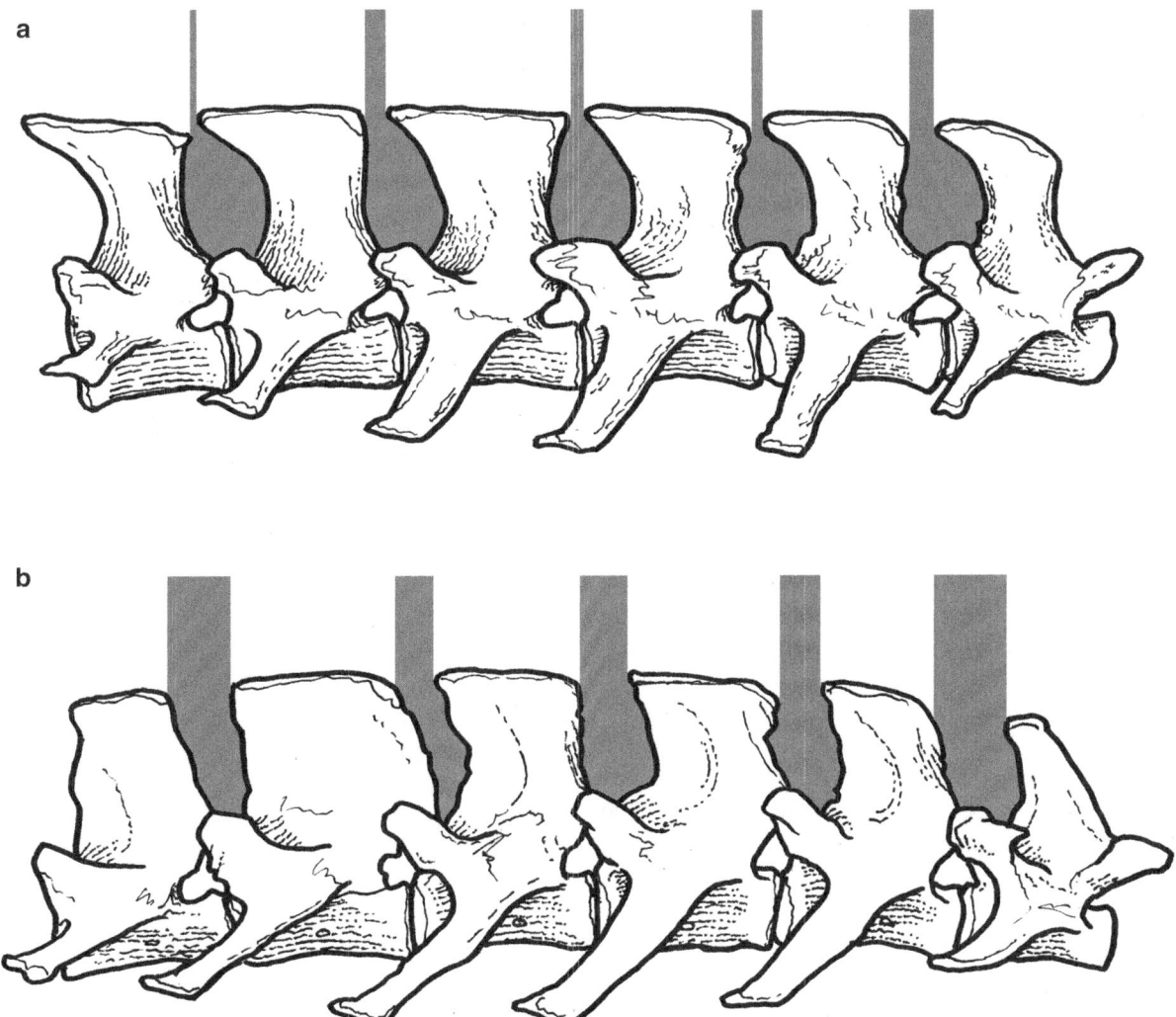

Fig. 6.2 Right lateral views of the lumbar vertebrae of (**a**) a Dorcas gazelle, *Gazella dorcas,* and (**b**) a gerenuk, *Litocranius walleri*, showing the enhanced spacing of the lumbar spinous processes (gray tone) that allows for lumbar lordosis in in the gerenuk. (After Richter 1970)

2. *Enlarged m. quadriceps femoris.* Horses can sleep standing up, in part because they have a "passive stay-apparatus" that can lock the partially extended knee joint in place, without any muscular exertion, to prevent it from flexing further under load. Similar morphology is found in the gerenuk and some other bovids. However, this mechanism probably does not actually serve a similar function in those animals (which sleep lying down), and its functional significance is unclear (Hermanson and MacFadden 1996). It cannot in any case be engaged in the fully extended knee (Richter 1970: 516), and thus cannot furnish the gerenuk with passive support for the extended knee in bipedal standing, as the collateral ligaments of the human knee do. The knee in a bipedally-standing gerenuk must therefore be held in extension by continual contraction of the knee extensor, the m. quadriceps femoris. Richter (1970: 534–535) found that this muscle is relatively larger in gerenuks than in the other gazelles he looked at. He interpreted this as a second adaptation to bipedality.

Apart from these two features (and the general elongation of the limbs and neck), Richter was unable to identify any peculiarity of the trunk and limbs in the gerenuk that could be functionally linked to its bipedal feeding behavior. "These investigations," he concluded, "… show clearly that facultative bipedality in quadrupeds need not be connected to any important morphological transformations in the locomotor apparatus" (*Die Untersuchungen … machen wahrscheinlich, daß fakultative Bipedie bei Quadrupeden nicht an wesentliche morphologische Umgestaltungen des Bewegungsapparates gebunden sein muß:* Richter 1970: 536).

This conclusion is unexpected. Because prolonged bipedal standing in a vertical posture imposes a 90° shift in the pull of gravity on the body, we might expect to find multiple differences in weight-bearing parts of the skeleton between gerenuks and their less habitually bipedal relatives, paralleling the well-known differences between humans and apes. Richter's negative findings therefore deserve reexamination. We undertook to test his conclusions further by looking at some features of the gerenuk hindlimb and vertebral column that he did not examine, and that might be expected to exhibit additional signs of adaptation to facultative bipedality.

The human hind limb differs from those of other hominoids in having been reshaped into a relatively inflexible, non-prehensile propulsive strut. The human tarsal region is elongated and rigid, forming a distinctive longitudinal arch. Fixed in an adducted position, the human hallux has lost its power to grasp. The mobility of many of the hindlimb joints is reduced in hominins (Tardieu 1979; Aiello and Dean 1990; DeSilva and Lovejoy 2009; DeSilva 2010). Hip abduction is especially restricted (MacLatchy and Bossert 1996; MacLatchy 1998), and the excursions of the limb in normal locomotion are largely restricted to swinging back and forth in a parasagittal plane. In these respects, humans can be thought of as converging to some extent with terrestrial quadrupeds; and all bovids have gone considerably further than humans have in these directions. In a cursorial unguligrade quadruped like the gerenuk, it would be fruitless to seek signals of bipedality in such features as tarsal rigidity or restriction of the locomotor excursions of the hind limb. We looked instead for differences between gerenuks and less bipedal bovids in a few features of the hip joint and vertebral column that might reflect the changes in the magnitude and direction of gravitational stress that come with standing upright on the hind legs.

Anatomy of the Bovid Hip

The pelvis of bovids (Fig. 6.3) looks uncomfortably spiky to an eye accustomed to looking at primate pelves. The ilia are long, pointed, and laterally flaring. As in most mammals, the iliac flare is more pronounced in larger species. This may reflect positive allometry of the limb muscles arising from the iliac blade (Elftman 1929) and/or allometric changes in the relative size of the gut. The pubic symphysis is long, immobile, and usually fused in adults. The ischium is long (over 70% as long as the ilium), affording an increased moment arm for the hamstrings in extending the protracted femur. Because bovids seldom sit on their haunches, the bovid ischial tuberosity is not a rounded, weight-bearing prominence like those of primates and many other mammals, but a laterally-projecting, generally pointed and triangular process, from which the hamstring muscles arise (Richter 1970).

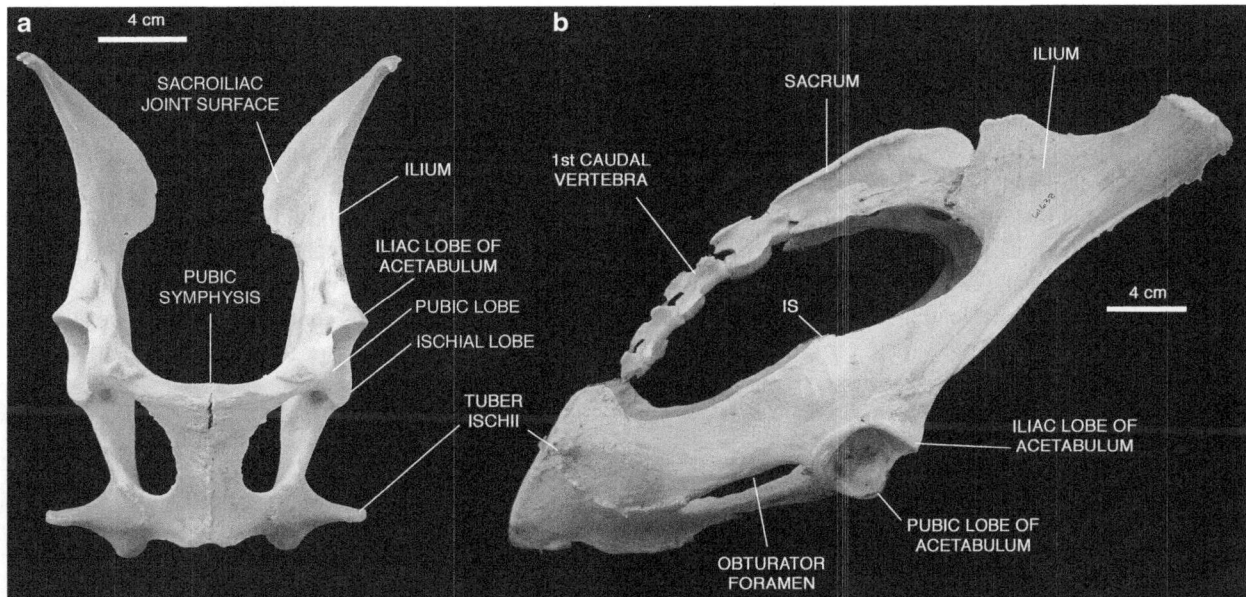

Fig. 6.3 a Ventral view of the hip bones of a gerenuk, *Litocranius walleri* (Harvard Mus. Comp. Zool. 13231) and **b** right lateral view of the pelvis of a yellow-backed duiker, *Cephalophus silvicultor* (MCZ 61638), showing the configuration and major landmarks of the bovid pelvis

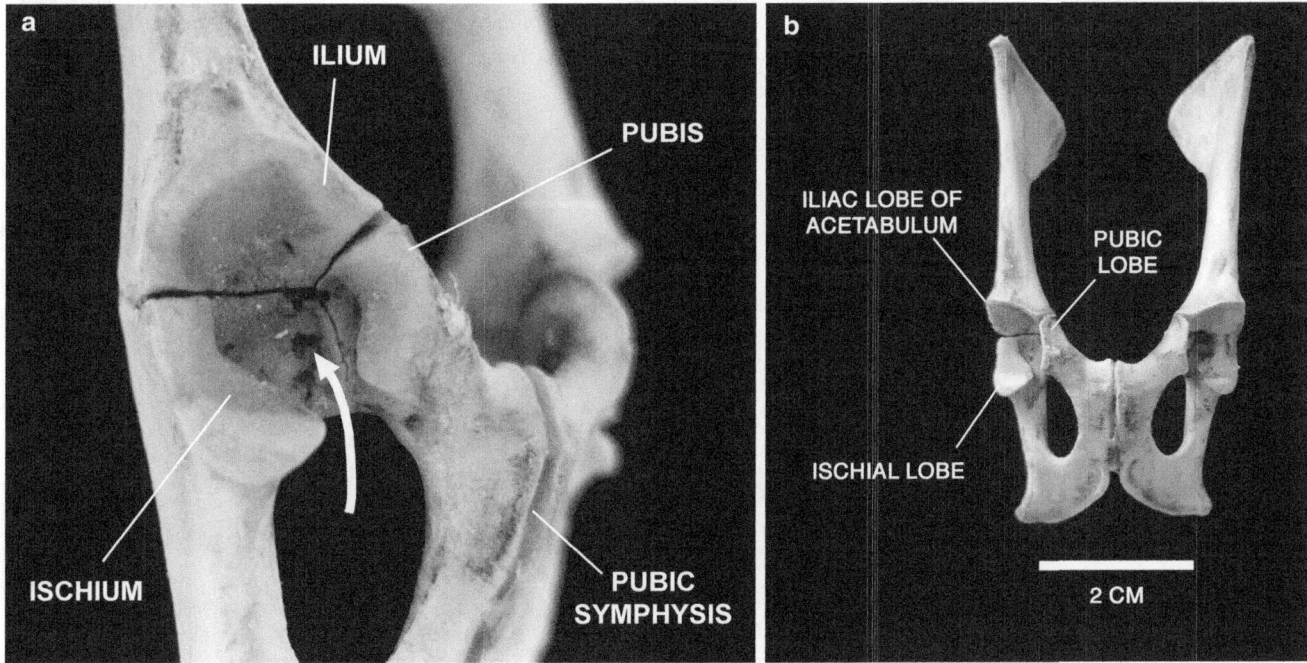

Fig. 6.4 Right ventrolateral (**a**) and ventral (**b**) views of the hip bones of a juvenile bovid (royal antelope, *Neotragus pygmaeus,* Harvard Mus. Comp. Zool. 38067), showing the unfused pubic symphysis and triradiate synchondroses. The gap between the ischial and pubic lobes of the acetabulum persists into adult life as an acetabular notch, equivalent to that in humans. Branches of the femoral vessels (usually supplemented by obturator branches in *Homo*) pass through this notch (white arrow in a) to supply the tissues of the hip joint and the proximal end of the femur (Chaveau and Arloing 1890; de Waal Malefijt et al. 1988)

Like that of other mammals, the bovid hip socket encompasses the intersection (the triradiate synchondrosis) of the three bones – ilium, ischium, and pubis – that fuse in adults to form the hip bone (Fig. 6.4). Each of the socket's three components bulges outward beyond the ends of its synchondroses with its two neighbors, so that the acetabular rim is not circular, but lobulated or trefoil-shaped. We can assume that the three lobes are loaded differently in different postural behaviors. In a standing human, most of the weight that is borne by the hindlimb is carried on the upper (dorsal) part of the socket (Hodge et al. 1986). We assume that this is also true in a bovid standing on all fours, and that this explains why the dorsal (iliac and ischial) lobes of the hip socket are larger and more protuberant than the ventrally situated pubic lobe (Figs. 6.3 and 6.4), reflecting the greater loads that they bear in quadrupedal postures.

In humans (Hodge et al. 1986), the loads borne by the different parts of the acetabulum fluctuate during locomotion. We assume that these findings apply to the bovid hip as well. Acetabular load is presumably least during swing phase. The posterodorsal (ischial) lobe of the hip socket probably receives its greatest load at the end of swing phase,

when the protracted hind limb touches down and thrusts upward and backward to check the descent of the pelvis. Conversely, the *antero*-dorsal (iliac) lobe probably bears the most load near the end of stance phase, when the retracted limb pushes off and thrusts upward and forward against the acetabulum.

In a gerenuk or other bovid standing upright on its hind legs only, the pelvis is rotated into a vertical position, and most or all of the weight of the entire upper part of the body – head, neck, forelimbs, and trunk – will presumably be

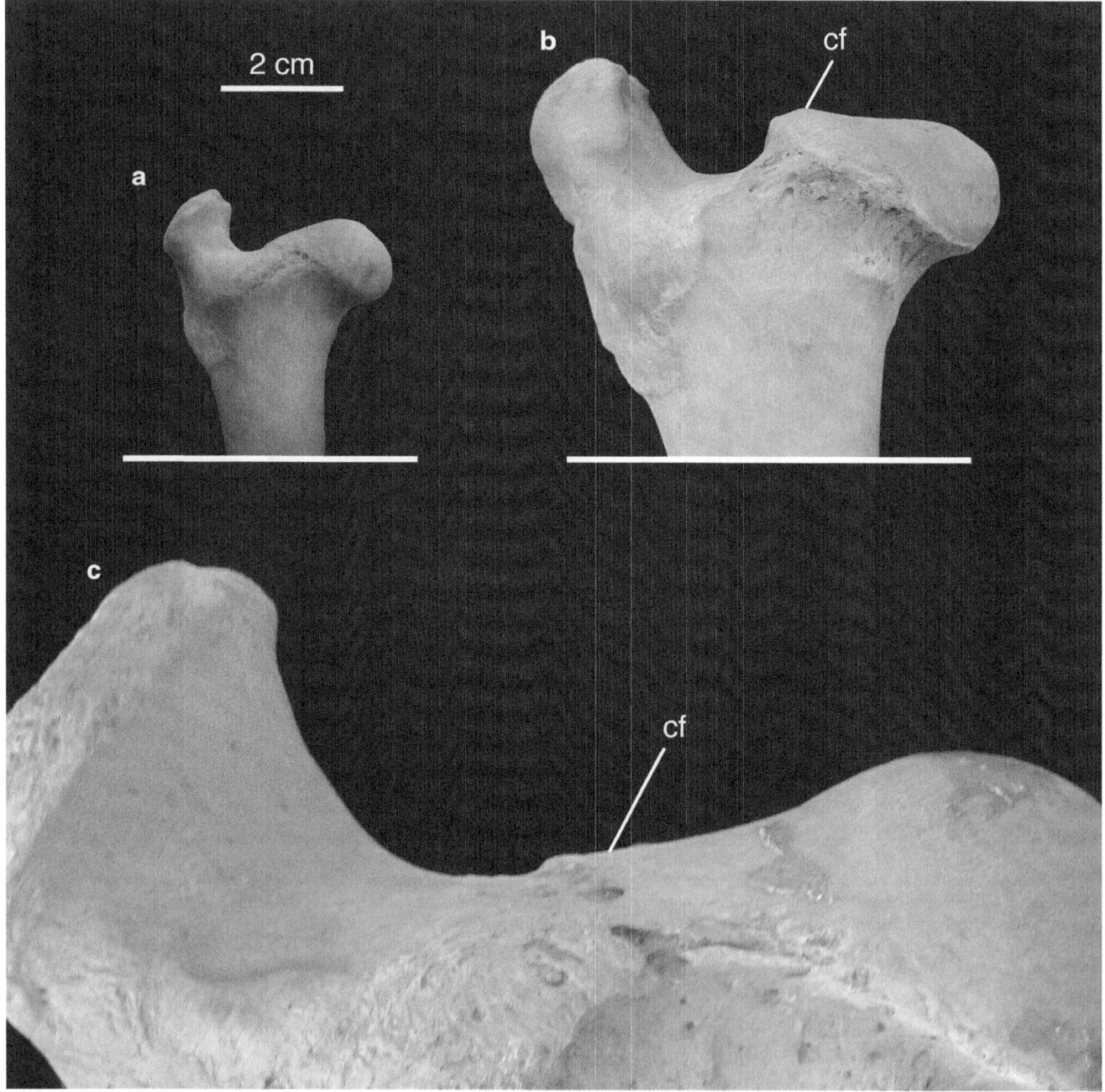

Fig. 6.5 Allometry of femoral head shape in bovids. Right femora, anterior views. **a** Klipspringer, *Oreotragus oreotragus*, 9–16 kg (Harvard Mus. Comp. Zool. 14555); **b** Grant's gazelle, *Gazella granti*, 45–65 kg (MCZ 13236); **c** American bison, *Bison bison*, 1100–2000 kg (MCZ 10). Abbreviation: cf, coronal flare. All to same scale

borne by the iliac lobes, at the cranial edges of the hip sockets. We might accordingly expect the iliac lobes to be relatively larger and more protuberant in gerenuks than in bovids that seldom or never stand on their hind legs, for the same sorts of reasons that the anterior horn of the semilunar surface of the acetabulum is relatively larger in humans than in African apes (Stern and Susman 1983).

In testing this expectation, allometry must be taken into account. The femoral head of bovids is not a sphere, but a transversely elongated ellipsoid (Fig. 6.5). Its non-spherical shape restricts movements other than flexion and extension, reflecting the hindlimb's mainly parasagittal excursions in walking and running. The head's articular surface extends onto the proximal end of the upper (dorsal) surface of the femoral neck. This dorsal expansion extends widely outside the acetabular socket when the femur is retracted (hip extension), and rotates fully into articulation (posteriorly, on the ischial lobe of the acetabulum) only when the femur is protracted – i.e., at the point of touchdown in the cycle of limb movements (hip flexion). The dorsal expansion appears to exhibit positive allometry. We have not attempted to measure this; but in larger bovids, it extends further onto the femoral neck, where its surface area is further increased by an outward flaring of its margins (coronal flare, "cf" in

Fig. 6.5), so that the upper edge of the articular surface is distinctly concave. The apparent positive allometry of the femoral head's articular surface is what we might expect on the basis of square-cube relationships, since the forces borne by joint areas are a function of the body's mass (which is roughly proportional to volume). The differential distribution of the articulation in different limb positions suggests that the hip joint sustains greater reaction forces at touchdown (ischial lobe) than at toe-off (iliac lobe), at least in large bovids. This implies that we might further expect the ischial lobe to be larger and more protuberant in larger bovids (the square-cube law again). This possibility needs to be taken into account in assessing the relative size of the ischial and iliac lobes of the acetabulum as a potential reflection of differences in feeding posture.

Materials and Methods

We measured the pelves (conjoined left and right ossa coxae) of 81 bovids, comprising 27 species belonging to 15 genera and 7 subfamilies of Bovidae (Table 6.1), including six gerenuks. The taxa sampled are distributed across a wide

Table 6.1 Sampled bovid taxa and numbers of specimens (N). N(ac) = number of specimens for which acetabular diameter was measured. Characterizations of species as browsers (b), grazers (g), or browser-grazers (gb) are based chiefly on Bodmer (1990) and Cerling et al. (2003)

Subfamily	Species	Common name	Diet	N	N(ac)
Cephalophinae:	Cephalophus maxwelli	Maxwell's duiker	b	1	
	Sylvicapra grimmia	Common duiker	b	7	5
Hippotraginae:	Oryx dammah	Scimitar oryx	g	2	2
	Oryx gazella	Gemsbok	g	5	5
Antilopinae:	Gazella dorcas	Dorcas gazelle	b	2	
	Gazella granti	Grant's gazelle	gb	8	5
	Gazella rufifrons	Red-fronted gazelle	gb	2	
	Gazella thomsoni	Thomson's gazelle	gb	5	4
	Litocranius walleri	Gerenuk	b	6	2
	Madoqua guentheri	Günther's dik-dik	b	2	2
	Madoqua kirki	Kirk's dik-dik	b	3	3
	Madoqua phillipsi	Phillips's dik-dik	b	2	
	Madoqua saltiana	Salt's dik-dik	b	2	
	Neotragus batesi	Bates's pygmy antelope	b	5	5
	Neotragus moschatus	Suni	b	4	4
	Oreotragus oreotragus	Klipspringer	b	5	4
	Ourebia montana	Mountain oribi	g	2	
	Ourebia ourebia	Oribi	g	7	7
Caprinae:	Capra hircus	Goat (domestic)	b	7	6
	Ovis aries	Sheep (domestic)	g	2	1
	Ovis canadensis	Bighorn sheep	gb	3	2
Reduncinae:	Redunca aridinium	Southern reedbuck	g	1	1
	Redunca fulvorufula	Mountain reedbuck	g	2	2
	Redunca redunca	Bohor reedbuck	g	3	1
Aepycerotinae:	Aepyceros melampus	Impala	gb	2	
Acelaphinae:	Acelaphus busephalus	Hartebeest	g	1	1
	Connochaetes taurinus	Blue wildebeest	g	1	1

range of body sizes, from the tiny *Neotragus batesi* (2–3 kg) up to the blue wildebeest (*Connochaetes taurinus*: 250–290 kg) and gemsbok (*Oryx gazella*: 100–300 kg). Using dial and spreading calipers, we took four measurements on the caudal half of each pelvis: the maximal transverse distances between the tips of each of the three pairs of contralateral acetabular lobes (interpubic breadth, interischial breadth, and interiliac breadth), and the maximum midsagittal length of the pubic symphysis (Fig. 6.6). The geometric mean of these four measurements was employed as a proxy for body size. To test the validity of this proxy, acetabular diameter was measured at the ischiopubic notch in the last 63 specimens that we examined, and then least-squares-regressed against the geometric mean of the four pelvic measurements. Acetabular diameter is highly correlated with body weight in quadrupedal primates, and is sometimes employed as a proxy for it (Steudel 1981; Jungers 1990). This diameter was very highly correlated ($r = 0.98$) with the geometric mean of the pelvic measurements in our bovid sample. We conclude that the combined pelvic measurements afford a sufficiently accurate indicator of body size for our purposes.

When plotted against the geometric mean of the pelvic measurements, interischial breadth (Fig. 6.6a-3) exhibits positive allometry over our entire bovid sample (Fig. 6.7a). This finding is compatible with our prediction that this lobe would be relatively larger and more protuberant in larger bovids. However, our hypothesis that the protrusion of the ischial lobe would increase more rapidly with body size than that of the iliac lobe is not borne out. Interiliac breadth (Fig. 6.6a-1) shows essentially the same positive allometry as interischial breadth (Fig. 6.7b). This means that the iliac lobe is also relatively larger and more protuberant in larger bovids. We interpret this finding to mean that these two lobes bear most of the body weight in standing and most of the locomotor forces in acceleration and deceleration (due to their dorsal position in the acetabulum). The linear dimensions of these lobes are therefore relatively larger in larger species. And because the interischial and interiliac breadths have similar allometries, their ratio does not covary with body size, as we had expected (Fig. 6.8).

The data plotted in Fig. 6.8 also refute our prediction that browsing bovids in general, and gerenuks in particular, would have relatively more protuberant iliac lobes (higher

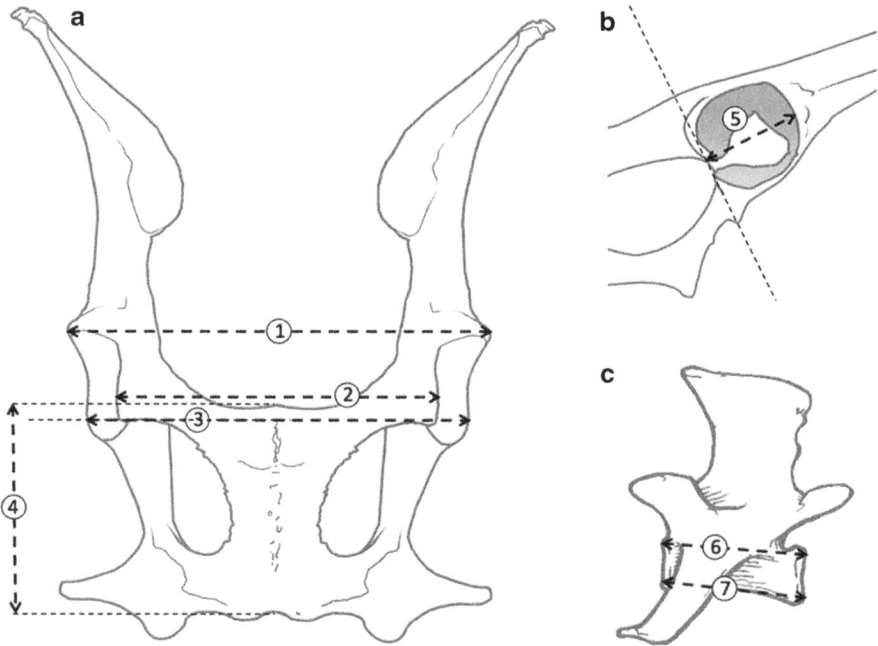

Fig. 6.6 Measurements employed in this study. **a** Ventral view of a bovid pelvis; **b** lateral view of acetabulum; **c** lateral view of a lumbar vertebra. *1* Maximum interacetabular breadth on ilium (interiliac breadth); *2* maximum interacetabular breadth on pubis (interpubic breadth); *3* maximum interacetabular breadth on ischium (interischial breadth); *4* craniocaudal length of pubic symphysis; *5* (maximum) acetabular diameter at the ischiopubic notch; *6* maximum dorsal length of vertebral body; *7* maximum ventral length of vertebral body

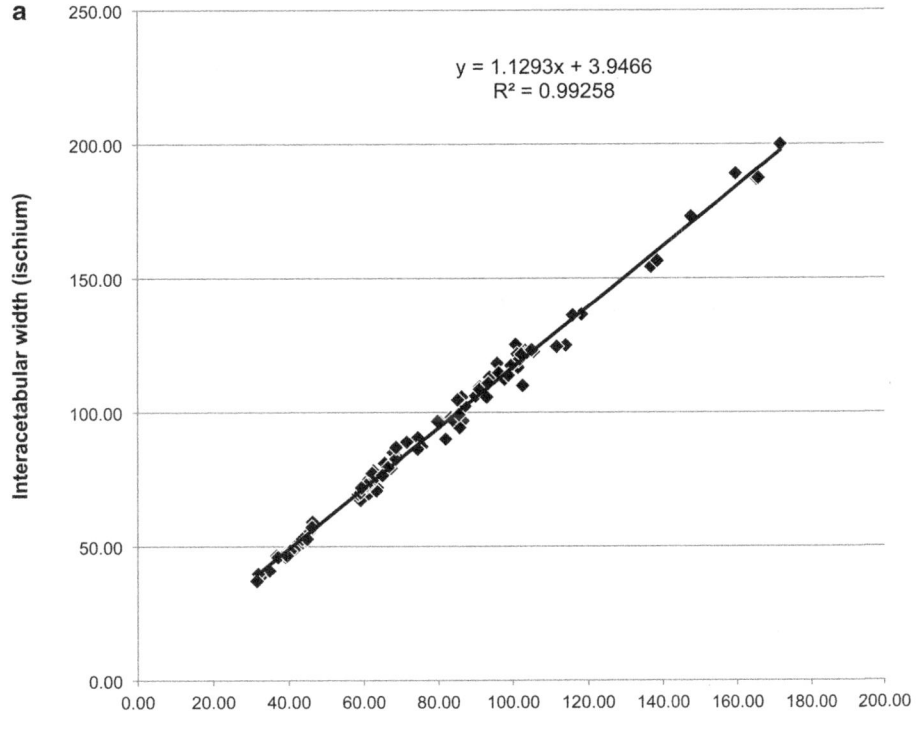

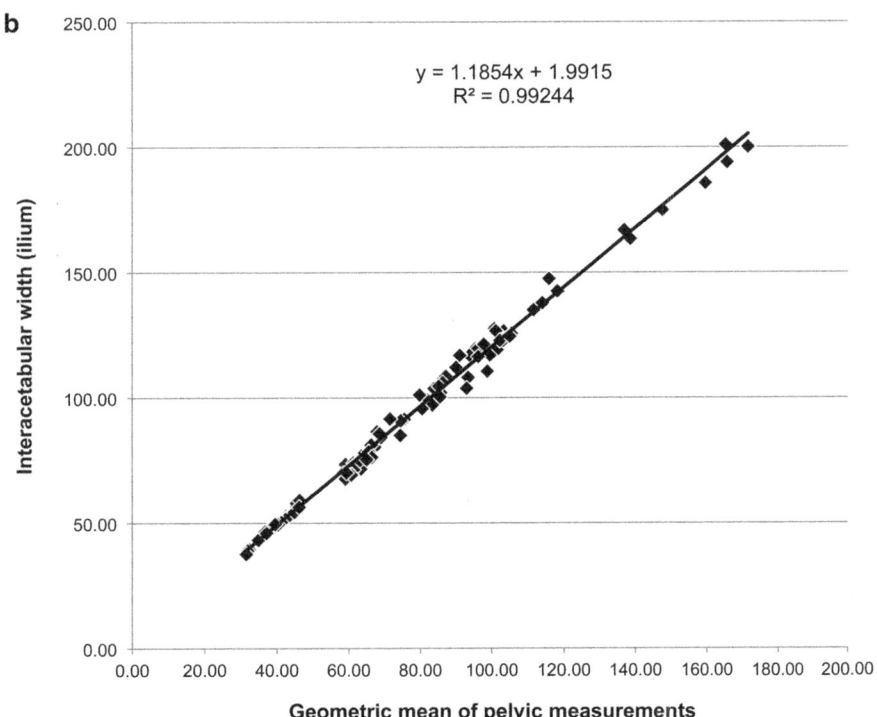

Fig. 6.7 Maximal interacetabular breadth of 81 bovids, measured at **a** the ischial lobes and **b** the iliac lobes of the acetabulum, regressed against a body-size proxy

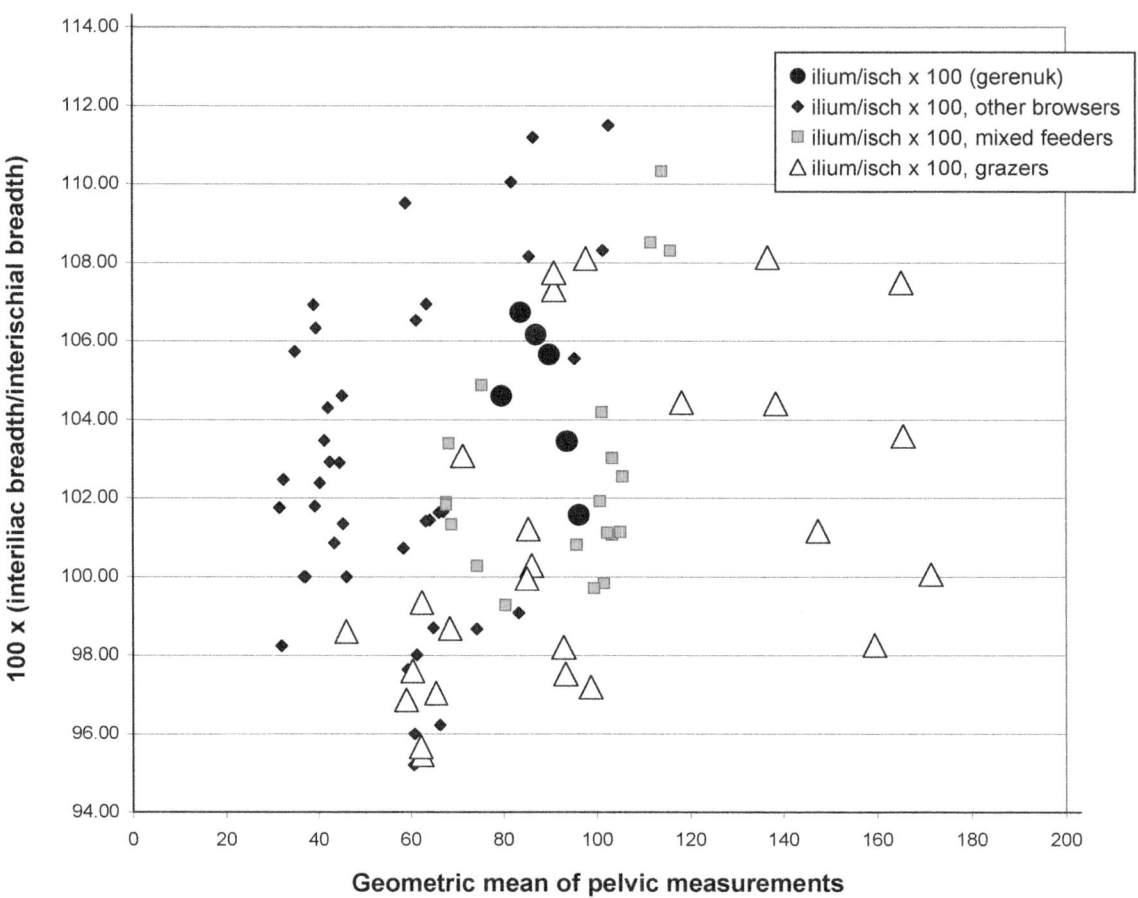

Fig. 6.8 Maximal iliac interacetabular breadth of 81 bovids, expressed as a percentage of maximal ischial interacetabular breadth (ilium/isch ×
100) and plotted against a body-size proxy. Higher index values (vertical axis) indicate relatively more protuberant iliac lobes of the acetabular
margin. White triangles, grazing species; gray squares, mixed feeders (grazer/browsers); black circles, gerenuk *(Litocranius walleri)*; black
diamonds, other browsing species

index values: vertical axis in Fig. 6.8) than grazers because
browsers more frequently stand on their hind legs. No
morphological signal of habitual postural bipedality, or of
browsing habits, is evident in these data. The gerenuks that
we measured resemble other bovids in their index values,
falling near the center of the bovid scatter.

In bovids as in humans and many other mammals, there is
a gap or notch in the acetabular rim at the site of the synchondrosis between the ischium and the ilium (Fig. 6.4).
Through this notch, blood vessels enter the hip socket to
supply its tissues and the head of the femur. In gerenuks,
uniquely among the bovids we have studied, this notch is
supplemented by a second notch, bridged over to form a

foramen, between the iliac and pubic lobes (Fig. 6.9). If this
notch also transmits blood vessels, it may betoken a different
pattern of blood supply to the hip socket and femoral head in
gerenuks. It might further be conjectured that this represents
some sort of vascular adaptation to prolonged bipedal
standing. However, we found no other signs of such adaptation in the morphology of the gerenuk acetabulum. The
increased loads borne by the human hip joint relative to
those of other apes are reflected in the relatively larger size
of the human femoral head and acetabulum (Jungers 1988);
but our measurements of acetabular diameter (Table 6.1)
revealed no differences in hip-socket size between gerenuks
and other bovids (Fig. 6.10).

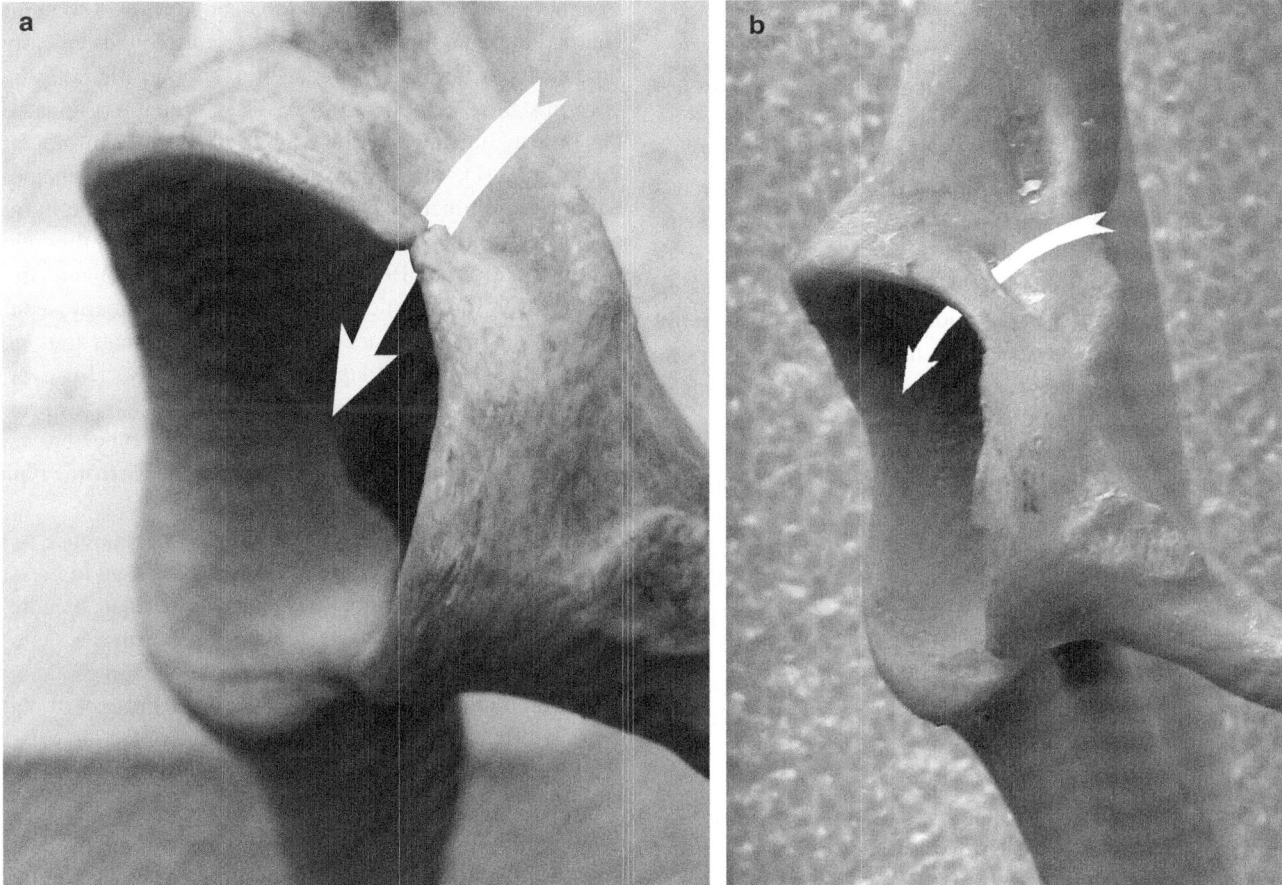

Fig. 6.9 Ventral (**a**: Harvard Mus. Comp. Zool. 12321) and ventromedial (**b**: MCZ 8734) views of the right acetabula of two gerenuks, showing the anterior acetabular foramen (arrows) in the bridged-over notch between the pubic and iliac lobes

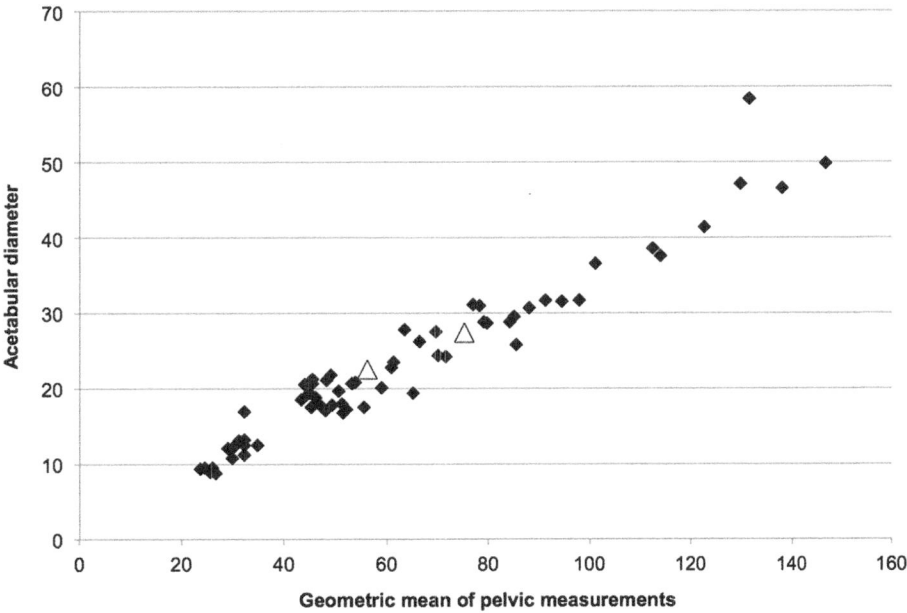

Fig. 6.10 Diameter of the acetabulum (maximal diameter measured at the ischiopubic notch) in the bovids sampled (Table 6.1), plotted against a body-size proxy. White triangles, gerenuks (*Litocranius*); black diamonds, other bovids. The acetabulum is not distinctively larger in the two gerenuks than in the others

Vertebral Morphology and Metrics

Like those of most quadrupedal mammals, the lumbar vertebrae of bovids are numerous (typically 6) and craniocaudally elongated. The lumbar vertebrae of humans and other hominoids are fewer (3–5), shorter, and broader. Humans differ from apes in having a characteristic lumbar lordosis. This is correlated with vertebral "wedging" – that is, with vertebral bodies and intervertebral disks that are craniocaudally shorter along the dorsal midline than along the ventral midline. It might be conjectured that the evident lordosis seen in a standing gerenuk would have similar correlates in the shape of the vertebral bodies.

Using sliding calipers, we measured the maximal mid-sagittal dorsal and ventral lengths of all six lumbar vertebral bodies of 12 bovid specimens belonging to 7 species: *Aepyceros melampus* (1)*, Gazella dorcas* (2)*, Gazella granti* (3)*, Gazella rufifrons* (1)*, Madoqua phillipsi* (1)*, Madoqua saltiana* (1), and *Litocranius walleri* (3). The ratio of dorsal to ventral length was calculated as a percentage for each vertebra. Values of this index below 100 reflect a lordotic excess of ventral length over dorsal length, as in humans.

All of the lumbar vertebrae measured exhibited index values exceeding 96.4, and the overwhelming majority (including all averages for each of the six lumbars across the gerenuk sample) slightly exceeded 100 (Table 6.2). Despite its habitual lumbar lordosis during bipedal feeding, *Litocranius* exhibits no lordotic wedging, either absolutely or by comparison with the other bovids measured, in the bodies of its lumbar vertebrae.

The endplates of human vertebral bodies increase steadily in size from the top of the neck down to the last lumbar vertebra. This size gradient is obviously adaptive for upright, bipedal posture and locomotion, in which each vertebra has

to carry more weight than the one above it. It is sometimes claimed that that this gradient is "a feature found typically in bipedal types because of the need to support the body in an upright stance" (Hooker 2007, p. 641), or that it originated in the human lineage as a bipedal adaptation (Stanford et al. 2009, p. 291). However, non-human primates, including baboons and other dedicated quadrupeds, resemble humans and differ from typical quadrupedal mammals in this regard (Latimer and Ward 1993; Shapiro 1993; Cartmill and Brown 2014). The monotonic head-to-tail size gradient seen in humans therefore appears to be an ancient primate trait, not a specifically hominin adaptation to bipedality. However, it may reflect an increased frequency of vertical postures in primates. If so, then gerenuks (and perhaps other browsing ungulates) might show similar differences from typical quadrupedal mammals.

We began by testing this hypothesis with reference to the lumbar vertebrae only. Sliding calipers were used to measure the maximal dorsoventral height and transverse breadth of the cranial endplate of each lumbar vertebra in the same 12 specimens used in measuring vertebral "wedging" (above). Vertebral body areas were estimated as the product of height times width and expressed as a percentage of the largest such product in each lumbar series. In all species (Fig. 6.11), vertebral body measurements exhibited a slight general caudad increase across the lumbar vertebral column, from around 80% at L.1 to 100% at L.5 or L.6. The gradient seen in gerenuks (*Litocranius*) varied slightly between specimens, but was no steeper than in the other bovids measured.

It might be objected that height times breadth is not a reliable estimator of actual vertebral body area. The endplates of bovid vertebrae deviate considerably in shape from a rectangle, and from one another as well (Fig. 6.12). In a study of the weight-bearing surfaces of vertebrae in humans and some other primates, Shapiro (1993) used the area of an

Table 6.2 Lumbar vertebral "wedging" index (midsagittal craniocaudal lengths of vertebral bodies: 100 × dorsal/ventral) values in African bovids. Index values below 100 reflect lordotic wedging; values over 100 indicate kyphotic wedging

	L.1	L.2	L.3	L.4	L.5	L.6
Litocranius mean =	**100.08**	**101.20**	**100.91**	**101.12**	**101.32**	**100.32**
range =	(96.5–103.0)	(100–103.1)	(100.1–101.9)	(100–103.1)	(100.6–102)	(98.2–102.1)
Gazella mean =	100.98	100.20	100.17	99.61	99.70	100.47
range =	(97.7–102.1)	(98.5–103.1)	(97.3–101.7)	(98–103.1)	(96.6–102.5)	(99.3–101.6)
Aepyceros (n = 1)	**101.14**	**102.87**	**104.81**	**104.88**	**104.21**	**100.05**
Madoqua mean =	**102.38**	**100.61**	**104.05**	**102.44**	**102.61**	**103.05**
range =	(101–103.7)	(98.3–102.9)	(101.6–106.5)	(101.5–103.4)	(101.8–103.4)	(101.4–104.7)

ellipse inscribed within the height-times-breadth rectangle as an estimate of endplate area. This probably affords a closer approximation to the actual value of the area. However, it adds nothing in comparing one vertebral area with another, since the ratio of any rectangle's area to that of the inscribed ellipse is a constant $4/\pi$.

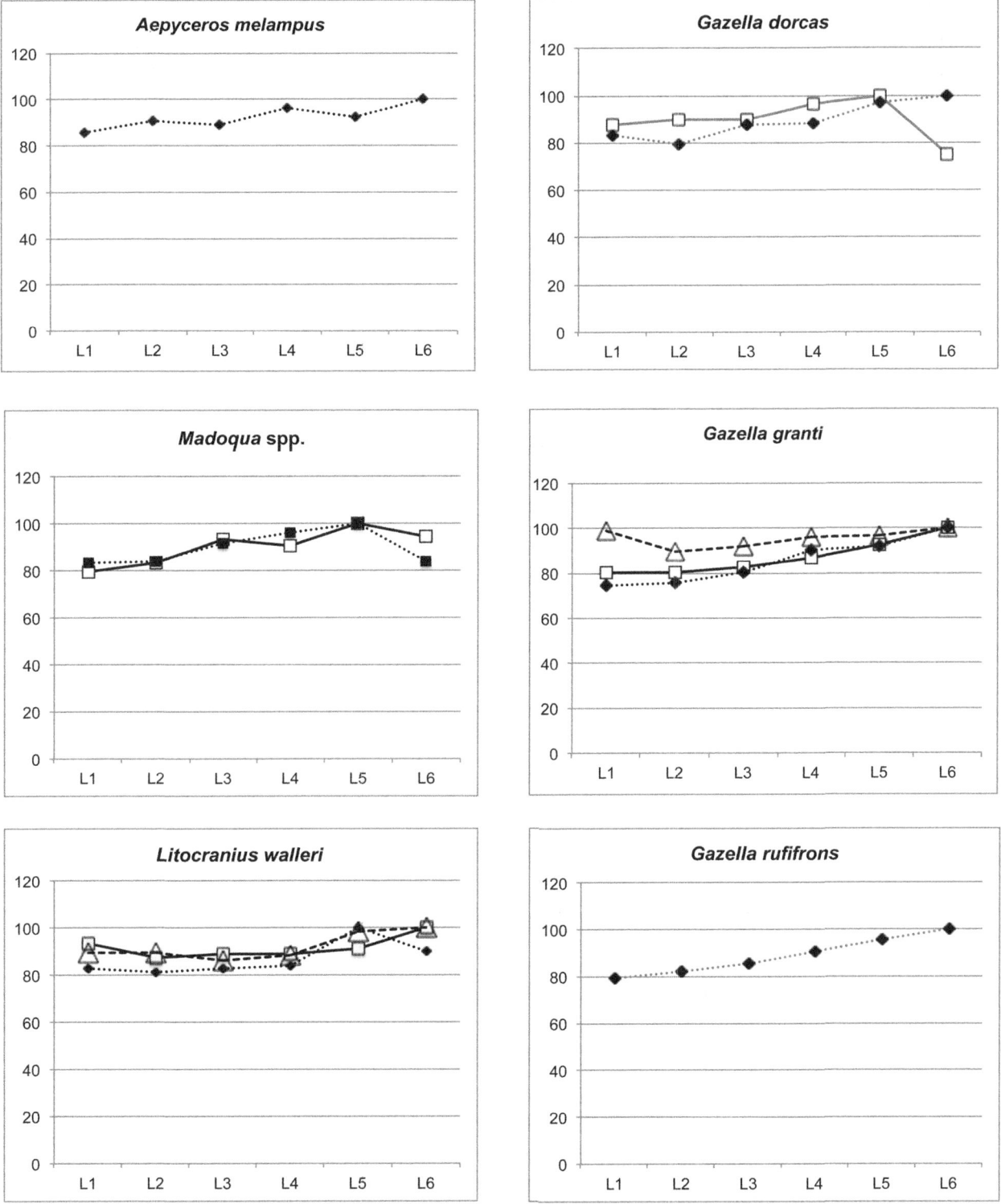

Fig. 6.11 Proxies of the areas of the cranial endplates of lumbar vertebrae (dorsoventral height times transverse breadth, expressed as a percentage of the value for the largest lumbar vertebra) in some African bovids. In all species, vertebral body measurements show a slight general caudad increase, from around 80% at the first lumbar vertebra (L.1) to 100% at L.5 or L.6. The gradient in gerenuks (*Litocranius*) is not steeper than in the other species

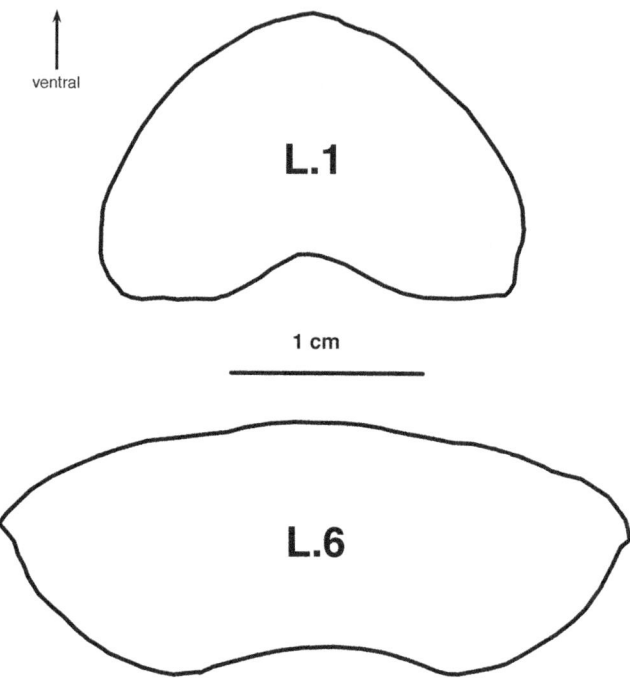

ventral

L.1

1 cm

L.6

Fig. 6.12 Outlines, traced from photographs, of the caudal endplates of the first and sixth lumbar vertebral bodies (L.1, L.6) in a gerenuk (Harvard Mus. Comp. Zool. 12321)

Because of the clinical importance of the lumbar spine, many investigators, going back at least to Davis (1961), have directly measured the weight-bearing surface areas of human lumbar vertebrae. Latimer and Ward (1993) measured the areas of the thoracic as well as the lumbar endplates in humans and other hominoids; and Pal and Routa (1986, 1987) measured areas of the caudal endplates of every alternate vertebra (C.2, C.4, etc.) from C.2 down to L.5. All these studies revealed the expected monotonic head-to-tail size increase in humans (Fig. 6.12). However, we have found no previous study in which the actual areas of vertebral-body endplates were measured for every vertebra in any mammal.

We are currently engaged in such a study of a wide sample of therian mammals (Cartmill and Brown 2014). For each specimen in this study, the caudal endplates of every vertebra from C.3 down to the last lumbar, and of the cranial endplate of S.1, are photographed together with a scale, sighting along a line tangent to the ventral midline of the vertebral foramen. The area of the endplate is then measured photogrammetrically using NIH ImageJ software. Although our data are only preliminary, they reveal informative comparisons. All the primates that we have examined, including the most quadrupedal (Fig. 6.14f), exhibit a humanlike profile, in which vertebral body area is least in the upper neck and increases monotonically in a head-to-tail

direction to peak in the lower lumbar section of the column (Fig. 6.13). A similar pattern is seen in macropodids. All the artiodactyls that we have measured so far (Fig. 6.14a–d) show an inverse pattern: vertebral body areas peak in the upper neck, fall to minimal values in mid-thorax, and rise again to a secondary peak in the lumbar region. Carnivorans (Fig. 6.14e) and *Thylacinus* show an intermediate pattern, with the primary peak in the lumbar region and a secondary peak in the neck. The gerenuk (Fig. 6.14a) does not deviate in a humanlike (i.e., primate-like) direction from the other artiodactyls thus far measured.

Conclusions

It remains possible that facultative bipedality has left its mark on other parts of the gerenuk body that have not yet been studied. Although we have not extended our investigation to the skull, gerenuks have a peculiarly elongated occiput (Elliot 1897; Gentry 1964), which may represent an adaptation to their head and neck postures or to their feeding behavior. We have not tried to evaluate or compare the weight-transmitting potential of the neural arches of gerenuk vertebrae, which may be important (especially in the lordotic lumbar region) when the trunk is held vertically (Sanders 1998). In a mammal that habitually holds its trunk in a vertical position, there are physiological reasons for expecting the cranial end of the thorax to be expanded (Chan 2014), and this may prove to be a feature of gerenuk anatomy. There are many questions left here for future research. But we can at any rate say that, apart from the modification of the lumbar spinous processes to permit lordosis – and the striking elongation of the limbs and neck – nothing so far established about the skeleton of gerenuks betrays their bipedal feeding habits. Although gerenuks spend a great deal of their feeding time standing on their hind legs in an upright posture, they do not show any distinctive enlargement of the hip socket as a whole or of the part of the acetabulum that presumably bears most of the weight in the upright position. The caudal parts of the gerenuk vertebral column are not distinctively enlarged for weight-bearing during bipedal postures, or "wedged" in connection with facultative lumbar lordosis. There is a very faint signal of bipedal behavior in gerenuk anatomy, but the hoped-for convergences with hominins are absent.

Two possible conclusions may be drawn from these facts. Either (A) bovid genes and anatomy are incapable of responding in a humanlike way to the selection pressures imposed by bipedal posture, or (B) those pressures are too slight to overcome stabilizing selection. We think that the gerenuk's lumbar lordosis argues against (A). More probably, strong selection favoring fast, efficient quadrupedal running – e.g., in

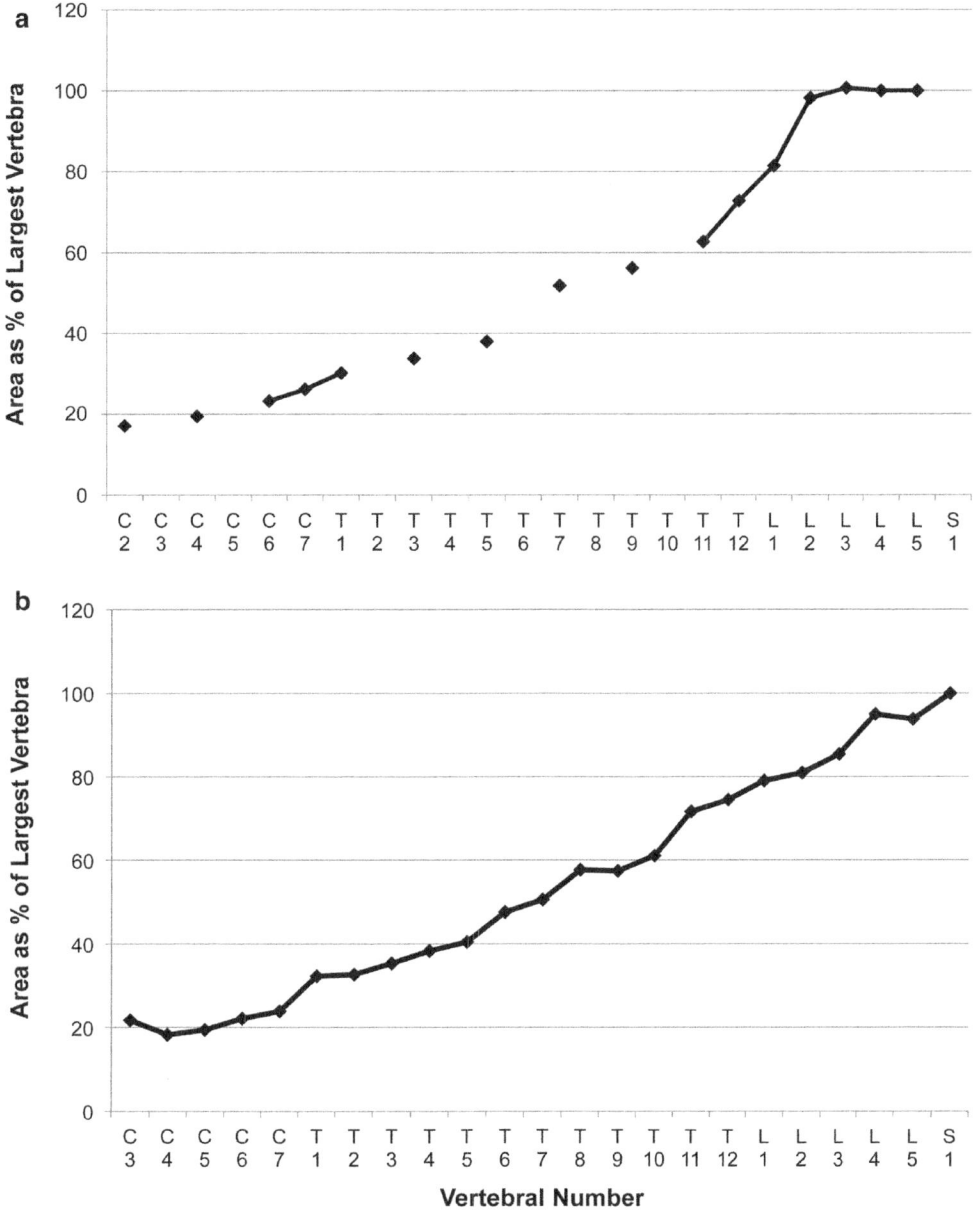

Fig. 6.13 Profiles of vertebral-body endplate areas for humans. **a** Averaged data for 44 male skeletons from Pal and Routa (1986, 1987). **b** Our data for one complete human vertebral column excluding the atlas and axis (BU 2, coll. Boston University Dept. of Anthropology). Endplate area was measured on the cranial surface of the first sacral vertebra and on the caudal surface of all other vertebrae. Gaps in the line connecting data points in (**a**) represent vertebrae not measured. C cervical; T, thoracic; L, lumbar; S, sacral (vertebrae)

escaping from predators – keeps the gerenuk's pelvic and lumbar morphologies from deviating much from those of other gazelles. We conclude that even in a habitual postural biped, marked anatomical adaptations to bipedality will not evolve until selection for effective quadrupedalism is relaxed. This conclusion reminds us that behavior is not so constrained by, or as predictable from, morphology as we might like.

The example of the gerenuk also suggests what sort of special morphology we should expect to see in an animal whose chief mode of feeding involves standing on the ground and reaching up – namely, adaptations for reaching

up, especially in the form of elongation of the limbs and/or the neck. But there is no sign that early hominids (*Ardipithecus, Australopithecus*) had relatively longer forelimbs or hindlimbs than earlier hominoids (Lovejoy et al. 2009). Although the available evidence does not conclusively rule this out, it does not favor the bipedal-feeding narrative about hominin origins.

Hooker (2007) suggested that several other features could be interpreted as bipedal adaptations in the skeleton of the gerenuk (and by inference in the fossil tylopod *Anoplotherium*), including "…flared ilia for muscle attachment

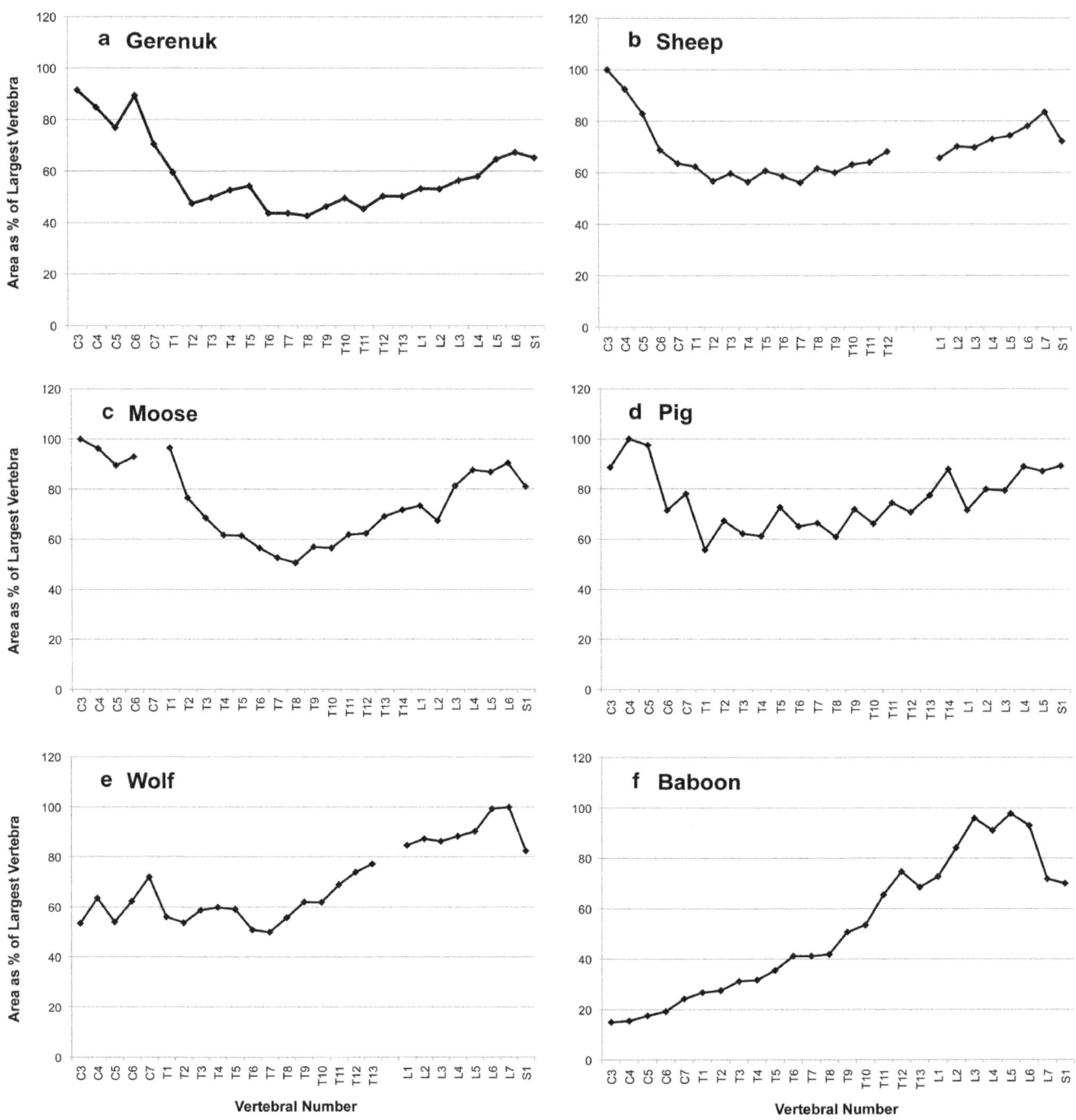

Fig. 6.14 Profiles of vertebral-body endplate areas for gerenuk (*Litocranius walleri,* averaged percentages for one male and one female), a domestic sheep (*Ovis aries,* Harvard Mus. Comp. Zool. 20975), a female moose (*Alces alces,* MCZ 64645), a domestic pig (*Sus scrofa,* MCZ 6246), a wolf (*Canis lupus,* MCZ 59176), and baboon (*Papio hamadryas,* averaged percentages for one male and one female). C, cervical; T, thoracic; L, lumbar; S, sacral (vertebrae). Endplate area was measured on the cranial surface of the first sacral vertebra and on the caudal surface of all other vertebrae. Gaps in lines connecting data points represent vertebrae missing in the specimen. In the artiodactyls, vertebral body area peaks in the cervical region, drops to a minimum in the thorax, and then rises again to a secondary peak at or near the last (sixth) lumbar vertebra. The gerenuk conforms to this pattern and does not deviate from the other artiodactyls in a humanlike direction

to raise the trunk vertically; long pubic symphysis to withstand the axial stresses under an erect stance; and trunk vertebrae enlarging in a posterior direction along the column to bear the increased weight of a vertical trunk region".

However, none of these supposed bipedal adaptations distinguish gerenuks from other bovids. Here again, the case of the gerenuk may provide a salutary warning for those seeking to explain the evolution of hominin bipedality. It may be that some of the features of human morphology that anthropologists like to explain as adaptations to bipedal posture and locomotion are nothing of the sort. For example, the head-to-tail gradient in vertebral size (from small cervicals to large lumbars) seen in the human vertebral column appears to be characteristic of all primates, including gorillas, baboons, and sloth lemurs. This gradient is evidently an ancient primate trait, originally evolved for reasons having nothing to do with bipedality. Its origins must be connected with some other factor(s) – e.g., with more habitually vertical trunk or neck postures in primates, or with reduced use of the head and neck in feeding and fighting, or with the hindlimb-dominated forms of locomotion that characterize primates (Demes et al. 1994; Kimura 2002; Schmitt 2009). Likewise, some other anatomical features that distinguish humans from apes and are commonly interpreted as bipedal adaptations may represent primate symplesiomorphies retained in humans, or apomorphies developed in the human ancestry for reasons having little or nothing to do with upright posture and locomotion (Lovejoy et al. 2009). We suggest that in the absence of decisive fossil evidence, the best way to sort out these issues is to extend our comparisons to non-hominoids and non-primates.

Acknowledgements We are grateful to Judith M. Chupasko and the rest of the staff at the Harvard Museum of Comparative Zoology for their unflagging help and support. We also thank the staff of the American Museum of Natural History, the Natural History Museum (London), and the Miami MetroZoo for their help. This research was financed by grants from Boston University.

References

Aiello, L., & Dean, C. (1990). *An introduction to human evolutionary anatomy*. London: Academic Press.

Alexander, R. (2004). Bipedal animals, and their differences from humans. *Journal of Anatomy, 204*, 321–330.

Bodmer, R. E. (1990). Ungulate frugivores and the browser-grazer continuum. *Oikos, 57*, 319–325.

Cartmill, M., & Brown, K. (2014). Vertebral body area profiles in primates and other mammals. *American Journal of Physical Anthropology, Supplement, 58*, 91.

Cerling, T. E., Harris, J. M., & Passey, B. H. (2003). Diets of East African bovidae based on stable isotope analysis. *Journal of Mammalogy, 84*, 456–470.

Chan, L. K. (2014). The thoracic shape of hominoids. *Anatomy Research International, 2014* (324850), 1–8.

Chaveau, A., & Arloing, S. (1890). *The comparative anatomy of the domesticated animals* (Trans. G. Fleming,). New York: Appleton.

Davis, P. R. (1961). Human lower lumbar vertebrae: Some mechanical and osteological considerations. *Journal of Anatomy, 95*, 337–344.

Demes, B., Larson, S. G., Stern, J. T., Jr., Jungers, W. L., Biknevicius, A. R., & Schmitt, D. (1994). The kinetics of primate quadrupedalism: "Hindlimb drive" reconsidered. *Journal of Human Evolution, 6*, 353–374.

DeSilva, J. M. (2010). Revisiting the "midtarsal break". *American Journal of Physical Anthropology, 141*, 245–258.

DeSilva, J. M., & Lovejoy, C. O. (2009). Functional morphology of the ankle and the likelihood of climbing in early hominins. *Proceedings of the National Academy of Sciences USA, 106*, 6567–6572.

de Waal Malefijt, J., Slooff, T. J. J., Huiskes, R., de Laat, E. A. T., & Barentsz, J. O. (1988). Vascular changes following hip arthroplasty: The femur in goats studied with and without cementation. *Acta Orthopedica Scandinavica, 59*, 643–649.

Elftman, H. O. (1929). Functional adaptations of the pelvis in marsupials. *Bulletin of the American Museum of Natural History, 58*, 189–232.

Elliot, D. G. (1897). List of mammals obtained by the Field Columbian Museum East African expedition to Somali-land in 1896. *Field Columbian Museum Publications (Zool.), 1*, 109–155.

Full, R. J., & Tu, M. S. (1991). Mechanics of a rapid running insect: Two-, four- and six-legged locomotion. *Journal of Experimental Biology, 156*, 215–231.

Gentry, A. W. (1964). Skull characters of African gazelles. *Annual Magazine of Natural History, 7*, 353–382.

Goetsch, A. L., Gipson, T. A., Askar, A. R., & Puchala, R. (2010). Feeding behavior of goats. *Journal of Animal Science, 88*, 361–373.

Grand, T. I. (1997). How muscle mass is part of the fabric of behavioral ecology in East African bovids (*Madoqua, Gazella, Damaliscus, Hippotragus*). *Anatomy and Embryology, 195*, 375–386.

Hermanson, J. W., & MacFadden, B. J. (1996). Evolutionary and functional morphology of the knee in fossil and extant horses (Equidae). *Journal of Vertebrate Paleontology, 16*, 349–357.

Hodge, W. A., Fijan, R. S., Carlson, K. L., Burgess, R. G., Harris, W. H., & Mann, R. W. (1986). Contact pressures in the human hip joint measured in vivo. *Proceedings of the National Academy of Sciences USA, 83*, 2879–2883.

Hooker, J. J. (2007). Bipedal browsing adaptations of the unusual late Eocene–earliest Oligocene tylopod *Anoplotherium* (Artiodactyla, Mammalia). *Zoological Journal of the Linnaean Society, 151*, 609–659.

Hunt, K. D. (1994). The evolution of human bipedality: Ecology and functional morphology. *Journal of Human Evolution, 26*, 183–202.

Hunt, K. D. (1996). The postural feeding hypothesis: An ecological model for the evolution of bipedalism. *South African Journal of Science, 92*, 77–90.

Jungers, W. L. (1988). Relative joint size and hominoid locomotor adaptations with implications for the evolution of hominid bipedalism. *Journal of Human Evolution, 17*, 247–265.

Jungers, W. L. (1990). Problems and methods in reconstructing body size in fossil primates. In J. D. Damuth & B. J. MacFadden (Eds.), *Body size in mammalian paleobiology: Estimation and biological implications* (pp. 103–118). Cambridge: Cambridge University Press.

Kimura, T. (2002). Primate limb bones and locomotor types in arboreal or terrestrial environments. *Morphological Anthropology, 83*, 201–219.

Kingdon, J. (2004). *The Kingdon pocket guide to African mammals*. London: A&C Black.

Köhler, M., & Moyà-Solà, S. (1997). Ape-like or hominid-like? The positional behavior of *Oreopithecus bambolii* reconsidered. *Proceedings of the National Academy of Sciences USA, 94*, 11747–11750.

Latimer, B., & Ward, C. (1993). The thoracic and lumbar vertebrae. In A. Walker & R. Leakey (Eds.), *The Nariokotome* Homo erectus *skeleton* (pp. 266–293). Cambridge: Harvard University Press.

Leuthold, W. L. (1978). On the ecology of the gerenuk *Litocranius walleri*. *Journal of Animal Ecology, 47*, 561–580.

Leuthold, W. L., & Leuthold, B. M. (1973). Notes on the behaviour of two young antelopes reared in captivity. *Zeitschrift für Tierzuchtung und Zuchtungsbiologie, 32*, 418–424.

Lovejoy, C. O., Suwa, G., Simpson, S. W., Matternes, J. H., & White, T. D. (2009). The great divides: *Ardipithecus ramidus* reveals the postcrania of our last common ancestors with African apes. *Science, 326*, 100–106.

Lydekker, R. (1908). *The game animals of Africa*. London: Rowland Ward.

MacLatchy, L. (1998). Reconstruction of hip joint function in extant and fossil primates. In E. Strasser, J. G. Fleagle, A. L. Rosenberger & H. M. Mchenry (Eds.), *Primate locomotion* (pp. 111–130). New York: Springer.

MacLatchy, L. M., & Bossert, W. H. (1996). An analysis of the articular surface distribution of the femoral head and acetabulum in anthropoids, with implications for hip function in Miocene hominoids. *Journal of Human Evolution, 31*, 425–453.

Pal, G. P., & Routa, R. V. (1986). A study of weight transmission through the cervical and upper thoracic regions of the vertebral column in man. *Journal of Anatomy, 148*, 245–261.

Pal, G. P., & Routa, R. V. (1987). Transmission of weight through the lower thoracic and lumbar vertebral regions in man. *Journal of Anatomy, 152*, 93–105.

Richter, J. (1970). Die fakultative Bipedie der Giraffengazelle *Litocranius walleri sclateri*. Ein Beitrag zur funktionellen Morphologie. *Morphologische Jahrbuch, 114*, 457–541.

Russo, G. A., & Shapiro, L. J. (2013). Reevaluation of the lumbosacral region of *Oreopithecus bambolii*. *Journal of Human Evolution, 65*, 253–265.

Sanders, W. J. (1998). Comparative morphometric study of the australopithecine vertebral series Stw-H8/H41. *Journal of Human Evolution, 34*, 249–302.

Schmitt, D. (2009). Primate locomotor evolution: Biomechanical studies of primate locomotion and their implications for understanding primate neuroethology. In M. L. Platt & A. A. Ghazanfar (Eds.), *Primate neuroethology* (pp. 31–63). Oxford: Oxford University Press.

Shapiro, L. (1993). Evaluation of "unique" aspects of human vertebral bodies and pedicles with a consideration of *Australopithecus africanus*. *Journal of Human Evolution, 25*, 433–470.

Stanford, C. B. (2002). Brief communication: Arboreal bipedalism in Bwindi chimpanzees. *American Journal of Physical Anthropology, 119*, 87–91.

Stanford, C. B., Allen, J. S., & Antòn, S. (2009). *Biological anthropology* (2nd ed.). Upper Saddle River, NJ: Pearson Prentice-Hall.

Stern, J. T. Jr., & Susman, R. L. (1983). The locomotor anatomy of *Australopithecus afarensis*. *American Journal of Physical Anthropology, 60*, 279–317.

Steudel, K. (1981). Body size estimators in primate skeletal material. *International Journal of Primatology, 2*, 81–90.

Tardieu, C. (1979). Aspects bioméchaniques de l'articulation du genou chez les primates. *Bulletins de la société anatomique de Paris, 4*, 66–86.

Thorpe, S. K. S., Holder, R. L., & Crompton, R. H. (2007). Origin of human bipedalism as an adaptation for locomotion on flexible branches. *Science, 316*, 1328–1331.

Tuttle, R. H. (1975). Parallelism, brachiation and hominoid phylogeny. In W. P. Luckett & F. S. Szalay (Eds.), *The phylogeny of the primates: A multidisciplinary approach* (pp. 447–480). New York: Plenum.

Vereecke, E. E., D'Août, K., & Aerts, P. (2006). Locomotor versatility in the white-handed gibbon (*Hylobates lar*): A spatiotemporal analysis of the bipedal, tripedal, and quadrupedal gaits. *Journal of Human Evolution, 50*, 552–567.

Chapter 7
Canine Height and Jaw Gape in Catarrhines with Reference to Canine Reduction in Early Hominins

William L. Hylander

Abstract Until recently, there has been little consensus as to the functional benefits of having vertically-shortened canines in the earliest humans. In an effort to resolve this problem, Hylander (2013) tested the hypothesis that canine height dimensions in catarrhines are linked to modifications in the amount of jaw gape. The data demonstrate that most adult male catarrhines have relatively larger canine overlap dimensions and relatively larger gapes than do conspecific females. Humans and hylobatids are the exceptions in that canine overlap is nearly the same between sexes, and so is relative gape, although humans have relatively small gape and hylobatids have relatively large gape. A correlation analysis demonstrated that a large portion of relative gape (maximum gape/projected jaw length) is predicted by relative canine overlap (canine overlap/jaw length). Relative gape is mainly a function of jaw muscle position and/or jaw muscle-fiber length. All things equal, more caudally positioned jaw muscles and/or longer muscle fibers increase the amount of gape. The net benefit for increasing gape in catarrhines is related to within species interactions as well as predation patterns. The cost, however, is to decrease bite force. In order to compensate for a decrease in bite force, jaw muscle mass must be increased so as to assure that the original bite force is maintained. On the other hand, and all things equal, more rostrally positioned jaw muscles and/or shorter muscle fibers decrease gape. The net benefit to decreasing gape is to increase bite force without a corresponding increase in muscle mass. Alternatively, the original bite force can be maintained whereas the costs of original muscle size can be reduced. Overall, the data support the hypothesis that canine reduction in early hominins is functionally linked to increased mechanical efficiency of the jaws. The purpose of this chapter is to review certain aspects of the original paper by Hylander (2013), as well as discussing additional implications of this study not previously considered. These include, but are not restricted to: (1) a review of recent developments about muscle mass and fiber lengths in a highly dimorphic model catarrhine primate, *Macaca fascicularis*; (2) a discussion of the fact that relative canine overlap in male catarrhines do not mirror those in female catarrhines; and (3) based on the catarrhine data, interpretations are advanced as to relevance of the functional significance of the high mandibular condyle position in certain catarrhines, with a particular emphasis on the high condyle of robust australopithecines.

Keywords Australopithecine • Functional analysis • Bite force • Canine reduction • Gape • Mandiblular corpus

Introduction

Arguably, there has been little consensus as to the functional benefits of having vertically-shortened canines in the earliest humans (e.g., Darwin 1871; Brace 1963; Holloway 1967; Washburn 1968; Jolly 1970; Greenfield 1992; Plavcan 2001; and many others). As cogently noted by Plavcan and van Schaik (1997: 369)… "Current models for the evolution of canine size in primates only suggest that when the canines are not used as weapons, some factor (as yet unknown) acts to quickly reduce canine size."

The main purpose of this paper is to: (1) review and discuss a data set designed to provide insights into the evolution of vertical canine size in catarrhine primates, with a particular focus on the functional benefits of reduced canine size in early hominins (Hylander 2013); (2) briefly discuss a recently published study on jaw muscle mass and fiber length of males and females in the sexually dimorphic model species, *Macaca fascicularis* (Terhune et al. 2015), and how this study supports predictions by Hylander (2013); (3) consider and discuss the puzzling fact that for catarrhines, when analyzing

W.L. Hylander (✉)
Department of Evolutionary Anthropology, Duke University, Durham, NC 27710, USA
e-mail: billhyl@duke.edu

© Springer International Publishing AG 2017
Assaf Marom and Erella Hovers (eds.), *Human Paleontology and Prehistory*,
Vertebrate Paleobiology and Paleoanthropology, DOI: 10.1007/978-3-319-46646-0_7

relative canine overlap, relative gape in male catarrhines do not mirror those in female catarrhines; instead, many females have larger than expected gapes; (4) evaluate presumed functional correlates of the highly positioned condyle in certain catarrhines, with special reference to robust australopithecines. In summary, the review of the Hylander paper (2013) can be found embedded throughout all sections of this particular manuscript, whereas the remaining topics to be considered can be found in the discussion section.

Functional Links Between Canine Height and Jaw Gape in Catarrhines

In 2004, I measured mandibular (jaw) length and maximum jaw gape in anesthetized adult male (n = 3) and female (n = 3) long-tailed macaques (*Macaca fascicularis*), and adult male (n = 3) Japanese macaques (*Macaca fuscata*). There were two interesting results. The first was that relative to jaw length, male long-tailed macaques opened their jaws much wider than do females. That is, when the mandible is projected onto the mid-sagittal plane (Fig. 7.1), these males opened their jaws over 110% of jaw length (about 74° of mandibular rotation), whereas the females opened their jaws somewhat less than 90% of jaw length (about 55° of rotation). These differences are important because maximum jaw opening is determined by how much the jaw muscles are able to stretch, rather than being restricted by accessory ligaments or bony structures.

The second interesting result occurred after measuring male Japanese macaques. Japanese macaques maximally opened their jaws both absolutely and relatively much less so than do male long-tailed macaques. Although male Japanese macaque mandibles are approximately 10% longer than those of male long-tailed macaques, Japanese macaques opened their jaws about 30% less, i.e., about 80% of jaw length (about 50° of rotation).

An additional important difference between these macaque species is related to the heights of their canine crowns. Previously published data (Plavcan 1990) demonstrate that the combined crown heights of the upper and lower canines (C^1 height + C_1 height = combined crown height) for male long-tailed and Japanese macaques are about 40 mm and 34 mm, respectively. Thus, the shorter jawed long-tailed macaques have absolutely larger vertical canine dimensions (and gapes) than Japanese macaques.

The preliminary data and known canine vertical dimensions for these two macaque species formed the basis of a working hypotheses predicting that among catarrhines, canine height dimensions and maximum gape are functionally linked, and that decreased gape promotes canine reduction, and more importantly, masticatory efficiency (Hylander 2013).

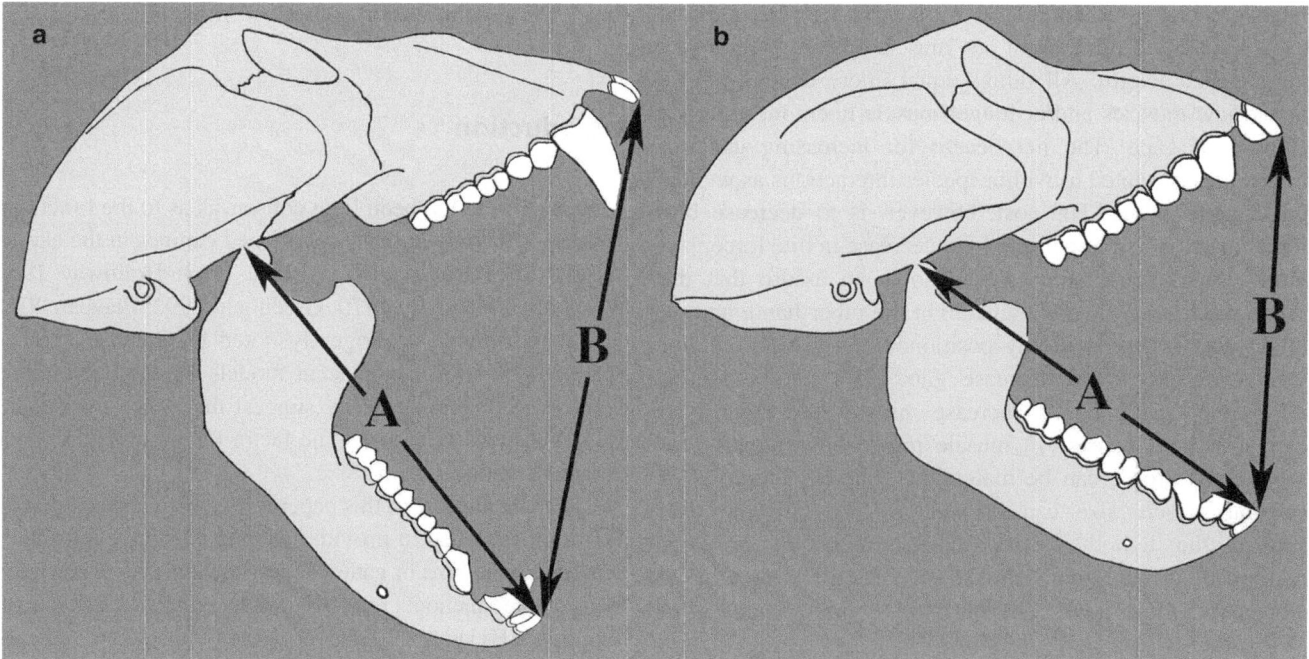

Fig. 7.1 Line drawings of male and female *Macaca fascicularis* during maximum jaw gape (lateral view). Measurement *A* is jaw length and *B* is maximum jaw gape. Male (**a**) is on the left and female (**b**) is on the right. Males maximally rotate their jaws open about 74°, whereas females do so about 55°. (Redrawn and modified from Hylander 2013)

Hypothesis and Predictions: Intraspecific Differences Between Adult Males and Females

Among conspecifics of highly dimorphic catarrhines for canine height, females with their shorter canine heights were predicted to have a *relatively* less gape. If true, this most likely means that females have relatively shorter jaw-closing muscle fibers and/or more rostrally positioned jaw-closing muscles. Although shorter muscle fibers and/or more rostrally positioned muscles have the disadvantage or cost of decreasing jaw gape, the advantage or benefit is to increase bite force (relative to muscle mass), and this in turn indicates that the jaws of females are mechanically more efficient than males during chewing and biting.

Conversely, if conspecific males with their vertically elongate canines have a much larger relative gape, a major benefit is that it facilitates a full display of their canines, as well as the ability to inflict deep wounds on conspecifics and predators, and this in turn enables them to more effectively compete with other adult males for increased access to females during breeding (cf. Leigh et al. 2008). Furthermore, having a relatively larger gape must be due mainly to having relatively longer jaw-closing muscle fibers and/or more caudally positioned jaw-closing muscles. Although jaw gape is increased, all things equal such as not changing the geometry of the mandible,[1] the mechanical cost is a decrease in bite force relative to muscle mass, and therefore the jaws of males are mechanically less efficient. In order for these males to maintain functional equivalence of bite force for chewing and biting, there must be a corresponding increase in relative jaw-closing muscle mass (another cost). Thus, compared to females, males must have relatively larger jaw muscles so as to maintain the necessary bite force.

Finally, the prediction for those catarrhines that exhibit minimal dimorphism for canine height dimensions (hylobatids and humans) was that there is little or no difference in relative gape between males and females, and therefore these conspecifics rotate their jaws open relatively more or less the same amount, and thus in terms of masticatory efficiency, conspecific males and females are near equivalent.

Hypothesis and Predictions: Interspecific Differences

The prediction for interspecific comparisons is that the amount of jaw gape in catarrhine primates is positively and intensely correlated with canine height and jaw length dimensions. Similarly, relatively longer vertically projecting canines are linked to relatively larger gapes, whereas relatively shorter projecting canines are linked to relatively smaller gapes. Thus, as in the above discussion of intraspecific differences, those species with vertically short canines and small gapes have a more mechanically efficient chewing apparatus. Furthermore, as with the intraspecific comparisons, it is likely that interspecific differences in gape are importantly linked to differences in muscle-fiber length, and/or relative jaw muscle position and masticatory efficiency.

Materials and Methods

As described in detail elsewhere (Hylander 2013), materials for this study were drawn from (1) Non-human living catarrhine primates housed at various zoos in Europe and the USA, as well as university animal-care facilities and regional primate centers in the USA; (2) Living human subjects residing mainly in Virginia and North Carolina; (3) Non-human catarrhine skulls housed at various natural history museums and universities in Europe and the USA. All living subjects and museum specimens are dental adults.

Living Subjects

Non-human living catarrhines. Jaw measurements were taken on 494 fully anesthetized subjects. At least two measurements were taken. These include jaw (mandibular) length and maximum jaw gape. Jaw length (A) is the linear distance between the posterior and lateral portion of the mandibular condyle to the mesial-incisal edge of the ipsilateral mandibular central incisor (Fig. 7.2). Jaw gape (B) is the linear distance between the incisal edges of the middle portion of the upper and lower central incisors when the jaws are opened maximally (Fig. 7.1).

During the measurement procedures, the upper and lower post-canine teeth were positioned in maximum occlusion so as to determine the relationship between the upper and lower central incisors. In most instances, the central incisors of non-human subjects exhibited an edge-to-edge bite (Fig. 7.3a). If there was overlap of the upper and lower central incisors, this overlap was determined by inscribing a line with a sharpened pencil along the labial surface of the lower incisor crown at the level of the incisal edge of the corresponding upper incisor. Following this, the linear distance between the pencil line and the incisal edge of the lower incisor was measured (b, variable X). The amount of incisor overlap was added to the variable jaw gape. This was done so as to fully account for the total amount of actual jaw opening. If there was no contact between the upper and

---

[1]Changing the geometry of the mandible includes altering the position of the mandibular condyle relative to the occlusal place, as well as, e.g., shifting the tooth row rostrally or caudally.

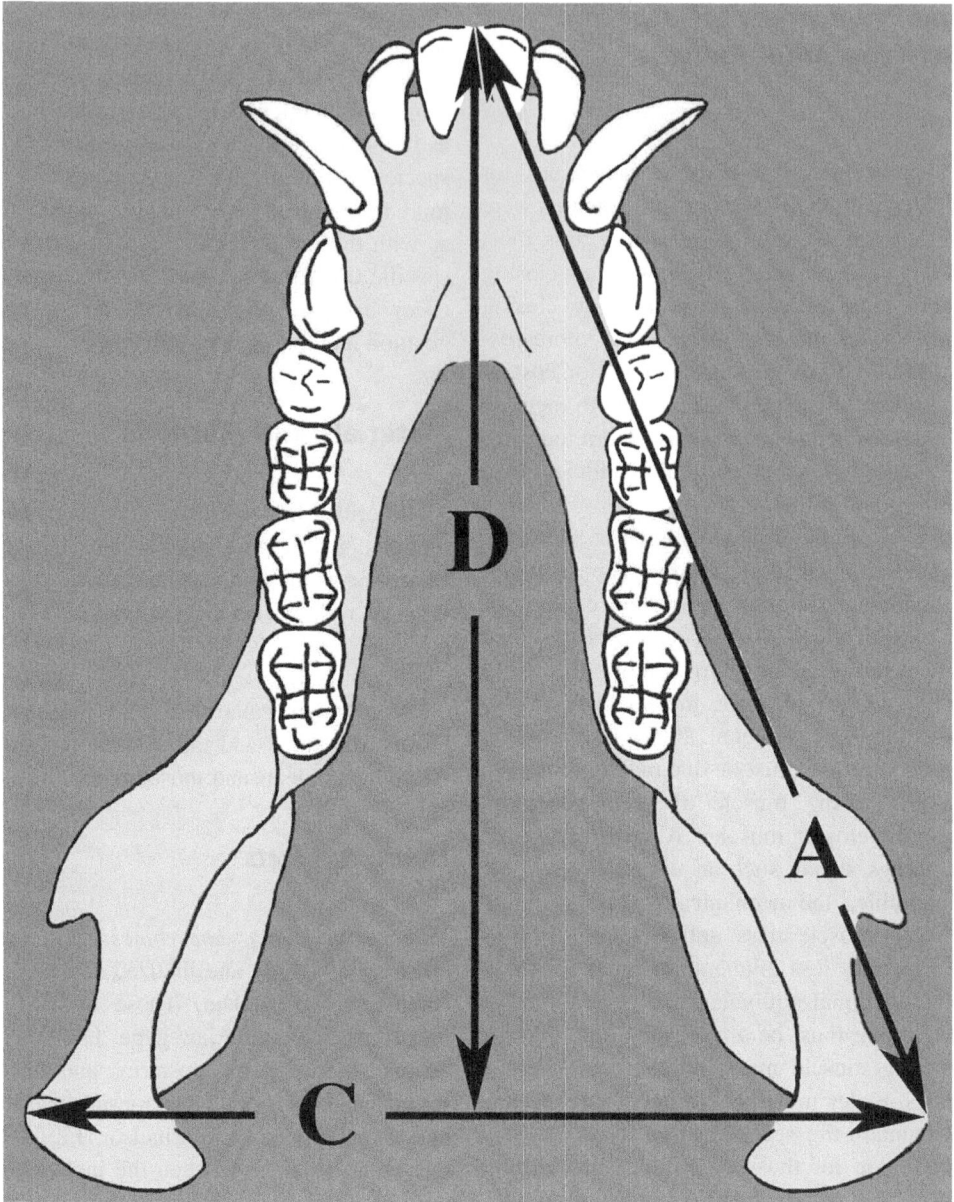

Fig. 7.2 Line drawing of a male macaque mandible (occlusal view). Measurement *A* is jaw length and *D* is jaw length projected onto the mid-sagittal plane. Measurement *C* is bicondylar width. (Redrawn and modified from Hylander 2013)

lower central incisors, the linear distance between the incisal edges of the upper and lower central incisors was measured (Fig. 7.3c, variable Y), and this amount was subtracted from the jaw gape value. Similar to when the incisors overlapped, this procedure ensured that the final gape measurement accounted for only gape due to jaw opening.

Humans. Overall, 45 subjects were selected to measure. All subjects ranged in age from 23–60 years old, and most were in their twenties and thirties. A total of 5 measurements were taken on each human subject. Three of these measurements and procedures are identical to those already

described for non-human subjects, i.e., jaw length (A), jaw gape (B) and the relationship between the upper and lower central incisors (Figs. 7.1, 7.2 and 7.3).

Two additional measurements were taken, including bicondylar width and canine overlap (Figs. 7.2 and 7.3). Bicondylar width (C) is the linear distance between the lateral poles of the left and right mandibular condyles (Fig. 7.2). So as to make the human data comparable to the museum specimens (see below) and based on anatomical dissections, 6 mm was subtracted from the initial bicondylar dimension. This more or less compensated for the thickness

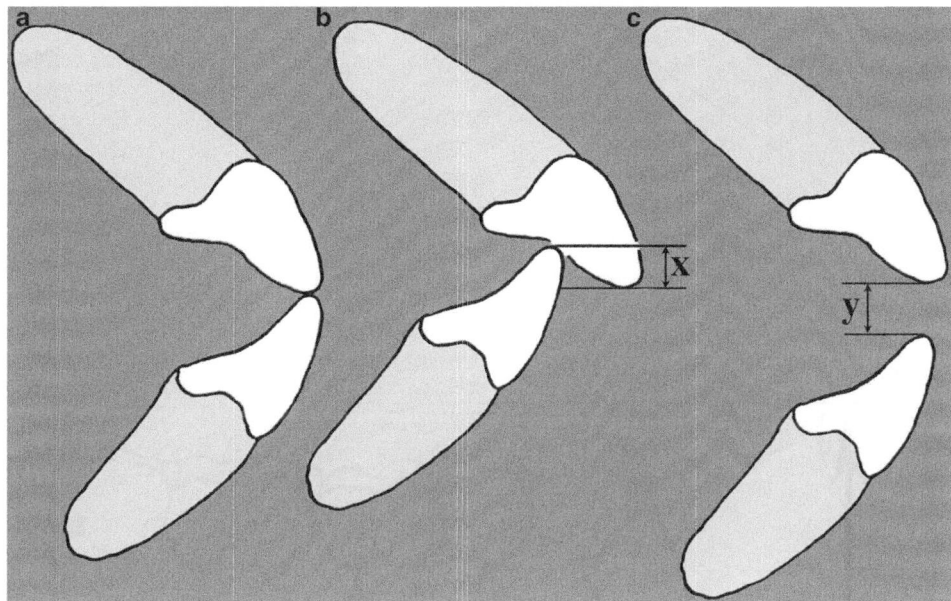

Fig. 7.3 Line drawings of the upper and lower central incisors in the sagittal plane (see text in Methods). **a** Edge-to-edge bite. **b** Overbite. **c** Open bite. (Redrawn and modified from Hylander 2013)

of the soft tissues overlying the human mandibular condyles. Canine overlap in humans was determined using procedures identical to measuring incisor overbite (Fig. 7.3b).

Non-human museum specimens. A total of 316 adult non-human catarrhine male and female skulls were measured, and these are of the same species as the living subjects. Two of the measurements were jaw length (A) and bicondylar width (C) (Figs. 7.1 and 7.2). Canine overlap was also measured. This was done as the calipers were positioned along the tips of the upper and lower canine crowns with the beaks of the calipers parallel to the edge of the alveolar bone of the upper molars (Fig. 7.4).

Derived Variables

There were a total number of 7 derived variables among the museum specimens and living subjects.

Non-human museum specimens. (1) Projected jaw length (Fig. 7.2 variable D) was determined based on jaw length (A) and ½ bicondylar width (Fig. 7.2 variable C) employing Pythagorean relations. Namely, $D = \sqrt{(A^2 - 0.5C^2)}$. Computing this variable is important for determining the amount of relative gape in the midsagittal plane. (2) Relative canine overlap is canine overlap (E)/jaw length (A); (3) Dimorphic canine overlap is male canine overlap divided by female canine overlap; (4) Dimorphic relative canine overlap is male relative canine overlap divided female relative canine overlap.

Non-human living catarrhines. (5) A correction factor was computed so as to determine projected jaw length in the

living subjects. The projected jaw length (Fig. 7.2, variable D) = jaw length (A) multiplied by the correction factor. This factor, based solely on the museum specimens = projected jaw length (D)/jaw length (A), was then multiplied by jaw length in the living subjects, with the assumption that these values are near identical; (6) Relative jaw gape is jaw gape (B)/jaw length (A); (7) Projected relative jaw gape is jaw gape (B)/projected jaw length (D).

Humans. For humans, projected jaw length (Fig. 7.2, variable D) was determined based on the procedure employed for non-human museum specimens. The remaining derived variables for humans were determined as for the museum and living non-human subjects.

Statistical Analysis

Descriptive statistics of all variables were determined. For the intraspecific comparisons (males versus females) of projected relative jaw gape, mean ratio values were tested for significance ($\alpha = 0.05$) using a nonparametric test (Mann-Whitney U Test, 2-sample, normal approximation). Finally, correlation procedures were used to ascertain the intensity (raw r^2 values) of the relationship of absolute and relative jaw gape (dependent variables) with various combinations of the following independent variables: jaw length, projected jaw length, canine overlap and relative canine overlap.

In addition to the above and because catarrhines differ in their phylogenetic relatedness, interspecific correlations were

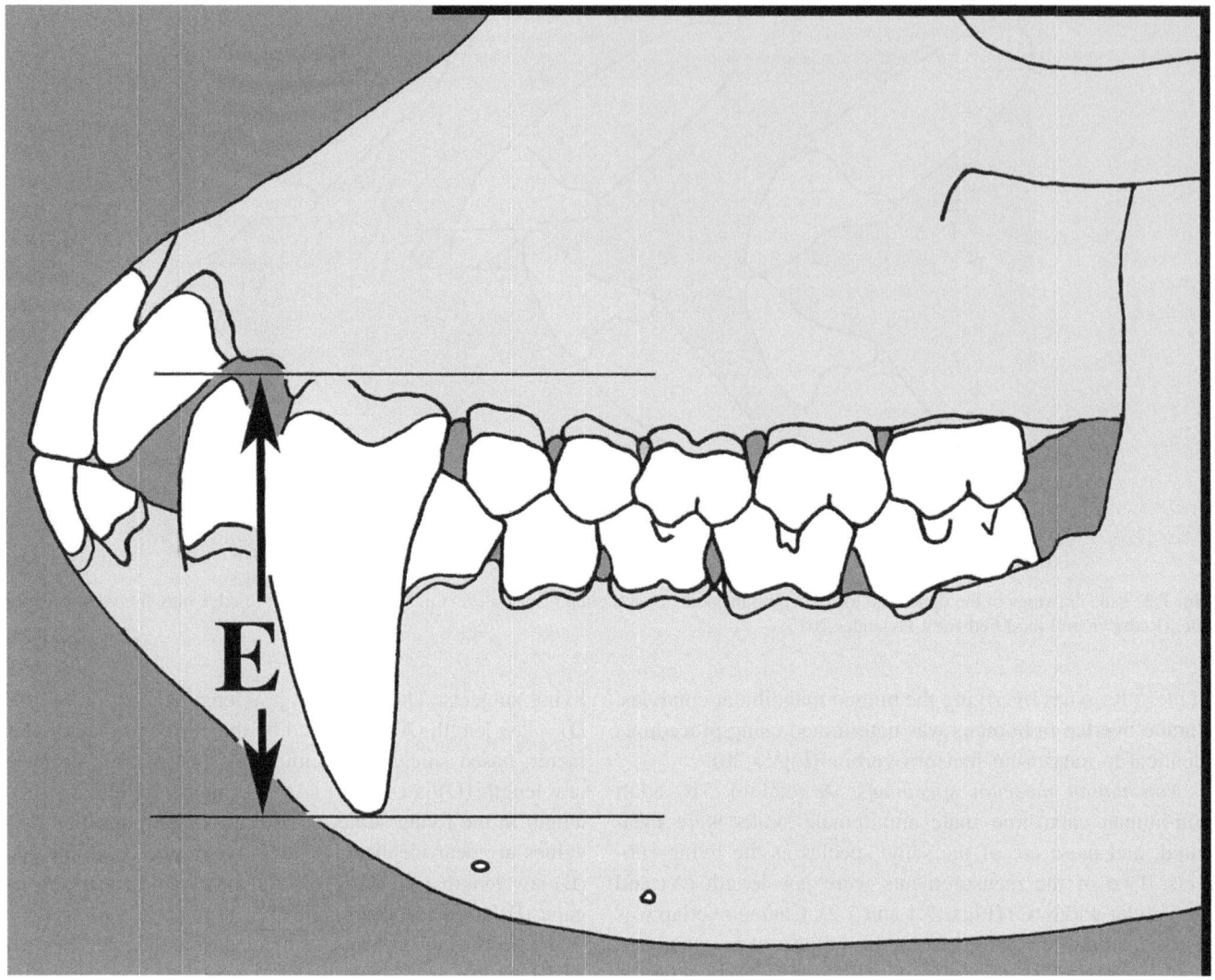

Fig. 7.4 Line drawing of the dentition of a male macaque with the upper and lower teeth in occlusion (lateral view). Measurement *E* is the variable canine overlap. (Redrawn and modified from Hylander 2013)

also calculated using independent contrasts (IC) (IC r^2 values). A consensus phylogeny was created using 10k Trees (Arnold et al. 2010), and phylogenetically informed correlations were conducted in Mesquite (Ver 2.73) using the PDAP: PDTree module (Maddison and Maddison 2010; Midford et al. 2005). For additional details, see Hylander (2013).

Results

Tables 7.1 and 7.2 summarize the descriptive statistics for the living subjects and the museum specimens, respectively.

All living cercopithecid species analyzed fit the predicted pattern, i.e., compared to females, conspecific males have a larger amount of projected relative gape (gape/projected jaw length (Table 7.1)). With the exception of those three species with insufficient sample sizes for gape, all males have

significantly larger values than for conspecific females ($p < 0.03$) (Table 7.1).

As with cercopithecids, the great apes also fit the predicted pattern. Moreover, with the exception of Japanese macaques, male great apes have less relative gape compared to male cercopithecids. Similarly, female great apes also tend to have relatively less gape than do female cercopithecids, although there are a few exceptions (Table 7.1). Most notably, female bonobos have more projected relative gape than do several female cercopithecids, as well as all other female great apes. Although female bonobos have relatively smaller gape values than do male bonobos, the male values are not quite significantly larger than females ($p < 0.075$). Interestingly, female bonobos have relatively larger gapes than do other female great apes. In contrast, male chimpanzees, gorillas and orangutans have significantly larger relative gapes than found in conspecific females ($p < 0.02$).

Table 7.1 Descriptive statistics of living catarrhines

Species	Sex	N	Jaw length	Projected jaw length	Gape	Relative gape	Projected relative gape	Males > females[a]
			Mean/SD	Mean/SD	Mean/SD	Mean/SD	Mean/SD	
Papio anubis	M	19	145.2/11.5	136.6/10.8	152.2/3.9	1.05/0.05	1.12/0.06	p < 0.0001
	F	12	121.7/8.3	113.0/7.7	97.3/8.2	0.80/0.09	0.87/0.10	
Papio hamadryas	M	7	149.2/11.5	140.4/10.8	144.2/8.4	0.97/0.03	1.03/0.04	p < 0.0001
	F	15	118.8/7.1	110.7/6.7	94.7/5.3	0.80/0.06	0.86/0.06	
Mandrillus sphinx	M	5	179.0/7.0	170.6/6.7	208.3/12.8	1.16/0.09	1.22/0.10	p < 0.005
	F	5	123.6/7.8	116.3/7.3	95.1/15.4	0.77/0.13	0.82/0.13	
Cercocebus atys	M	17	100.1/5.9	92.5/5.4	103.7/9.9	1.04/0.09	1.12/0.09	p < 0.0004
	F	23	87.2/5.1	80.1/4.7	68.3/8.1	0.79/0.11	0.86/0.12	
Colobus guereza	M	8	100.2/7.7	92.9/7.1	90.8/7.1	0.91/0.10	0.98/0.11	p < 0.015
	F	6	85.9/5.9	79.1/5.4	68.6/4.6	0.80/0.04	0.87/0.04	
Theropithecus gelada	M	7	131.3/4.8	123.4/4.5	129.0/9.4	0.98/0.08	1.05/0.08	p < 0.003
	F	5	102.7/4.0	95.3/3.7	85.4/3.3	0.83/0.01	0.90/0.01	
Cercopithecus neglectus	M	2	82.2/3.1	76.4/2.9	80.4/1.6	0.98/0.06	1.05/0.06	Insufficient sample size
	F	2	70.8/1.1	64.8/1.0	61.1/5.4	0.86/0.05	0.94/0.07	
Chlorocebus ethiops	M	14	81.8/3.6	76.1/3.3	94.1/5.7	1.15/0.06	1.24/0.06	p < 0.0004
	F	7	73.1/1.9	67.8/1.8	71.7/5.3	0.98/0.07	1.06/0.08	
Cercopithecus diana	M	1	87.3	81.9	101.5	1.16	1.24	Insufficient sample size
	F	3	68.5/0.8	63.6/0.7	60.6/5.2	0.88/0.07	0.95/0.08	
Erythrocebus patas	M	8	102.4/5.9	96.4/5.6	130.1/10.8	1.27/0.05	1.35/0.06	p < 0.0001
	F	11	81.8/4.8	76.2/4.5	83.2/7.7	1.02/0.12	1.10/0.12	
Macaca fuscata	M	23	101.6/7.5	93.4/6.9	73.9/7.8	0.73/0.06	0.79/0.06	p < 0.0001
	F	20	86.6/5.0	78.8/4.5	51.0/4.9	0.59/0.06	0.65/0.06	
Macaca mulatta	M	26	98.0/5.0	90.4/4.6	90.9/8.9	0.93/0.09	1.00/0.10	p < 0.0001
	F	24	81.5/5.3	74.9/4.9	60.1/6.7	0.74/0.07	0.80/0.08	
Macaca fascicularis	M	18	92.1/4.6	86.2/4.3	103.6/8.3	1.12/0.06	1.20/0.06	p < 0.0001
	F	23	72.1/3.9	66.5/3.6	57.3/5.4	0.80/0.06	0.86/0.07	
Macaca silenus	M	5	101.8/3.2	95.2/3.0	103.4/3.6	1.02/0.06	1.09/0.06	p < 0.003
	F	7	85.5/3.0	78.3/2.7	63.0/3.2	0.74/0.04	0.81/0.05	
Macaca nemestrina	M	12	114.7/6.1	107.3/5.7	122.2/10.3	1.07/0.07	1.14/0.08	p < 0.0001
	F	26	90.0/4.4	82.7/4.1	69.5/8.7	0.77/0.08	0.84/0/09	
Trachypithecus cristatus	M	8	75.7/3.6	68.5/3.3	62.5/3.9	0.83/0.06	0.91/0.06	p < 0.02
	F	3	71.0/2.4	63.7/2.2	53.1/0.5	0.75/0.03	0.83/0.04	
Trachypithecus francoisi	M	2	73.6/4.0	64.8/3.5	62.0/11.8	0.84/0.11	0.95/0.13	Insufficient sample size
	F	2	71.8/2.9	63.9/2.6	53.7/0.5	0.75/0.04	0.84/0.04	
Lophocebus aterrimus	M	4	99.7/6.1	91.8/5.6	85.2/9.9	0.86/0.14	0.93/0.07	p < 0.03
	F	3	90.3/1.5	82.4/1.3	56.7/3.1	0.63/0.04	0.69/0.02	
Hylobates syndactylus	M	4	96.5/4.4	89.5/4.1	101.6/2.8	1.05/0.03	1.14/0.03	p < 0.20
	F	4	91.2/6.5	84.9/6.1	93.5/7.1	1.03/0.04	1.10/0.04	
Hylobates leucogenys	M	6	77.5/6.5	71.6/6.0	82.9/7.0	1.07/0.08	1.16/0.08	Females larger P < 0.46[b]
	F	4	71.9/3.5	65.5/3.2	77.9/2.3	1.09/0.05	1.19/0.05	
Hylobates lar	M	5	69.8/5.0	64.6/4.6	76.3/3.7	1.09/0.05	1.18/0.05	Males and females identical
	F	6	66.8/3.2	60.9/2.9	71.9/1.6	1.08/0.05	1.18/0.05	
Pan troglodytes	M	10	143.5/8.9	131.8/8.1	114.8/12.8	0.80/0.05	0.87/0.06	p < 0.0002
	F	10	144.8/6.9	133.9/6.4	87.2/14.2	0.60/0.10	0.65/0.11	
Pan paniscus	M	8	113.2/6.2	102.3/5.6	86.6/7.1	0.76/0.04	0.85/0.04	p < 0.075
	F	9	114.6/5.6	104.2/5.1	84.1/5.0	0.73/0.05	0.81/0.06	
Gorilla gorilla	M	9	195.0/11.0	180.4/10.2	154.0/7.4	0.79/0.04	0.85/0.04	p < 0.0001
	F	9	166.1/9.8	153.3/9.0	107.2/15.6	0.65/0.09	0.70/0.09	
Pongo abelii	M	7	187.7/16.7	174.6/15.6	133.8/17.2	0.71/0.05	0.77/0.05	p < 0.0025
	F	10	156.5/11.7	143.7/10.8	94.8/11.9	0.61/0.06	0.66/0.06	
Pongo pygmaeus	M	5	193.9/16.7	178.4/15.4	121.5/16.4	0.63/0.05	0.68/0.05	p < 0.02
	F	5	153.7/7.1	142.2/6.6	87.2/5.4	0.57/0.02	0.61/0.02	

(continued)

Table 7.1 (continued)

Species	Sex	N	Jaw length	Projected jaw length	Gape	Relative gape	Projected relative gape	Males > females[a]
Homo sapiens	M	24	111.8/8.1	91.2/8.6	55.3/7.0	0.50/0.06	0.61/0.09	Females larger
	F	21	102.1/6.5	83.1/8.0	53.3/6.5	0.52/0.07	0.64/0.09	p < 0.14[b]

M and F are males and females. SD is the standard deviation. N is number of living subjects measured. Relative gape is gape/jaw length; Projected relative gape is gape/projected jaw length

[a]Mann-Whitney U Test to determine if males are significantly larger than females for the variable Projected Relative Gape (1-tailed test)

[b]White-cheeked gibbons and humans are the only catarrhines in which females have larger mean values for the variable projected relative gape, although the mean values are not significantly different from one another (2-tailed test)

Table 7.2 Descriptive statistics of non-human museum catarrhines and living humans

Species	Sex	N	Jaw length	Bicondylar width	Correction factor	Projected jaw length	Canine overlap	Relative canine overlap	Dimorphic canine overlap	Dimorphic relative canine overlap
			Mean/SD	Mean/SD	Mean/SD	Mean/SD	Mean/SD	Mean/SD	Male/female	Male/female
Papio anubis	M	8	152.8/6.7	102.9/6.0	0.941/0.007	143.9/6.8	43.4/2.7	0.285/0.012	4.52	3.48
	F	5	117.1/5.4	84.3/3.7	0.933/0.004	109.2/5.2	9.6/2.9	0.082/0.024		
Papio hamadryas	M	6	133.3/9.4	90.2/3.5	0.941/0.005	125.4/9.6	29.0/4.2	0.216/0.012	3.26	2.88
	F	7	118.5/6.9	85.9/4.1	0.932/0.006	110.4/6.8	8.9/1.5	0.075/0.019		
Mandrillus sphinx	M	9	169.2/9.3	102.8/2.5	0.952/0.006	161.1/9.8	55.3/4.5	0.328/0.028	12.96	9.11
	F	5	116.4/8.5	78.4/3.5	0.940/0.006	108.0/7.9	4.3/2.2	0.036/0.046		
Cercocebus atys	M	5	100.2/6.4	76.6/2.6	0.924/0.006	92.6/6.5	25.2/1.6	0.252/0.017	3.76	3.27
	F	5	87.3/6.2	70.0/5.0	0.917/0.005	80.1/5.8	6.7/0.6	0.077/0.009		
Colobus guereza	M	5	89.6/4.6	67.1/2.8	0.927/0.005	83.1/4.6	23.9/2.5	0.267/0.029	1.73	1.60
	F	5	82.3/2.9	64.0/1.9	0.921/0.003	75.8/2.8	13.8/4.8	0.167/0.057		
Theropithecus gelada	M	5	132.9/5.4	90.9/3.7	0.940/0.002	124.9/5.2	39.5/2.9	0.297/0.013	6.08	5.12
	F	5	110.8/3.7	82.4/2.2	0.928/0.003	102.9/3.6	6.5/3.1	0.058/0.027		
Cercopithecus neglectus	M	6	81.9/7.0	60.7/3.7	0.928/0.008	76.0/7.0	23.1/3.8	0.285/0.013	2.06	1.70
	F	5	67.1/4.1	53.6/3.1	0.916/0.009	61.5/4.0	11.2/1.5	0.168/0.012		
Chlorocebus aethiops	M	5	73.2/2.2	55.8/2.2	0.930/0.008	68.0/2.4	21.4/2.7	0.293/0.043	2.21	2.00
	F	5	66.1/2.6	49.3/1.3	0.930/0.003	61.3/2.6	9.7/1.7	0.146/0.022		
Cercopithecus diana	M	5	82.0/3.7	56.9/2.0	0.937/0.007	76.9/3.9	25.6/3.8	0.311/0.033	2.12	1.77
	F	7	68.7/2.8	51.4/1.8	0.927/0.005	63.7/2.8	12.1/1.6	0.176/0.021		
Erythrocebus patas	M	5	99.3/5.1	66.2/2.1	0.942/0.004	93.6/5.2	33.9/4.2	0.341/0.033	3.23	2.60
	F	8	80.1/4.6	57.8/2.7	0.932/0.004	74.7/5.0	10.5/1.1	0.131/0.013		
Macaca fuscata	M	11	97.8/6.6	79.0/4.3	0.914/0.009	89.4/6.6	15.0/3.0	0.154/0.027	2.73	2.44
	F	9	87.1/3.9	72.2/2.8	0.910/0.006	79.2/3.8	5.5/1.2	0.063/0.014		
Macaca mulatta	M	5	94.0/2.0	73.0/5.4	0.922/0.009	86.7/1.2	18.5/2.1	0.198/0.022	3.43	3.14
	F	6	86.7/6.1	69.0/3.3	0.917/0.009	79.5/6.2	5.4/0.9	0.063/0.012		
Macaca fascicularis	M	8	86.2/2.1	61.7/1.6	0.934/0.005	80.5/2.3	19.8/3.0	0.230/0.037	4.40	3.83
	F	7	74.4/4.7	57.1/3.7	0.923/0.004	68.7/4.4	4.5/2.1	0.060/0.026		
Macaca silenus	M	3	105.9/4.2	75.0/1.4	0.935/0.003	99.0/4.3	24.0/3.7	0.227/0.029	3.87	2.87
	F	5	79.6/1.5	62.4/1.3	0.920/0.005	73.2/1.8	6.2/1.8	0.079/0.023		
Macaca nemestrina	M	6	108.5/4.5	77.9/3.8	0.933/0.007	101.2/4.6	29.6/2.4	0.273/0.034	6.17	5.06
	F	8	86.6/6.5	67.2/1.8	0.921/0.009	79.8/6.8	4.8/2.5	0.054/0.024		
Trachypithecus cristatus	M	5	76.5/1.9	65.0/0.8	0.905/0.007	69.3/2.2	18.2/1.7	0.238/0.020	2.33	2.11
	F	5	69.6/1.1	61.0/2.1	0.898/0.007	62.5/1.2	7.8/0.8	0.113/0.006		
Trachypithecus francoisi	M	4	71.2/1.5	66.1/2.4	0.886/0.006	63.0/1.2	15.9/3.9	0.223/0.057	2.21	2.19
	F	5	70.0/3.8	64.0/3.1	0.889/0.002	62.3/3.5	7.2/1.3	0.102/0.018		
Lophocebus aterrimus	M	7	92.5/4.1	71.6/1.9	0.921/0.009	85.2/4.5	15.1/2.8	0.164/0.030	3.21	2.73
	F	6	78.0/0.7	64.4/1.3	0.911/0.003	71.0/1.7	4.7/0.8	0.060/0.011		
Hylobates syndactylus	M	5	88.6/2.8	66.2/2.8	0.927/0.008	82.2/3.1	20.8/1.7	0.236/0.024	1.18	1.13
	F	5	84.0/3.2	67.0/1.7	0.911/0.003	77.0/3.1	17.6/3.8	0.209/0.038		
Hylobates leucogenys	M	5	75.1/5.6	57.5/5.2	0.923/0.010	69.3/5.3	22.4/5.4	0.297/0.066	1.17	1.73
	F	5	72.7/1.2	59.8/2.8	0.911/0.010	66.2/1.6	19.2/4.3	0.264/0.057		
Hylobates lar	M	6	71.5/2.3	57.2/3.7	0.916/0.008	65.5/2.0	19.6/2.3	0.274/0.034	1.26	1.20
	F	6	68.5/3.1	55.7/2.3	0.913/0.008	62/5.3.2	15.6/4.0	0.228/0.059		

(continued)

Table 7.2 (continued)

Species	Sex	N	Jaw length	Bicondylar width	Correction factor	Projected jaw length	Canine overlap	Relative canine overlap	Dimorphic canine overlap	Dimorphic relative canine overlap
Pan troglodytes	M	5	137.3/8.5	107.6/7.7	0.919/0.013	126.3/8.7	22.5/1.5	0.164/0.010	2.30	2.22
	F	6	130.3/11.2	98.1/4.1	0.925/0.011	120.1/11.6	9.8/2.2	0.074/0.013		
Pan paniscus	M	6	115.0/4.1	94.4/3.3	0.912/0.007	104.8/4.1	12.0/0.8	0.105/0.009	2.32	1.99
	F	7	113.4/4.9	93.7/3.3	0.909/0.007	103.0/5.1	9.4/1.3	0.079/0.011		
Gorilla gorilla	M	8	182.3/3.5	138.8/6.4	0.924/0.005	168.5/9.6	28.8/5.1	0.157/0.023	2.32	1.99
	F	9	158.2/2.5	121.4/5.8	0.923/0.002	146.2/7.9	12.4/1.6	0.079/0.011		
Pongo abelii	M	6	174.9/16.4	125.7/9.8	0.933/0.008	163.2/16.1	31.1/8.0	0.176/0.033	2.14	1.78
	F	7	146.6/5.8	121.7/4.8	0.909/0.011	133.2/6.5	14.5/0.8	0.099/0.006		
Pongo pygmaeus	M	7	183.0/5.1	143.6/9.5	0.920/0.010	168.3/4.8	33.1/3.9	0.181/0.022	3.06	2.45
	F	8	145.6/7.4	110.9/7.6	0.924/0.007	134.6/6.9	10.8/2.7	0.074/0.019		
Homo sapiens	M	24	111.8/8.1	129.0/8.3	0.815/0.026	91.2/8.6	4.0/1.5	0.036/0.014	1.29	1.20
	F	21	102.1/6.5	117.9/5.7	0.813/0.030	83.1/8.0	3.1/1.3	0.030/0.013		

M and F are males and females. N is the number of museum specimens and living humans

Relative Canine Overlap = Canine Overlap/Jaw Length

Dimorphic Canine Overlap = male Canine Overlap/female Canine Overlap

Dimorphic Relative Canine Overlap = male Relative Canine Overlap/female Relative Canine Overlap

Hylobatids and humans also fit the predicted pattern, although in this case the prediction is that projected relative gape values in males are not significantly different than in conspecific females. The data in Table 7.1 support this prediction.

Although hylobatids and humans have little dimorphism for canine overlap, hylobatids have a large amount of relative gape whereas humans have a small amount of relative gape. These predicted differences are linked to the relative size of their canines (Table 7.2), i.e., humans have relatively vertically short canines and hylobatids have relatively vertically long canines. Also, relative gape values for male and female hylobatids are very similar to what is seen for male cercopithecids, and humans are much more similar to female great apes (and female Japanese macaques) (Table 7.1). The one notable exception here is the relatively large gape values for female bonobos.

Predicting Gape: More on Interspecific Analyses of Catarrhines

Figure 7.5 is a bivariate plot of jaw gape versus jaw length. As expected, a large portion of the variation in gape is explained (raw r^2 value = 0.55). Accounting for phylogenetic relationships indicates an even larger value (IC r^2 value = 0.76).

Nevertheless, and as indicated earlier, it was originally thought that a more functionally relevant independent variable is *projected* jaw length, rather than simply jaw length. As it turned out, this correction had relatively minor significance for most catarrhines, although it had its most significant effect on humans as humans have a very broad (wide) cranial base, and therefore a relatively large bicondylar dimension.

Figure 7.6 is a bivariate plot of jaw gape versus projected jaw length. Again, a large portion of the variation in gape is explained (raw r^2 value = 0.60), (IC r^2 value = 0.73). Both raw and IC values are significantly different from zero (p < 0.0001).

Figure 7.7 is a bivariate plot of jaw gape versus canine overlap. This figure indicates that there is a strong correlation between jaw gape and canine overlap (raw r^2 = 0.74). Similarly, accounting for phylogenetic relationships indicates a nearly identical value. All of these r^2 values are significantly different from zero (p < 0.0001).

Figure 7.8 is a plot of actual jaw gape versus predicted jaw gape. Here the gape predictions are based on a multiple correlation of the two independent variables, projected jaw length and canine overlap. This figure demonstrates that a surprisingly large amount of gape is predicted (raw r^2 = 0.89 and IC r^2 = 0.87), and these values are significantly different from zero (p < 0.0001).

Multiple correlations were also performed for predicting catarrhine gapes separately by sex. For male catarrhines, the raw and IC r^2 values are 0.90 and 0.86, respectively, whereas for females they are 0.73 and 0.71. All r^2 values are significantly different from zero (p < 0.0001). Figure 7.9 is a plot of *relative* projected gape versus *relative* canine overlap. Again there are strong correlations (raw and IC r^2 = 0.71 and 0.57, respectively), and these values are significantly different from zero (p < 0.0001). Similarly, males and females were also analyzed separately. The male r^2 values are 0.71 and 0.60, respectively, whereas the female values are 0.64 and 0.33, respectively. All values are significantly different from zero (p < 0.002), although surprisingly, compared to males, female r^2 values for the independent contrasts are considerably less (i.e., males = 0.60, females = 0.33).

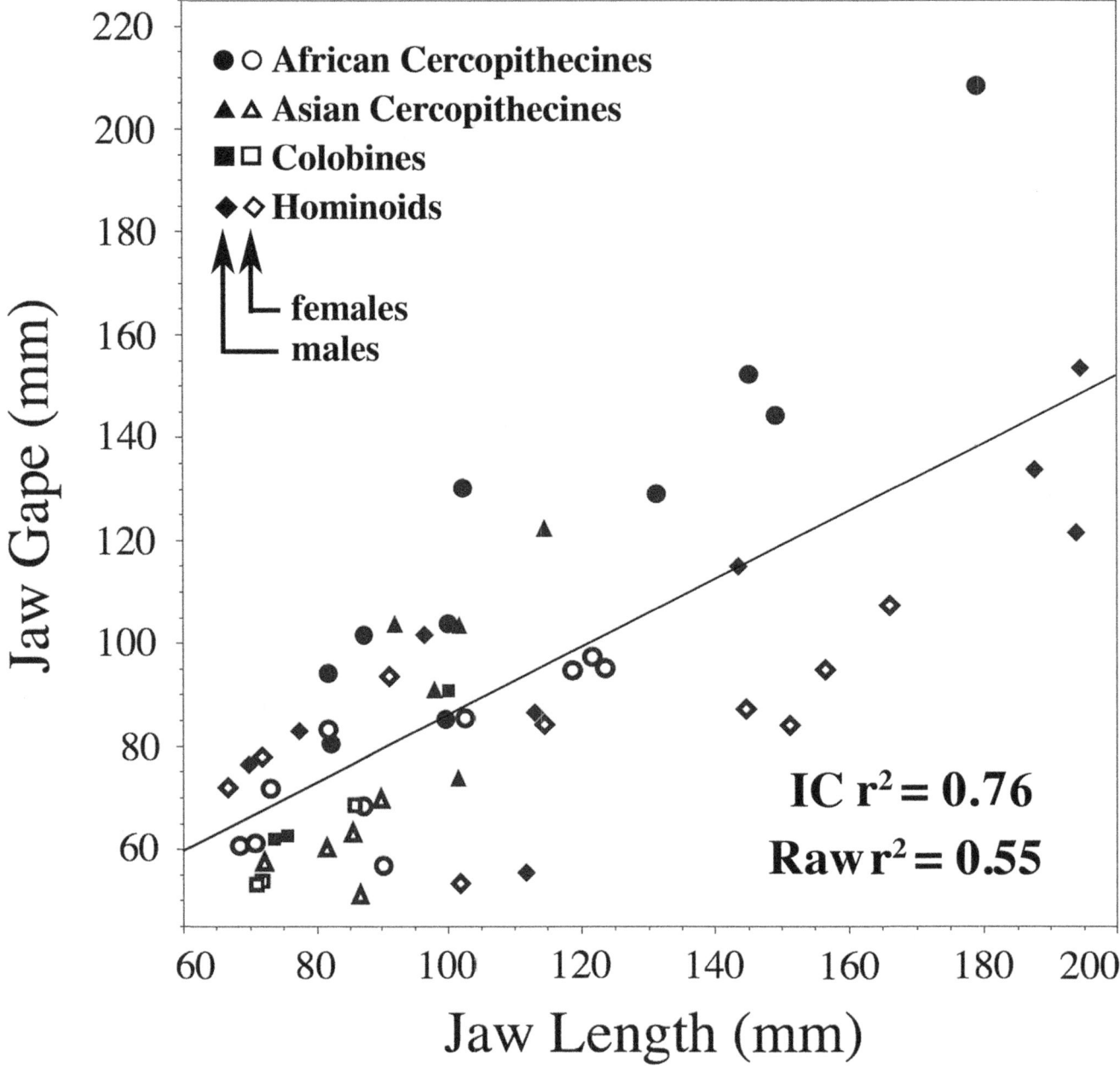

Fig. 7.5 Bivariate plot of jaw gape and jaw length. *Note that males are solid symbols and females are open symbols.* Least-squares regression line is based on raw data; r^2 values are based on raw values and independent contrasts (IC)

Discussion

Intraspecific Predictions

The data in Tables 7.1 and 7.2 strongly support the intraspecific predictions. Although the amount of projected relative gape varies considerably among various catarrhine species, within those species that are highly dimorphic for canine overlap dimensions, males invariably have relatively larger gapes than do females. This includes all 18 species of Old World monkeys analyzed, as well as the five species of great apes.

Note in Table 7.2 that with the exception of hylobatids and humans, bonobos have the least amount of dimorphic relative canine overlap values. The mean values in humans

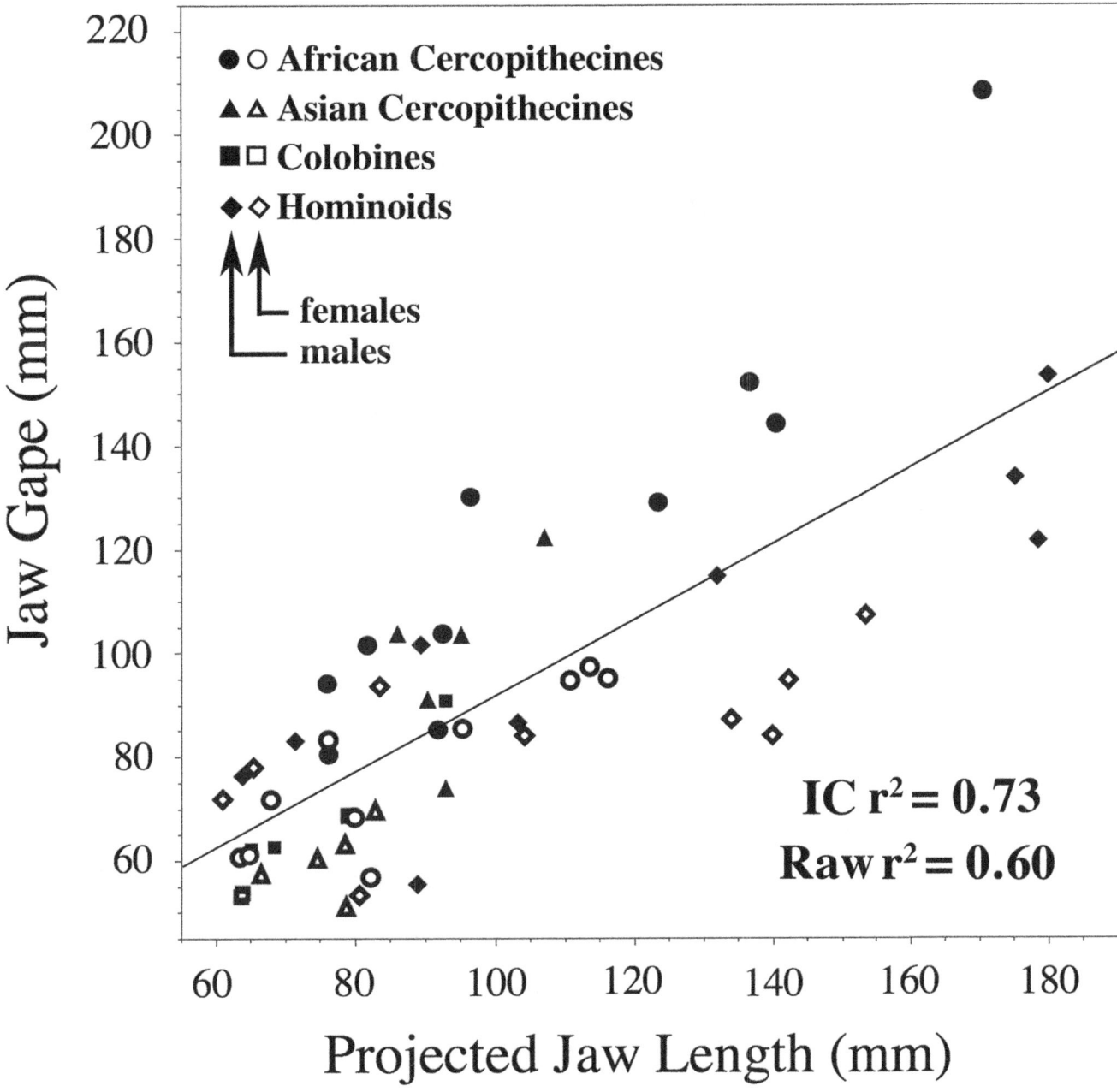

Fig. 7.6 Bivariate plot of jaw gape and projected jaw length. See legend for Fig. 7.5. (Redrawn and modified from Hylander 2013)

and hylobatids range from 1.20 (humans and *Hylobates lar*) to 1.13 (*Hylobates leucogenys* and *Hylobates syndactylus*), whereas the remaining catarrhines range from 9.11 (*Mandrillus sphinx*) to 1.27 (*Pan paniscus*).

The intraspecific predictions for humans and the three hylobatids species are also supported in that the conspecifics exhibit little or no difference in relative gape. Although overall, males have slightly more relative canine overlap than do females (Table 7.2), the projected relative gape values are not significantly larger in males. Mean values for male and female *Hylobates lar* are the same; male *Hylobates*

leucogenys and male humans are slightly smaller than in conspecific females; and male *Hylobates syndactylus* are slightly larger than females (p < 0.15).

Interspecific Predictions

As with the intraspecific predictions, the data in Tables 7.1 and 7.2 as well as the data in Fig. 7.8, strongly support the interspecific prediction. That is, the amount of jaw gape is strongly correlated to the two independent variables,

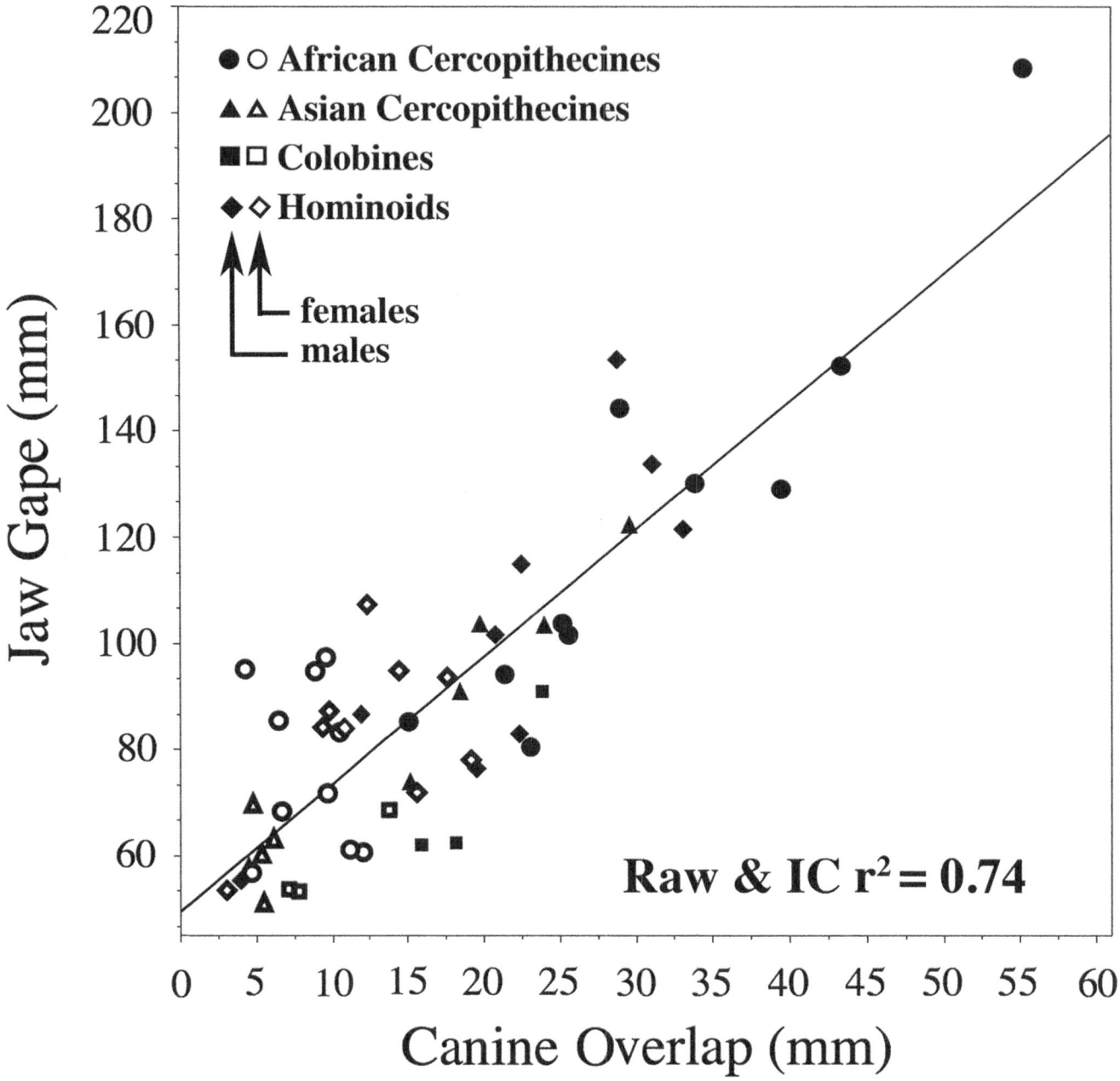

Fig. 7.7 Bivariate plot of jaw gape and canine overlap. See legend for Fig. 7.5. (Redrawn and modified from Hylander 2013)

projected jaw length and canine overlap for both the raw and IC values ($r^2 = 0.89$ and 0.87, respectively).

Most importantly, note in Fig. 7.9 that after correcting for jaw length, there continues to be a strong relationship between projected relative gape and relative canine overlap for the raw and IC values ($r^2 = 0.71$ and 0.57, respectively). As before, this result coincides with the initial predictions, and therefore provides persuasive evidence that maximum jaw gape is strongly linked to canine height (canine overlap). That being the case, a more detailed discussion of catarrhine jaw mechanics will follow.

Prior to a discussion of catarrhine jaw mechanics, however, additional comments about bonobos are in order. First, female bonobos appear to deviate from the overall pattern of relative projected gape compared to that seen in female great apes and all humans in that female bonobos have relatively large gapes. That is, female chimps, gorillas, orangs and all humans have relative projected gape values ranging from 0.61 to 0.70, whereas female bonobos are 0.81 (male bonobos = 0.85) (Table 7.1).

Second, of all of the supposed canine dimorphic catarrhines (cercopithecoids and great apes), bonobos have the

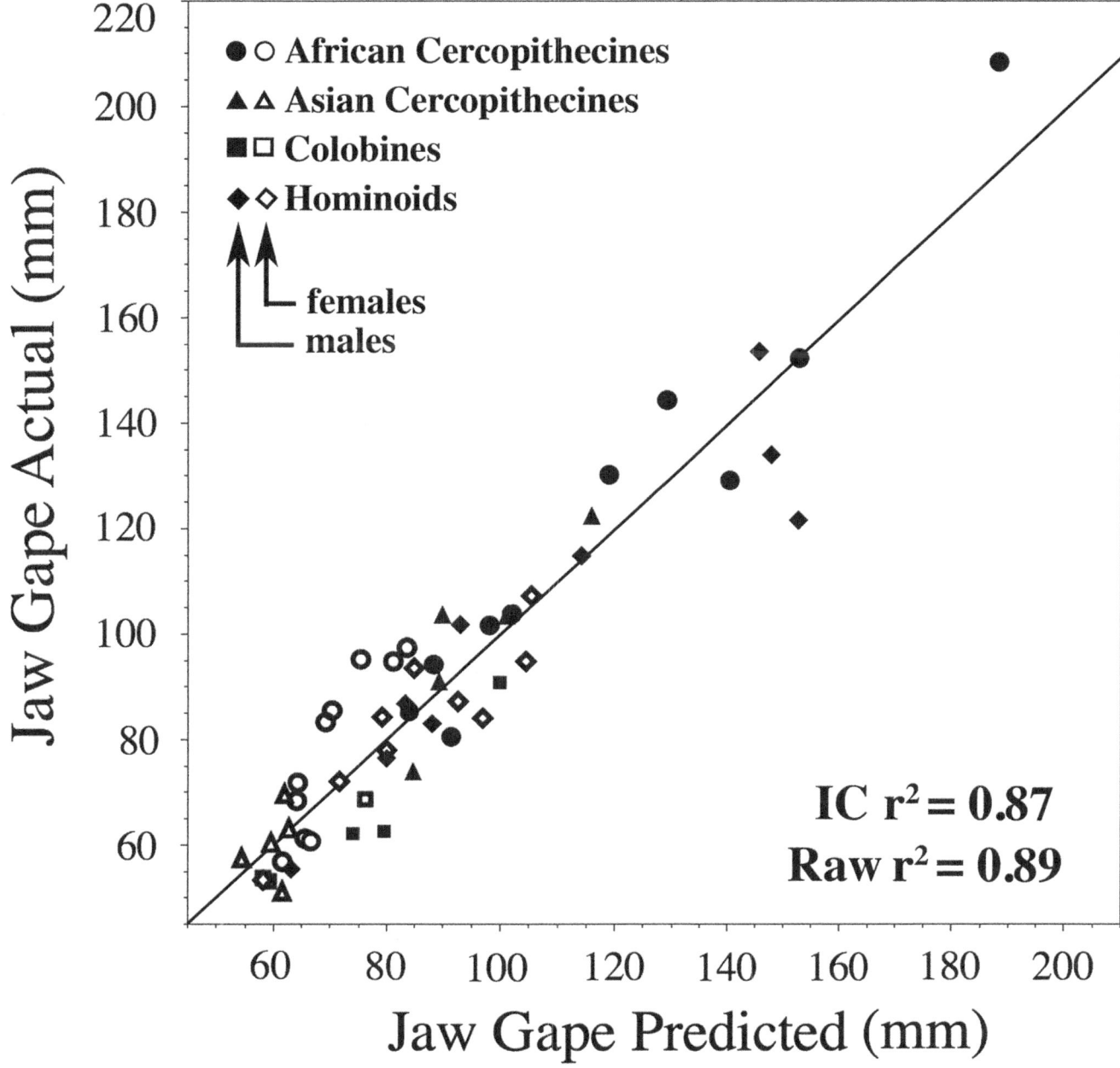

Fig. 7.8 Bivariate plot of actual and predicted jaw gape. Predicted jaw gape (dependent variable) is based on a multiple correlation of independent variables projected jaw length and canine overlap. See legend for Fig. 7.5. (Redrawn and modified from Hylander 2013)

smallest estimate for the variable *dimorphic relative canine overlap* (bonobos = 1.27). Moreover, my sample of wild shot bonobos comes from the Tervuren collection in Belgium (Royal Museum of Central Africa) (measured in 2007). Most importantly, after considering the nature of this collection based on my 2007 observations, I came to the conclusion that perhaps my canine overlap values may not be reasonable estimates. Why do I say that? It is simply because I was surprised to learn that many of the adult bonobo skulls

were unsexed. That is, frequently the museum card for bonobo skeletons indicated a question mark for sex.

At that time I suspected that the variable *dimorphic relative canine overlap* for bonobos might be less than 1.27, and that the true value is more similar (but not identical) seen in humans and hylobatids. By way of explanation, I had originally assumed in 2007 that these wild shot animals had been sexed in the field. Much to my surprise during a return visit to Tervuren in 2011, the museum curator (Dr. Wim Wendelen,

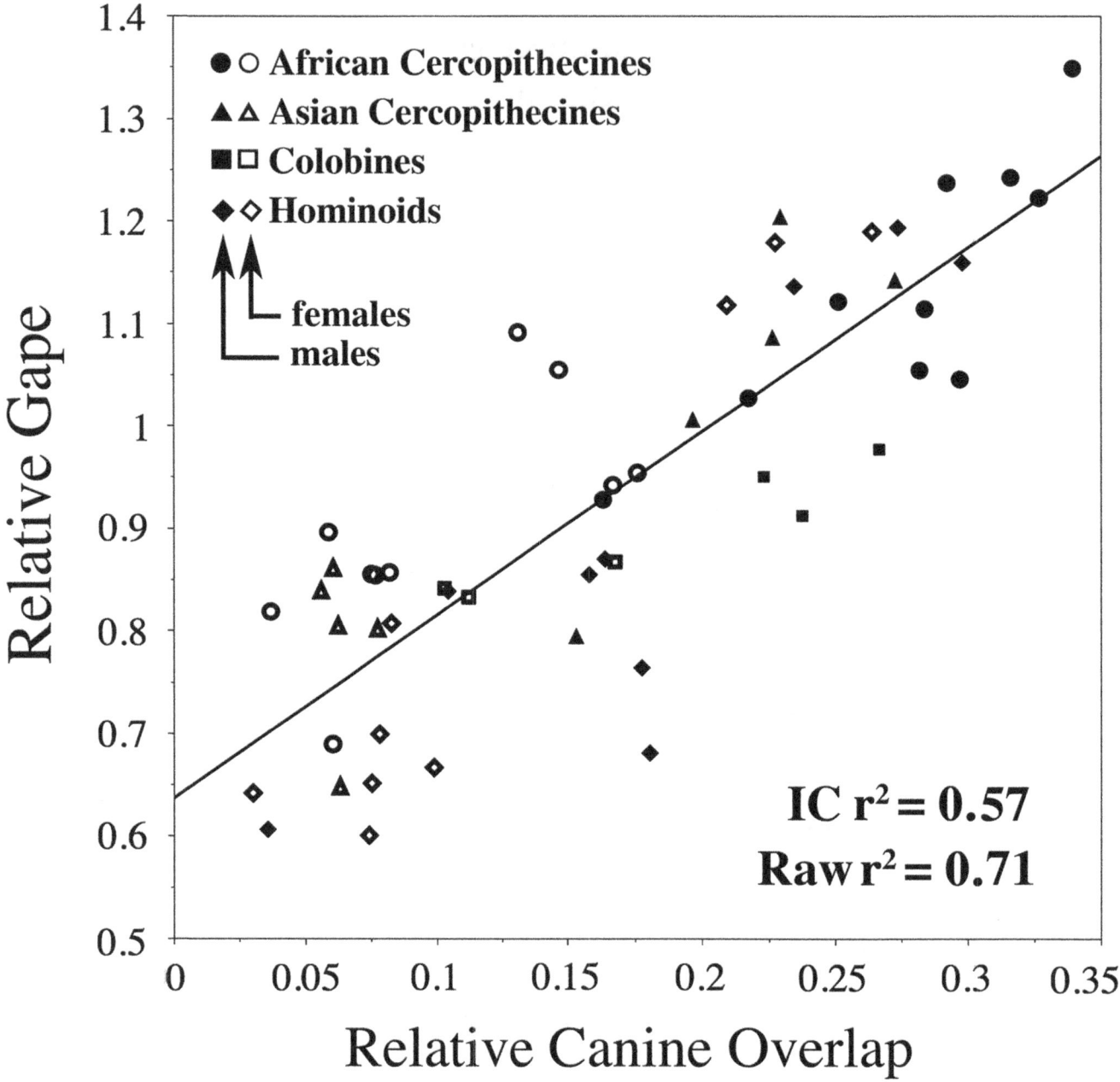

Fig. 7.9 Bivariate plot of relative gape and relative canine overlap. Relative gape = jaw gape/projected jaw length. Relative canine overlap = Canine overlap/jaw length. See legend for Fig. 7.5. (Redrawn and modified from Hylander 2013)

personal communication) informed me that these skulls were sexed after the bodies were skeletonized. That said, although my sample in Table 7.2 may have been correctly sexed, the unsexed skulls that I omitted from my sample would likely have reduced the differences between male and female mean canine overlap values as these unsexed skulls (based largely on canine size) were some combination of males and females. If so, the unsexed skulls would have extended the range of overlap between males and females. Finally, this may explain why although relative gape values for males are somewhat larger, male and female bonobos are not quite significantly

different from one another (p < 0.075).[2] Whether or not they are truly significantly different is not the issue. Instead, the data suggest compared to other great apes, male and female bonobos are more similar to one another, and that female bonobos have an unusually large gape.

---

[2]Although canine height values for male bonobos are much larger than females, as defined by Plavcan (1990), and that overall male canines are much more tusk-like, male and female canine overlap values are more similar to one another (Hylander 2013).

Costs and Benefits of Modifying Bite Force, Gape and Canine Overlap in Catarrhines

Both jaw mechanics and the various costs and benefits are illustrated in Fig. 7.10. Figure 7.10a is a drawing of a male catarrhine mandible in the lateral projection. Here there are three forces applied to the jaw during biting on the mandibular second molar, and these forces are the combined left and right condylar reaction force (F_c), the combined left and right jaw-muscle resultant force (F_m) and the unilateral bite reaction force (F_b). The actual location and direction of these forces are first-approximations, and their precise characteristics are immaterial for this discussion. In addition, the approximate relevant moment arms (X, Y and Z) about these forces are also included.

F_m pulls the mandible upwards and slightly forward, whereas the equal and opposite reaction forces ($F_c + F_b$) push the mandible downwards and backwards. In order to achieve static equilibrium and as indicated by the length of the black arrows, the combined condylar and bite reaction forces ($F_c + F_b$) are about 60% and 40% (respectively) of F_m.

The evolution of increased bite force. Starting with Fig. 7.10a, assume that there is increased selection for an increase in F_b due to a dietary shift requiring more forceful chewing. All other things equal, such as not changing the overall geometry of the mandible[1], in order to increase F_b, there are three options. Option 1 (Fig. 7.10b) is to increase the overall mass of the jaw-closing muscles so as to increase its physiological cross-sectional area (PCSA), which results in a corresponding (proportional) increase of F_m, F_b and F_c. These increases are indicated by the circled gray solid extensions to the black arrows in this and other subsequent figures. Option 2 (Fig. 7.10c) is to maintain the same amount of jaw muscle mass but shorten its muscle fibers. This modification also results in an increase in the PCSA, and this in turn results in a proportional increase in F_m, F_b and F_c. Option 3 (9D) is to maintain the same muscle mass and muscle fiber lengths but shift F_m more rostrally (forwards). As F_m is shifted rostrally, the magnitude of F_m remains unchanged, F_b is increased and F_c is decreased. The decrease in F_c is indicated by the circled gray dots at the end of the black arrow.

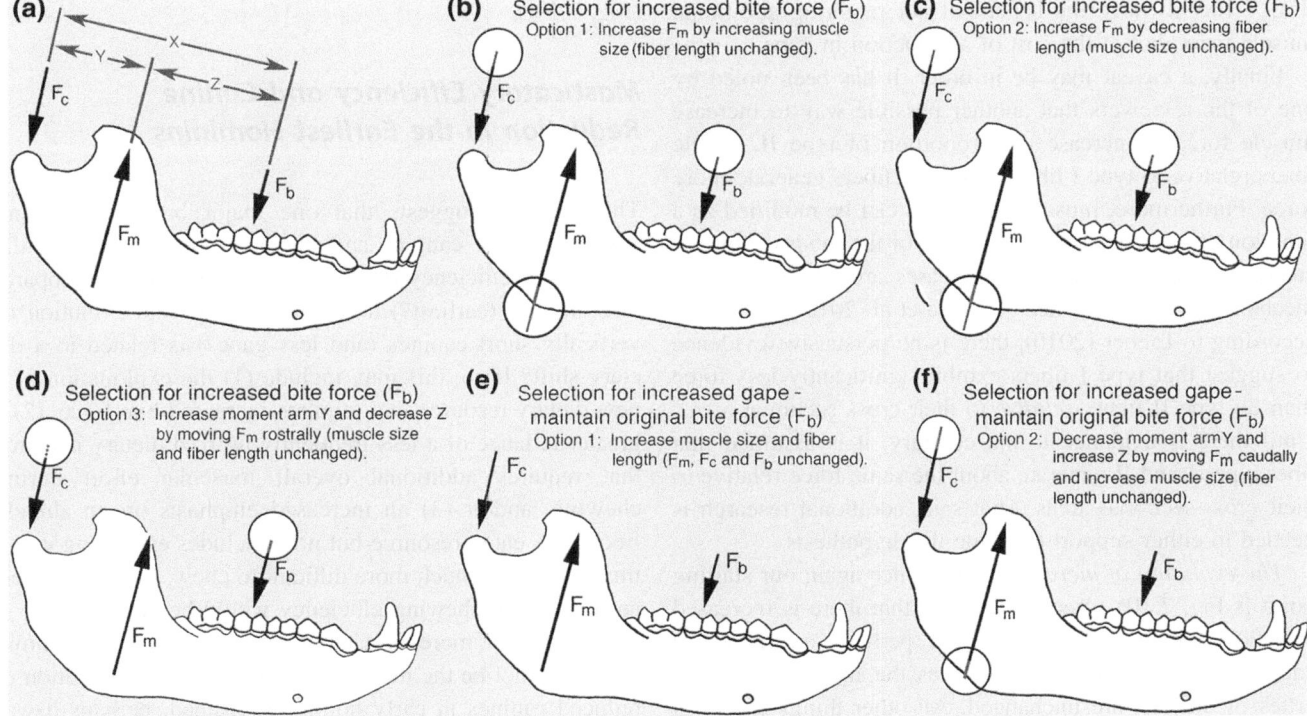

Fig. 7.10 Jaw mechanics. **a.** Starting point. Variables *X, Y and Z* are relevant moment arm variables. **b.** Selection for increased bite force by increasing muscle mass. Variables X, Y and Z remain constant. **c.** Selection for increased bite force by decreasing muscle fiber length. Variables X, Y and Z remain constant. **d.** Selection for increased bite force by shifting the resultant muscle force rostrally. Variable X remains constant whereas Y is increased and Z is decreased. **e.** Selection for increased gape by increasing muscle fiber length. Variables X, Y and Z remain constant. **f.** Selection for increased gape by shifting the resultant jaw-closing muscle force caudally. Variable X remains constant whereas Y is decreased and Z is increased. (Redrawn and modified from Hylander 2013)

Although all of these options have the benefit of increasing F_b, their costs differ. The cost of Option 1 (Fig. 7.10b) requires the additional growth and maintenance of more muscle mass. Another cost may be the increased wear and tear of the articular tissues of the TMJ due to an increase in F_c. Although Options 2 and 3 (Fig. 7.10c and d) are spared the cost of increasing muscle mass, the cost is a reduction in jaw gape. This cost can be tolerated so long as it does not compromise canine function for gape displays and/or biting large objects such as conspecifics or predators. If importantly compromised, Option 1 is the only viable strategy. Furthermore, as in Option 1, an additional cost for Option 2 is that F_c is increased with the associated wear and tear of the joint articular surfaces. In Option 3, there is the cost of changing muscle attachment areas by shifting the jaw muscles rostrally. In this option, moments arms Y and Z from Fig. 7.10a are increased and decreased, respectively.

In summary, when the overall geometry of the mandible is unchanged, there are three options to increase bite force, and each is associated with various costs and benefits. Most importantly, Option 1 (Fig. 7.10b) has the benefit of maintaining the original gape, but at the cost of developing and maintaining additional muscle mass. Options 2 and 3 (Fig. 7.10c, d) have the benefit of not requiring additional muscle mass, but at the cost of a reduction in gape.

Finally, a caveat may be in order. It has been noted by one of the reviewers that another possible way to increase muscle force is increase the proportion of type II muscle fibers relative to type I fibers as type II fibers generate more force. Furthermore, muscle fiber types can be modified as a function of the mechanical properties of the foods ordinarily eaten, and that type II fiber increases are linked to more mechanically resistant diets (Ravosa et al. 2010). That said, according to Lieber (2010), there is no persuasive evidence to suggest that type I fibers exhibit significantly less force than do type II fibers *relative* to their cross sectional areas. Until there is evidence to the contrary, it is assumed that fiber types I and II generate about the same force relative to their cross-sectional areas. That said, additional research is needed to either support or refute this hypothesis.

The evolution of increased gape. Once again our starting point is Fig. 7.10a. Assume for now that there is increased selection for increased gape (and perhaps an increased canine overlap dimensions) whereas the mechanical properties of the diet are unchanged. All other things equal, in order to increase gape, there are two options. One is to increase the length of its jaw closing muscle fibers whereas the other is to shift F_m caudally. In both options, the cost is a decrease in F_b. This is because in the first option the PCSA of jaw muscle mass has been decreased due to increased muscle fiber length. Therefore, F_m, F_b and F_c are all reduced.

In the second option, F_b has also decreased because although the magnitude of F_m is unchanged, F_m has been shifted caudally, resulting in a smaller moment arm Y and a larger moment arm Z (see Fig. 7.10a). In this option, although F_b + F_c still equals F_m, F_b has decreased and F_c has increased.

Assume for now that a decrease in F_b is unacceptable because the diet is unchanged. In order to avoid a decrease in F_b, F_m must be increased, and this can only be accomplished by the cost of increasing additional muscle mass. In Option 1 (increasing muscle fiber length and muscle mass), the newly added muscle mass now preserves the original F_m, F_b and F_c (Fig. 7.9e) by maintaining the original PCSA. In Option 2 (shifting F_m caudally and increasing muscle mass), the newly added muscle mass increases the original PCSA and F_m, preserves the original F_b and increases F_c (Fig. 7.9f).

In summary, in order to increase gape as well as preserve the original F_b, there are two options, and both have the cost of developing and maintaining additional muscle mass. Option 1 (lengthen the jaw-muscle fibers and increase muscle mass) causes the magnitude of F_m, F_b and F_c to be identical to the starting point of Fig. 7.10a. Option 2 (shift F_m caudally and increase muscle mass) causes F_m and F_c to be increased, as well as the preservation of the original F_b (Fig. 7.10f).

Masticatory Efficiency and Canine Reduction in the Earliest Hominins

This analysis suggests that one major benefit to having vertically short canines and decreased gape increases the mechanical efficiency of the catarrhine masticatory apparatus. In early (earliest?) hominins, perhaps the evolution of vertically short canines (and less gape?) is related to a dietary shift? If so, this may include (1) the exploitation of a new dietary resource that requires increased bite force, (2) a greater reliance of a less frequently utilized dietary resource that requires additional overall muscular effort during chewing, and/or (3) an increased emphasis on an already frequently eaten resource but now includes exploiting it at a time when it is much more difficult to chew. In all instances, an increase in chewing efficiency would be beneficial.

Evolving a more efficient masticatory apparatus, however, may not be the ultimate factor driving the evolution of reduced canines in early hominins. Instead, perhaps it was driven by a shift in mating patterns that caused reduced competition with conspecifics, or a favorable shift in a decrease in predation. If so, a more efficient masticatory apparatus would be the same end result. This is because the original equilibrium between those selective forces favoring increased gape/elongated canines as opposed to those selective forces favoring decreased gape/shortened canines is disrupted. Relaxed selection for maintaining large canine overlap and gape in combination with the ever-present selection for a more efficient masticatory apparatus (presumably desirable for all mammals) leads (directional selection) to a new state of equilibrium, resulting in canine reduction, decreased gape and a more efficient masticatory apparatus.

There are, however, other factors possibly influencing canine reduction in early hominins. For example, perhaps the increased use of unmodified rocks or stones, as well as wooden clubs, served as weapons. Under this scenario, the routine use of these objects reduced selection to maintain the original canine overlap dimensions and gape, and thus tipped the balance in favor of reduced canines, decreased gape and increased chewing efficiency.

For obvious reasons, the fossil record does not allow us to choose unambiguously the most likely ultimate reason for canine reduction in the earliest hominins. Some would argue that this ambiguity is linked to whether or not *Ardipithecus ramidus* (as well as earlier hypothesized hominins such as *Sahelanthropus tchadensis, Ardipithecus kadabba*, and *Orrorin tugenensis*) (Pickford et al. 2002; Brunet et al. 2002, 2005; Haile-Selassie et al. 2004; White et al. 2009; Wood and Harrison 2011) is a basal hominin. If they are not hominins, the combination of reduced canines, transversely thick mandibular corpora and increased enamel thickness of the *Australopithecus anamensis/Australopithecus afarensis* lineage arguably provides support for a dietary shift involving the exploitation of foods that are much more difficult to chew.

It seems quite likely that *Ar. ramidus* is indeed an early (earliest?) hominin, as cogently and persuasively argued by White and colleagues, and if the last common ancestor of panins and hominins had relatively similar amounts of enamel thickness and transversely thick mandibular corpora (but not canine height reduction) as *Ar. ramidus* (White et al. 2009), then the ultimate cause of canine reduction in the earliest hominins is unclear. More to the point, however, is that Lovejoy's argument that canine reduction and pair bonding (monogamy) are functionally linked (Lovejoy 2009), is contradicted by what we see in "pair bonded" hylobatids in that both sexes have very elongated canines, and therefore pair bonding is not necessarily linked to canine reduction in males.

While a shift in diet, mating/predation patterns, or the habitual use of various objects as weapons have been previously proposed as important influences on canine reduction, my data showed suggest how a diet composed of very difficult-to-chew food items may have promoted canine reduction (and reduced canine sexual dimorphism). For example, if the earliest hominins prove to have been committed to a diet of underground storage organs (USOs), as many have argued (see Dominy et al. 2008), my analysis clearly indicates that reduced gape and then canine reduction would follow, leading to a mechanically more efficient chewing apparatus. Of course if canine reduction and decreased gape were incompatible with mating/predation patterns in the earliest hominins, then arguably the only option would have been to increase the amount of jaw muscle mass (and PCSA) so as to increase bite force for masticating these tougher or more obdurate food objects.

Parenthetically, in an important study by Taylor and Vinyard (2009), compared to untufted capuchins (*Cebus capucinus* and *C. albifrons*), *Cebus apella* (a tufted capuchin) are said to solve the problem when ingesting and biting hard objects without decreasing gape by simply increasing the relative amount of temporalis (and masseter?) force. In this study, however, there are no measures of maximum gape for capuchins. Moreover, these authors have combined male and female capuchins in their analysis of muscle morphology. Making the distinction between males and females would have been helpful so as to determine if there are predictable differences in gape (and canine overlap) and muscle fiber length within and between these capuchin species. Of course these criticisms are after the fact as my analysis of gape and canine overlap was published several years later (Hylander 2013).

My unpublished data indicate that male and female *Cebus apella* exhibit the predicted differences in gape. That is, *Cebus apella* males with their greater relative canine overlap values (0.25; n = 5), open their jaws about 88% of jaw length (n = 5), whereas female relative canine overlap values (0.15; n = 6), open their jaws (n = 3) about 81% of jaw length. Thus, apparently there are differences in relative gape and perhaps muscle architecture differences between male and female *Cebus apella*. Overall, I am convinced that Taylor and Vinyard (2009) are correct in stating that *Cebus apella* has relatively larger jaw-closing muscles compared to other capuchins. On the other hand, linking relative jaw gape and relative canine overlap with their muscle morphology data would perhaps have been even more insightful.

In summary, the data and analysis presented here suggests that the relatively short vertical canine dimensions for the earliest hominins are arguably linked to small gape, increased mechanical advantage of the jaw closing muscles and increased masticatory efficiency. Furthermore, presumably increased selective pressures for decreased gape (and increased masticatory efficiency) in the earliest portion of the hominin lineage is the driving force for canine reduction. That is, canine reduction in its vertical dimensions (along

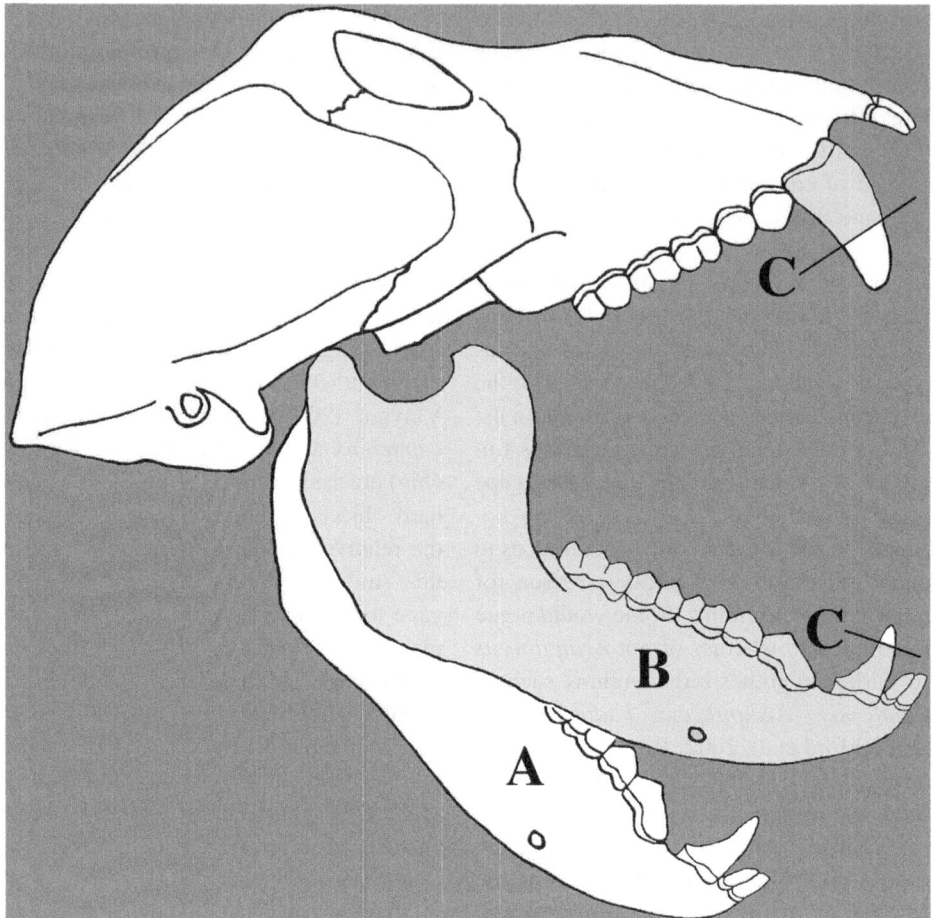

Fig. 7.11 Decreased gape and canine height reduction in catarrhines. *A*. Average maximum gape in a model species (e.g., long-tailed macaques). *B*. Average maximum gape in a second model species (e.g., Japanese macaques). For purposes of increasing masticatory efficiency, assume that species A reduces its maximum gape to that seen in B. If this were to occur, it follows that the gape between upper and lower canines would now be much smaller, and arguably less suitable for inflicting a punishing bite on both conspecifics and predators. One way to partially counter the newly restricted canine gape dimension in B is to reduce its upper and lower canine heights. The line *C* indicates the new location of the tips of the now reduced upper and lower canines. The differences in canine height and maximum gape here closely follows that seen in long-tailed macaques compared to Japanese macaques. See text for further discussion

with decreased gape) is advantageous so as to maintain sufficient space between the tips of the upper and lower canines during wide opening so as to facilitate and continue to inflict a punishing bite on conspecifics and/or predators. Furthermore, and as an example, assume for now that the primitive condition for macaques is to have very elongated canines as in male long-tailed macaques and pigtailed macaques, and the derived condition is in male Japanese macaques (Fig. 7.11).

Finally, this overall proximate explanation is not mutually exclusive relative to other suggestions for additional benefits of canine reduction, although some are more plausible than others (see review in Plavcan 2001). One of the more plausible suggestions for canine reduction in early hominins is often referred to as the "rotary chewing" hypothesis (Jolly 1970). This hypothesis states that in early hominins, canine height reduction allows or facilitates an increase in transverse movements of the teeth and jaws during the power stroke of mastication, and this in turn may result in a more effective mechanical solution for increasing particle-size reduction of newly added difficult-to-chew food items. *Australopithecus boisei* is perhaps a good example of the culmination of this trend in having very small non-projecting canines, flat postcanine occlusal surfaces, large jaw muscles, very thick enamel, robust jaws and rostrally positioned jaw-closing muscles.

The rotary chewing hypothesis, however, cannot account for much of the variability seen in canine vertical (overlap) dimensions in catarrhines. For example, although male pig-tailed macaque mandibles are about 10% longer than in male Japanese macaques, the amount of canine overlap in pigtails is almost 100% larger (Table 7.2). It is unlikely that the much lesser (but substantial) amount of canine overlap in male Japanese macaques is somehow linked to an increase in transverse movements of the teeth and jaws during chewing, simply because their canines still extend well beyond the occlusal plane. On the other hand, perhaps once the canines of early hominins reached a certain level of canine overlap reduction, selection could then proceed to favor an additional reduction of canine overlap (as in *Au. boisei*) so as to increase the amount of transverse chewing movements (Rak and Hylander 2008), as well as continuing to select for decreased gape and increased masticatory efficiency.

Relative Muscle Size, Position and Fiber Length in a Model Dimorphic Catarrhine Species: Macaca fascicularis (long-tailed macaques)

In 2013 (Hylander), I noted *"It is clear that there is a considerable amount of research to be done. For example, in order to understand better why female catarrhines have less relative gape than do conspecific males, more work needs to be done on analyzing jaw muscle architecture and position."*

With that in mind, Terhune et al. (2015) elected to use *Macaca fascicularis* as a model species so as to ask the following question. How do male long-tailed macaques relative to female long-tailed macaques manage to have increased gapes while presumably at the same time do not sacrifice the amount of bite force magnitude? That is, do the males shift their jaw closing muscles relatively more caudally towards the jaw joint so as to increase gape, or do they increase their jaw muscle fiber lengths so as to increase jaw gape, or both? Of course either of these strategies requires the hypothesized obligatory addition of more muscle mass and force so as to maintain the equivalent and necessary amount of bite force (see Fig. 7.10 and previous discussion).

As it turns out, the biomechanical solution for male long-tailed macaques is to opt for both strategies for muscle position and fiber lengths, along with the obligatory increase in the amount of muscle mass. Most interestingly, the masseter muscle is shifted caudally (to increase gape) whereas the temporalis muscle fibers are lengthened as well as a large increase in temporalis muscle mass (to increase gape and force, respectively). Although not analyzed, presumably the medial pterygoid is behaving in a fashion similar to the masseter. Finally, as outlined earlier (Fig. 7.10), this is not the only possible strategy available to macaques and other catarrhines. Whatever the case, perhaps the long-tailed macaque strategy is typical for cercopithecoids, as well as all great apes? Only additional research can shed light on this matter.

Why Is It that Relative Gape in Male Catarrhines does not Correspond to Expected Relative Gape in Conspecific Female Catarrhines?

As noted in the results section, Fig. 7.8 is a plot of actual jaw gape versus predicted jaw gape. Recall here that gape predictions are based on a multiple correlation of two independent variables, projected jaw length and canine overlap. This figure (males and females combined) demonstrates that a surprisingly large amount of gape is predicted (raw $r^2 = 0.89$ and IC $r^2 = 0.87$). Importantly for here, multiple correlations were also performed for predicting catarrhine gapes separately by sex. For male catarrhines, the raw and IC r^2 values are 0.90 and 0.86, respectively, whereas for females they are 0.73 and 0.71. Although all values are statistically significant, note that males consistently have larger r^2 values than do females.

Also as noted in the RESULTS section, Fig. 7.9 is a plot of *relative* projected gape versus *relative* canine overlap. Again, there are strong and significant correlations (raw and IC $r^2 = 0.71$ and 0.57, respectively. Similarly, males and females were also analyzed separately. In this case, the male r^2 values are 0.71 and 0.60, respectively, whereas the female values are 0.64 and 0.33, respectively. As before, all values are significantly different from zero ($p < 0.002$).

For here, in the interests of avoiding repetition, let's simply focus on the raw and IC (independent contrasts) r^2 values for data from Fig. 7.9. The interesting question here is why is it that the r^2 values for catarrhine males (0.71 and 0.57, respectively) are so much larger than for females (0.64 and 0.33, respectively)? Importantly, a perusal of the data in Tables 7.1 and 7.2 indicate that for catarrhine females there is a greater amount of variation for relative gape versus relative canine overlap, and this is reflected by the lower r^2 values. For example, *Papio hamadryas* females maximally open their jaws 86% of jaw length, whereas *Pan troglodytes* females open their jaws only 65% of jaw length. On the other hand, for these same female species, they both have near identical values of relative canine overlap (i.e., 7.5% and 7.4% of jaw length, respectively). Interestingly, many other female catarrhines also have relatively small values of relative canine overlap (see *Papio anubis*, *Mandrillus sphinx*, *Cercocebus atys*, all 5 species of macaques analyzed, etc.), and most of these females have relatively large gape.

As these female catarrhines have small relative values of canine overlap, do these differences in gape reflect differences in relative food object size during ingestion? That is, for example, do female *Papio hamadryas* baboons ingest much larger food objects than do female *Pan troglodytes*, and therefore this accounts for their relatively large gape? I suspect that this is not the case. Instead, rather than invoking dietary issues, there may be other reasons why females appear to be tracking gapes similar to what is seen in their

conspecific males. That said, perhaps what we are seeing here can be referred to as a "correlated response" (cf. Lande 1980; Plavcan 1998). Although this explanation strikes me as plausible, it is not very satisfying as testing this hypothesis is difficult (and unclear to me as to how to do so).

Reconsidering the High Mandibular Condyle of Catarrhines (and Robust australopiths)

Years ago, Smith and Savage (1959) suggested that high mandibular condyles of herbivores are more mechanically efficient for the medial pterygoid and masseter muscles than are low-positioned condyles, because the high condyle is linked to larger jaw-closing moment arms (and moments) of these muscles during chewing. Conversely, carnivorans are

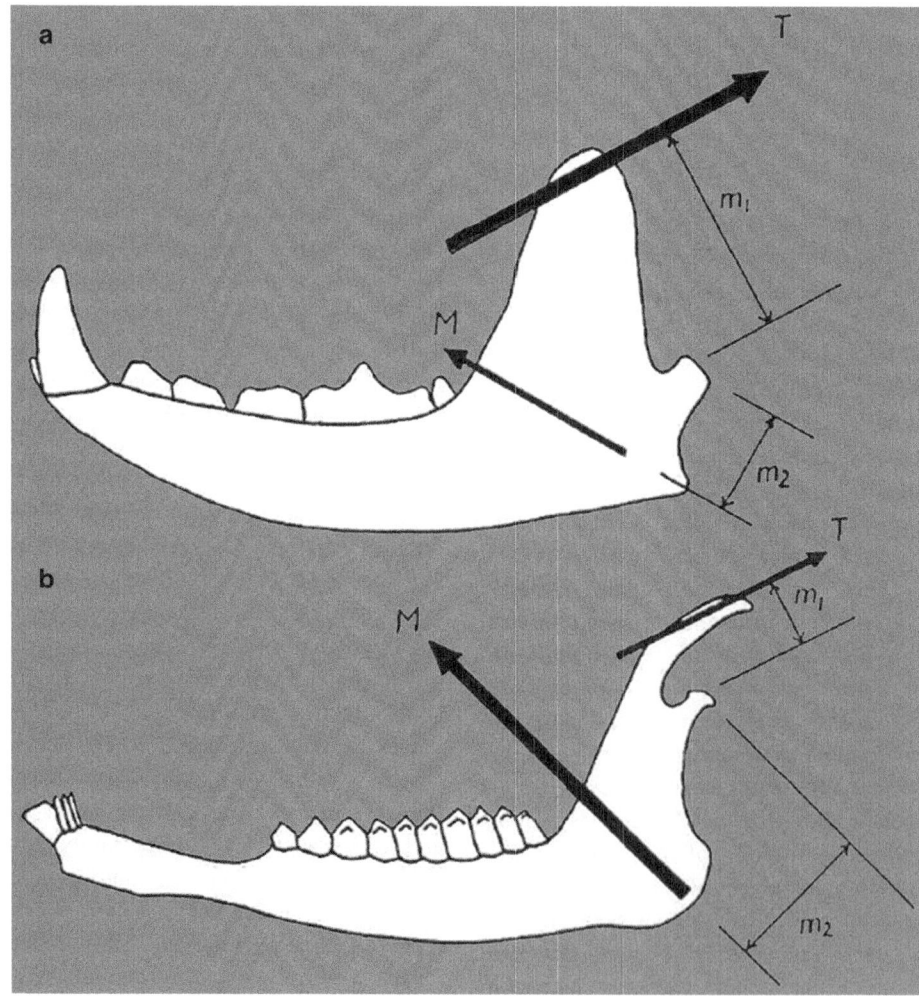

Fig. 7.12 A typical carnivoran in **a** and a typical bovid or cervid in **b**. *M* and *T* indicate resultant force for the masseter and temporalis muscles, respectively; m_1 and m_2 indicate moment arms for the temporalis and masseter muscles, respectively. Redrawn and modified from Smith and Savage (1959)

said to be more mechanically efficient for the temporalis muscles because of their relatively high coronoid processes well above the condyle (which is located at or near the occlusal plane). They go on to argue that although the dominant muscle mass for herbivores is the medial pterygoid and masseter muscles, the dominant muscle mass of carnivorans is the temporalis muscle. Furthermore, they imply whereas the arrangement of the jaw muscles in herbivores are well suited for chewing on their postcanine teeth, the arrangement of the jaw muscles in carnivorans are well suited for canine biting while subduing a struggling prey (Fig. 7.12).

As a beginning graduate student many years ago, all of the above made sense. Upon years of further reflection, however, it is clear that there are muddles in the Smith and Savage models. That is, carnivorans surely have large gapes, and although their moment arms may be large for the temporalis, clearly these same muscles must be designed for considerable stretch so as to facilitate a large gape. That said, it must take substantial increases in temporalis muscle mass so as to increase PCSA to increase muscle and bite force. This in turn brings us back to the necessity of considering muscle architecture, and not just muscle moment arms and muscle mass (Taylor and Vinyard 2009).

More recently and many others (see Rak and Hylander 2008 for references) have considered additional competing hypotheses regarding condylar position, with a particular emphasis on robust australopiths (and other catarrhines). Nevertheless, my purpose here is not to review various competing hypotheses for condylar position in various mammals. Instead, I'll simply focus on observations regarding maximum relative gape data for baboons and geladas (Hylander 2013), and these observations are arguably relevant for understanding why or why not robust australopiths have highly positioned mandibular condyles.

It is well known that relative to the occlusal plane, baboons have low-positioned condyles, whereas geladas have high-positioned condyles (Fig. 7.13). Following conventional wisdom, high condyles are presumably linked to larger muscle moments or moment arms for the medial pterygoid and masseter muscles as well as increased bite force, whereas low condyles are linked to smaller muscle moments or moment arms and decreased gape and bite force. Importantly, and all things considered equal, larger moments or moment arms for the medial pterygoid and masseter muscles should be associated with *less gape*, whereas smaller ones should be associated with *more gape*. Of course for carnivorans, the converse must also be the case. That is, higher coronoids well above the occlusal plane among carnivorans presumably are also linked to less gape. Intuitively, considering what carnivorans do with their teeth and jaws, that doesn't make sense. Whatever the case, lets move on to catarrhines.

Surprisingly, relative gape (maximum gape/projected jaw length) in baboons and geladas are near identical (Table 7.1), in spite of very different condylar positions. Relative gape values for male and female baboons and geladas are as follows: *Papio anubis* 1.12 and 0.87, *Papio hamadryas* 1.03 and 0.86, and *Theropithecus gelada* 1.05 and 0.90, respectively. Contrary to expectations, relative gape values (for each sex separately) are more or less similar. That is, among these 3 species, although *Papio anubis* males have relatively larger relative gapes than the other males, female *Theropithecus gelada* have larger relative gapes than the other females. Thus, among these catarrhines, these data do not support the hypothesis that high condyles are necessarily linked to decreased gape (and mechanical efficiency).

As noted earlier, among catarrhines, relative canine height above the occlusal plane (canine overlap) is intensely

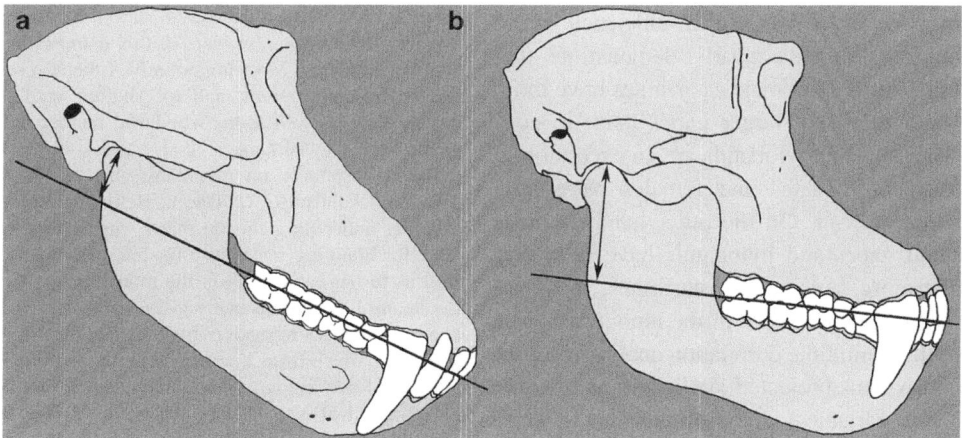

Fig. 7.13 Condylar position in baboons (**a**) and geladas (**b**). *Note that condyle heights about the occlusal plane* vary between **a** and **b**. That is, condylar height in geladas (**b**) is about twice as large as in baboons (**a**)

linked to relative gape (Hylander 2013). That said, perhaps there are other benefits for the high condylar position of robust australopithecines (and other mammals?) (Hylander 2015). Furthermore, as discussed earlier, it is interesting that although *Theropithecus* and *Papio* females (and macaques) have very little canine overlap dimensions, they have a relatively large amount of relative gape. Most importantly for this discussion, it is quite clear that most ideas about condylar position and jaw mechanics have been based on the erroneous and implicit assumption that muscle fiber lengths are more or less equal (Hylander 2015; Terhune et al. 2015). Until we have a clear idea about jaw-muscle fiber length, PCSA and muscle geometry among extant catarrhines, we are not in a strong position to make inferences about the details of their jaw mechanics based on jaw morphology alone. That said, catarrhine jaw mechanics analyses based on fossil or museum osteological specimens should be considered with a distinct warning label. That is, until we know more about muscle morphology (jaw muscle position), architecture (fiber length and angles of pinnation), PCSA and jaw gape in extant primates, all bets are off when dealing with fossil primates. This suggestion has been implied by others, but most cogently expressed by Taylor and Vinyard (2009: 718) when they noted, "Finally, this implication reinforces the utility and importance of analyses comparing living species that can incorporate soft tissues as well as behavioral information for inferring past life ways in extinct taxa known only from skeletal morphology (e.g., Witmer 1995)."

Conclusions

This study reviews the hypothesis that canine reduction in living catarrhines is linked to the amount of jaw gape, and therefore has an impact on the mechanical efficiency of the masticatory apparatus. The data clearly demonstrate that relative to jaw length, most adult male catarrhines have more elongate canines and relatively larger gapes than do conspecific females. Humans and hylobatids are an exception to this rule, i.e., conspecific males and females have little canine and gape dimorphism. On the other hand, humans have relatively small gapes and hylobatids have relatively large gapes. Furthermore, among all catarrhines, there are considerable interspecific differences in the amount of gape relative to jaw length. A multiple correlation analysis of gape (dependent variable) versus projected jaw length and canine overlap (independent variables) demonstrates that a large amount of gape is predicted by these two independent variables. Similarly, a large portion of projected relative gape is predicted by relative canine overlap.

Relative maximum gape (gape/projected jaw length) must be largely a function of jaw-adductor muscle position and/or muscle-fiber length, and if so, there are important costs and benefits linked to modifying these muscle characteristics. All things equal, more rostrally positioned jaw muscles and/or shorter muscle fibers decrease gape, and the benefit is to increase bite force and therefore, the mechanical efficiency of the jaws. In contrast, more caudally positioned jaw muscles and/or longer muscle fibers increase the amount of gape (for elongate canines), but at the cost of reducing mechanical efficiency. Overall, this analysis provides support for the hypothesis that a major proximate benefit for canine height reduction in early hominins is functionally linked to increased mechanical efficiency of the jaws. In addition, as canine height has been reduced in early hominins, this arguably is due to either reduced selection for canine height, positive selection for reduced gape, or some combination of these two factors. Furthermore, it is hypothesized that the driving force for canine height reduction is gape reduction. If canine reduction in height does not occur, this may constrain inflicting deep, painful and discouraging bites on conspecifics and predators. The ultimate reason for reduced canines in the earliest hominins, however, is unclear. If the *Australopithecus anamensis/afarensis* lineage is representative of the earliest hominins, the data and analysis here arguably favors the dietary-shift hypothesis. On the other hand, if *Ardipithecus ramidus* is an early hominin, the ultimate reason for canine reduction is likely due to some combination of a shift in diet, mating patterns, predation patterns, or the habitual use of various objects (but not recognizably stone tools) as weapons.

Acknowledgments I thank Matt Cartmill, Yoel Rak, Mike Plavcan, Carel van Schaik, Chris Vinyard, Christine Wall, Tim White and Richard Wrangham for their comments and suggestions on an earlier published version of this study (Hylander 2013). I also thank Andrea Taylor and Chris Vinyard for help and comments, as well as Matt Ravosa, Erella Hovers and Assaf Marom for their editorial suggestions, as well as the other reviewers of this manuscript for their helpful and cogent comments. Most importantly, I thank Yoel Rak for his continued friendship, support, and for sharing with me his keen eye for morphology, as well as his wonderful memory of his many and hilarious Israeli (Jewish) jokes. I also thank him for his wonderful and repeated hospitality on my many visits to Israel. His gracious and wonderful family (Ricka, Carmi, Benjamin and Ariel) always seemed to suffer under the delusion that I was a *"real mensch"*. I also thank Yoel for honoring my family for his many visits to Durham (NC), as well as to our cattle farm in the mountains of Virginia. My dear wife Linda and I, during his many visits, frequently wonder about how long one human needs to recover from jet lag. Over the past 15 years I have learned so much from Yoel. In addition, on many of our research trips, Yoel and I may not have not gotten much work done, but we did have a lot of fun. I also thank him for teaching me about Neandertal morphology, most notably during our many trips together to central and western Europe. Probably the best thing he taught me on these trips was where to find the best prosciutto and wine. Thanks Yoel, you dah man. We all love you.

References

Arnold, C., Matthews, L. J., & Nunn, C. L. (2010). The 10K trees website. A new online resource for primate phylogeny. *Evolutionary Anthropology, 19*, 114–118.

Brace, C. L. (1963). Structural reduction in evolution. *American Naturalist, 97*, 39–49.

Brunet, M., Guy, F., Pilbeam, D., Lieberman, D., Likius, A., Mackaye, H., et al. (2002). A new hominid from the Upper Miocene of Chad, Africa. *Nature, 418*, 145–152.

Brunet, M., Guy, F., Pilbeam, D., Lieberman, D., Likius, A., Mackaye, H., et al. (2005). New material of the earliest hominid from the Upper Miocene of Chad. *Nature, 434*, 752–755.

Darwin, C. (1871). *The descent of man, and selection in relation to sex.* New York: D. Appleton and Co.

Dominy, N. J., Vogel, E. R., Yeakel, J. D., Constantino, P., & Lucas, P. W. (2008). Mechanical properties of plant underground storage organs and implications for dietary models of early hominins. *Evolutionary Biology, 35*, 159–175.

Greenfield, L. O. (1992). Origin of the human canine. A new solution to an old enigma. *Yearbook of Physical Anthropology, 35*, 153–185.

Haile-Selassie, Y., Suwa, G., & White, T. (2004). Late Miocene teeth from Middle Awash, Ethiopia, and early hominid dental evolution. *Science, 303*, 1503–1505.

Holloway, R. L., Jr. (1967). Tools and teeth: Some speculations regarding canine reduction. *American Anthropologist, 69*, 63–67.

Hylander, W. L. (2013). Functional links between canine height and jaw gape in Catarrhines with special reference to early hominins. *American Journal of Physical Anthropology, 150*, 247–259.

Hylander, W. L. (2015). Reconsidering the high mandibular condyle of robust australopiths. *American Journal of Physical Anthropology, Supplement, 60*, 174.

Jolly, C. J. (1970). The seed-eaters: A new model of hominid differentiation based on a baboon analogy. *Man, 5*, 5–26.

Lande, R. (1980). Sexual dimorphism, sexual selection and adaptation in polygenic characteristics. *Evolution, 34*, 292–307.

Leigh, S. R. Setchel, J. M., Charpentier, M., Knapp, L. A., & Wickings, E. J. (2008). Canine tooth size and fitness in mandrills (*Mandrillus sphinx*). *Journal of Human Evolution, 55*, 75–85.

Lieber, R. L. (2010). *Skeletal muscle function, structure and plasticity* (3rd ed.). Philadelphia: Lippincott Williams and Wilkins, Wolters Kluwer Co.

Lovejoy, C. O. (2009). Reexamining human origins in light of *Ardipithecus ramidus*. *Science, 326*, 74.

Maddison, W. P., & Maddison, D. R. (2010). http://mesquiteproject.org.

Midford, P. E, Garland, T. Jr., & Maddison, W. P. (2005). PDAP Package of Mesquite. Version 1.07.

Pickford, M., Sénut, B., & Treil, J. (2002). Bipedalism in *Orrorin tugenensis* revealed by its femora. *Comptes Rendus Palevol, 1*, 191–203.

Plavcan, J. M. (1990). Sexual dimorphism in the dentition of extant anthropoid primates. PhD Dissertation, Duke University.

Plavcan, J. M. (1998). Correlation response, competition and female canine size in Primates. *American Journal of Physical Anthropology, 107*, 401–416.

Plavcan, J. M. (2001). Sexual dimorphism in primate evolution. *Yearbook of Physical Anthropology, 44*, 25–53.

Plavcan, J. M., & van Schaik, C. P. (1997). Interpreting hominid behavior on the basis of sexual dimorphism. *Journal of Human Evolution, 32*, 345–374.

Rak, Y., & Hylander, W. L. (2008). What else is the tall mandibular ramus of the robust australopiths good for? In C. J. Vinyard, M. J. Ravosa, & C. E. Wall (Eds.), *Primate craniofacial function and biology* (Developments in Primatology Series) (pp. 431–442). New York: Springer.

Ravosa, M. J., Nin, G. J., Costley, D. B., Daniel, A. N., Stock, S. R., & Stack, M. S. (2010). Masticatory biomechanics and masseter fiber-type plasticity. *Journal of Musculoskeletal and Neuronal Interactions, 10*, 46–55.

Smith, J. M., & Savage, R. J. G. (1959). The mechanics of mammalian jaws. *School Science Review, 40*, 389–401.

Taylor, A. B., & Vinyard, C. J. (2009). Jaw-muscle fiber architecture in tufted capuchins favors generating relatively large muscle forces without compromising gape. *Journal of Human Evolution, 57*, 710–720.

Terhune, C. E., Hylander, W. L., Vinyar, C. J., & Taylor, A. B. (2015). Jaw-muscle architecture and mandibular morphology influence relative maximum gapes in sexually dimorphic *Macaca fascicularis*. *Journal of Human Evolution, 82*, 145–158.

Washburn, S. L. (1968). On Holloway's "Tools and teeth". *American Anthropologist, 70*, 97–101.

White, T. D., Asfaw, B., Beyene, Y., Hailie-Selassie, Y., Lovejoy, C. Suwa, O., et al. (2009). *Ardipithecus ramidus* and the paleobiology of early hominids. *Science, 326*, 64–86.

Witmer, L. M. (1995). The extant phylogenetic bracket and the importance of reconstructing soft tissues in fossils. In J. J. Thomason (Ed.), *Functional morphology in vertebrate paleontology* (pp. 19–33). Cambridge: Cambridge University Press.

Wood, B., & Harrison, T. (2011). The evolutionary context of the first hominins. *Nature, 470*, 347–352.

Chapter 8
Paranthropus: Where Do Things Stand?

Bernard Wood and Kes Schroer

Abstract In 1960 John Robinson suggested that the newly defined species *Zinjanthropus boisei* should be transferred to the genus *ParanthropusParanthropus* (Broom 1938) as *Paranthropus boisei* (Leakey 1959). Since then fossil evidence of two hyper-megadont early hominin taxa has come to light. One of these taxa, *Paraustralopithecus aethiopicus* (Arambourg and Coppens 1968), has been added to the *Paranthropus* genus, whereas the second taxon, *Australopithecus garhi* (Asfaw et al. 1999), has been included in a different taxon, *Australopithecus*. This contribution will tease out why different alpha-taxonomic decisions were made about the generic affinities of *Paraustralopithecus aethiopicus* and *Australopithecus garhi*. It will also review the types of data that are now available for generating and testing hypotheses about the relationships of megadont and hyper-megadont hominins. On the basis of this review, in this paper we will suggest a hypothesis, or hypotheses, that are most consistent with the current fossil and contextual data from East and southern Africa.

Keywords Analogy • Biogeography • Convergence • Homoplasy • Eastern Africa • Megadontia • Southern Africa

Introduction

In the 1970s *Paranthropus* had been all but abandoned as a hominin taxon. Many researchers familiar with the early hominin fossil record, including the dedicatee of this volume (e.g., Rak et al. 2007), do not recognize a separate genus for hypodigms they refer to as *Australopithecus robustus* and *Australopithecus boisei sensu lato* [i.e., the combined hypodigms of *Australopithecus boisei* (Leakey 1959) *sensu stricto* and *Australopithecus aethiopicus* (Arambourg and Coppens 1968)]. But some researchers, including the authors, maintain that the morphologies of these early hominins cannot be comfortably accommodated within the genus *Australopithecus*. This contribution reviews the fossil evidence for early hominins with wide faces and especially large postcanine tooth crowns [hereafter referred to as 'megadont' (i.e., *Paranthropus robustus*) and 'hyper-megadont' (i.e., *Paranthropus boisei* and *P. aethiopicus*)] hominins, examines why and how the genus *Paranthropus* was established and why some researchers have revived it, and, finally, the strengths and weaknesses of the case for continued use of the genus *Paranthropus*. We have not provided citations for the section covering the fossil evidence; the relevant references can be found in Wood and Constantino (2007) and Wood and Schroer (2013).

Fossil Evidence

Southern Africa

The first evidence of hominins with wide, flat faces, large and robust mandibular corpora and especially large (i.e., megadont) postcanine tooth crowns was the TM 1517 cranium recovered in 1938 from the cave site of Kromdraai in the Blaauwbank Valley, South Africa. The first discoveries of similar-looking hominins from Swartkrans, another breccia-filled cave complex close by in the same valley,

B. Wood (✉)
CASHP, George Washington University, Washington, DC 20052, USA
e-mail: bernardawood@gmail.com

K. Schroer
Neukom Institute Fellow in Computational Sciences and Department of Anthropology, Dartmouth College, Hanover, NH 03755, USA
e-mail: kes.schroer@gmail.com

© Springer International Publishing AG 2017
Assaf Marom and Erella Hovers (eds.), *Human Paleontology and Prehistory*,
Vertebrate Paleobiology and Paleoanthropology, DOI: 10.1007/978-3-319-46646-0_8

were made in 1948 and since then, more than 400 hominin fossil specimens representing ca. 150 individuals have been recovered in breccia dumps, or *in situ*, at Swartkrans. A third cave, Drimolen, is close by and is the second largest source of megadont hominins in southern Africa after Swartkrans. The Drimolen hominin sample includes a well-preserved skull, DNH 7, a mandible with an almost complete dentition, DNH 8, and an unusual number of immature individuals. The non-metrical morphology of the Drimolen dental remains has been interpreted as being intermediate between that of Swartkrans and Kromdraai. Two other sites in the Blaauwbank Valley, Cooper's Cave and Gondolin, have also yielded evidence of megadont early hominins. It has been suggested that the same hominin taxon, or its precursor, has been sampled at Sterkfontein, also in the Blaauwbank Valley, but other researchers who have carried out a careful analysis of the collection disagree (Table 8.1).

The best estimates of the first and last appearance dates of the megadont hominins from southern Africa comes from Swartkrans. Direct uranium-lead dating of the flowstone layers above and below the Hanging Remnant and Lower Bank deposits at Swartkrans gives an age of ca. 2 Ma for Member 1 at Swartkrans and contemporaneous deposits across the sites, thus providing a first appearance date for

megadont hominins. The most recent evidence of megadont hominins in southern Africa comes from Member 3 at Swartkrans, and faunal and other evidence suggests a last appearance datum of ca. 1 Ma.

East Africa and Malawi

In East Africa, there is evidence of early hominins with even larger postcanine tooth crowns than *P. robustus*, so large that we refer to them as hyper-megadont. These unusually large tooth crowns are combined with small incisors, a small canine, and especially large and robust mandibular bodies. The first evidence of these hyper-megadont hominins consisted of a large deciduous molar, OH 3, recovered in 1955 from locality BK in Lower Bed II at Olduvai Gorge in Tanzania. It puzzled researchers, but its significance became clearer in 1959 when a well-preserved sub-adult cranium, OH 5, with massive postcanine tooth crowns and diminutive anterior teeth was recovered from locality FLK in Bed I at Olduvai Gorge. Four years later, a well-preserved adult mandible whose dentition, based on absolute and relative size, matched the dentition of OH 5 was recovered from Peninj just north of Olduvai Gorge, also in Tanzania.

Table 8.1 Timeline of important events in the discovery and analysis of the fossil evidence of *Paranthropus aethiopicus*, *Paranthropus boisei*, and *Paranthropus robustus*. After Wood and Schroer (2013)

1938	Recovery of TM 1517 from Kromdraai and its publication by Robert Broom as the holotype of *Paranthropus robustus*	1967	Revecory of the first hyper-megadont postcanine teeth from the Omo-Shungura Formation, in the following year Arambourg and Coppens assign to the new taxon *Paraustralopithecus aethiopicus*aethiopicus
1939	A single tooth was found at Cooper's Cave. Fossils found since have been assigned to *P. robustus*	1969	Recovery of the KNM-ER 406 cranium from the site that was then known as East Rudolf
1949	Recovery of SK 6 from Swartkrans and its publication by Robert Broom as the holotype of *Paranthropus crassidens*	1971	A partial face from Chesowanja (KNM-CH 1)is categorized as a possible female specimen of *P. boisei*
1952	Publication of the *Swartkrans Ape-Man* monograph by Robert Broom and John Robinson	1973	A partial cranium from Koobi Fora (KNM-ER 732) is recognized as confirmatory evidence of substantial size and shape sexual dimorphism in *P. boisei*
1955	Recovery of OH 3 from Olduvai Gorge, with hindsight the first *Paranthropus* specimen to be discovered in East Africa	1985	Recovery of the first and only well-preserved crania of *P. aethiopicus* (KNM-WT 17000) from West Turkana in Kenya
1958	Publication of John Robinson's monograph the *Dentition of the Australopithecinae* that spelt out the dental differences between *P. robustus* and *Australopithecus africanus*	1993	Publication of Bob Brain's monograph on the site and hominin fossil evidence from Swartkrans. Recovery of the first well-preserved skull of *P. boisei* (KGA 10-525) from Konso in Ethiopia, published in 1997
1959	Mary Leakey discovers the remains of OH 5 at FLK in Olduvai Gorge and its publication by Louis Leakey as the holotype of *Zinjanthropus boisei*	1994	Recovery of the first well-preserved skull of *P. robustus* (DNH 7) from Drimolen in South Africa
1960	John Robinson first uses the name combination *Paranthropus boisei*	1999	Publication of the first hominid teeth recovered from Gondolin
1964	Kamoya Kimeu recovers a remarkably well preserved mandible from Peninj that matches the OH 5 cranium	1999	A maxillary fragment from the site of Malema, Malawi is provisionally assigned to *P. boisei*, greatly expanding the known range of this taxon
1967	Publication of Phillip Tobias' seminal analysis of the OH 5 cranium		

Fig. 8.1 Map of the sites that contribute to the hypodigms of *Paranthropus aethiopicus*, *Paranthropus boisei* and *Paranthropus robustus*. Redrawn after Wood and Schroer (2013)

In 1967, a mandible, Omo 18-1967-18, with alveoli that suggested the postcanine teeth were large was recovered from Member C in the Shungura Formation in southern Ethiopia, and since then a fragmentary hyper-megadont cranium and several hyper-megadont mandibles and numerous isolated teeth have been recovered from the Shungura Formation. However, the largest collection of hyper-megadont crania and mandibles in East Africa comes from sites nearby on the eastern and western shores of Lake Turkana in northern Kenya. Two hemi-mandibles with robust bodies, KNM-ER 403 and 404, plus an abraded and edentulous palate, KNM-ER 405, were collected in 1968, and since then a succession of crania and calvariae (e.g., KNM-ER 406, 407, 732, 733, 13750, 23000) and mandibles (e.g., KNM-ER 729, 3230) have been recovered from what

was then known as East Rudolf and what is now called Koobi Fora, or East Lake Turkana. Morphologically similar cranial remains have also been found in sediments across that lake in a region known as West Turkana (e.g., KNM-WT 16005, 17000, 17400) (Fig. 8.1).

The next East African site to yield evidence of a hyper-megadont hominin was Chesowanja in Kenya, where in 1970 a right hemiface and anterior cranial base, KNM-CH 1, was recovered from the Chemoigut Formation. The morphology of the face and the absolute size and proportions of the dentition were judged to be similar to those of OH 5 and the Koobi Fora fossils. Further evidence from the Horn of Africa came in the early 1990s when a well-preserved skull, KGA 10-525, was recovered at Konso (initially called Konso Gardula) in Ethiopia, and subsequently a maxilla was

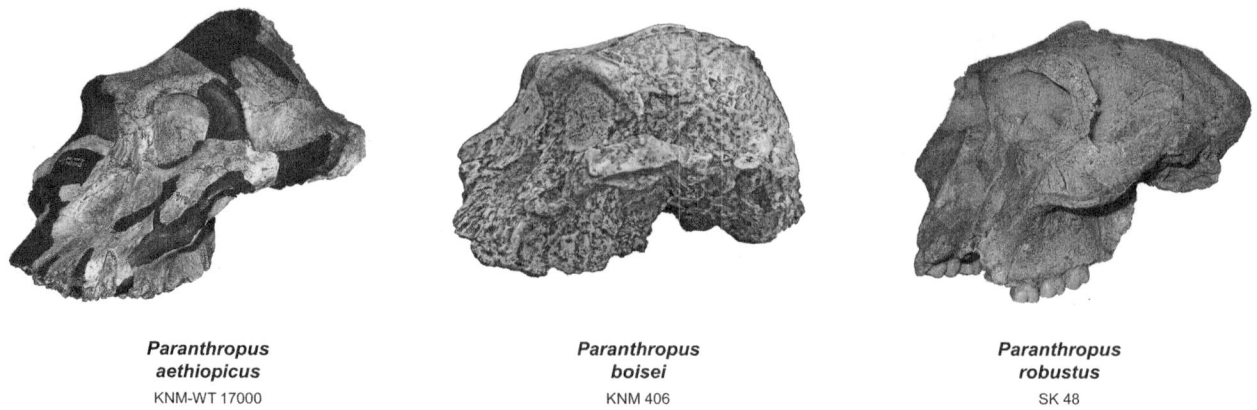

Paranthropus	Paranthropus	Paranthropus
aethiopicus	boisei	robustus
KNM-WT 17000	KNM 406	SK 48

Fig. 8.2 Left lateral views of the well-preserved holotype cranium of *Paranthropus aethiopicus*, and representative crania of *Paranthropus boisei* and *Paranthropus robustus*. Not to scale. Redrawn after Wood and Schroer (2013)

discovered at Malema in Malawi. The latter discovery was significant from a biogeographical standpoint because it extended the southern extent of the range of hyper-megadont hominins by more than five hundred miles.

The oldest well-preserved evidence of hyper-megadont hominins from East Africa comes from ca. 2.7−2.6 Ma strata in Member C at Omo-Shungura, and if a maxilla from the Ndolanya Beds at Laetoli is included in the hypodigm, then this would also point to an estimated first appearance date between 2.7 and 2.5 Ma. The lack of hyper-megadont hominins with small incisors and canines in the older sediments at Omo-Shungura and in the lower Lomekwi Member at West Turkana suggests that the ca. 2.7 Ma first appearance date of these hominins is likely to be close to the time of origin, or immigration, of the East African hyper-megadont hominins. The youngest known remains are most likely two isolated teeth recovered from Olduvai Gorge (OH 3 and 38) dating to ca. 1.3 Ma, or the remains from Konso in Ethiopia dated to ca. 1.4 Ma. However, because there are no major East African hominin sites in the period between ca. 1.3 and 1.0 Ma, we have no reliable information about how long these hominins might have persisted in East Africa beyond these last appearance dates.

From time to time, researchers have suggested that megadont hominins with large, robust mandibles have been found outside of Africa (e.g., Robinson 1954), but none of the candidates have turned out to match the distinctive morphology seen in early hominins found at sites in southern and eastern Africa (Fig. 8.2).

Taxonomy

When Broom (1938) announced and described the TM 1517 cranium from Kromdraai, he claimed that its shorter, flatter face, its small canines and incisors, and the differences in the size and shape of its molars and premolars compared to those of *Australopithecus africanus* from Taung and *Australopithecus transvaalensis* from Sterkfontein, were worthy of recognition at the generic level, so Broom designated TM 1517 as the holotype of a new genus and species, *Paranthropus Paranthropus robustus*. When the first megadont hominins were recovered in November 1948 from what was then called the "pink breccia" at Swartkrans, Broom (1949) designated the SK 6 mandible as the holotype of *Paranthropus crassidens*, but he gave no morphological reasons for making a specific distinction between the hominins from Swartkrans and Kromdraai. The initial species-level distinction between *P. crassidens* and *P. robustus* was soon amended to the subspecific level (Robinson 1954, 1956, 1968; Campbell 1963), and although Howell (1978) restored the specific distinction between the Kromdraai and Swartkrans samples and Grine (1985) described differences between the deciduous dentitions of the two samples, most researchers view the differences between the megadont hominins recovered from the two sites as consistent with variation *within* a single species rather than the type of variation found *between* species.

As for the taxonomy of the initial fossil evidence from East Africa, although OH 5 was initially placed in a novel genus and species, *Zinjanthropus boisei* Leakey 1959, five years later Louis Leakey and colleagues, without explanation, demoted *Zinjanthropus* to the level of a subgenus as *Australopithecus* (*Zinjanthropus*) (see Leakey and Leaky 1964), and not long afterwards one of those authors abandoned any generic distinction between *Zinjanthropus* and *Australopithecus* (Tobias 1967). Researchers now refer to the taxon as *Australopithecus boisei* or *Paranthropus boisei* (Table 8.2).

In his "preliminary diagnosis" of OH 5, Leakey (1959) drew attention to twenty distinctive features (e.g., malar morphology, the anterior accentuation of the sagittal crest, and the imbalance between the diminutive canines and the

Table 8.2 List of the sites and a summary of what fossil evidence they contribute to the hypodigms of *Paranthropus aethiopicus*, *Paranthropus boisei*, and *Paranthropus robustus*. After Wood and Schroer (2013)

Region	Site	Formation	Age of remains (Ma)	Dating method	Nature of the evidence	Taxa
Eastern Africa	Laetoli, Tanzania	Ndolanya	2.7–2.5	Radiometric	EP 1500/01 (maxilla)	*P. aethiopicus*
	Omo, Ethiopia	Shungura	2.6–2.3 2.3–1.2	Radiometric, magnetostratigraphy, tephrostratigraphy	Omo 18-18 (edentulous mandible; holotype of *P. aethiopicus*) and others, mostly isolated teeth Various specimens, mostly teeth	*P. aethiopicus* *P. boisei*
	West Turkana, Kenya	Nachukui	2.5–2.35	Radiometric, magnetostratigraphy, tephrostratigraphy	KNM-WT 17000 (cranium) KNM-WT16005 (mandible) Various specimens	*P. aethiopicus* *P. boisei*
	Malema, Malawi	Chiwondo	2.5–2.3	Biostratigraphy	HCRP-RC-911 (maxilla)	*P. boisei*
	Koobi Fora, Kenya	Koobi Fora	2.2–1.88 1.88–1.65 1.65–1.39	Radiometric, tephrostratigraphy	KNM-ER 1500 (partial skeleton) and others KNM-ER 406, 407, 732 (all crania) and others KNM-ER 729, 3230 (both mandibles) and others	*P. boisei*
	Chesowanja, Kenya	Chemoigut	2.0–1.5	Biostratigraphy, radiometric dating of capping layer	KNM-CH1 (partial cranium), other fragments	*P. boisei*
	Olduvai, Tanzania	Olduvai	1.9–1.7 1.7–1.2	Biostratigraphy, radiometric	OH 5 (cranium; holotype of *P. boisei*) Various specimens	*P. boisei*
	Peninj, Tanzania	Humbu	1.7–1.3	Radiometric; magnetostratigraphy	Mandible	*P. boisei*
	Konso, Ethiopia	Konso	1.45–1.3	Radiometric, tephrostratigraphy	KGA 10–525 (skull) and others	*P. boisei*
Southern Africa	Kromdraai, South Africa	Monte Cristo	2.0–1.5	Biostratigraphy, magnetostratigraphy	Close to 30 *Paranthropus* specimens, including TM 1517 (holotype of *P. robustus*)	*P. robustus*
	Drimolen, South Africa	Monte Cristo	2.0–1.6	Overall faunal assemblage composition; no absolute dates	>80 hominins, including DNH 7 (nearly complete female skull) and DNH 8 (male mandible)	*P. robustus*
	Gondolin, South Africa	Eccles	1.9–1.5	Biostratigraphy, magnetostratigraphy	GA 1 and GA 2 (isolated teeth)	*P. robustus*
	Cooper's Cave, South Africa	Monte Cristo	1.9–1.4	Biostratigraphy, uranium-lead dating	Various specimens, mostly isolated dental specimens	*P. robustus*
	Swartkrans, South Africa	Monte Cristo	1.8–1.0	Biostratigraphy	>300 *Paranthropus* specimens total, many isolated dental remains, including SK 6	*P. robustus*

massive postcanine dentition) that he felt justified naming a novel genus and species for the cranium. When Tobias (1967) presented his detailed analysis of OH 5, he concluded that it showed affinities with *Australopithecus africanus* and more closely with *P. robustus* (he referred to the latter as *Australopithecus robustus*), but he also detailed a suite of characters in which OH 5 differed from the *P. robustus* hypodigm. Tobias' interpretation of these differences is best put in context by the following quotation, "the Olduvai australopithecine differs from *Australopithecus robustus* in a

manner similar to that in which the latter differs from *Australopithecus africanus*" (Tobias 1967:233). Tobias went on to conclude that "the australopithecines had differentiated into a series of taxa, characterized by differing degrees of enlargement of the cheek teeth and naturally, of the supporting structures, muscular prominences, masticatory stress columns, and so on...." (Tobias 1967:228). Yet, as painstaking and detailed as Tobias' analysis was it was based on a single specimen and the results must be affected by the limitations that attend any study of one fossil

(Smith 2005), no matter how careful the study and how well-preserved the fossil.

Despite that caveat, for most researchers discoveries of ca. 2.3–ca. 1.3 Ma fossils at East African sites around Lake Turkana have been consistent with recognizing a single hyper-megadont species. The exceptions are Delson's (1997) suggestion that the evidence from Konso (Suwa et al. 1997) might justify a reassessment of *Paranthropus* taxonomy, and the possibility that discoveries at Gondolin (Menter et al. 1999) and Drimolen (Keyser 2000; Keyser et al. 2000) may help close the morphological gap between *P. robustus* and *P. boisei*. However, Wood and Lieberman (2001) concluded "the Konso specimens fit within the population parameters of *P. boisei* predicted by the 'pre-Konso' hypodigm" (p. 20), and when Constantino and Wood (2004) compared the regional hypodigms of *Paranthropus* before and after the addition of the new material from Drimolen and Gondolin, they found that the number of significant metrical differences between the postcanine dentition from eastern and southern Africa had increased rather than decreased. The balance of the evidence suggests a single hyper-megadont taxon inhabited East Africa between ca. 2.3 and ca. 1.3 Ma; the only evidence we presently have for *Australopithecus garhi* (see below), which also has large premolars and molars, is ca. 200 kyr earlier.

The pre-2.3 Ma evidence of hyper-megadont hominins from East Africa presents a more complex story. Arambourg and Coppens (1968) had made the ca. 2.6 Ma Omo 18-1967-18 mandible the holotype of a new species and genus, *Paraustralopithecus aethiopicus*. Few researchers now recognize *Paraustralopithecus* as a separate genus, but many consider that the pre-2.3 Ma hyper-megadont hominins from Omo-Shungura and West Turkana (e.g., KNM-WT 17000) belong to a species distinct from *A.* or *P. boisei*, and they refer to the taxon as either *Australopithecus aethiopicus* or *Paranthropus aethiopicus*. The hypodigm of this species would include a well-preserved adult cranium from West Turkana (KNM-WT 17000) together with mandibles (e.g., KNM-WT 16005) and isolated teeth from the Shungura Formation (Suwa et al. 1994, 1996). Some would also include the L. 338y-6 juvenile cranium in Member E of the Shungura Formation, and a maxilla from the ca. 2.3 Ma Ndolanya Beds at Laetoli, in the taxon.

The cranial evidence for *P. aethiopicus* resembles that of *P. boisei*, but the face of the former taxon is more prognathic, the cranial base is less flexed, the inferred size of the incisors and canines is larger, and the postcanine teeth are not quite so large or derived. But there is only one relatively complete *P. aethiopicus* cranium, and so the warnings of Smith (2005) about making taxonomic inferences based on small samples are especially relevant. Some researchers who are prepared to accept that a species may evolve significantly

over time (e.g., Walker et al. 1986) do not recognize *P. aethiopicus* as a separate taxon and instead include the hypodigm of *P. aethiopicus* within *Paranthropus Paranthropus boisei sensu lato*.

A novel hominin species, *Australopithecus garhi*, was established by Asfaw et al. (1999) to accommodate a fragmented cranium recovered from the ca. 2.5 Ma Hatayae Member of the Bouri Formation in the Middle Awash study area in Ethiopia. *Australopithecus garhi* combines a primitive cranial morphology with large-crowned postcanine teeth; the crowns of the anterior premolars are especially large. However, unlike *P. boisei*, the canines are large and the enamel apparently lacks the extreme thickness seen in the latter taxon. A partial skeleton combining a long femur with a long forearm was found nearby, but is not associated with the type cranium (Asfaw et al. 1999) and these fossils have not been formerly assigned to *A. garhi*. Yet, despite its large postcanine tooth crowns, the cranium of *A. garhi* lacks the derived features of *Paranthropus*. Asfaw et al. (1999) suggested *Au. garhi* may be ancestral to *Homo*, but the results of phylogenetic analyses of the limited fossil evidence are not consistent with this hypothesis. The morphology of the mandibles reported in the same publication as the cranium of *Au. garhi* is in some respects like that of the mandibles associated with *P. aethiopicus*, but in some ways the dental morphology is more similar to non hyper-megadont specimens from the Shungura Formation. If it is demonstrated that the type specimen of *P. aethiopicus*, Omo 18-18, belongs to the same taxon as the mandibles that appear to match the *A. garhi* cranium, then *P. aethiopicus* would have priority as the name for the *A. garhi* hypodigm.

Australopithecus or *Paranthropus*?

Prior to the discovery of *P. boisei*, Robinson (1954) reviewed the australopith remains from southern Africa and set out the morphological features that distinguish what others have referred to as the "robust" and "gracile" remains. These were incorporated into taxonomic definitions (Robinson 1954:198) that were subsequently amended and augmented (Robinson 1968:169). These features, together with a series of detailed dental characters extracted from Robinson (1956), constitute the characters that, prior to the publication of the detailed analysis of OH 5, were claimed to distinguish *Paranthropus* (i.e., *P. robustus* plus *P. crassidens*) from *Australopithecus* (i.e., *A. africanus*).

In his detailed review of the same fossil evidence considered by Robinson, Tobias (1967) did not so much deny the existence of character differences between "robust" and "gracile" australopiths, as place a different interpretation on them. He, in common with Leakey (1959), suggested that the differences between what later became known as the "robust" and "gracile" australopiths from southern Africa were equal to, if not exceeded by, the differences between the "robust" forms from southern Africa and OH 5. But Tobias rejected the notion that the two groups should be put in separate genera and in a later paper he argued that the southern African and East African "robust" forms were allopatric populations of a single "superspecies" within the genus *Australopithecus* (Tobias 1973). Subsequent authors such as Pilbeam and Gould (1974) and Corruccini and Ciochon (1979) effectively endorsed Tobias' decision to sink *Paranthropus* into *Australopithecus* by advancing the argument that australopiths were allometrically "scaled variants" of the same morphotype. Today, many researchers follow Tobias (1967), who followed Washburn and Patterson (1951), and subsume *Paranthropus* within the genus *Australopithecus*. The term "robust" australopith is widely used to informally identify the megadont and hypermegadont taxa within *Australopithecus sensu lato*.

Studies of the mandible, cranial base, endocranium, face, adult dentition, deciduous dentition, enamel microstructure

Table 8.3 Shared, and distinguishing, features of *Paranthropus boisei* and *Paranthropus robustus*

Trait	*Paranthropus robustus*	*Paranthropus boisei*
Face	Flat, wide, and dished compared to *Australopithecus* External anterior pillar is present, but internal structure is similar to *P. boisei*	Especially flat, wide, and dished External anterior pillar is absent, but internal structure is similar to *P. robustus*
Cranial features related to mastication	Large infratemporal fossa compared to *Australopithecus* Pronounced ectocranial cresting compared to *Australopithecus* *Gorilla*-like ramus that includes a coronoid process higher than the condylar process	Especially large infratemporal fossa Especially pronounced ectocranial cresting, suggesting greater development of the temporalis Especially wide mandibular ramus that includes a coronoid process higher than the condylar process Extensive overlap of the parieto-temporal suture
Dentition	Large, molarized postcanines, extra distal cusps, thick enamel, and accelerated development compared to *Australopithecus* Higher incidence of pitting compared to *Australopithecus* suggesting a hard and tough diet	Very large, hyper-molarized postcanines including an increased number of molar roots, higher frequency of molar roots, higher frequency of extra distal cusps, hyper-thick enamel, and similar accelerated development Lower complexity, suggesting a less mechanically challenging diet
Isotopic signal	C_3	C_4

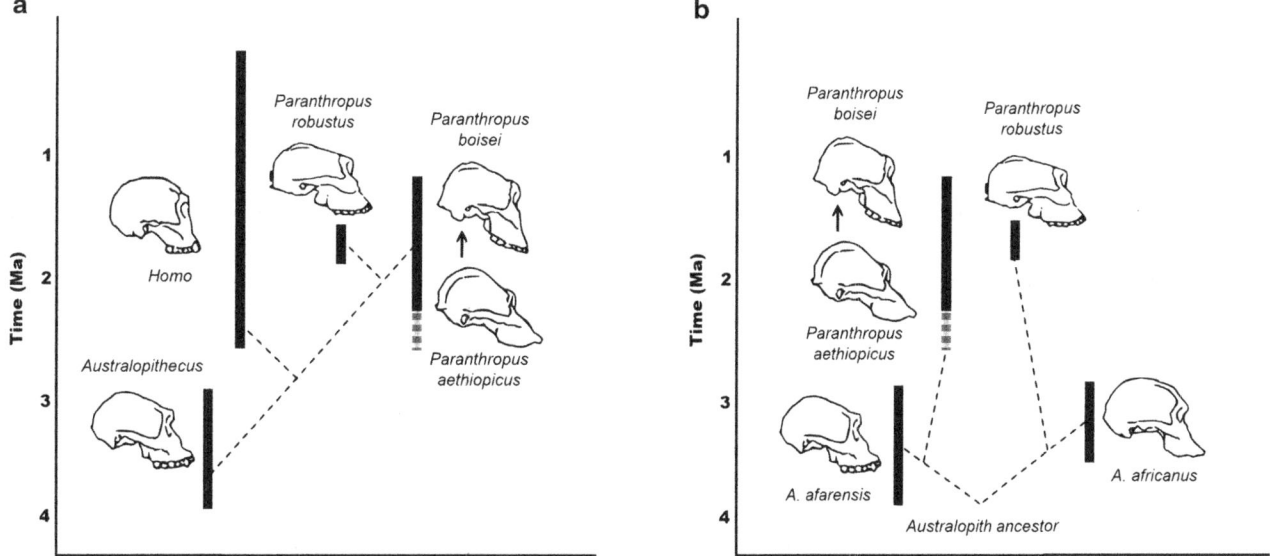

Fig. 8.3 Hypotheses about the relationships among *Paranthropus* considered in this review: **a** *Paranthropus* monophyly, **b** *Paranthropus* paraphyly. Redrawn after Strait et al. (1997)

and dental eruption pattern have all suggested ways in which the "robust" australopiths show either unique morphology or distinctive combinations of morphologies that individually are more widely distributed among early hominins. Some of these claims have been contested, but probable derived features of the skull of "robust" australopiths include a particularly thick mandibular corpus, apparently unique patterns of facial buttressing, and peculiar sutural and endocranial morphology. Dental characters special to the group include molarized mandibular premolars, preferentially enlarged mandibular molar talonids and a concomitantly high incidence of distal accessory cusps in the mandibular molars (Wood and Constantino 2007 and references therein). The case for retaining *Paranthropus* as a phenetically-distinct genus for the "robust" australopiths was cogently put by Robinson when he suggested that such a course would aptly reflect the "different adaptive patterns" (Robinson 1972:251) of the "gracile" (i.e., *Au. africanus*) and "robust" australopiths (Fig. 8.3, Table 8.3).

Phylogenetic Relationships of *Paranthropus robustus* and *Paranthropus boisei*

How are the two major taxa we have been considering, *P. robustus* and *P. boisei*, related? First, could they be so closely related that they do not deserve to be recognized as separate species? Eldredge and Tattersall (1975) suggested that the issue was "highly debatable" and some early cladistic analyses (e.g., Delson et al. 1977; Johanson and White 1979 and Skelton et al. 1986) made no distinction between the two taxa, but Johanson and White subsequently revised their position (White et al. 1981). Olson (1978) cited basicranial and dental characters that are unique to *P. boisei*, and suggested that an excessively overlapping squamosal suture may be a peculiarity of that taxon. Rak (1983) identified features of the mandible and face, respectively, that may be peculiar to *P. boisei*, and Grine (1984) listed apomorphies of the latter species' deciduous dentition. In short, these studies have supported Tobias (1967) in his assessment that OH 5 and its ilk are specifically distinct from *P. robustus*. Subsequent additions to the *P. boisei* hypodigm have underscored the current conventional wisdom that while *P. boisei* shares derived characters with *P. robustus*, even more characters set *P. boisei* apart and support its status as a taxon distinct from *P. robustus* (Wood and Constantino 2007).

But did the eastern and southern African "robust" taxa evolve from a most recent common ancestor (MRCA), exclusive to themselves, and thus form a monophyletic group, or did the megadont and hyper-megadont taxa in the two

regions evolve independently – *P. robustus* from *A. africanus*, and *P. boisei* from *P. aethiopicus*? This is not a trivial question for if the two forms evolved from a most recent common ancestor, then because the less derived "robust" form (*P. robustus*) is apparently more recent than the more derived form (*P. boisei*), this would either imply several reversals in cranial morphology, or that *P. robustus* existed for several hundred thousand years prior to its known first appearance datum. Alternatively, if the two regional variants arose independently, it would be a striking example of homoplasy for at least two, and probably more, hominin lineages would have independently acquired a suite of morphology that includes postcanine megadontia and robust mandibles.

The Case for and against *Paranthropus* Monophyly

Most cladistic analyses of early hominins have found support for *Paranthropus* monophyly. Wood (1988) reviewed fifteen studies that treated the eastern and southern African "robust" taxa separately in phylogenetic analyses, all of which concluded that the two regional variants were sister taxa (though some of the studies used the same data sets, so these results are not quite as impressively consistent as they appear). Subsequently, Corruccini (1994) reviewed the results of early hominin cladistic analyses and also concluded that one of the few reliable parts of the hominin cladogram was the *Paranthropus* clade. Strait et al. (1997) subjected 60 raw and adjusted traits from five previous studies to eight parsimony analyses, and in all cases the "robust" taxa formed a single clade. Strait and Grine (2004) combined 109 non-metrical traits with 89 traits based on linear measurements and, using two differently composed in-groups, also found that the three "robust" taxa (*P. robustus*, *P. boisei* and *P. aethiopicus*) consistently formed a monophyletic group, a result also reached by the cladistic analysis by Kimbel et al. (2004). Other studies that focused on specific morphology also support the conclusions of these global phylogenetic analyses. For example, LaCruz (2007) suggested that details of the enamel cap of *A. africanus* and *P. robustus* are too dissimilar for them to be sister taxa, Villmoare and Kimbel (2011) suggested that the internal structure of the circumnasal region of the maxilla is a synapomorphy of *P. robustus* and *P. boisei*, and Gunz et al. (2012) showed that a *P. robustus* cranium, SK 48, is more likely to be a scaled variant of *P. boisei* than a scaled variant of *A. africanus*.

In the face of all this analytical support for a "robust" australopith clade, why should *Paranthropus* monophyly be doubted? The reason, in a word, is homoplasy. The term was introduced by Ray Lankester who wrote that "when identical or nearly similar forces, or environments, act on two or more parts of an organism… the resulting correspondences called forth in the several parts in the two organisms will be nearly or exactly alike. I propose to call this kind of agreement *homoplasis* or *homoplasy*" (Lankester 1870, p. 39). Homoplasy, which refers to any resemblances between taxa that were not inherited from their most recent common ancestor, comes in several forms. Two types of homoplasy, analogy and convergence, are both caused by adaptation to similar environments. A third type of homoplasy, parallelism, is a by-product of development, not adaptations. Convergence usually occurs across greater phylogenetic distances than parallelism. Most cases of a fourth type of homoplasy, reversal (e.g., brain size increases and then decreases), are probably the result of natural selection, but recent work on silenced gene reactivation suggests that some reversals may also be neutral with regard to adaptation. The last type of homoplasy, homoiology, is attributed to non-genetic factors (e.g., activity-induced bone remodeling). In each case, homoplasies can be mistaken for shared derived similarities (i.e., synapomorphies), which are the principal evidence for phylogeny. As such, homoplasy complicates attempts to estimate phylogenetic relationships. Indeed, if homoplasies are sufficiently numerous, they can prevent a reliable phylogeny from being generated.

The first reason to suspect that homoplasy occurs anywhere in the hominin clade is comparative evidence from other mammalian groups evolving in Africa during the same time period, and in similar paleoenvironments, as hominins. Phylogenetic studies of bovids (Gatesy et al. 1997), hippos (Boissiere 2005), carnivores (van Valkenburgh 2007), Old World monkeys (Jablonski and Leakey 2008), elephants (Todd 2010) and equids (Bernor et al. 2010) all suggest that the evolutionary history of these groups shows evidence of substantial homoplasy during the period of time spanned by the megadont and hyper-megadont hominins. This comparative evidence does not mean that hominins *must* also have been affected by homoplasy, but it suggests it would be unwise to rule it out. Substantial homoplasy is also explicit in interpretations of the evolutionary history of non-hominin hominoids (Pilbeam 2002).

The second reason to suspect homoplasy in the hominin clade is that if consistency indices (CI) are any guide to the prevalence of homoplasy, then the ca. 0.65 average CI for hominin cladistic analyses means that approximately 35% of the characters used in the analyses must have been independently acquired (i.e., they are homoplasies) (Wood 1988).

In the *Paranthropus* clade specifically, another reason to suspect homoplasy is that many, but by no means all, of the characters that link *Paranthropus* taxa in the same clade are related to the masticatory system. There is empirical evidence that these characters are likely to be functionally integrated, thus potentially they are non-independent and if so, they should not be coded as individual independent characters in a cladistic analysis (Gunz et al. 2012). There is also some comparative evidence from other groups of mammals (e.g., Maglio 1975; Vrba 1979, 1984) to suggest that the masticatory system might be the equivalent of a "homoplasy ghetto." Another reason to question the hypothesis of "robust" australopith monophyly is because there is circumstantial evidence of homoplasy in traits related to the masticatory apparatus in other parts of the hominin fossil record. For example, the faces of *Kenyanthropus platyops* and *Homo rudolfensis* are, like *P. boisei*, both orthognathic relative to earlier hominins, but whereas the former have small or moderately sized postcanine teeth the latter shows extreme postcanine megadontia. Since *K. platyops* and *H. rudolfensis* are generally not considered to be closely-related to *P. boisei*, the cited similarities among these taxa must be due to homoplasy.

Other reasons to suspect that homoplasy may impede our ability to reconstruct a reliable phylogeny for *Paranthropus* are the results of three studies that looked in detail at the dental evidence for the *Paranthropus* clade. The results of all three studies were in support of falsifying the hypothesis of *Paranthropus* monophyly. The first of the three tests involved the relative size of the areas of the cusps of the mandibular postcanine tooth crowns. Wood (1988) reasoned that if *P. robustus* and *P. boisei* were sister taxa, then it is likely they would share a common pattern of dental development and would be expected to conform to the same scaling relationships; Gunz et al. (2011) used similar logic in their investigation of overall cranial shape. According to Wood's model, differences in cusp morphology between the taxa would be predictable from a combination of size differences and the extrapolation of any scaling relationships present in the smaller-toothed taxon. But in only two of the ten analyses involving the relative size of the whole or parts of the crowns of mandibular postcanine teeth did such a scaling relationship explain the observed differences. In the other eight analyses, there was either insufficient correlation between the variables to make any allometric prediction, or the observed differences between the two taxa were not the same as those predicted by the allometric relationships observed in the smaller-crowned (*P. robustus*) taxon.

The second test considered whether *P. robustus*, the less derived of the proposed sister taxa, was closer to the primitive state of a character morphocline. The root system of the P_3 is one of the few systems where the morphoclines have been worked out in any detail (Wood 1988; Emonet et al. 2012). In hominins, two distinct morphoclines lead from the inferred primitive condition (Wood et al. 1988). One, towards P_3 root reduction and simplification, culminates in modern humans. The second, towards greater root complexity, culminates in molar-like P_3 roots. The two *Paranthropus* taxa are *not* on the same morphocline. Instead, the roots of *P. robustus* correspond to one of the character states along the morphocline that leads towards reduced root complexity relative to the inferred primitive condition, whereas the P_3 roots of *P. boisei* correspond to the character state that shows the greatest root complexity (Wood 1988).

The third study uses the inhibitory cascade, a model that interprets the relative size of the occlusal surface of mammalian molars in terms of developmental mechanisms (Kavanagh et al. 2007), to test for similarities in the relative size of the occlusal surfaces of the postcanine teeth of the two *Paranthropus* taxa. The inhibitory cascade model has detected derived developmental conditions in the dentitions of rodents, ungulates, carnivores, and platyrrhines, and Schroer and Wood (2015) applied it to the postcanine dentition of a sample of catarrhine taxa, including fossil hominins. Extant congeners shared their fit to the inhibitory cascade of the molars; the only exception among the extant catarrhine taxa considered in this study was *Papio*, which may itself be paraphyletic (Zinner et al. 2009). When the same model was applied to *Paranthropus*, the differences in the relative size relationships observed in the molar and premolar-molar cascades in *P. robustus* and *P. boisei* were not consistent with the hypothesis that they belonged to the same genus.

To be considered within the same genus, taxa do not just have to be monophyletic, but they should also be in the same grade. That is, taxa within a genus should have an adaptive regime that is more similar to the type species of that genus than it is to the type species of another genus (Wood and Collard 1999). The different relative postcanine tooth sizes of *P. robustus* and *P. boisei* suggest that their diets may not have been the same, and support for such a dietary difference has come from recent studies of stable isotopes preserved in the teeth of the two taxa. The $^{13}C/^{12}C$ signal recovered from *P. boisei* specimens from Olduvai Gorge (van der Merwe et al. 2008; Cerling et al. 2011), Chesowanja, Koobi Fora, and Peninj (Cerling et al. 2011) and from a larger sample from the Turkana Basin (Cerling et al. 2013) suggest a C_4-dominated diet for *P. boisei*. A C_4-dominated diet is fundamentally different from that of all known living and fossil hominoids, which vary from nearly pure C_3 consumers like gorillas and chimpanzees, to a diet like that of *A. africanus* and *P. robustus* (Sponheimer et al. 2006, 2013) that is dominated by, but not confined to, C_3 foods. The primate whose carbon isotope composition best matches that of *P. boisei* is the extinct baboon *Theropithecus oswaldi*, whose preferred food was most likely grass! A diet of grasses or sedges is also consistent with the dental macrowear of *P. boisei*, for the sand and grit that is inevitably included in unwashed grasses or sedges would account for the high degree of macrowear on the postcanine teeth in that taxon.

The dissimilar dental microwear signals for *P. boisei* and *P. robustus* (Scott et al. 2005; Ungar et al. 2012) adds to the evidence that there are differences in the adaptive regimes of *P. robustus* and *P. boisei*.

The final reason to question *Paranthropus* monophyly concerns biogeography. Turner and Wood (1993b) assessed the probability of monophyly by examining the biogeographic patterns of African Plio-Pleistocene large mammals. They concluded that during the time range of *Paranthropus*, there was evidence in at least one mammalian group of faunal dispersal between regions, with several monophyletic groups having representatives in both regions. They suggested that while this lends credibility to the hypothesis of *Paranthropus* monophyly, it does not refute a polyphyletic origin for this group. In a second study, Turner and Wood (1993a) worked on the assumption that the well-developed masticatory system of *Paranthropus* was an adaptation to enable the consumption of tough food items in response to environmental aridity. They found that similar trends were detectable in the craniodental anatomy of other terrestrial mammals from this time period, and parallels in lineage turnover suggest that a large-scale response to environmental changes was occurring. Although this second study by Turner and Wood did not contradict the first one, it did suggest there are comparative precedents for regional mammalian lineages independently evolving similar masticatory adaptations in response to changing environmental conditions.

Differences in geological context, taphonomic history and collection methods, as well as a lack of a precise chronology in one of the regions, complicate attempts to compare the faunas of eastern and southern Africa, but access to new comprehensive datasets encouraged Patterson et al. (2014) to re-examine this critical time period in the African paleontological record. They investigated the biogeographic histories of three terrestrial African mammalian families whose fossil records span the past 3 million years to provide a comparative test of the hypothesis of *Paranthropus* monophyly. They used presence/absence data from 52 eastern African and 40 southern African fossil localities. These localities contain data for 117 species from 38 genera within the family Bovidae, and 34 species from 15 genera within the families Hyaenidae and Felidae. These assemblages were placed into 500 ka time slices and compared at both the genus and species level using the Jaccard index of faunal similarity. Results show that sampling biases have more effect on the patterns of interchange between eastern and southern African bovids than they do on the patterns of interchange seen in the Hyaenidae and Felidae. However, even when these biases are taken into account, there are persistent differences in the degree of interchange within and between these families. These findings suggest that mammalian groups (including hominins) can have very different histories of exchange between eastern and southern Africa over the past 3 million years. If these three families, especially Bovidae, are suitable proxies for the southern and eastern African megadont and hyper-megadont hominin taxa, then the results of this biogeographic comparative study are consistent with relatively independent evolutionary trajectories for the hominins in the two regions.

Conclusions

Most of the present cladistic evidence is in favor of monophyly and if one is comfortable with the conclusion that hard-tissue morphology *is* capable of recovering sound hypotheses about phylogenetic relationships established on the basis of independent genetic evidence (e.g., Strait and Grine 2004), then *Paranthropus* monophyly must be the null hypothesis. But if one is more skeptical about the ability of hard-tissue morphology to recover phylogenetic relationships (e.g., Collard and Wood 2000) or about the non-independence of traits used in cladistics analysis, then what to many researchers seems to be overwhelming evidence for *Paranthropus* monophyly seems less compelling.

According to the results of phylogenetic analyses the question of *Paranthropus* monophyly looks to be resolved, but future research must strive to determine whether the superficial and detailed similarities seen in the hard-tissue morphology of eastern and southern African *Paranthropus* taxa is due to their sharing a most recent common ancestor, or due to one or more types of homoplasy.

Much new fossil and other evidence has been accumulated since Grine's (1988) *Evolutionary History of the "Robust" Australopithecines*, but despite these developments, we are not obviously closer to resolving the conundrum of *Paranthropus*. There are rays of hope, however, in that we may be closer to reconstructing the diet of *P. boisei* and closer to understanding more about its postcranial skeleton, always assuming that it can provide reliable evidence regarding monophyly (e.g., Pilbeam 2002). What is not in doubt however, is that it is *very* unlikely that *any Paranthropus* taxon was the direct ancestor of modern humans. For many, this lessens their appeal, but to others, including the dedicatee and the authors, this makes their paleobiology more, not less, intriguing.

Acknowledgments BW would like to thank the Provost of George Washington University for his continuing support for the GW University Professorship of Human Origins and via the GW Signature Program for his support of CASHP. KS was supported by the NSF-IGERT DGE-0801634 and NSF-GRFP and is currently supported by Neukom Institute for Computational Science, Dartmouth. We are grateful to Bill Kimbel and an anonymous reviewer for excellent suggestions that improved the manuscript.

References

Arambourg, C., & Coppens, Y. (1968). Decouverte d'un australopithecien nouveau dans les Gisements de L'Omo (Ethiopie). *South African Journal of Science, 64*, 58–59.

Asfaw, B., White, T., Lovejoy, O., Latimer, B., Simpson, S., & Suwa, G. (1999). *Australopithecus garhi*: A new species of early hominid from Ethiopia. *Science, 284*, 629–635.

Bernor, R. L., Armour-Chelu, M., Gilbert, H., Kaiser, T., & Schulz, E. (2010). Equidae. In W. B. Sanders (Ed.), *Cenozoic mammals of Africa* (pp. 685–721). Berkeley: University of California Press.

Boissiere, J.-R. (2005). The phylogeny and taxonomy of Hippopotamidae (Mammalia: Artiodactyla): A review based on morphology and cladistic analysis. *Zoological Journal of the Linnean Society, 143*, 1–26.

Broom, R. (1938). The Pleistocene anthropoid apes of South Africa. *Nature, 142*, 377–379.

Broom, R. (1949). Another new type of fossil ape-man (*Paranthropus crassidens*). *Nature, 163*, 57.

Campbell, B. (1963). Quantitative taxonomy and human evolution. In S. L. Washburn (Ed.), *Classification and human evolution* (pp. 50–74). Chicago: Aldine.

Cerling, T. E., Mbua, E., Kirera, F. M., Manthi, F. K., Grine, F. E., Leakey, M. G., et al. (2011). Diet of *Paranthropus boisei* in the early Pleistocene of East Africa. *Proceedings of the National Academy of Sciences USA, 108*, 9337–9341.

Cerling, T. E., Manthi, F. K., Mbua, E. N., Leakey, L. N., Leakey, M. G., Leakey, R. E., et al. (2013). Stable isotope-based diet reconstructions of Turkana Basin hominins. *Proceedings of the National Academy of Sciences USA, 110*, 10501–10506.

Collard, M. C., & Wood, B. A. (2000). How reliable are human phylogenetic hypotheses? *Proceedings of the National Academy of Sciences USA, 97*, 5003–5006.

Constantino, P., & Wood, B. (2004). *Paranthropus* paleobiology. In *Miscela´nea en homenaje a Emiliano Aguirre, Volumen III. Paleoantropologia* (pp. 136–151). Madrid: Museo Arqueolo´cico Regional.

Corruccini, R. S. (1994). How certain are hominoid phylogenies? The role of confidence intervals in cladistics. In R. S. Corruccin & R. L. Ciochon (Eds.), *Integrative paths to the past: Paleoanthropological advances in honor of F Clark Howell* (pp. 167–183). Englewood Cliffs: Prentice Hall.

Corruccini, R. S., & Ciochon, R. L. (1979). Primate facial allometry and interpretations of australopithecine variation. *Nature, 281*, 62–64.

Delson, E. (1997). One skull does not a species make. *Nature, 389*, 445–446.

Delson, E., Eldredge, N., & Tattersall, I. (1977). Reconstruction of hominid phylogeny: A testable framework based on cladistic analysis. *Journal of Human Evolution, 6*, 263–278.

Eldredge, E., & Tattersall, I. (1975). Evolutionary models, phylogenetic reconstruction, and another look at hominid phylogeny. In F. S. Szalay (Ed.), *Approaches to primate Paleobiology* (pp. 218–242). Basel: Karger.

Emonet, E. G., Tafforeau, P., Chaimanee, Y., Guy, F., de Bonis, L., Koufos, G., et al. (2012). Three-dimensional analysis of mandibular dental root morphology in hominoids. *Journal of Human Evolution, 62*, 146–154.

Gatesy, J., Amato, G., Vrba, E., Schaller, G., & DeSalle, R. (1997). A cladistics analysis of mitochondrial ribosomal DNA from the Bovidae. *Molecular Phylogenetics and Evolution, 7*, 303–319.

Grine, F. E. (1984). Comparison of the deciduous dentitions of African and Asian hominids. *Cour Forschungsinst Senckenb, 69*, 69–82.

Grine, F. E. (1985). Australopithecine evolution: The deciduous dental evidence. In E. Delson (Ed.), *Ancestors: The hard evidence* (pp. 153–167). New York: Liss.

Grine, F. E. (1988). Evolutionary history of the "robust" australopithecines: A summary and historical perspective. In F. E. Grine (Ed.), *Evolutionary history of the "robust" australopithecines* (pp. 509–520). New York: Aldine de Gruyter.

Gunz, P., Neubauer, S., Maureille, B., & Hublin, J.-J. (2011). Virtual reconstruction of the Le Moustier 2 newborn skull. Implications for Neanderthal ontogeny. *Paleo Revue d'Archéologie Préhistorique, 22*, 155–172.

Gunz, P., Neubauer, S., Golovanova, L., Doronichev, V., Maureille, B., & Hublin, J.-J. (2012a). A uniquely modern human pattern of endocranial development. Insights from a new cranial reconstruction of the Neandertal newborn from Mezmaiskaya. *Journal of Human Evolution, 62*, 300–313.

Gunz, P., Ramsier, M., Kuhrig, M., Hublin, J.-J., & Spoor, F. (2012b). The mammalian bony labyrinth reconsidered, introducing a comprehensive geometric morphometric approach. *Journal of Anatomy, 220*, 529–543.

Howell, F. C. (1978). Hominidae. In V. J. Maglio & H. B. S. Cooke (Eds.), *Evolution of African mammals* (pp. 154–248). Cambridge: Harvard University Press.

Jablonski, N. G., & Leakey, M. G. (2008). *Koobi Fora research project vol. 6. The fossil monkeys*. California Academy of Sciences: San Francisco.

Johanson, D. C., & White, T. D. (1979). A systematic assessment of early African hominids. *Science, 202*, 321–330.

Kavanagh, K. D., Evans, A. R., & Jernvall, J. (2007). Predicting evolutionary patterns of mammalian teeth from development. *Nature, 449*, 427–432.

Keyser, A. W. (2000). The Drimolen skull: The most complete australopithecine cranium and mandible to date. *South African Journal of Science, 96*, 189–197.

Keyser, A. W., Menter, C. G., Moggi-Cecchi, J., Pickering, T. R., & Berger, L. R. (2000). Drimolen: A new hominid-bearing site in Gauteng, South Africa. *South African Journal of Science, 96*, 193–197.

Kimbel, W., Rak, Y., & Johanson, D. C. (2004). The skull of *Australopithecus afarensis*. New York: Oxford University Press.

Lacruz, R. S. (2007). Enamel microstructure of the hominid KB 5223 from Kromdraai, South Africa. *American Journal of Physical Anthropology, 132*, 175–182.

Lankester, E. R. (1870). *On comparative longevity in man and the lower animals*. London: Macmillan.

Leakey, L. S. B. (1959). A new fossil skull from Olduvai. *Nature, 184*, 491–493.

Leakey, L. S. B., & Leakey, M. D. (1964). Recent discoveries of fossil hominids in Tanganyika, at Olduvai and near Lake Natron. *Nature, 202*, 5–7.

Maglio, V. J. (1975). Origin and evolution of the Elephantidae. *Transactions of the American Philosophical Society, 63*, 1–149.

Menter, C. G., Kuykendall, K. L., Keyser, A. W., & Conroy, G. C. (1999). First record of hominid teeth from the Plio-Pleistocene site of Gondolin, South Africa. *Journal of Human Evolution, 37*, 299–307.

Olson, T. R. (1978). Hominid phylogenetics and existence of *Homo* in Member 1 of the Swartkrans Formations, South Africa. *Journal of Human Evolution, 7*, 159–178.

Patterson, D. B., Faith, J. T., Bobe, R., & Wood, B. (2014). Regional diversity patterns in African bovids, hyaenids, and felids during the past 3 million years: The role of taphonomic bias and implications for the evolution of *Paranthropus*. *Quaternary Science Reviews, 96*, 9–22.

Pilbeam, D. R. (2002). Perspectives on the Miocene Hominoidea. In W. Hartwig (Ed.), *The primate fossil record* (pp. 303–310). Cambridge: Cambridge University Press.

Pilbeam, D., & Gould, S. J. (1974). Size and scaling in human evolution. *Science, 186*, 892–901.

Rak, Y. (1983). *The australopithecine face*. New York: Academic Press.

Rak, Y., Ginzburg, A., & Geffen, E. (2007). Gorilla-like anatomy on *Australopithecus afarensis* mandibles suggests *Au. afarensis* link to robust australopiths. *Proceedings of the National Academy of Sciences USA, 104*, 6568–6572.

Robinson, J. T. (1954). The genera and species of the Australopithecinae. *American Journal of Physical Anthropology, 12*, 181–200.

Robinson, J. T. (1956). The dentition of the Australopithecinae. *Transvaal Museum Memoirs, 9*, 1–179.

Robinson, J. T. (1968). The origin and adaptive radiation of the australopithecines. In G. Kurth (Ed.), *Evolution und hominisation* (2nd ed., pp. 150–175). Stuttgart: Fischer.

Robinson, J. T. (1972). The bearing of East Rudolf fossils on early hominid systematics. *Nature, 240*, 239–240.

Scott, R. S., Ungar, P. S., Bergstrom, T. S., Brown, C. A., Grine, F. E., Teaford, M. F., et al. (2005). Dental microwear texture analysis reflects diets of living primates and fossil hominins. *Nature, 436*, 693–695.

Schroer, K., & Wood, B. (2015). Modeling the dental development of fossil hominins through the inhibitory cascade. *Journal of Anatomy, 226*, 150–162.

Skelton, R. R., McHenry, H. M., & Drawhorn, G. M. (1986). Phylogenetic analysis of early hominids. *Current Anthropology, 27*, 21–43.

Smith, R. J. (2005). Species recognition in paleoanthropology: Implications of small sample sizes. In D. E. Lieberman, R. J. Smith, & J. Kelley (Eds.), *Interpreting the past: Essays on human, primate, and mammal evolution in honor of David Pilbeam* (pp. 207–219). Boston: Brill Academic Publishers.

Sponheimer, M., Passey, B. H., de Ruiter, D. J., Guatelli-Steinberg, D., Cerling, T. E., & Lee-Thorp, J. (2006). Isotopic evidence for dietary variability in the early hominin *Paranthropus robustus*. *Science, 314*, 980–982.

Sponheimer, M., Alemseged, Z., Cerling, T. E., Grine, F. E., Kimbel, W. H., Leakey, M. G., et al. (2013). Isotopic evidence of early hominin diets. *Proceedings of the National Academy of Sciences USA, 110*, 10513–10518.

Strait, D. S., & Grine, F. E. (2004). Inferring hominoid and early hominid phylogeny using craniodental characters: The role of fossil taxa. *Journal of Human Evolution, 47*, 399–452.

Strait, D. S., Grine, F. E., & Moniz, M. A. (1997). A reappraisal of early hominid phylogeny. *Journal of Human Evolution, 32*, 17–82.

Suwa, G., Wood, B. A., & White, T. D. (1994). Further analysis of mandibular molar crown and cusp areas in Pliocene and early Pleistocene hominids. *American Journal of Physical Anthropology, 9*, 407–426.

Suwa, G., Asfaw, B., Beyene, Y., White, T., Katoh, S., Nagaoka, S., et al. (1997). The first skull of *Australopithecus boisei*. *Nature, 389*, 489–492.

Tobias, P. V. (1967). Olduvai Gorge. *The cranium and maxillary dentition of Australopithecus (Zinjanthropus) boisei* (Vol 2). Cambridge: Cambridge University Press.

Tobias, P. V. (1973). Darwin's prediction and the African emergence of the genus *Homo*. *Accademia Nazionale dei Lincei, Quaderno, 182*, 63–85.

Todd, N. E. (2010). New phylogenetic analysis of the family Elephantidae based on cranial-dental morphology. *The Anatomical Record, 293*, 74–90.

Turner, A., & Wood, B. A. (1993a). Taxonomic and geographic diversity in "robust" australopithecines and other African Plio-Pleistocene mammals. *Journal of Human Evolution, 24*, 147–168.

Turner, A., & Wood, B. A. (1993b). Comparative palaeontological context for the evolution of the early hominid masticatory system. *Journal of Human Evolution, 24*, 301–318.

Ungar, P. S., Krueger, K. L., Blumenschine, R. J., Njau, J., & Scott, R. S. (2012). Dental microwear texture analysis of hominins recovered by the Olduvai landscape paleoanthropology project, 1995–2007. *Journal of Human Evolution, 63*, 429–437.

van der Merwe, N., Masao, F., & Bamford, M. (2008). Isotopic evidence for contrasting diets of early hominins *Homo habilis* and *Australopithecus boisei* of Tanzania. *South African Journal of Sciences, 104*, 153–155.

van Valkenburgh, B. (2007). Déjà vu: The evolution of feeding morphologies in the Carnivora. *Integrative and Comparative Biology, 47*, 147–163.

Villmoare, B. A., & Kimbel, W. H. (2011). CT-based study of internal structure of the anterior pillar in extinct hominins and its implications for the phylogeny of robust *Australopithecus. Proceedings of the National Academy of Sciences USA, 108*, 16200–16205.

Vrba, E. S. (1979). Phylogenetic analysis and classification of fossil and recent Alcelaphini (Mammalia: Bovidae). *Biological Journal of the Linnean Society, 11*, 207–228.

Vrba, E. S. (1984). Evolutionary pattern and process in the sister group Alcelaphini-Aepycerotini (Mammalia: Bovidae). In N. Eldredge & S. M. Stanley (Eds.), *Living fossils* (pp. 62–79). New York: Springer.

Walker, A., Leakey, R. E., Harris, J. M., & Brown, F. H. (1986). 2.5-Myr *Australopithecus boisei* from west of Lake Turkana, Kenya. *Nature, 322*, 517–522.

Washburn, S. L., & Patterson, B. (1951). Evolutionary importance of the South African "man-apes". *Nature, 167*, 650–651.

White, T. D., Johanson, D. C., & Kimbel, W. H. (1981). *Australopithecus africanus*: Its phylogenetic position reconsidered. *South African Journal of Science, 77*, 445–470.

Wood, B. A. (1988). Are 'robust' australopithecines a monophyletic group? In F. E. Grine (Ed.), *Evolutionary history of the "robust" australopithecines* (pp. 269–284). New York: Aldine de Gruyter.

Wood, B. A., Abbott, S. A., & Uytterschaut, H. (1988). Analysis of the dental morphology of Plio-Pleistocene hominids. IV. Mandibular postcanine root morphology. *Journal of Anatomy, 156*, 107–139.

Wood, B., & Collard, M. (1999). The human genus. *Science, 284*, 65–71.

Wood, B., & Constantino, P. (2007). *Paranthropus boisei*: Fifty years of fossil evidence and analysis. *Yearbook of Physical Anthropology, 50*, 106–132.

Wood, B., & Lieberman, D. E. (2001). Craniodental variation in *Paranthropus boisei*: A developmental and functional perspective. *American Journal of Physical Anthropology, 116*, 13–25.

Wood, B., & Schroer, K. (2013). *Paranthropus*. In D. Begun (Ed.), *Companion to paleoanthropology* (pp. 457–478). New York: Wiley-Blackwell.

Zinner, D., Groeneveld, L. F., Keller, C., & Roos, C. (2009). Mitochondrial phylogeography of baboons (*Papio* spp.) – Indication for introgressive hybridization? *BMC Evolutionary Biology, 9*, 83.

Chapter 9
Feeding Behavior and Diet in *Paranthropus boisei*: The Limits of Functional Inference from the Mandible

David J. Daegling and Frederick E. Grine

Abstract The craniofacial morphology of *Paranthropus boisei* is highly derived, representing the evolutionary culmination of one robust australopith lineage. Following its discovery, OH 5 was popularly described as "Nutcracker Man," and this image of the East African robust australopiths as hard-object feeders has persisted for the last half-century. Emerging lines of evidence, however, suggest that the diet of this species was not primarily comprised of hard foods. An alternative view is that *P. boisei* consumed a relatively tough diet of grasses and sedges. As the covariation of diet and/or feeding behavior with mandibular morphology has been the focus of a voluminous literature, this paper evaluates whether the jaws of *P. boisei* – interpreted within a framework of masticatory mechanics in particular and bone biology in general – can be interpreted as functionally coherent with a herbivorous diet that lacked a significant component of durophagy. In terms of proportion and geometry, australopith mandibles have no parallel among living primates and *P. boisei* represents the extreme expression of this morphotype. From the perspective of primate masticatory biomechanics, the inference of loading regimes experienced in this fossil species is speculative, and subsequent inference of diet from corpus geometry should be regarded with skepticism. However, in terms of overall mandibular architecture, and from what is known about the biomechanical influences governing bone hypertrophy, competing hypotheses of dietary specialization are equally plausible on morphological criteria. Mandibular hypertrophy is an expected outcome of a fibrous diet requiring extensive and prolonged mastication, especially in a taxon in which occlusal morphology is suboptimal for the breakdown of fibrous foods.

Keywords Australopith • Dentition • Diet • Durophagy • Early *Homo* • Mandibular hypertrophy • Masticatory biomechanics • Stable isotopes

Introduction

Paranthropus boisei is commonly seen as representing the pinnacle of robust australopith evolution in terms of its craniodental morphology. The mandibular and postcanine tooth hypertrophy that characterizes *P. boisei* is seemingly unparalleled among australopiths. Under a credulous application of the adaptationist paradigm, these features must be explicable in terms of feeding behavior and ecology. More deliberate consideration would allow that these features could be correlated responses to other craniofacial adaptations which may or may not be related to feeding ecology, but for purposes of discussion we proceed from the assumption that masticatory/ingestive loads and mandibular morphology are functionally linked. Current views on the question of diet in *P. boisei* are discordant with respect to microwear and isotope data on the one hand and biomechanical analyses on the other. Microwear analyses suggest a general absence of hard objects in the diet (Walker 1981; Ungar et al. 2008, 2012), and stable isotope data suggest sedges and grasses were eaten (van der Merwe et al. 2008; Cerling et al. 2011). Conversely, biomechanical models suggest specialization for hard-object feeding (Rak 1983; Strait et al. 2008, 2013), and comparative studies of dental enamel embrace this view (Constantino et al. 2010, 2011). This paper explores whether the mandibular remains attributed to *P. boisei* can productively inform these alternative interpretations. In particular, does the mandible accord with durophagous inferences from the cranium, or does its structure align better with a diet that did not necessarily comprise hard objects, but is functionally responsive to

D.J. Daegling (✉)
Department of Anthropology, University of Florida, 1112
Turlington Hall, Gainesville, FL 32611-7305, USA
e-mail: daegling@ufl.edu

F.E. Grine
Departments of Anthropology and Anatomical Sciences,
Stony Brook University, Stony Brook, NY 11794-4364, USA
e-mail: frederick.grine@stonybrook.edu

© Springer International Publishing AG 2017
Assaf Marom and Erella Hovers (eds.), *Human Paleontology and Prehistory*,
Vertebrate Paleobiology and Paleoanthropology, DOI: 10.1007/978-3-319-46646-0_9

repetitive loading. Alternatively, is this dichotomous interpretation itself misguided; i.e., is the functional morphology of *P. boisei* mandibles consistent with both regimens?

There are nearly 40 mandibular specimens that are likely attributable to *P. boisei*, but only four are relatively complete (Wood and Constantino 2007). Collectively, the sample provides an instructive, if incomplete, glimpse into population variation in mandibular form. One may safely say that *P. boisei* mandibles are exceptionally broad, display a high degree of sexual dimorphism, and represent the most derived morphology of mandibular form among the australopiths (Fig. 9.1).

Postcanine megadontia defines *P. boisei* as well, although these large teeth do not by themselves explain the size of the

mandible (Daegling and Grine 1991; Plavcan and Daegling 2006).

The Interpretive Dilemma of "Robust" Mandibles

The functional significance of *P. boisei* mandibular morphology has been inferred from corpus size and geometry as well as the size and height of the ramus. The relatively large corpus dimensions in *P. boisei* have been postulated to be due to large bending and particularly large twisting moments acting on the mandible during mastication (Hylander 1979, 1988). The tall mandibular ramus of *P. boisei* has certain

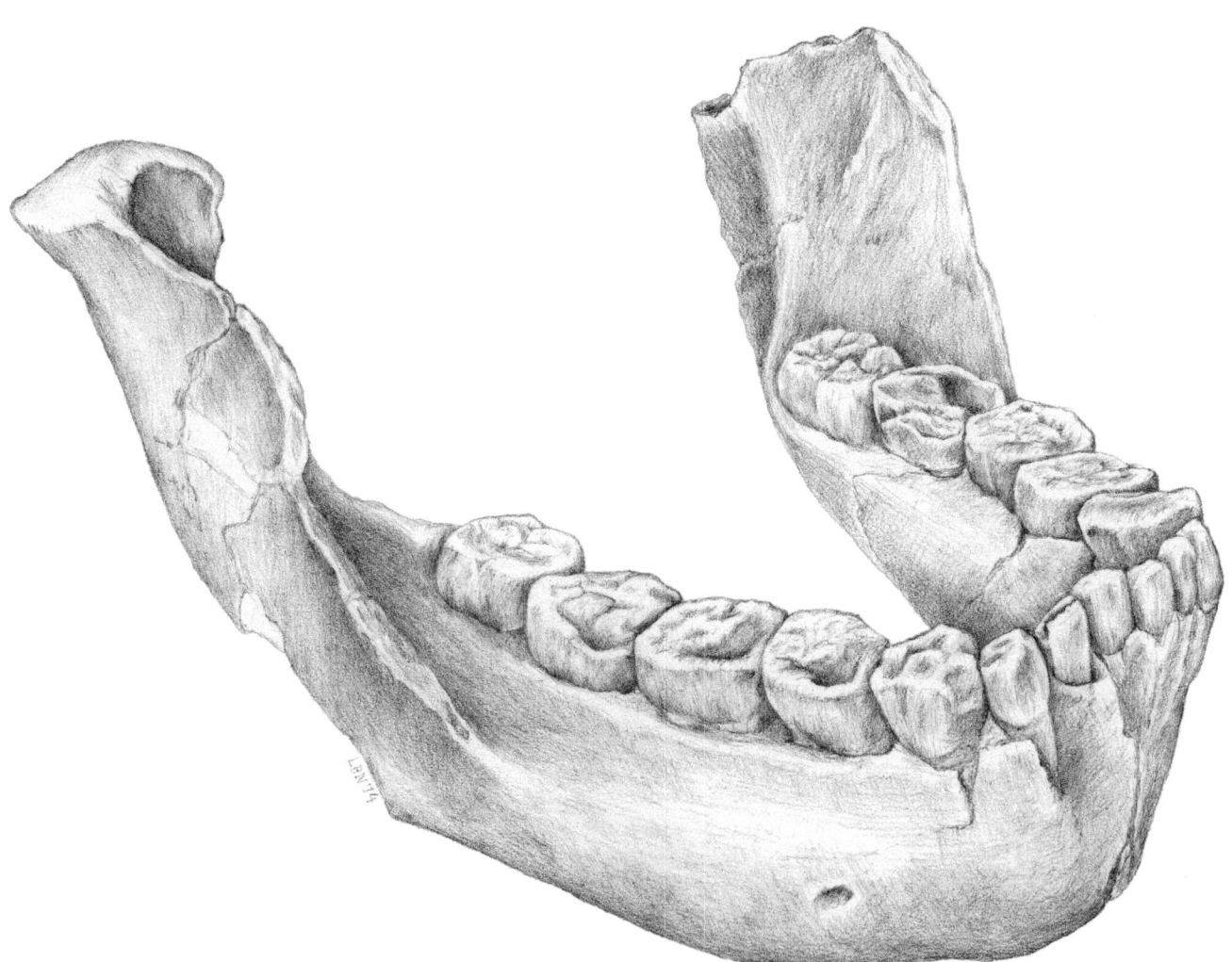

Fig. 9.1 The Peninj or Natron *Paranthropus boisei* mandible is one of the most complete jaws attributable to this species. This oblique anterolateral view highlights the megadont postcanine dentition and massive mandibular corpus that is typical of the species. Drawing by Luci Betti-Nash

mechanical consequences, including improved adductor leverage (DuBrul 1977), and an increased anterior-posterior component to occlusal forces (Rak and Hylander 2008). These observations by themselves tell us little about the particulars of diet, but considered in the context of megadontia and molarization of the premolars, they collectively suggest a mechanically demanding diet requiring a great deal of effort to process. The tall ramus also results in greater simultaneity of occlusal contact along the tooth row (Ward and Molnar 1980), which suggests bulk feeding – perhaps on low-quality foods that required a good deal of work to process.

One approach to deciphering the functional significance of mandibular variation may be to relate diet and feeding behavior relative to a concept of work. The term "work" is defined physically as force times distance (*Fd*) which itself relates to energetics of feeding, since energy can be conceived of as the capacity to do work. With respect to the mechanics of feeding, work can be invoked in various contexts: e.g., muscular activity, jaw movement, the fracturing of food during occlusion, etc. With respect to mandibular corpus mechanical properties, one is presumably interested in the work done on the mandible itself (specifically, internal work manifested as strain energy).

In mastication, work is a cyclical phenomenon regardless of how the variable may be conceived or measured. Under a naive assumption that masticatory movements, occlusal contacts and loads are constant across cycles,[1] one can consider whether a bite of 500 N in a single masticatory cycle is equivalent to 10 cycles involving a peak occlusal force of 50 N in each. Put another way, is chewing vigorously for an hour comparable to chewing relatively effortlessly all day long in terms of the structural response of mandibular bone to these distinct load histories? Whether these represent "equivalent" strain histories in terms of evolutionary or functional morphology depends on how bone reacts to them.

This problem of interpretation has persisted since Hylander (1979) first posited that intermittent high strains as well as long-term cyclical strains may each challenge the integrity of mandibular bone. The reason is that bone is not only susceptible to fail under a singular loading event, but can also fail by fatigue, in which repetitive loads of initially tolerable magnitude eventually weaken and fracture the bone. The structural solution to both threats – high loads versus repeated loads – is to increase the mass of mandibular bone. This can lower peak stress to a value below ultimate (i.e., failure) stress in the former case, and expand fatigue life by lowering net cyclical stress in the latter. Remodeling may also serve to increase fatigue life by removing accumulated damage, although remodeled bone itself appears to be less strong and have reduced fatigue life relative to primary bone (Martin et al. 1998).

In this context, a "massive" mandible may represent a structural solution to either forceful biting and chewing, persistent and prolonged mastication, or both. What kinds of foods are implicated by these alternatives? Stiff, stress-limited or "hard" foods require high forces to initiate fracture, whereas tough or displacement-limited foods require more constant force application to drive fractures through them (Williams et al. 2005). Relative to foods such as apple or carrot, both tough and hard foods engender higher bite forces in experimental animals (Weijs and deJongh 1977; Hylander 1979).

The logic behind the proposition that jaw morphology reflects diet is straightforward. Since certain foods are harder/tougher than others, processing such foods requires more work and this additional work translates into more stress. Reducing this stress is a simple matter of adding bone, or distributing existing bone mass more efficiently relative to physiological loads. In effect, the implicit assumption in comparative research on mandibular geometry is that bone metabolic activity is geared toward reaching some equilibrium state of stress. While this chain of reason is internally coherent, we have surprisingly little empirical corroboration of what this equilibrium state is, except to note that it is not constant throughout the skeleton (Hylander and Johnson 1992; Hsieh et al. 2001). In the context of this discussion, stress and strain are being treated as interchangeable concepts (i.e., in experimental studies strain is measured to infer states of stress). This follows from the definition of Young's modulus, although it is important to recognize that spatial heterogeneity in this variable is the rule rather than the exception for mandibular bone (Schwartz-Dabney and Dechow 2003; Dechow and Hylander 2000; Daegling et al. 2014). This means that under physiological loads stress will be proportional to strain, although the precise nature of this proportionality may differ among locations.

Paranthropus boisei Mandibles in Comparative Context

Paranthropus is distinguished by its "robust" mandibles, expressed as an index of breadth over height (Fig. 9.2). While this leads to the perception that *P. boisei* corpora are unusually thick (Chamberlain and Wood 1985), what is actually most remarkable by australopith standards is the relative height of *P. boisei* mandibles (cf. Figs. 9.3 and 9.4). This suggests that these jaws are exceptional in terms of their ability to counter primarily vertical (parasagittal) bending moments. Among extant anthropoids, relatively deep corpora are associated with both hard-object feeding (e.g., *Lophocebus albigena and* pithecines) and folivory (some colobines), albeit not universally (Hylander 1979; Bouvier 1986a, b).

---

[1]This is an assumption of convenience. In reality, masticatory excursions vary even within a single chewing sequence. A convincing demonstration of this can be had by attentively eating an apple.

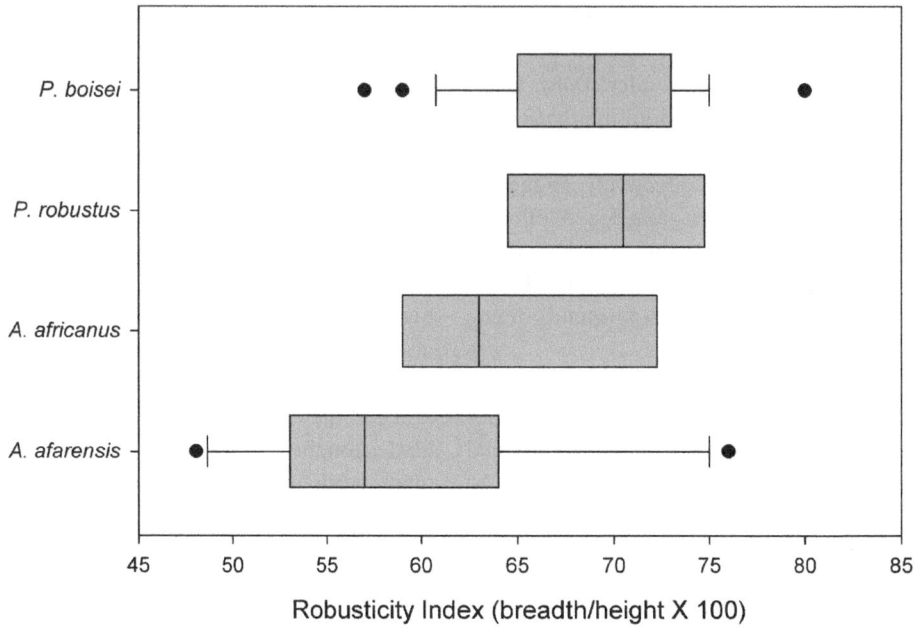

Fig. 9.2 The "robusticity index" is the ratio of corpus breadth to corpus height. Higher index values indicate a relatively broad corpus. Because this is an index of shape, the biomechanical significance of any particular value is not reliably inferred in the absence of other data. The shape of the corpus in *Paranthropus boisei* is not exceptional among australopiths in general, with median values falling within the ranges of other taxa. Sample sizes: *Australopithecus afarensis* (N = 11), *A. africanus* (N = 6), *P. robustus* (N = 8), *P. boisei* (N = 25). Whiskers represent 10th and 90th percentiles (not defined for N < 9), boxes represent 25th and 75th percentiles, and the line within the box is the median. Outliers (when present) are plotted individually. Corpus breadth is measured at midcorpus along the minimum plane of section at M_1; corpus height is measured as the minimum height from the lateral alveolar margin of M_1 to the corpus base

Sagittal bending, however, is not the only important load acting on the mandibular corpus in anthropoids (including *Paranthropus*): Hylander (1988) noted that mastication in general, and biting on the premolars in particular, also results in torsion of the mandibular corpus. Tension and compression arising from occlusal loads are also important sources of stress on the working-side corpus (Hirabayashi et al. 2002; Ishigaki et al. 2003; Benazzi et al. 2012), and the vertical resultants of occlusal, joint reaction and muscular forces create direct shear stress (Hylander 1979).

Corpus breadth in *Paranthropus* follows an australopith scaling pattern, with two dramatic exceptions: SK12, representing *P. robustus*, and Omo 7A-125, representing *P. boisei* (Fig. 9.3). These two specimens possess unusually broad corpora by comparison. Otherwise, the genus is unremarkable in this dimension by early hominin standards. Because the australopiths in general have broad corpora (Fig. 9.5), however, the significance of variation in this metric is of interest for functional/mechanical interpretation. A transversely thick corpus has been interpreted as a structural solution to counter increased transverse bending and axial torsion during mastication (Hylander 1979, 1988). Although the effect of increased transverse dimensions on bending rigidity is transparent, experimental work has since indicated that underlying variations in cortical thickness

condition torsional rigidity as much as external geometry (Daegling and Hylander 1998). In addition, comparative analysis indicates that corpus breadth is probably a relatively poor indicator of torsional resistance (Daegling 2007a), with overall corpus size being the best proxy in the absence of subperiosteal contour data. By this criterion *P. boisei* is exceptional, having massive corpora in terms of areal dimensions (Chamberlain and Wood 1985). Among australopiths, then, *P. boisei* mandibles are well-suited for resistance of both parasagittal bending and axial torsion (Hylander 1988).

Information on cortical bone distribution within mandibular sections can provide a more precise assessment of structural competence. No such data have been collected for *P. boisei*, although it may be predicted with reasonable confidence that whatever that pattern might be, the status of the hypodigm as a biomechanical outlier will not change appreciably. The reason is that overall size of the section is the most important determinant of its structural rigidity. Cortical geometry has been examined in both *Australopithecus africanus* and *Paranthropus robustus*, and relative cortical thickness is less in the former (Daegling and Grine 1991). *Australopithecus africanus*, however, does not reveal any compromised mechanical competence as a result: given equivalent loads, its mandibles are as rigid in parasagittal

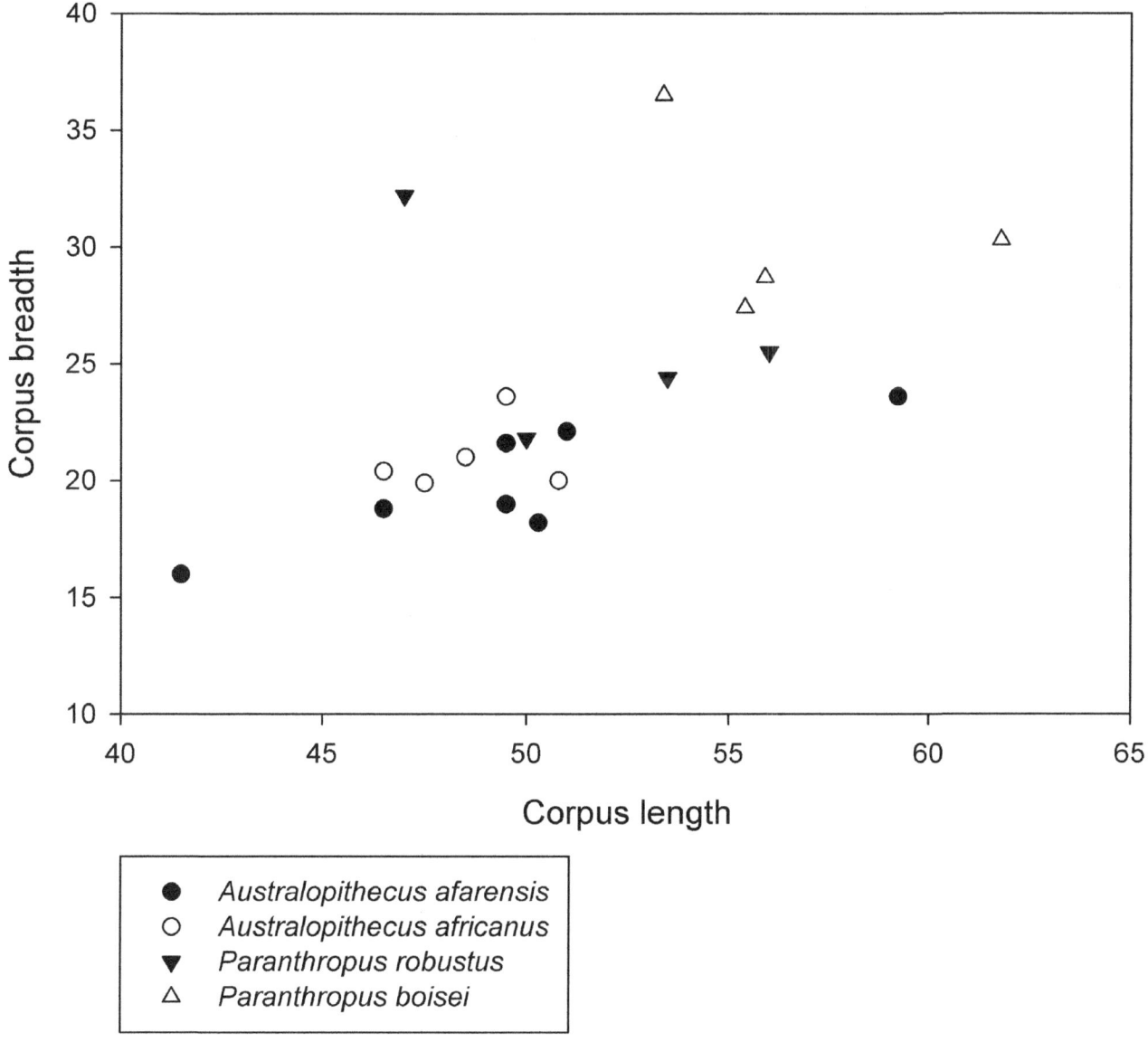

Fig. 9.3 Corpus breadth interpreted against corpus length provides a general, if imprecise, measure of transverse bending stiffness. It is a less reliable measure of torsional strength and rigidity (Daegling 2007a). For the most part, *Paranthropus* mandibles follow a general australopith pattern. The *P. robustus* outlier is SK 12; the *P. boisei* outlier is Omo 7A-125. The length of the corpus is measured from infradentale to the distal margin of M_2

bending as those of *P. robustus* (Daegling and Grine 2007). This indicates that *A. africanus* is simply more economical in the use of cortical bone; i.e., it achieves the same relative structural rigidity with less material.

This finding may be interesting in and of itself, but for dietary inference it is of uncertain utility as there is no clear relationship between cortical packing (i.e., the ratio of cortical area to the entire subperiosteal area) and diet in anthropoids. Orangutans, being amply documented as consumers of both hard and tough foods (summarized in Taylor 2006), have relatively thin cortices relative to African apes

and humans (Daegling 2007b). *Cebus apella* is the most notorious hard-object feeder among platyrrhines, but utilizes the same proportion of bone in cross-section as its congener *C. capucinus,* which does not consume hard objects (Daegling 1992). Species that represent different subfamilies of cercopithecids do not display any apparent differences in cortical packing (Daegling 2002). However, because these comparative data are meager in terms of taxonomic breadth, at this point it is little more than a guess that a null hypothesis of no interspecific differences in cortical packing is the rule rather than the exception. Cortical packing is,

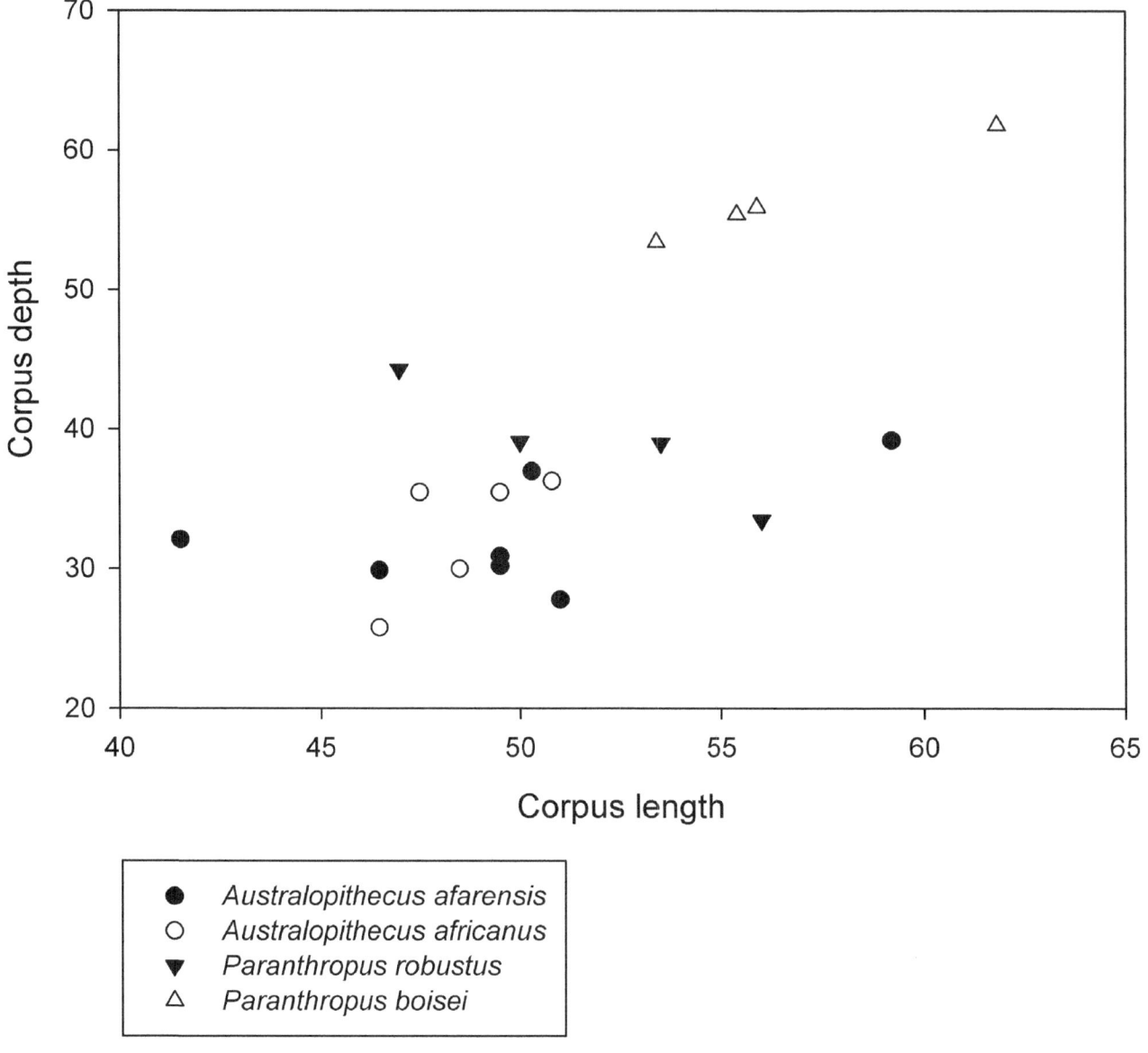

Fig. 9.4 Corpus depth interpreted relative to corpus length is a reasonable proxy for corpus stiffness under parasagittal bending. *Paranthropus boisei* is distinct from other australopiths in having exceptionally deep mandibles. Thus, the large corpora in *P. boisei* are primarily due to their great depth rather than exceptional breadth. The covariation between corpus depth and length is weak in the non-*boisei* forms

however, a relative measure of cortical bone area and relates more to the economical use of cortical bone in resistance of bending and torsion. For other sources of stress (e.g., compression and shear), absolute bone area is the critical quantity for assessing stress-resisting capacity. Corpus geometry (shape) is of far less concern in such cases.

A separate, but no less relevant consideration is whether the mandibles of *P. boisei* were large relative to body size. Smith (1993) and Pilbeam and Gould (1974) suggested that robust australopiths (i.e., *Paranthropus*) simply represented a scaled-up form of *Australopithecus*. The implication is that mandibular proportions are effects of somatic size rather than

trophic adaptation.[2] A definitive assessment of this proposition is not possible for the simple reason that body size is effectively unknown for most australopiths. While methods for estimating body size in extinct taxa from postcranial elements exist (Jungers 1988, 1990), "there is no sign of a well-authenticated

_____

[2]If this postulate is to have any explanatory force, it must assume that body size and diet are independent. In fact, body size has predictable effects on diet in mammals, including primates (Kay 1975; Ravosa 1999; Sailer et al. 1985). Consequently, the assumption that body size and feeding mechanics represent alternative explanations for craniofacial morphology is untenable.

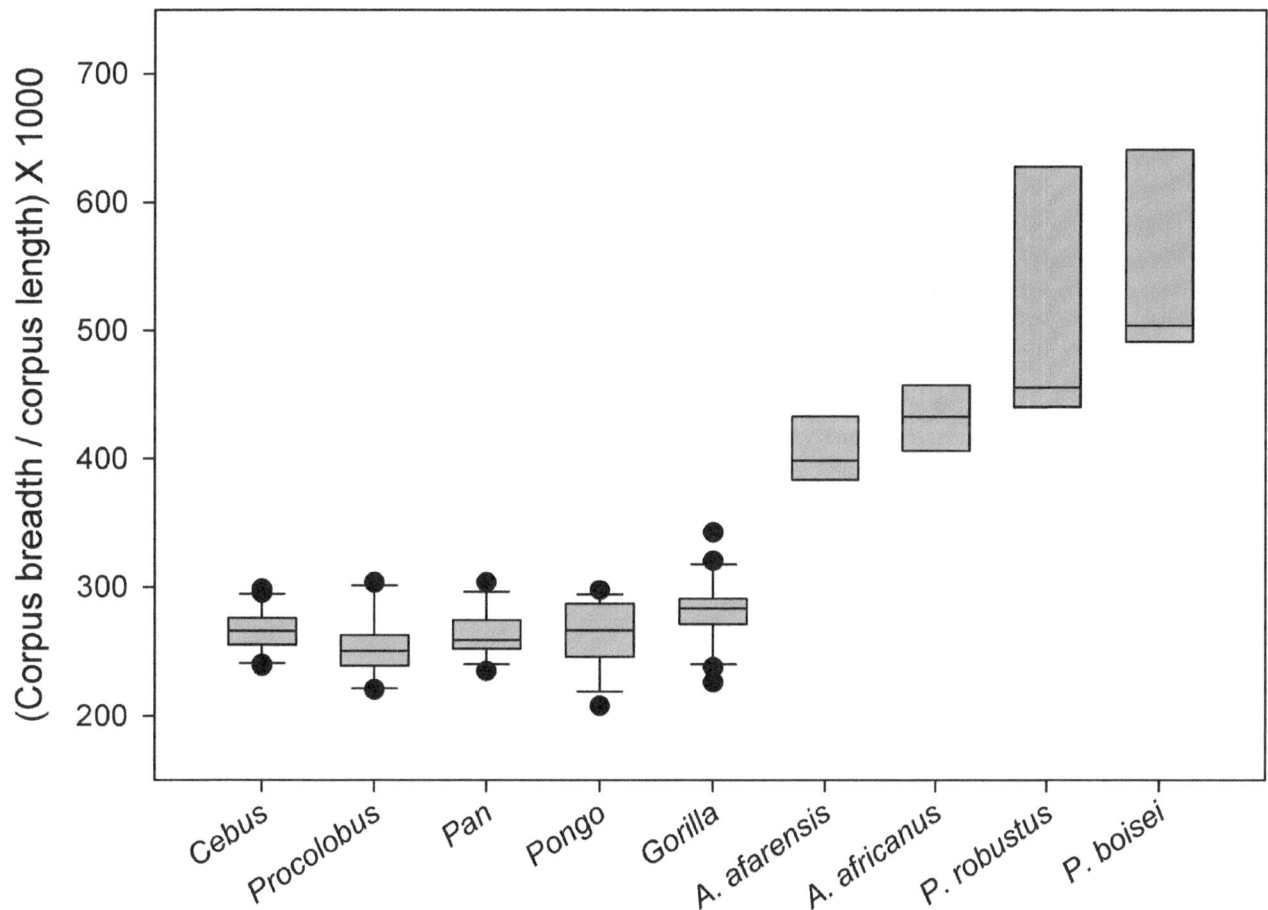

Fig. 9.5 Relatively broad mandibles characterize the australopith radiation in comparison to extant anthropoids. A thick corpus is an ideal structural solution for strengthening a mandible under transverse bending, but also contributes to structural rigidity under loading regimes of torsion, direct shear and parasagittal bending, albeit this does not necessarily represent an optimal structural solution in these cases. The large variation in the *Paranthropus* samples is potentially attributable to a high degree of sexual dimorphism, but taphonomic factors may also contribute to the observed variance. As Wood (1991) notes with respect to East African *Paranthropus*, mandibular dimensions may be diminished by erosion of periosteal surfaces, but they may also be artificially expanded by infilling of post-mortem cracks with matrix. Sample sizes for the australopith samples as in Fig. 9.4; among anthropoids 10 males and females are sampled from each taxon with the exceptions of *Pan* (8 females, 10 males) and *Pongo* (7 females, 6 males)

P. boisei skeleton" (Wood and Constantino 2007: 111) from which to estimate body size in this taxon. Indeed, the only postcranial skeletal remains that have been attributed to *P. boisei* on the basis of associated cranial remains is the poorly preserved KNM-ER 1500 specimen (Grausz et al. 1988). According to Grausz and colleagues (1988), this specimen represents a small female individual, which they attributed to *P. boisei* on the basis of a tiny fragment of the inferior margin of a mandibular corpus that was seen to be similar in size and morphology to the homologous region of the KNM-ER 15930 mandible. The latter was regarded by Walker and Leakey (1988) and by them as a presumptive female specimen of *P. boisei*. Both fossils exhibit a thick corpus below the mental foramen and a blunt inferior marginal crest. Although Wood (1991) has observed that these features are also exhibited by mandibles (e.g., KNM-ER 1802) that are generally attributed

to early *Homo*, the KNM-ER 1500 knee joint does not exhibit the enlarged epiphyses that characterize species of *Homo* (e.g., *Homo erectus*, *H. neanderthalensis* and *H. sapiens*) for which these elements are known with certainty.

Wood and Aiello (1998) suggest that *P. boisei* mandible size is not expected on the basis of body size, but this argument is based on body size estimates derived from cranial remains (specifically, orbit dimensions). This is less than ideal because cranial features may not covary very well with body mass (Hylander 1985) and they may not be independent of mandibular size if these features are influenced by masticatory activity.

Provisionally assuming that KNM-ER 1500 does belong within the *P. boisei* hypodigm, a coarse-grained assessment (not without sampling and methodological issues) can be undertaken. Taking average corpus size and body mass

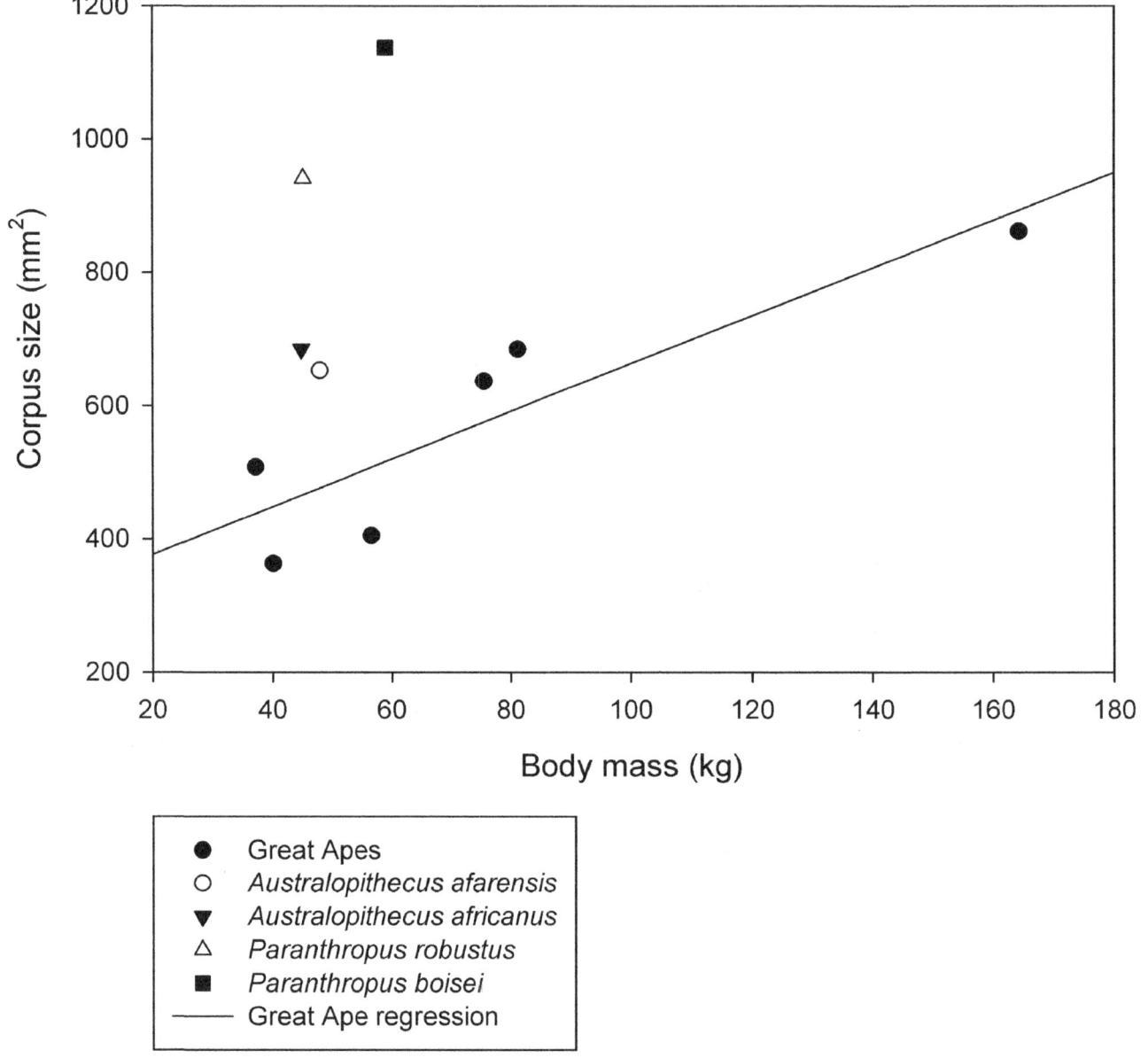

Fig. 9.6 Corpus size in australopiths is not a simple covariate of body size, particularly in *Paranthropus*. For this to be the case in *P. boisei*, one must postulate a body size in excess of that of male gorillas. Points represent mean corpus size (calculated as the product of breadth versus height) and mean body weights determined separately for male and female *Pan*, *Pongo* and *Gorilla*, with corpus size derived from samples represented in Fig. 9.5. Body weights are derived from the literature (Jungers 1988; McHenry 1992) and for the fossil taxa are based on regressions utilizing postcranial elements. An important caveat is that the individuals used to calculate corpus size are not the same as those for which body size data were collected or inferred. Yet only by circular reasoning would one conclude that *Paranthropus* jaw size is a simple correlate of body size (Wood and Aiello 1998)

estimates derived from postcranial elements (Jungers 1988; McHenry 1992), australopiths have mandibles that are unexpectedly large given body size (Fig. 9.6). For *Paranthropus* in particular, their mandibles are relatively large even if these hominins had body sizes approaching those of male gorillas.

The term "robusticity" is frequently invoked in descriptions of hominin mandibles: its meaning is not always clear. It has

been formally defined as the ratio of corpus breadth to corpus height, the "robusticity index" (e.g., Tobias 1966; Leakey et al. 1970; Chamberlain and Wood 1985; Rosas 1995). High index values distinguish australopith jaws relative to others hominins. The term is also invoked informally to mean large or strong, with no particular shape configuration implied. This latter use is really a statement about size alone, and implies (correctly) that australopiths have mandibles that are relatively stronger than

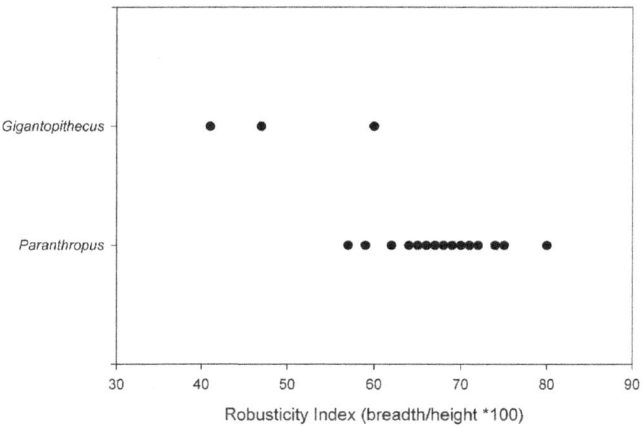

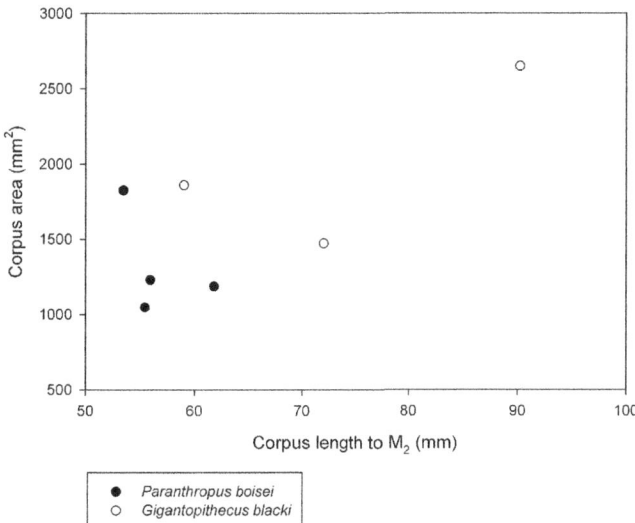

Fig. 9.7 The robusticity index of two "megadont" large-bodied hominoids. *Gigantopithecus blacki* has a large mandible, the shape of which is distinct from *Paranthropus boisei* and recalls living hominoids to a greater degree. The megadont status of *G. blacki* is questionable, since there are no non-gnathic fossils of this taxon from which to independently estimate body mass, although any size smaller than extant gorillas is unlikely (Johnson 1979). The relative narrowness of the *G. blacki* corpora argues against mandibular "robusticity" being allometrically driven. The specimen of *G. blacki* in the *Paranthropus* range is mandible II, a subadult individual. Based on what is known of mandibular growth in *Pan* and *Gorilla*, it is likely that the index value in this individual would have shifted toward the adult *G. blacki* specimens at maturity (Taylor 2002)

Fig. 9.8 The corpus of *Gigantopithecus blacki* is as large as that of *Paranthropus boisei*, even though its cross-sectional shape is different (see Fig. 9.7). Although both taxa's mandibles are accurately described as mechanically robust, it is unclear what the shape differences and size comparability mean in terms of diet. Independent evidence of microwear suggests neither was a hard-object specialist. If both forms represent specialization for herbivory, the challenge is then understanding the mechanical significance of the profound shape differences between them. Phylogenetic inertia may be invoked by arguing that *G. blacki* being a pongine and *Paranthropus* derived from an australopith precursor has canalized their mandibular evolution. While this is reasonable, it is an untested assertion. The larger implication for functional inference is that there may be multiple morphological solutions to the same dietary challenge

those of modern hominoids (Daegling and Grine 2007). The functional meaning of different values of the robusticity index, by contrast, is mysterious. By this ratio, *Paranthropus* has robust mandibles, but by this criterion so does *Lemur* relative to *Pongo*. Regardless of whether the term is used in the sense of shape or size, both australopiths as a group and anthropoids as a whole show increasing robusticity with size (Chamberlain and Wood 1985; Wood and Aiello 1998; Smith 1983).

Observation of these trends raises anew the question of whether what we see in *P. boisei* represents a kind of universal primate solution for having a large body and a low-quality diet that is difficult to process. Alternatively, the trend might have nothing to do with diet or ecology, but rather reflects some sort of allometric imperative that relates to chewing mechanics (Ross et al. 2008; Ravosa et al. 2010), muscular recruitment, or some other intrinsic factor. *Gigantopithecus* provides an instructive counterpoint in this regard. This ape has jaws that are not robust in the shape sense (Fig. 9.7), but in terms of relative size and mechanical rigidity are comparable to those of *P. boisei* (Fig. 9.8). Like *P. boisei*, microwear on *Gigantopithecus* molars provides no evidence of specialization on hard objects (Daegling and Grine 1994). Megadontia and corpus hypertrophy presumably reflect an unusual amount of masticatory effort, but this biomechanical assessment brings us no closer to understanding the particulars of their respective feeding behaviors, even if the isotopic evidence indicates little dietary

overlap. These considerations suggest that while corpus shape and size certainly inform questions of biomechanical performance, diet *per se* is probably inaccessible from the comparative morphology of the mandible. More sensible inferences are probably those that try to understand how differences in masticatory mechanics influence mandibular bone in ontogenetic and evolutionary contexts.

The Perspective from Bone Biology

Experimental research has established that bone is more responsive to dynamic loads than static ones (Rubin et al. 1990). Strain magnitude is but one component of the dynamic environment, and load frequency has been clearly established as an important determinant of bone metabolic activity (Ozcivici et al. 2010). Provided strain magnitude are high enough, as few as four loading cycles a day are sufficient to prevent bone loss, and only 36 cycles per day can engender bone apposition at supraphysiological strains (Rubin et al. 1990). Identical physiological strains of 500 $\mu\epsilon$ induce different responses in bone depending on frequency: low frequencies (1 Hz) do not prevent bone loss, whereas significant formation occurs under frequencies of 15 Hz (McCleod and Rubin 1989). The implication of a vast body of research is that the skeleton's response to strain appears to be dose-mediated (Rubin and Lanyon 1985). An important corollary finding, however, is that patterns of bone apposition do not serve to minimize or homogenize peak strains. Rather, minimization of peak strain does not appear to be the goal of functional adaptation of bone tissue (Rubin et al. 1990). This finding complicates biomechanical interpretation of skeletal tissue in extant organisms, to say nothing of fossils.

What emerges from the literature on bone biology is that bone metabolic activity appears to be tied to the interactive effects of load frequency and load magnitude. Such a finding echoes Hylander (1979) original concern that a few large loads may be as important as a multitude of small ones for functional adaptation ("norms of reaction") during development. It is clear that there are limits to what functional adaptation can achieve in terms of ontogenetic morphological change; raising a chimpanzee on a cooked human diet will not yield an adult mandible that is recognizably human in form. One might then ask whether interspecific differences in mandibular form are consequently inexplicable in terms of functional adaptation. If the answer is in the affirmative, then there has been a colossal misappropriation of research effort over the last 30 years. This seems unlikely. Norms of reaction to alterations in the mandibular stress environment are well known and are manifested ontogenetically (Bouvier and Hylander 1981; Beecher and Corruccini 1981; Corruccini and Beecher 1984; Ciochon et al. 1997; Ravosa et al. 2008), but this does not mean that variation in these norms is not heritable. Genetic assimilation of these reaction norms is an implicit assumption of past and present comparative research; indeed, conceptualizing an intelligible relationship of evolution to development requires it (Via and Lande 1985; Badyaev 2005; Nussey et al. 2005; Pigliucci et al 2006). The alternative is a purely structuralist perspective (*sensu* O'Grady 1986) that in extremis denies that ontogenetic functional activity has any significant role in morphological evolution.

Scott et al. (2014) have provided compelling experimental evidence which cogently argues that hard-object feeding – whether habitual or on a "fall-back" resource basis – may not explain the extreme dentognathic morphology of *P. boisei*. The source of their inference was a comparison of dietary treatments among cohorts of white rabbits which included control and tough diets as well as "annual" and "seasonal" variation. Their experimental analysis suggests *P. boisei* morphology is adequately explained by reliance on tough foods that required prolonged postcanine processing and "concomitantly elevated masticatory stresses owing to higher repetitive loading and longer load durations resulting from extended bouts of milling and grinding" (Scott et al. 2014: 4).

How do developments in bone mechanobiology research change our interpretations of australopith foraging and feeding behavior relative to, for example, Robinson's (1954) "dietary hypothesis" of the ecological differences between *Australopithecus* and *Paranthropus*? At least in terms of the mandibular evidence, our uncertainty has ironically increased with our better understanding of bone function and development. Specifically, we have no clear strategy for distinguishing the morphological "outcome" of a diet of exceptionally hard foods from one that is exceptionally tough, whether that is reckoned on evolutionary or developmental time scales. In terms of overall masticatory work, these diets may not be very different, and if mandibular bone is responsive to some aspect of the stress environment that is sensitive to a general parameter of work, the resultant configurations of the corpus may be indistinct. Hard foods logically require higher bite forces to initiate fracture, but mastication of tough foods appears to engender relatively high levels of mandibular bone strain as well (Weijs and deJongh 1977; Hylander 1979). Thus, the greater number of cycles likely to be required to process tough foods may actually produce more internal work on the mandible than a hard-object diet, especially in an animal saddled with a dentition generally devoid of occlusal relief.

With respect to the question of differentiating high versus repetitive loading environments, one promising line of investigation may be the analysis of secondary remodeling in mandibular bone. Thin sections from a *P. robustus* mandible (SKX 5013) indicate secondary osteonal densities which are relatively low by modern primate standards (Daegling and Grine 2007), including hard-object specialists. Bouvier and Hylander (1981) demonstrated that a hard diet could engender more remodeling activity in macaques, which seemingly throws into question the consensus view that *P. robustus* relied on hard objects to at least some extent. Rejecting durophagy in *P. robustus*, however, may be premature for two reasons. First, the protocol used by Bouvier and Hylander (1981) almost certainly ensured that the control group (with its diet "the consistency of fudge") had fewer daily masticatory cycles than the hard object treatment group. Second, investigation of osteonal densities from mandibles of West African monkeys (Daegling and McGraw 2012; Lad et al. 2013) indicate that species with hard diets (*Cercocebus atys*) have less osteonal bone than species that have fewer hard objects in the diet but engage in more masticatory cycles when feeding (e.g., *Colobus polykomos*). If, in fact, secondary remodeling is more sensitive to accumulated cyclical loads than load intensity, then low incidence of osteonal bone in *P. robustus* remains consistent with interpretations that this taxon was a hard-object feeder. Assuming this to be a valid and generalizable conclusion, one would predict that if *P. boisei* was processing a tough diet requiring a large number of daily masticatory cycles, then the eventual investigation of its mandibular microanatomy will reveal high density of secondary osteonal bone. This prediction rests on the assumption that dynamic strain similarity (Rubin and Lanyon 1984), or the existence of a strain interval that corresponds to a "lazy zone" of bone metabolic activity (Frost 2003), is operative. These models suggest that bone may be locally "tuned" to model and remodel in order to maintain an optimum strain interval during normal activity. Under such assumptions, safety factors in bone do not vary appreciably. The veracity of this last assumption is open to question with respect to the primate skull (Hylander and Johnson 1992). On the mandibular evidence overall, the diet of *Paranthropus boisei* is uncertain. However, it would appear to be safe to assume whatever was ingested required an unusual amount of masticatory effort to process.

Independent Evidence for Diet: Stable Light Isotopes and Dental Microwear

One of the most widely utilized and perhaps best understood of the biogeochemical approaches to palaeodietary reconstruction relies upon stable carbon isotopes. The underlying principle for stable carbon isotope analysis relates to the differences between plants that follow different photosynthetic pathways. Of particular relevance to reconstructing early hominin diets is the observation that in tropical African environments, virtually all trees, bushes, and forbs utilize C_3 photosynthesis, while grasses and sedges use the C_4 pathway. In comparison to C_4 plants, C_3 plants are strongly depleted in ^{13}C relative to atmospheric CO_2. Consequently, C_3 plants have distinctly lower $\delta^{13}C$ values than C_4 plants (Grine et al. 2012). The carbon isotopes in these plants are ultimately incorporated into the tissues of the animals that consume them, with the result that the tooth enamel of dedicated C_4 grass consumers has a distinctive carbon isotope signature from that of animals consuming C_3 vegetation.

Carbon isotope data has been gathered from a substantial number of *P. boisei* specimens (n = 27) that span nearly the entire known temporal depth of the species (ca. 2.2–1.4 Ma) (van der Merwe et al. 2008; Cerling et al. 2011, 2013; Wynn et al. 2013). The sample indicates a diet dominated (ca. 80%) by ^{13}C-enriched foods, and there is no suggestion of any temporal trend within this species sample. Of course, the limitation of these carbon isotope data is that they cannot distinguish between a plant-based diet, a meat-based diet (where, in this case, C_4-consuming animals such as wildebeest or insects were themselves consumed by *P. boisei* individuals), an aquatic diet comprising algae, or an

omnivorous diet that included both the basal herbaceous items as well as potential prey consumed by *P. boisei*. Nevertheless, and despite the observation that the teeth of *P. boisei* were not particularly well-suited to selenodont/ruminant grazing (Kay 1985; Teaford and Ungar 2000), there is no compelling morphological or ancillary evidence for carnivory.

The isotopic evidence for what appears to be a fairly strong reliance on grasses (leaves and/or seeds) and/or sedges by *P. boisei* is perhaps best contextualized with respect to the $\delta^{13}C$ values that have been recorded for specimens of its presumptive lineal ancestor, *P. aethiopicus*. Four specimens attributed to this species have been sampled in the temporal range of 2.5–2.3 Ma (Cerling et al. 2013). Only the molar of KNM-WT 17000 exhibits a value that falls within the observed *P. boisei* range; those for the other *P. aethiopicus* individuals indicate a somewhat stronger C_3 component to the diet. Although the *P. aethiopicus* sample is rather paltry, these data suggest that P. aethiopicus had a diet with a C_4 component of around 50% (or greater). In other words, on average, *P. aethiopicus* consumed substantially more C_3 resources than *P. boisei*, but it had already begun to show a greater consumption of ^{13}C-enriched foods than the majority of earlier East African australopiths – e.g., *Au. afarensis* (Wynn et al. 2013), *Kenyanthropus platyops* (Cerling et al. 2013) and *Au. anamensis* (Cerling et al. 2013). Of particular relevance to the current discussion of the mandibular evidence, Sponheimer et al. (2013) found a significant relationships between early hominin $\delta^{13}C$ values and postcanine occlusal tooth area ($r^2 = 0.86$, t = 5.50, P < 0.01), and the cross-sectional area of the mandibular corpus at the level of the first molar ($r^2 = 0.83$, t = 4.91, P < 0.01 - see Sponheimer et al. [2013: fig. S4B and table S2]). There is a trend for mandibular size to increase through time with an increase in the levels of C_4 consumption. Thus, it would appear that to *P. aethiopicus* had already begun to embark on the road to a dietary shift that entailed a greater consumption of C_4 items, and that *P. boisei* simply continued and culminated this trend. It would then seem that the real task at hand is the explanation of the dished face, anteriorly positioned zygomatics, thick palate, large and robust mandibular corpora and the large cheek teeth with hyperthick and minimally decussated enamel in *P. aethiopicus*.

If the australopiths trended towards increasing reliance on C_4 resources (e.g., increasing folivory) rather than C_3-based foods (e.g., frugivory), they would be expected to exhibit craniodental features that emphasize the generation of and resistance to highly repetitive (but not necessarily low-magnitude) loads during prolonged periods of mastication. Certainly, australopith cranial and mandibular architecture is consistent with this scenario (Daegling et al. 2011, 2013). Thus, there is no compelling reason to invoke hard-object feeding as an explanatory paradigm for "robust" australopith adaptations in the absence of isotopic evidence for the consumption of foods such as nuts, hard fruits and fruit seeds.

Dental microwear as an indicator of diet has been extensively studied in a variety of extant mammal taxa (e.g., Grine et al. 2012 and references therein; DeSantis et al. 2013; Purnell et al. 2013; Schultz et al. 2013; Withnell and Ungar 2014). The relationship between occlusal surface wear textures and the properties of dietary items has been amply documented, such that microwear is capable of distinguishing among broad dietary categories when their constituent items differ in their fracture properties. What would seem to be abundantly clear is that microwear fabrics reflect occlusal movements that relate directly to the fracture properties of the foods being chewed. A recent attempt to demonstrate that siliceous plant phytoliths are incapable of scratching tooth enamel (e.g., Lucas et al. 2013), and that exogenous grit rather than diet is implicated in the formation of microwear, has been effectively rebutted (Rabenold and Pearson 2014; Borrero-Lopez et al. 2015). Indeed, Rabenold and Pearson (2011) have provided evidence in support of their notion of overall 'phytolith load' as a factor in the evolution of thick tooth enamel.

There is no evidence from occlusal microwear of *P. boisei* teeth that this species processed hard objects regularly (Ungar et al. 2008, 2012; Grine et al. 2012). Interestingly, however, a variety of microwear analyses have noted a higher incidence of pitting and greater average texture complexity in *P. robustus* compared to *Au. africanus* and *Au. afarensis* (Grine 1981, 1986; Grine and Kay 1988; Scott et al. 2005), which provides comparative evidence that *P. robustus* consumed more hard and brittle items than earlier hominins. Scott et al. (2005) suggested that this variation compares favorably with extant primate species such as *Lophocebus albigena* and *Cebus apella* that consume hard objects as fallback foods. Whether or not these hard items were consumed regularly or only intermittently, their presence in the diet suggests that the cranial, mandibular and dental characteristics that define *Paranthropus* as a genus are associated with potentially very different species-specific diets.

In summary, evidence from stable isotope analysis (van der Merwe et al. 2008; Cerling et al. 2011) and dental microwear (Ungar et al. 2008, 2012) are consistent with an interpretation of a tough diet in *P. boisei* and do not conform with one of hard objects. Since *Paranthropus* has a craniofacial morphology unlike any living animal (making predictions of biomechanical performance from finite element models [e.g., Dzialo et al. 2014; Smith et al. 2015] somewhat tenuous), it is difficult to argue that their skulls compel an interpretation of any particular dietary specialization. Indeed, paleoecological evidence can be marshaled to argue that *P. boisei* was more of a generalist (Wood and Strait 2004).

Evidence from Dental Morphology and Structure

The thick enamel, postcanine megadontia and bunodonty of australopith (especially *Paranthropus*) dentitions have been invoked as *prima facie* evidence for hard-object feeding (Strait et al. 2008, 2013). The proposal that *Paranthropus* subsisted on tough foods may seem to be nonsensical given the morphology of its dentition (assuming, for example, that fibrous, essentially two-dimensional foods are most efficiently processed by sharp shearing crests [Ungar 2007]). By a criterion of optimal utility, this is correct, but it also dismisses the reality of phylogenetic or developmental constraint. It is hardly unreasonable to posit that *P. boisei's* phylogenetic heritage ensured that it would be endowed with a large, bunodont, and thickly enamelled postcanine battery. Though these teeth would by ideally suited for crushing hard objects, it does not necessarily follow that this is what they were used for. Indeed, Berthaume et al. (2010) showed that the teeth of P. boisei would not have been not particularly efficient at fracturing hard food items, and comparative microwear would appear to rule this out (Walker 1981; Ungar et al. 2008, 2012).

Thick enamel has been repeatedly cited as an attribute that coincides with durophagy in primates (Kay 1981; Dumont 1995; Lambert et al. 2004; Lucas et al. 2008; Vogel et al. 2008; Constantino et al. 2009). The fact that *P. boisei* sports "hyperthick" enamel (Beynon and Wood 1986; Grine and Martin 1988; Skinner et al. 2015) can lead to the conclusion that hard objects "explain" the craniodental complex of these hominins. Yet broader comparative datasets suggest that thick enamel in primates may be equally well explained as an adaptation to prevent attrition owing to abrasive diets that may not have hard objects as a major component (Rabenold and Pearson 2011; Lucas et al. 2013). This finding implicates longevity as an important ecological covariate of enamel thickness variation (Pampush et al. 2013). Indeed, Galbany et al. (2014) have suggested that not only the physical properties of the foods consumed, but the underlying soil composition – particularly its quartz richness – are factors that significantly impact tooth wear.

Considering enamel thickness on its own, the nature of its "adaptive signal" in *P. boisei*, or any other hominin, remains unsettled. However, if the hyper-thick enamel of *P. boisei* is taken at face value as an adaptation to hard-object processing, as posited by Lucas et al. (2008: 383), where "the mastication of hard objects was important enough to select for a thick enamel cap, and … an increased enamel thickness of the postcanine dentition resulted in relatively greater fitness and hence was under positive selection pressure," then one might also expect the structure of the enamel to be subjected to the same selective force. However, the strength and distribution of prism decussation in enamel is recognized as being a critical mechanism to prevent crack propogation (Spears and Macho 1998; Macho et al. 2003, 2005; Popowics et al. 2004; Shimizu et al. 2005; Shimizu and Macho 2008; Xie et al. 2008; Bajaj and Arola 2009; Yahyazadehfar et al. 2013; Yilmaz et al. 2015). Yet, the hyper-thick enamel of *P. boisei* is curiously devoid of strong decussation (Beynon and Wood 1986; Ramirez-Rozzi 1998; Macho 2014). As such, the thickly enameled, bunodont cheek teeth of *P. boisei* are clearly capable of resisting wear but they are not structurally optimized to resist fractures engendered by hard object feeding.

Occlusion and Bite Forces

Paranthropus boisei had tall mandibular rami. As noted above, this has important functional consequences. The tall ramus should reduce the occlusal force gradient along the tooth row (Ward and Molnar 1980), effectively equalizing bite forces across the postcanine dentition. The anterior component of the power stroke that is enhanced by a tall ramus may have combined with mediolateral jaw movements to create a "rotary" masticatory action (Rak and Hylander 2008). This may account for the nearly flat pattern of attrition that characterizes *Paranthropus* cheek teeth (Grine 1981). The mechanical consequence of this masticatory configuration was likely an increase in torsional moments acting on the mandibular corpus (Rak and Hylander 2008). A large corpus is an effective solution for minimizing the stresses associated with these moments, to some degree independent of geometry.

Once again, these observations do not resolve the question of diet in *P. boisei*. With the observation of megadontia and extremely thick enamel, they suggest bulk processing of low-quality foods in a high-attrition environment. Whether this means particularly hard or tough foods is an open question, but given that either requires enhanced bite force, adductor hypertrophy has to accompany large expansion of occlusal area to maintain equivalent bite "pressures" (Demes and Creel 1988; Wroe et al. 2010). Big teeth are not necessary for imparting stress to foods – indeed, for large items they could be detrimental in this regard – but they provide insurance against wear (Lucas et al. 1986).

Conclusion

Paranthropus boisei mandibles indicate that the species had a diet requiring a large amount of masticatory work. Beyond this, their jaws tell us little about the foods that were eaten. A "robust" (i.e., thick) mandible is not the inevitable outcome of dietary specialization in large-bodied hominoids, as

Gigantopithecus mandibles may represent a heavy masticatory workload, but one using a jaw of distinct structural geometry. The combination of postcanine megadontia, mandibular corpus hypertrophy and thick enamel does not compel an interpretation that *Paranthropus boisei* was adapted for consumption of hard objects. It is plausible that the bulk processing of low quality fibrous foods was the target of natural selection in this lineage. This hypothesis requires the assumption that bunodonty in *P. boisei* was not the optimal solution for the comminution of its food. Phylogenetic considerations (viz., brachydont, bunodont, thick-enameled precursors) permit the recognition that an optimal occlusal solution (e.g., some form of hypsodonty or lophodonty) may not have been available to this species.

Because *Paranthropus* has a craniofacial morphology unlike any living animal, predictions of biomechanical performance from finite element or more elementary models are, at best, tentative. After all, biomechanical modelling also has been employed to argue that the *Paranthropus boisei* cranium represents defensive adaptations for interspecific agonism. Thus, craniofacial morphology – analyzed in isolation – can lend itself to a multitude of adaptive just-so stories.

Acknowledgments We thank Erella Hovers and Assaf Marom for the invitation to contribute to this volume. We are honored to be included in a volume that recognizes the enduring contributions of Yoel Rak to paleoanthropology. Yoel has been that rare combination of an individual who managed to be a perpetual voice of reason while at the same time encouraging ideas to challenge the status quo. The professional enrichment that Yoel has provided to our careers is exceeded only by the personal enrichment that he brought to our lives in our varied, but always memorable, interactions over the years. Michael Berthaume and an anonymous reviewer provided cogent critiques that improved the manuscript.

References

Badyaev, A. V. (2005). Stress-induced variation in evolution: From behavioural plasticity to genetic assimilation. *Proceedings of the Royal Society B, 272*, 877–886.

Bajaj, D., & Arola, D. D. (2009). On the R-curve behavior of human tooth enamel. *Biomaterials, 30*(23), 4037–4046.

Beecher, R. M., & Corruccini, R. S. (1981). Effects of dietary consistency on craniofacial and occlusal development in the rat. *The Angle Orthodontist, 51*, 61–69.

Benazzi, S., Kullmer, O., Grosse, I. R., & Weber, G. W. (2012). Brief communication: Comparing loading scenarios in lower first molar supporting bone structure using 3D finite element analysis. *American Journal of Physical Anthropology, 147*, 128–134.

Berthaume, M., Grosse, I. R., Patel, N. D., Strait, D. S., Wood, S., & Richmond, B. G. (2010). The effect of early hominin occlusal morphology on the fracturing of hard food items. *The Anatomical Record, 293*, 594–606.

Beynon, A. D., & Wood, B. A. (1986). Variations in enamel thickness and structure in East African hominids. *American Journal of Physical Anthropology, 70*, 177–193.

Borrero-Lopez, O., Pajares, A., Constantino, P. J., & Lawn, B. R. (2015). Mechanics of microwear traces in tooth enamel. *Acta Biomaterialia, 14*, 146–153.

Bouvier, M. (1986a). A biomechanical analysis of mandibular scaling in Old World monkeys. *American Journal of Physical Anthropology, 69*, 473–482.

Bouvier, M. (1986b). Biomechanical scaling of mandibular dimensions in New World monkeys. *International Journal of Primatology, 7*, 551–567.

Bouvier, M., & Hylander, W. L. (1981). The effect of strain on cortical bone structure in macaques (*Macaca mulatta*). *Journal of Morphology, 167*, 1–12.

Carrier, D. R., & Morgan, M. H. (2014). Protective buttressing of the hominin face. *Biological Review, 90*, 330–346.

Cerling, T. E., Mbua, E., Kirera, F. M., Manthi, F. K., Grine, F. E., Leakey, M. G., et al. (2011). Diet of *Paranthropus boisei* in the early Pleistocene of East Africa. *Proceedings of the National Academy of Sciences USA, 108*, 9337–9341.

Cerling, T. E., Manthi, F. K., Mbua, E. N., Leakey, L. N., Leakey, M. G., Leakey, R. E., et al. (2013). Stable isotope-based diet reconstructions of Turkana Basin hominins. *Proceedings of the National Academy of Sciences USA, 110*, 10501–10506.

Chamberlain, A. T., & Wood, B. A. (1985). A reappraisal of variation in hominid mandibular corpus dimensions. *American Journal of Physical Anthropology, 66*, 399–405.

Ciochon, R. L., Nisbett, R. A., & Corruccini, R. S. (1997). Dietary consistency and craniofacial development related to masticatory function in minipigs. *Journal of Craniofacial Genetics and Developmental Biology, 17*, 96–102.

Constantino, P. J., Lucas, P. W., Lee, J. J. W., & Lawn, B. R. (2009). The influence of fallback foods on great ape tooth enamel. *American Journal of Physical Anthropology, 140*, 653–660.

Constantino, P. J., Lee, J. J. W., Chai, H., Zipfel, B., Ziscovici, C., Lawb, B. R., et al. (2010). Tooth chipping can reveal the diet and bite forces of fossil hominins. *Biology Letters, 6*, 826–829.

Constantino, P. J., Lee, J. J. W., Morris, D., Lucas, P. W., Hartstone-Rose, A., Lee, W.-K., et al. (2011). Adaptation to hard-object feeding in sea otters and hominins. *Journal of Human Evolution, 61*, 89–96.

Corruccini, R. S., & Beecher, R. M. (1984). Occlusofacial morphological integration lowered in baboons raised on soft diet. *Journal of Craniofacial Genetics and Developmental Biology, 4*, 135–142.

Daegling, D. J. (1992). Mandibular morphology and diet in the genus *Cebus*. *International Journal of Primatology, 13*, 545–570.

Daegling, D. J. (2002). Bone geometry in cercopithecoid mandibles. *Archives of Oral Biology, 47*, 315–325.

Daegling, D. J. (2007a). Morphometric estimation of torsional stiffness and strength in primate mandibles. *American Journal of Physical Anthropology, 132*, 261–266.

Daegling, D. J. (2007b). Relationship of bone utilization and biomechanical competence in hominoid mandibles. *Archives of Oral Biology, 52*, 51–63.

Daegling, D. J., & Grine, F. E. (1991). Compact bone distribution and biomechanics of early hominid mandibles. *American Journal of Physical Anthropology, 86*, 321–339.

Daegling, D. J., & Grine, F. E. (1994). Bamboo feeding, dental microwear, and diet of the Pleistocene ape *Gigantopithecus blacki*. *South African Journal of Science, 90*, 527–532.

Daegling, D. J., & Grine, F. E. (2007). Mandibular biomechanics and the paleontological evidence for the evolution of human diet. In P. S. Ungar (Ed.), *The evolution of human diet: The known, the Unknown, and the unknowable* (pp. 77–105). New York: Oxford University Press.

Daegling, D. J., & Hylander, W. L. (1998). Biomechanics of torsion in the human mandible. *American Journal of Physical Anthropology, 105*, 73–87.

Daegling, D. J., McGraw, W. S., Ungar, P. S., Pampush, J. D., Vick, A. E., & Bitty, E. A. (2011). Hard-object feeding in sooty mangabeys

(*Cercocebus atys*) and interpretation of early hominin feeding ecology. *PLoS ONE, 6*, e23095.

Daegling, D. J., & McGraw, W. S. (2012). Osteonal bone density in the mandibles of West African colobines. *American Journal of Physical Anthropology, Supplement, 54*, 124.

Daegling, D. J., Judex, S., Ozcivici, E., Ravosa, M. J., Taylor, A. B., Grine, F. E., et al. (2013). Viewpoints: Feeding mechanics, diet, and dietary adaptations in early hominins. *American Journal of Physical Anthropology, 151*, 356–371.

Daegling, D. J., Granatosky, M. C., & McGraw, W. S. (2014). Ontogeny of material stiffness heterogeneity in the macaque mandibular corpus. *American Journal of Physical Anthropology, 153*, 297–304.

Dechow, P. C., & Hylander, W. L. (2000). Elastic properties and masticatory bone stress in the macaque mandible. *American Journal of Physical Anthropology, 112*, 553–574.

Demes, B., & Creel, N. (1988). Bite force, diet, and cranial morphology of fossil hominids. *Journal of Human Evolution, 17*, 657–670.

DeSantis, L. R. G., Scott, J. R., Schubert, B. W., Donohue, S. L., McCray, B. M., Van Stolk, C. A., et al. (2013). Direct comparisons of 2D and 3D dental microwear proxies in extant herbivorous and carnivorous mammals. *PLoS ONE, 8*, e71428.

DuBrul, E. L. (1977). Early hominid feeding mechanisms. *American Journal of Physical Anthropology, 47*, 305–320.

Dumont, E. R. (1995). Enamel thickness and dietary adaptation among extant primates and chiropterans. *Journal of Mammalogy, 76*, 1127–1136.

Dzialo, C., Wood, S. A., Berthaume, M., Smith, A., Dumont, E. R., Benazzi, S., et al. (2014). Functional implications of squamosal suture size in *Paranthropus boisei*. *American Journal of Physical Anthropology, 153*, 260–268.

Frost, H. M. (2003). Bone's mechanostat: A 2003 update. *The Anatomical Record A: Discoveries in Molecular, Cellular and Evolutionary Biology, 275*, 1081–1101.

Galbany, J., Romero, A., Mayo-Alesón, M., Itsoma, F., Gamarra, B., Pérez-Pérez, A., et al. (2014). Age-related tooth wear differs between forest and savanna primates. *PLoS ONE, 9*, e94938.

Grausz, H. M., Leakey, R. E., Walker, A. C., & Ward, C. V. (1988). Associated cranial and postcranial bones of *Australopithecus boisei*. In F. E. Grine (Ed.), *Evolutionary history of the "robust" australopithecines* (pp. 127–132). New York: Aldine de Gruyter.

Grine, F. E. (1981). Trophic differences between 'gracile' and 'robust' australopithecines: A scanning electron microscope analysis of occlusal events. *South African Journal of Science, 77*, 203–230.

Grine, F. E. (1986). Dental evidence for dietary differences in *Australopithecus* and *Paranthropus*: A quantitative analysis of permanent molar microwear. *Journal of Human Evolution, 15*, 783–822.

Grine, F. E., & Kay, R. F. (1988). Early hominid diets from quantitative image analysis of dental microwear. *Nature, 333*, 765–768.

Grine, F. E., & Martin, L. B. (1988). Enamel thickness and development in *Australopithecus* and *Paranthropus*. In F. E. Grine (Ed.), *Evolutionary history of the "robust" australopithecines* (pp. 3–42). New York: Aldine de Gruyter.

Grine, F. E., Sponheimer, M., Ungar, P. S., Lee-Thorp, J., & Teaford, M. F. (2012). Dental microwear and stable isotopes inform the paleoecology of extinct hominins. *American Journal of Physical Anthropology, 148*, 285–317.

Hirabayashi, M., Motoyoshi, M., Ishimaru, T., Kasai, K., & Namura, S. (2002). Stresses in mandibular cortical bone during mastication: Biomechanical considerations using a three-dimensional finite element method. *Journal of Oral Science, 44*, 1–6.

Hsieh, Y. F., Robling, A. G., Ambrosius, W. T., Burr, D. B., & Turner, C. H. (2001). Mechanical loading of diaphyseal bone in vivo: The strain threshold for an osteogenic response varies with location. *Journal of Bone Mineral Research, 16*, 2291–2297.

Hylander, W. L. (1979). The functional significance of primate mandibular form. *Journal of Morphology, 160*, 223–240.

Hylander, W. L. (1985). Mandibular function and biomechanical stress and scaling. *American Zoologists, 25*, 315–330.

Hylander, W. L. (1988). Implications of *in vivo* experiments for interpreting the functional significance of "robust" australopithecine jaws. In F. E. Grine (Ed.), *Evolutionary history of the "robust" australopithecines* (pp. 55–83). New York: Aldine de Gruyter.

Hylander, W. L., & Johnson, K. R. (1992). Strain gradients in the craniofacial region of primates. In Z. Davidovitch (Ed.), *Biological mechanisms of tooth movement and craniofacial adaptation* (pp. 559–569). Columbus, OH: Ohio State University College of Dentistry.

Ishigaki, S., Nakano, T., Yamada, S., Nakamura, T., & Takashima, F. (2003). Biomechanical stress in bone surrounding an implant under simulated chewing. *Clinical Oral Implants Research, 14*, 97–102.

Johnson, A. E., Jr. (1979). Skeletal estimates of *Gigantopithecus* based on a gorilla analogy. *Journal of Human Evolution, 8*, 585–587.

Jungers, W. L. (1988). New estimates of body size in australopithecines. In F. E. Grine (Ed.), *Evolutionary history of the "robust" australopithecines* (pp. 115–125). New York: Aldine de Gruyter.

Jungers, W. L. (1990). Problems and methods in reconstructing body size in fossil primates. In J. Damuth & B. J. MacFadden (Eds.), *Body size in mammalian paleobiology: Estimation and biological implications* (pp. 103–118). Cambridge: Cambridge University Press.

Kay, R. F. (1975). The functional adaptations of primate molar teeth. *American Journal of Physical Anthropology, 43*, 195–215.

Kay, R. F. (1981). The nut-crackers–a new theory of the adaptations of the Ramapithecinae. *American Journal of Physical Anthropology, 55*, 141–151.

Kay, R. F. (1985). Dental evidence for the diet of Australopithecus. *American Journal of Physical Anthropology, 14*, 15–41.

Lad, S. E., Daegling, D. J., & McGraw, W. S. (2013). Mandibular remodeling in sympatric West African cercopithecids. *American Journal of Physical Anthropology, Supplement, 56*, 176.

Lambert, J. E., Chapman, C. A., Wrangham, R. W., & Conclin-Brittain, N. L. (2004). Hardness of cercopithecine foods: Implications for the critical function of enamel thickness in exploiting fallback foods. *American Journal of Physical Anthropology, 125*, 363–368.

Leakey, M., Tobias, P. V., Martyn, J. E., & Leakey, R. E. (1970). An Acheulean industry with prepared core technique and the discovery of a contemporary hominid mandible at Lake Baringo, Kenya. *Proceedings of the Prehistoric Society (New Series), 35*, 48–76.

Lucas, P. W., Corlett, R. T., & Luke, D. A. (1986). Postcanine tooth size and diet in anthropoid primates. *Zeitschrift Morphologie Anthropologie, 76*, 253–276.

Lucas, P. W., Constantino, P., Wood, B., & Lawn, B. (2008). Dental enamel as a dietary indicator in mammals. *BioEssays, 30*, 374–385.

Lucas, P. W., Omar, R., Al-Fadhalah, K., Almusallam, A. S., Hebry, A. G., Michael, S., et al. (2013). Mechanisms and causes of wear in tooth enamel: Implications for hominin diets. *Journal of the Royal Society Interface, 10*. doi:10.1098/rsif.2012.0923.

Macho, G. A. (2014). Baboon feeding ecology informs the dietary niche of *Paranthropus boisei*. *PloS ONE, 9*, e84942.

Macho, G. A., Jiang, Y., & Spears, I. R. (2003). Enamel microstructure - a truly threedimensional structure. *Journal of Human Evolution, 45*, 81–90.

Macho, G. A., Shimizu, D., Jiang, Y., & Spears, I. R. (2005). *Australopithecus anamensis*: A finite-element approach to studying the functional adaptations of extinct hominins. *The Anatomical Record, 283A*, 310–318.

Martin, R. B., Burr, D. B., & Sharkey, N. A. (1998). *Skeletal tissue mechanics*. New York: Springer.

McHenry, H. M. (1992). Body size and proportions in early hominids. *American Journal of Physical Anthropology, 87*, 407–431.

McLeod, K., & Rubin, C. (1989). Predictions of osteogenic mechanical loading paradigms from electrical response data. *Transmission Biology Growth Reproduction Society, 9*, 20.

Nussey, D. H., Postma, E., Gienapp, P., & Visser, M. E. (2005). Selection on heritable phenotypic plasticity in a wild bird population. *Science, 310*, 304–306.

O'Grady, R. T. (1986). Historical processes, evolutionary explanations, and problems with teleology. *Canadian Journal of Zoology, 64*, 1010–1020.

Ozcivici, E., Luu, Y. K., Adler, B., Qin, Y.-X., Rubin, J., Judex, S., et al. (2010). Mechanical signals as anabolic agents in bone. *Nature Review of Rheumatology, 6*, 50–59.

Pampush, J. D., Duque, A. C., Burrows, B. R., Daegling, D. J., Kennedy, W. F., & McGraw, W. S. (2013). Homoplasy and thick enamel in primates. *Journal of Human Evolution, 64*, 216–224.

Pigliucci, M., Murren, C. J., & Schlichting, C. D. (2006). Phenotypic plasticity and evolution by genetic assimilation. *Journal of Experimental Biology, 209*, 2362–2367.

Pilbeam, D., & Gould, S. J. (1974). Size and scaling in human evolution. *Science, 186*, 892–901.

Plavcan, J. M., & Daegling, D. J. (2006). Interspecific and intaspecific relationships between tooth size and jaw size in primates. *Journal of Human Evolution, 51*, 171–184.

Popowics, T. E., Rensberger, J. M., & Herring, S. W. (2004). Enamel microstructure and microstrain in the fracture of human and pig molar cusps. *Archives of Oral Biology, 49*, 595–605.

Purnell, M. A., Crumpton, N., Gill, P. G., Jones, G., & Rayfield, E. J. (2013). Within-guild dietary discrimination from 3-D textural analysis of tooth microwear in insectivorous mammals. *Journal of Zoology, 291*, 249–257.

Rabenold, D., & Pearson, O. M. (2011). Abrasive, silica phytoliths and the evolution of thick molar enamel in primates, with implications for the diet of Paranthropus boisei. *PLoS ONE, 6*, e28379.

Rabenold, D., & Pearson, O. M. (2014). Scratching the surface: A critique of Lucas et al. (2013) conclusion that phytoliths do not abrade enamel. *Journal of Human Evolution, 74*, 130–133.

Rak, Y. (1983). *The australopithecine face*. New York: Academic Press.

Rak, Y., & Hylander, W. L. (2008). What else is the tall mandibular ramus of the robust australopiths good for? In C. J. Vinyard, M. J. Ravosa, & C. E. Wall (Eds.), *Primate craniofacial function and biology* (pp. 431–442). New York: Springer.

Ramirez-Rozzi, F. (1998). Can enamel microstructure be used to establish the presence of different species of Plio-Pleistocene hominids from Omo, Ethiopia? *Journal of Human Evolution, 35*, 543–576.

Ravosa, M. J. (1999). Anthropoid origins and the modern symphysis. *Folia Primatologia, 70*, 65–78.

Ravosa, M. J., Ross, C. F., Williams, S. H., & Costley, D. B. (2010). Allometry of masticatory loading parameters in mammals. *The Anatomical Record, 293*, 557–571.

Ravosa, M. J., Lopez, E. K., Menegaz, R. A., Stock, S. R., Stack, M. S., & Hamrick, M. W. (2008). Adaptive plasticity in the mammalian masticatory complex: You are what, and how, you eat. In C. J. Vinyard, M. J. Ravosa, & C. E. Wall (Eds.), *Primate craniofacial function and biology* (pp. 293–328). New York: Springer.

Robinson, J. T. (1954). Prehominid dentition and hominid evolution. *Evolution, 8*, 324–334.

Rosas, A. (1995). Seventeen new mandibular specimens from the Atapuerca/Ibeas Middle Pleistocene hominids sample (1985–1992). *Journal of Human Evolution, 28*, 533–559.

Ross, C. F., Reed, D. A., Washington, R. L., Eckhardt, A., Anapol, F., & Shahnoor, N. (2008). Scaling of chew cycle duration in primates. *American Journal of Physical Anthropology, 138*, 30–44.

Rubin, C. T., & Lanyon, L. E. (1984). Dynamic strain similarity in vertebrates; an alternative to allometric limb bone scaling. *Journal of Theoretical Biology, 107*, 321–327.

Rubin, C. T., & Lanyon, L. E. (1985). Regulation of bone mass by mechanical loading: The effect of peak strain magnitude. *Calcified Tissue International, 37*, 441–447.

Rubin, C. T., McLeod, K. J., & Bain, S. D. (1990). Functional strains and cortical bone adaptation: Epigenetic assurance of skeletal integrity. *Journal of Biomechanics, 23*, 43–54.

Sailer, L. D., Gaulin, S. J. C., Boster, J. S., & Kurland, J. A. (1985). Measuring the relationship between dietary quality and body size in primates. *Primates, 26*, 14–27.

Schwartz-Dabney, C. L., & Dechow, P. C. (2003). Variations in cortical material properties throughout the human dentate mandible. *American Journal of Physical Anthropology, 120*, 252–277.

Schultz, E., Piotrowski, V., Clauss, M., Mau, M., Merceron, G., & Kaiser, T. M. (2013). Dietary abrasiveness is associated with variability of microwear and dental surface texture in rabbits. *PLoS ONE, 8*, e56167.

Scott, J. E., McAbee, K. R., Eastman, M. M., & Ravosa, M. J. (2014). Experimental perspectives on fallback foods and dietary adaptations in early hominins. *Biology Letters, 10*, 20130789.

Scott, R. S., Ungar, P. S., Bergstrom, T. S., Brown, C. A., Grine, F. E., Teaford, M. F., et al. (2005). Dental microwear texture analysis reflects diets of living primates and fossil hominins. *Nature, 436*, 693–695.

Shimizu, D., & Macho, G. A. (2008). Effect of enamel prism decussation and chemical composition on the biomechanical behaviour of dental tissue: A theoretical approach to determine the loading conditions to which modern human teeth are adapted. *The Anatomical Record, 291*, 182–208.

Shimizu, D., Macho, G. A., & Spears, I. R. (2005). Effect of prism orientation and loading direction on contact stresses in prismatic enamel of primates: Implications for interpreting wear patterns. *American Journal of Physical Anthropology, 126*, 427–434.

Skinner, M. M., Alemseged, Z., Gaunitz, C., & Hublin, J.-J. (in press, 2015). Enamel thickness trends in Plio-Pleistocene hominin mandibular molars. *Journal of Human Evolution*.

Smith, A. L., Benazzi, S., Ledogar, J. A.,Tamvada, K., Pryor Smith, L. C., Weber, G. W. et al. (2015). The feeding biomechanics and dietary ecology of *Paranthropus boisei*. *The Anatomical Record, 298*, 145–167.

Smith, R. J. (1983). The mandibular corpus of female primates: Taxonomic, dietary and allometric correlates of interspecific variations in size and shape. *American Journal of Physical Anthropology, 61*, 315–330.

Smith, R. J. (1993). Categories of allometry: Body size versus biomechanics. *Journal of Human Evolution, 24*, 173–182.

Spears, I. R., & Macho, G. A. (1998). Biomechanical behaviour of modern human molars: Implications for interpreting the fossil record. *American Journal of Physical Anthropology, 106*, 467–482.

Sponheimer, M., Alemseged, Z., Cerling, T. E., Grine, F. E., Kimbel, W. H., Leakey, M. G., et al. (2013). Isotopic evidence of early hominin diets. *Proceedings of the National Academy of Sciences, 110*(26), 10513–10518.

Strait, D. S., Wright, B. W., Richmond, B. G., Ross, C. F., Dechow, P. C., Spencer, M. A., et al. (2008). Craniofacial strain patterns during premolar loading: Implications for human evolution. In C. J. Vinyard, M. J. Ravosa, & C. E. Wall (Eds.), *Primate craniofacial function and biology* (pp. 173–198). New York: Springer.

Strait, D. S., Constantino, P., Lucas, P. W., Richmond, B. G., Spencer, M. A., Dechow, P. A., et al. (2013). Viewpoints: Diet and dietary adaptations in early hominins: The hard food perspective. *American Journal of Physical Anthropology, 151*, 339–355.

Taylor, A. B. (2002). Masticatory form and function in the African Apes. *American Journal of Physical Anthropology, 117*, 133–156.

Taylor, A. B. (2006). Feeding behavior, diet, and the functional consequences of jaw form in orangutans, with implications for the evolution of *Pongo*. *Journal of Human Evolution, 50*, 377–393.

Teaford, M. F., & Ungar, P. S. (2000). Diet and the evolution of the earliest human ancestors. *Proceedings of the National Academy of Sciences USA, 97*, 13506–13511.

Tobias, P. V. (1966). A re-examination of the Kedung Brubus mandible. *Zoologische Mededelingen, 41*, 307–320.

Ungar, P. S. (2007). Dental functional morphology. In P. S. Ungar (Ed.), *The evolution of human diet: The known, the unknown, and the unknowable* (pp. 39–55). New York: Oxford University Press.

Ungar, P. S., Grine, F. E., & Teaford, M. F. (2008). Dental microwear and diet of the Plio-Pleistocene hominin *Paranthropus boisei*. *PLoS ONE, 3*, e2044.

Ungar, P. S., Krueger, K. L., Blumenschine, R. J., Njau, J., & Scott, R. S. (2012). Dental microwear texture analysis of hominins recovered by the Olduvai Landscape Paleoanthropology Project, 1995–2007. *Journal of Human Evolution, 63*, 429–437.

van der Merwe, N. J., Masao, F. T., & Bamford, M. K. (2008). Isotopic evidence for contrasting diets of early hominins *Homo habilis* and *Australopithecus boisei* of Tanzania. *South African Journal of Science, 104*, 153–155.

Via, S., & Lande, R. (1985). Genotype environment interaction and the evolution of phenotypic plasticity. *Evolution, 39*, 505–522.

Vogel, E. R., van Woerden, J. T., Lucas, P. W., Utami Atmoko, S. S., van Schaik, C. P., & Dominy, M. J. (2008). Functional ecology and evolution of hominoid molar enamel thickness: *Pan troglodytes schweinfurthii* and *Pongo pygmaeus wurmbii*. *Journal of Human Evolution, 55*, 60–74.

Walker, A. C. (1981). Diet and teeth: Dietary hypotheses and human evolution. *Philosophical Transactions of the Royal Society London B, 292*, 57–64.

Walker, A. C., & Leakey, R. E. (1988). The evolution of *Australopithecus boisei*. In F. E. Grine (Ed.), *Evolutionary history of the "robust" australopithecines* (pp. 247–258). New York: Aldine de Gruyter.

Ward, S. C., & Molnar, S. (1980). Experimental stress analysis of topographic diversity in early hominid gnathic morphology. *American Journal of Physical Anthropology, 53*, 383–395.

Weijs, W. A., & deJongh, H. J. (1977). Strain in mandibular alveolar bone during mastication in the rabbit. *Archives of Oral Biology, 22*, 667–675.

Williams, S. H., Wright, B. W., Truong, V. D., Daubert, C. R., & Vinyard, C. J. (2005). Mechanical properties of foods used in experimental studies of primate masticatory function. *American Journal of Primatology, 67*, 329–346.

Withnell, C. B., & Ungar, P. S. (2014). A preliminary analysis of dental microwear as a proxy for diet and habitat in shrews. *Mammalia, 78*, 409–415.

Wood, B. A. (1991). *Koobi Fora research project IV. Hominid cranial remains from Koobi Fora*. Oxford: Clarendon.

Wood, B. A., & Aiello, L. C. (1998). Taxonomic and functional implications of mandibular scaling in early hominins. *American Journal of Physical Anthropology, 105*, 523–538.

Wood, B. A., & Constantino, P. (2007). *Paranthropus boisei*: Fifty years of evidence and analysis. *Yearbook of Physical Anthropology, Supplement, 45,* 106–132.

Wood, B., & Strait, D. (2004). Patterns of resource use in early *Homo* and *Paranthropus*. *Journal of Human Evolution, 46,* 119–162.

Wroe, S., Ferrara, T. L., McHenry, C. R., Curnow, D., & Chamoli, U. (2010). The craniomandibular mechanics of being human. *Proceedings of the Royal Society B.* doi:10.1098/rsif.2012.0923.

Wynn, J. G., Sponheimer, M., Kimbel, W. H., Alemseged, Z., Reed, K., Bedaso, Z. K., et al. (2013). Diet of *Australopithecus afarensis* from the Pliocene Hadar Formation, Ethiopia. *Proceedings of the National Academy of Sciences USA, 110,* 10495–10500.

Xie, Z., Swain, M., Munroe, P., & Hoffman, M. (2008). On the critical parameters that regulate the deformation behavior of tooth enamel. *Biomaterials, 29,* 2697–2703.

Yahyazadehfar, M., Bajaj, D., & Arola, D. D. (2013). Hidden contributions of the enamel rods on the fracture resistance of human teeth. *Acta Biomaterialia, 9,* 4806–4814.

Yilmaz, E. D., Schneider, G. A., & Swain, M. V. (2015). Influence of structural hierarchy on the fracture behavior of tooth enamel. *Philosophical Transactions of the Royal Society A, 373,* 20140130. doi:10.1098/rsta.2014.0130.

Zhao, L. X., & Zhang, L. Z. (2013). New fossil evidence and diet analysis of *Gigantopithecus blacki* and its distribution and extinction in South China. *Quaternary International, 286,* 69–74.

Chapter 10
Aspects of Mandibular Ontogeny in *Australopithecus afarensis*

Halszka Glowacka, William H. Kimbel, and Donald C. Johanson

Abstract Human and ape mandibles differ in the proportion of adult size attained at equivalent dental emergence stages; for most dimensions human mandibles are more advanced. These dissimilarities in pattern of growth underlie the vastly different adult mandibular morphologies of these taxa. *Australopithecus* mandibles represent a third distinctive mandibular morphology, but the pattern of its mandibular growth remains underexplored. The *Australopithecus afarensis* sample from the Hadar site, Ethiopia, ca. 3.4–3.0 Ma, is represented by three infant (pre-M_1 emergence) and two juvenile (pre-M_3 emergence) mandibles. A recently recovered mandible, A.L. 1920-1, though edentulous, appears to capture an *A. afarensis* individual during M_2 emergence, thus bridging these developmental stages. In this chapter, we (1) describe three new infant/juvenile *A. afarensis* mandibles and confirm that the suite of features used to distinguish *A. afarensis* from other taxa is present early in ontogeny, and (2) investigate how the *A. afarensis* mandible changes in size and shape throughout growth in comparison to humans and chimpanzees. Our results indicate that *A. afarensis* resembles humans more than chimpanzees in its percentage of adult corpus breadth attained at successive stages of dental emergence. *A. afarensis* is also more similar to humans in corpus cross-sectional shape changes throughout ontogeny. We suggest that canine reduction may have had an important influence on the growth trajectory of the *A. afarensis* mandibular corpus such that, as in humans, it achieved adult values relatively early. Our results underscore the importance of considering the influence of the developing dentition on both juvenile and adult mandibular morphology.

Keywords *Australopithecus afarensis* • Chimpanzee • Hadar • *Homo* • Mandibular growth • Tooth eruption

Introduction

Because mandibular remains are commonly used to diagnose hominin species, understanding when diagnostic features develop during mandibular ontogeny of hominins can help identify nonadult material taxonomically and also contribute to an understanding of how novel morphological features arise in relation to hypothesized selective forces. While the ontogeny of extant hominid (great ape + human) mandibles has been well characterized (e.g., Björk 1963; Humphrey 1999; Boughner and Dean 2008; Coquerelle et al. 2011; Singh 2014; Terhune et al. 2014), little is known about how early hominin mandibles grow to attain their adult forms (but see Cofran (2014), on the growth of the *Australopithecus robustus* mandible). This is largely due to the paucity of juvenile remains in the fossil record of early hominins.

Shape differences in the adult mandibular corpus have been linked to the mechanical demands of different diets (e.g., Hylander 1979; Bouvier and Hylander 1981; Ravosa 2000; Ross et al. 2011; Scott et al. 2014). Experimental and comparative studies of primates do not, however, support a consistent relationship between mandibular morphology and dietary category (Ross et al. 2012). An alternative hypothesis is that adult mandibular morphology reflects, in part, the size and shape of the developing dentition that the mandibular corpus houses during its growth and development (Dean and Beynon 1991). Plavcan and Daegling (2006), for example, found that mandibular corpus depth is influenced by the size of the permanent canine, but that postcanine tooth size does not covary in a systematic way with corpus shape across primates.

Loss of the large honing canine is a principal hominin apomorphy. Canine-crown height reduction occurred early in hominin evolution as evidenced by the presence of

H. Glowacka (✉) · W.H. Kimbel · D.C. Johanson
Institute of Human Origins, School of Human Evolution and Social Change, Arizona State University, Tempe, AZ 85287, USA
e-mail: halszka.glowacka@asu.edu

W.H. Kimbel
e-mail: wkimbel.iho@asu.edu

D.C. Johanson
e-mail: johanson.iho@asu.edu

reduced canines in late Miocene-early Pliocene *Orrorin tugenensis, Sahelanthropus tchadensis, Ardipithecus kadabba*, and *Ardipithecus ramidus* (Brunet et al. 2002; Haile-Selassie 2001; Haile-Selassie et al. 2004; Senut et al. 2001; Suwa et al. 2009; White et al. 1994). Although the canine of both *Ardipithecus* and early *Australopithecus* have basal crown dimensions that overlap those of bonobos (*Pan paniscus*) and female chimpanzees (*Pan troglodytes*), they are, on average, more similar in size to human canines than they are to chimpanzee canines and have relatively small unworn crown heights (Kimbel and Delezene 2009; Suwa et al. 2009; Ward et al. 2013). The reduction in canine-root size, on the other hand, lagged behind crown-size reduction (Suwa et al. 2009; Ward et al. 2013). The canine roots in early *Australopithecus* show size reduction between *Australopithecus anamensis* (ca. 4.2–3.9 Ma) and its likely phyletic descendant *Australopithecus afarensis* (3.8–3.0 Ma), especially in the mandible (Ward et al. 2013). *Australopithecus afarensis* is the earliest known hominin species to possess both a reduced canine crown and root.

Canine-crown formation time in early hominins is more similar to that of humans than chimpanzees (Table 10.1; Dean et al. 1993, 2001), in which the large, sexually dimorphic canines take up to seven years to form (Table 10.1; Schwartz and Dean 2001). If mandibular corpus morphology reflects the size of the adult canine crown and root (Plavcan and Daegling 2006), then growth of the canine can be expected to influence the morphology of the mandibular corpus during ontogeny. Based on the size and duration of formation of chimpanzee canines, and the reduced canines in the hominin lineage, we might expect to see differences between apes and humans in the pattern of corpus growth, at least in the region of the antemolar dentition. Following this logic, *A. afarensis* should resemble humans more than apes in its pattern of mandibular corpus growth.

In this chapter we have two goals. First, we provide morphological descriptions of three recently recovered infant and juvenile mandibular specimens of *A. afarensis* from Hadar, Ethiopia, and describe how these differ from infant/juvenile mandibles of other hominin taxa. At the time of the description of *A. afarensis*, only a single infant mandible was known from Hadar Formation deposits (A.L. 333-43; White and Johanson 1982). The expanded mandibular sample allows us to ask whether or not diagnostic mandibular morphology is present in nonadult mandibles attributed to *A. afarensis*. Second, the

recently enlarged *A. afarensis* sample allows us to conduct a quantitative assessment of ontogenetic patterns of change in corpus size and shape in comparison to humans and chimpanzees. We ask the question: Given its reduced canines, does *A. afarensis* more closely resemble apes or humans in the pattern of mandibular corpus growth?

Fossil Descriptions and Comparisons

Our use of anatomical terminology for the mandible follows that of Weidenreich (1936). All measurements are in millimeters unless otherwise noted.

A.L.1920-1

A.L. 1920-1 is a partial mandible of a juvenile preserving the entire right corpus with the root of the ascending ramus and symphyseal region with the adjacent left anterior corpus (Fig. 10.1). The specimen, recovered in 2012, is from the surface of lower Sidi Hakoma Member sediments of the Hadar Formation (ca. 3.35 Ma).

Preservation

No tooth crowns are preserved; only the alveoli of LI1 to mesial LC and RI1 to mesial RM2 are present. The partial crypt of the unerupted RM3 is visible at the broken posterior edge of the ramus. Broken roots of RP4-RM1 are preserved at or just below the alveolar margins but the other alveoli are empty. The corpus is well preserved with very good surface detail in most areas. Anteriorly, the break cuts vertically from the mesial edge of the left canine alveolus anteriorly to the midline posteriorly. About ¾ of the way to the basal margin it deviates to the right beneath the left incisors and takes away a flake of cortical bone just above the base. On the right, the basal margin is intact posteriorly to mid-M2. The right buccal alveolar margin is broken at C-P3 and the incisors; the right lingual alveolar margin is intact except at P4-M1, where it is broken down to the level of the tooth roots. Posteriorly on the right, the ramus root is broken away below mid-corpus.

Table 10.1 Mandibular canine crown formation times (yrs) in chimpanzees, humans, and *Australopithecus*

Taxon	Sex	Mean	Source
Pan troglodytes	Male	6.81	Schwartz and Dean (2001)
	Female	5.85	Schwartz and Dean (2001)
Homo sapiens	Male	4.58	Schwartz and Dean (2001)
	Female	3.98	Schwartz and Dean (2001)
Australopithecus	?	3.73–4.51*	Dean et al. (2001)

*Range of mean crown formation times using a periodicity of 8–10 days; data for specimens attributed to the genus *Australopithecus*

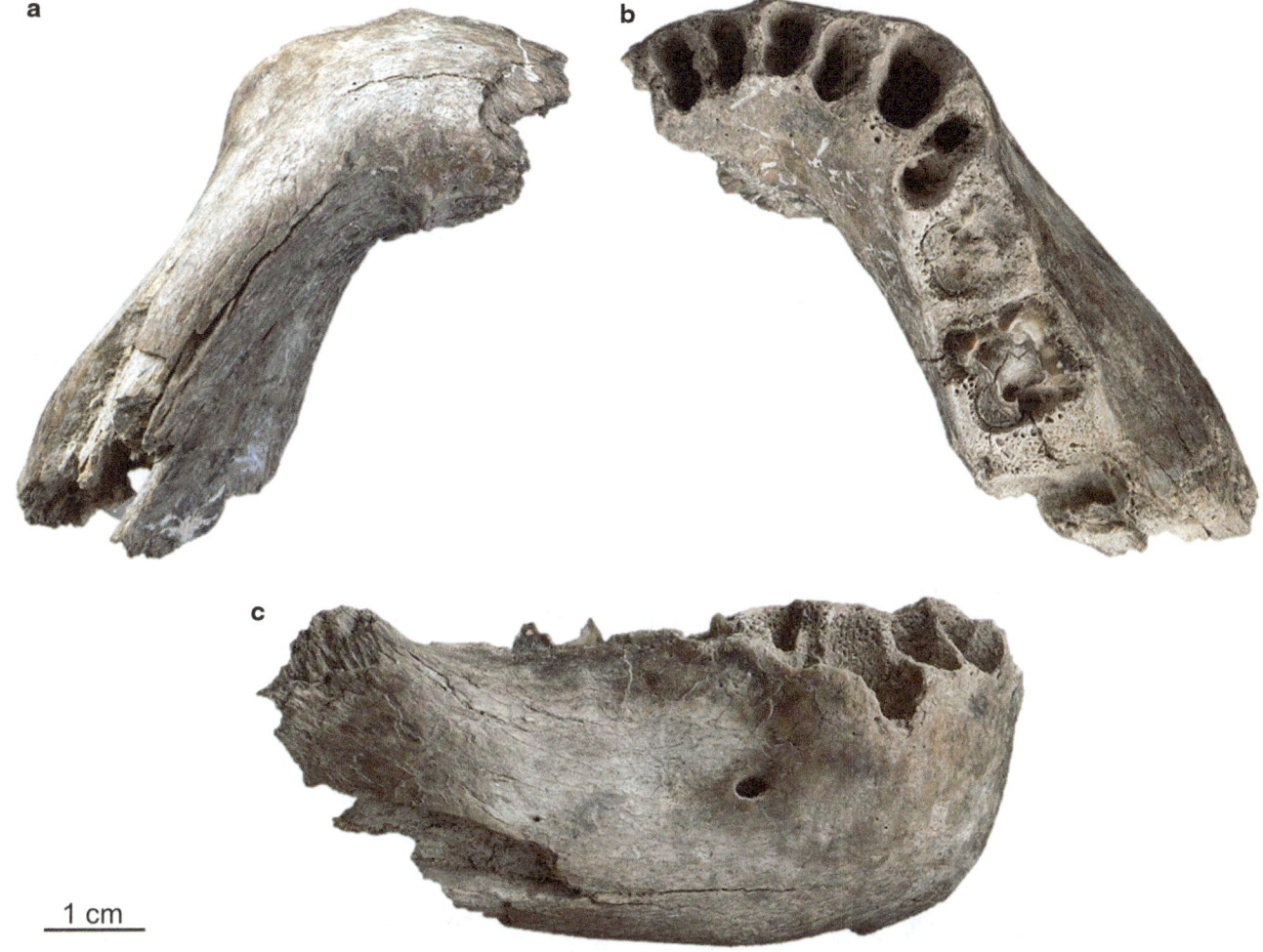

Fig. 10.1 **a** A.L. 1920-1, inferior; **b** A.L. 1920-1, occlusal; **c** A.L. 1920-1, lateral

Morphology

Lateral aspect. With the mandible oriented on the alveolar plane, the slightly convex basal margin deviates superiorly to the rear. Corpus depth (26.5-26.1) is greatest under the canine to P3. A prominent C/P3 jugum dominates the superior half of the lateral corpus, swelling the demarcation from the anterior corpus. Posterior to the jugum, a shallow, horizontally elongated hollow extends posteroinferiorly toward the root of the ramus. For much of its length the hollow is bounded above by a low, rounded *torus lateralis superior* and below by thickened bone above the base, the *torus marginalis*. The deepest part of the hollow is below P4. At its anteroinferior end, below P3/P4, is a single mental foramen. Situated just above midcorpus, the circular foramen has a diameter of 2.5 and opens anterosuperiorly. Posteriorly, the *torus lateralis* superior merges with a modestly projecting lateral prominence, denoting the root of the ascending ramus. The surface here is fairly flat and the ramus root is located high on the corpus below the distal

portion of M1. As the anterior margin of the ramus arises from the corpus, it cuts across the mesial half of the M2 alveolus.

Anterior aspect. The anterior corpus is convex mediolaterally across the incisor row and weakly bulbous inferosuperiorly. A slight median ridge, more easily palpable than visible, runs inferiorly from the I1/I1 interdental septum for 7.5, then divides into two faintly expressed limbs that diverge across the anterior corpus before fading near the base. On either side of the median ridge, the surface is slightly flattened. A single pin-prick-size foramen sits in the midline just below the dividing ridge. The surface along the incisors' labial alveolar margin is decorated with tiny foramina.

Basal aspect. The preserved segment of the basal margin is thick and rounded. As it passes posteriorly, it deviates laterally from the medially protruding alveolar prominence of the medial corpus, exposing an extensive subalveolar plane in this view. Anteriorly, on the medial side of the basal margin, a narrow pitted zone marks the anterior digastric insertion along the curve of the base below the canine alveolus. In the midline the anterior corpus curves continuously and smoothly inward to the basal margin. A weak but slightly abraded mental spine runs from the inner aspect of the basal margin for ca. 5.2 along the posterior aspect of the symphysis.

Medial aspect. A thick alveolar prominence, most pronounced below M1, dominates the medial corpus. From its inferior side, a barely perceptible mylohyoid line runs a short distance anteroinferiorly before hooking down to the medially thickened base below P4/M1. Posterior and inferior to the line the subalveolar surface bears a shallow fossa that extends to the broken posterior edge of the specimen. Although the anterior break is not on midline, the slightly bulbous anterior symphyseal contour is evident. The postincisive planum is short and inclined at about 45° to the basal plane. A suggestion of the superior transverse torus's lateral reflection reaches the medial corpus opposite the canine alveolus. The inferior transverse torus projected further posteriorly than the superior torus and is situated just above the basal plane.

Occlusal aspect. The hollowed area of the lateral corpus below P3/P4 is accentuated by the lateral prominence at M1/M2 and especially by the pronounced bulge of the canine jugum anteriorly. As defined by the alveoli, the anterior dental row forms a smooth arc. The P3 was two-rooted, with the buccal root's jugum merging with that of the canine.

The state of the M2 crypt suggests that the crown had breached the alveolar margin. The anterior wall of the crypt bears a clear impression of a grooved mesial root, and the buccal wall curvature conforms to the inter-radicular space between the mesial and distal roots. At the base of the crypt, the size of the impressions for the root apices indicate 1/2 to 3/4 root closure. We therefore conclude that the crown was in the final stages of eruption at death. A small (roughly 7 × 7) patch of the mesial wall of the RM3 crypt is exposed in the root of the ramus. It is not possible to determine the state of the crown's development.

A.L. 1030-1

The specimen is a left mandibular corpus fragment of an infant with erupted dp3 and dp4 (Fig. 10.2). At the anterior break, which approximates the midline, the unerupted I1 and I2 crowns are exposed in situ. Posteriorly, the specimen preserves the root of the ascending ramus. A.L. 1030-1 was

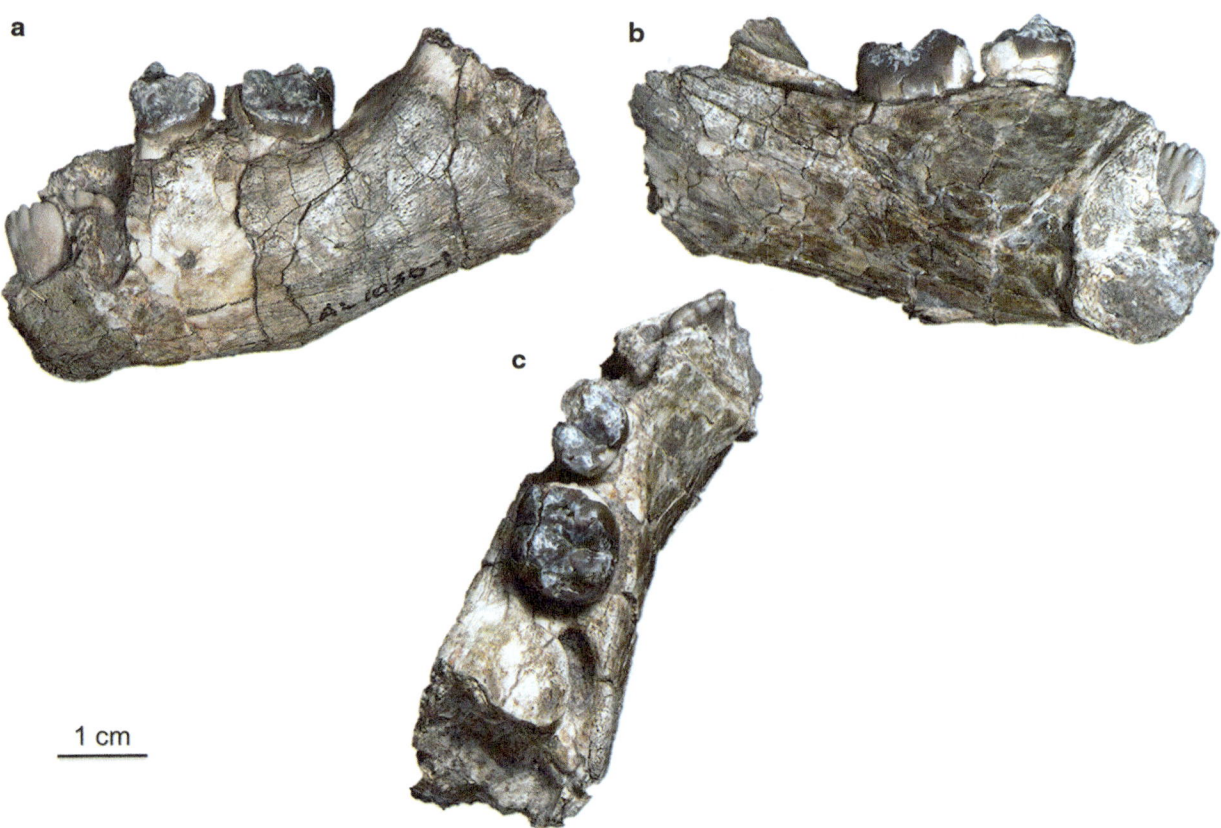

Fig. 10.2 **a** A.L. 1030-1, lateral; **b** A.L. 1030-1, medial; **c** A.L. 1030-1, occlusal

recovered in 2002 from the surface of upper Denen Dora (DD) Member deposits (ca. 3.2 Ma) of the Hadar Formation.

Preservation

Most surfaces are traversed by fine, superficial, mostly vertical cracks that have a minor effect on corpus dimensions. A rectangular (13.8 × 15.3) section of buccal cortical plate below dp3 is slightly displaced medially by crushing. Anterior to this disturbance, the superior half of the buccal plate is missing, exposing the labial face of I1 and a bit of the incisal edge of I2 in their crypts. Crushing has also displaced a triangular (15 × 9.6) segment of bone on the inferior half of the medial corpus below dc. In the midline cross section the lingual and mesial surfaces of the unerupted I1 crown are exposed. Posteriorly, the specimen is broken at the root of the ramus, posterior to which the M1 crypt and the abraded mesial wall of the M2 crypt can be seen.

Morphology

Lateral aspect. Corpus depth increases anteriorly and, prior to damage, was likely maximal anterior to dp3. The mental foramen is situated just above midcorpus below dp3. The foramen is circular (1.7 × 1.7) and opens anterosuperiorly. Above the foramen and passing posterosuperiorly up to the alveolar margin at mesial dp4 is a very mild hollow from which the low but well defined oblique line slopes gently posteriorly immediately below the alveolar margin to the root of the ramus. Otherwise, the lateral face of the corpus is fairly flat, with only hints of the *torus marginalis* and *prominentia lateralis*. The latter is expressed as a very weak vertical convexity below the ramus root. The extramolar sulcus is very narrow (3.5) and situated high on the corpus. The basal margin is strongly undulating, with an inferior inflexion point, accentuated by slight displacement of cortex along a crack, below mesial dp4.

Anterior aspect. The surface of the anterior corpus is vertically convex, with a posteriorly retreating inferior segment. The symphyseal cross section is ovoid; its vertical height is roughly twice its maximum anteroposterior dimension (14.5). The symphyseal axis is weakly inclined, forming an angle of ca. 85° with the basal plane.

Medial aspect. The crypt for the M1 is exposed in medial view; the alveolar bone is interrupted distal to dp4 by the opening of the crypt. The medial corpus bears a pronounced alveolar prominence adjacent to the M1 crypt. The prominence quickly loses definition anteriorly, fading below dp3/dp4 to meet the lateral reflection of the post-incisive planum. Beneath the prominence the medial surface recedes

laterally to the basal margin. Surface cracking makes it difficult to discern the *fossa subalveolaris*, which, if present, was very weak. The slightly hollowed post-incisive planum is inclined about 61° to the basal plane. The posterior edge of the symphyseal cross section is damaged, but in true medial view, a small but distinct inferior transverse torus projects posteriorly about 3 above the basal margin and slightly beyond the post-incisive planum.

Basal aspect. The strong lateral offset of the basal margin relative to the alveolar contour is evident. Posteriorly, the basal margin deviates from the midline more than the alveolar contour. The basal margin is sharp posteriorly but becomes blunt and rounded under the deciduous premolars. Anteriorly, the base is flattened across the midline, where very shallow but palpable bilateral oval depressions mark the insertions of the anterior bellies of the digastric muscles. These impressions are divided by a barely raised crest that passes posteriorly and then superiorly into the genioglossal fossa on the lingual symphyseal surface.

Occlusal aspect. Crushing of the lateral and medial corpus (see above) makes it difficult to evaluate the natural contours in this view. The buccal alveolar contour appears to be smoothly convex, consistent with the relatively weak topography of the lateral corpus described above. A hint of buccal deviation corresponding to a modest jugum is evident at the mesiobuccal dp3 root.

A.L. 333n-1

This specimen is a right hemi-mandible of an infant preserving the corpus from the di2 alveolus anteriorly to the M1 crypt posteriorly and a complete mandibular ramus, with intact angle, condyle, and coronoid process (Fig. 10.3). Recovered in 1999, it derives from the surface of a gully draining DD-2 sub-member deposits at the A.L. 333 hominin locality (ca. 3.22 Ma).

Preservation

The corpus surface is well preserved, with few superficial cracks and no distortion. The corpus is broken anteriorly, lateral to the midline, preserving the distal alveolus of di1 and exposing the mesial interproximal surface of what appears to be the unerupted I2 crown. The ramus has been displaced laterally, posteriorly, and slightly inferiorly along a horizontal crack at the level of the projected occlusal plane. The medial and lateral poles of the condyle are broken and the superior margin of the coronoid process is lightly abraded.

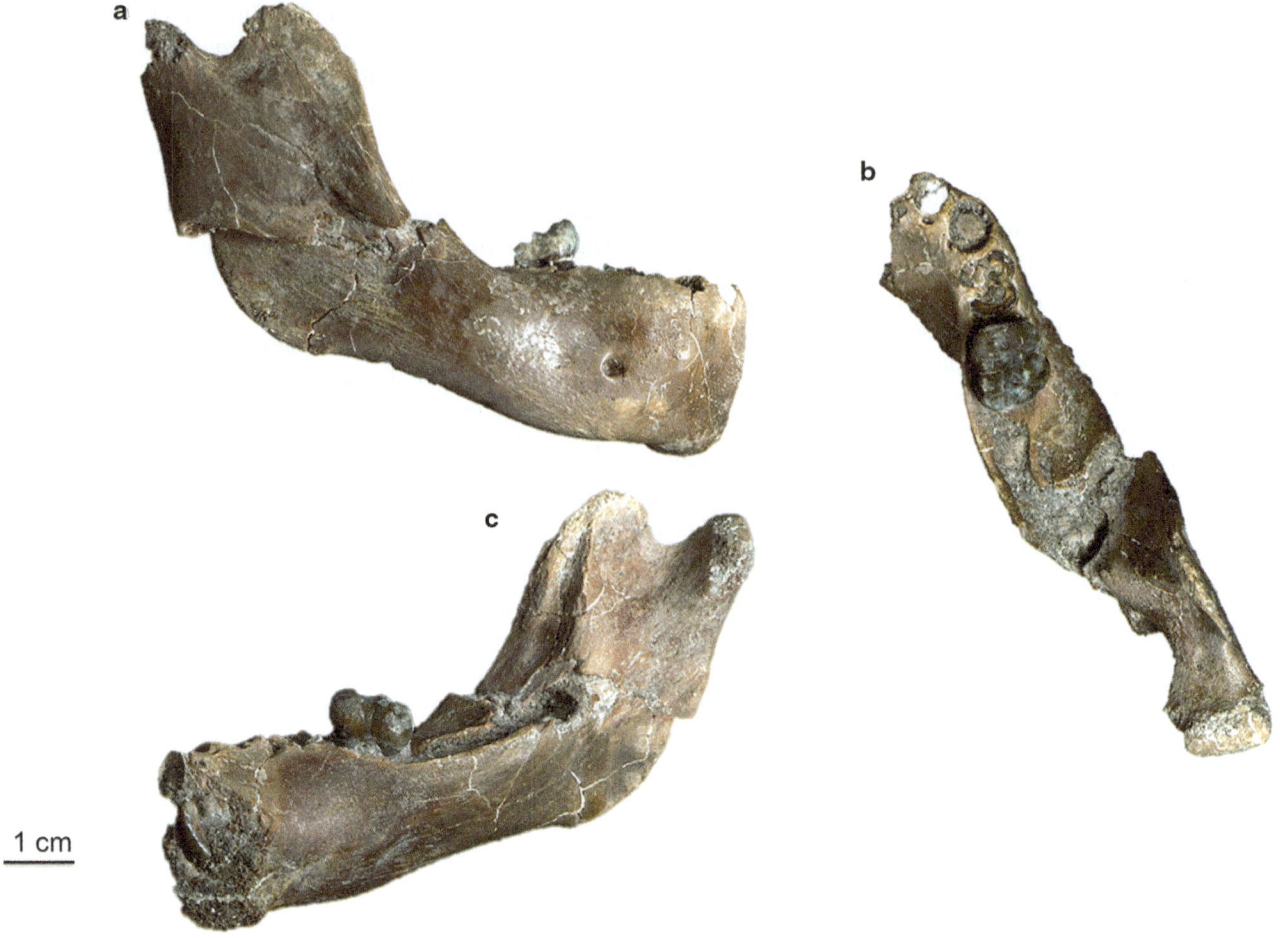

1 cm

Fig. 10.3 **a** A.L. 333n-1, lateral; **b** A.L. 333n-1, occlusal; **c** A.L. 333n-1, medial

Morphology: The Corpus

Lateral aspect. The corpus is deepest at dp3, becoming substantially shallower posteriorly. The mental foramen, situated inferior to midcorpus level beneath the mesial root of dp3, is ovoid, with an anteroposterior long axis and anterosuperiorly directed opening. The root of the ramus is situated high on the corpus, defining a narrow extramolar sulcus. The anterior margin of the ramus arises free of the corpus posterior to dp4. The oblique line, running a short distance anteriorly at 55° to the basal margin, merges with a low, diffuse *torus lateralis superior* at mesial dp4. Below the oblique line, the root of the ramus is marked by a swollen *prominentia lateralis*, from which a strong *torus marginalis* sweeps anteroinferiorly to merge with the basal margin at distal dp3. Together, the *torus lateralis* and *torus marginalis* define the upper and lower boundaries, respectively, of an extensive teardrop-shaped hollow, widest and deepest anteriorly, that extends posteriorly to the lateral prominence. Anteriorly, the hollow terminates abruptly at a prominent

button-like canine jugum; the jugum spreads posterosuperiorly to blend with the *torus lateralis* at dp3.

Anterior aspect. Anterior to the canine jugum, the surface is flat inferosuperiorly and mediolaterally, with only modest development of deciduous incisor juga. The basal margin here is concave.

Medial aspect. The anterior break, a vertical section through the di1/di2 inter-alveolar space, exposes what is probably the mesial aspect of the developing I2 crown. The plane of the break, though not on midline, suggests a symphyseal cross section that is about twice as tall as it is wide and only slightly inclined from the vertical, with a barely convex anterior contour. The posterior contour juts out prominently to define an extensive *planum alveolare* set at 58° to the basal plane. The lateral reflection of the well developed superior transverse torus intersects the medial corpus at dp3/dp4. Below the torus, the right genioglossal fossa is a shallow pit. The inferior transverse torus is slightly less posteriorly projecting than the superior torus and is situated right on the basal plane. The mylohyoid line runs anteroinferiorly from the damaged mandibular foramen to the inferior transverse torus. It delimits a well defined subalveolar fossa, which forms an elongate oval depression between a swollen area corresponding to the M1 crypt posteriorly and the basal margin below mesial dp4.

Basal aspect. This view emphasizes the thickness of the corpus at the level of the M1 crypt; it thins quickly anterior and posterior to this point. The basal margin of the corpus is thick and rounded. Posteriorly, it thins dramatically towards the gonial angle. Anteriorly, a long, curved anterior digastric muscle insertion flattens the base at the level of dc. This area corresponds to the concave basal margin in anterior view. The digastric fossa is marked by a small deep pit near its anterior extremity beneath di2. In the midline, the spinelike edge of the inferior transverse torus projects posteriorly to obscure the glenioglossal pit.

Occlusal aspect. The protrusion of the *prominentia lateralis* posteriorly and the sudden bulge of the dc jugum anteriorly delimit the lateral corpus hollow and contribute to a very obvious mediolateral thinning of the corpus at dp3.

Morphology: The Ramus

Lateral aspect. Despite the displacement of its superior half (see above), the ramus can be visually projected back into its anatomical position to determine the slope of its anterior border, which is set at 55° to the basal plane. Anteroposterior ramus breadth is ca. 28 at the level of the alveolar plane. With the mandible oriented on the alveolar plane, the coronoid process is about 2.0 taller than the condyle. Although its uppermost edge is slightly abraded, the coronoid process has a rounded, posteriorly extended tip. The mandibular notch is shallow and slightly wider than it is deep. The deepest point is approximately centered in the notch. The ectocondyloid buttress, a thickening of bone running from the condyle anteroinferiorly to the level of the projected occlusal plane, is weakly expressed. Anterior to the buttress, the ramal surface is mildly depressed, presumably reflecting the insertion of m. masseter. The masseteric fossa is delimited anteriorly by the thickened anterior ramal border. The gonial angle is very weakly everted and slightly thickened, marking the inferiormost insertion of the masseter.

Anterior aspect. The anterior border of the ramus deviates medially as it rises superiorly and becomes vertical along the anterior edge of the coronoid. The anterior margin is sharp.

Posterior aspect. The gonial angle is thick but with a sharp edge. Despite displacement, it is clear that, as the ramus rises, both the posterior margin and the condyle tilt medially.

Medial aspect. The posterior margin of the ramus is straight, with only a hint of an *incisura suprangularis* below

the neck of the condyle. A shallow mylohyoid groove is well marked, running across the entire medial surface, from the damaged but robust *torus triangularis* anteroinferiorly to the lateral reflection of the inferior transverse torus below dp3. Three pterygoid tuberosities are discernible at the gonial angle. The inferiormost is the strongest and runs about 4.5 anterosuperiorly from the inner edge of the gonial angle; the superiormost is weakest and originates at about 1/3 the height of the ramus. A raised, flat roughened area extends the medial pterygoid insertion about halfway up the posterior margin of the ramus. The endocoronoid ridge is powerfully developed, running from the anterior margin of the coronoid process anteroinferiorly to the *torus triangularis*. Anterior to the ridge lies a deep, smooth sulcus, which is bounded anteriorly by a strong ridge that parallels and thickens the anterior border of the ramus, especially superiorly. This area marks the probable insertion of the temporalis muscle. The endocondyloid ridge extends anteroinferiorly from the anteromedial side of the condyle. It is crestlike and better developed than the endocoronoid ridge superiorly, but low, rounded, and less prominent than the latter inferiorly, where they converge in the *torus triangularis*. The two ridges and the coronoid notch define the boundaries of a deep, fossalike *planum triangularis*. Posterior to the endocondyloid crest is a deep, 7-wide groove, the *sulcus colli*, which runs out to the posterior edge of the condylar neck and is bounded below this point by the elevated area marking the superior extent of the medial pterygoid insertion. Damage to the *torus triangularis* precludes description of the area around the mandibular foramen. The medial pole of the condyle is eroded, exposing trabecular bone.

Comparative Morphology

Based on juvenile Hadar specimen A.L. 333-43, White et al. (1981) observed that mandibles of *A. afarensis* and *A. africanus* (e.g., Taung) could be distinguished taxonomically at relatively early ontogenetic stages. The recently expanded sample of infant mandibles from Hadar, described here, supports this contention. Specimens A.L. 1030-1, A.L. 333n-1, and A.L. 1920-1 join A.L. 333-43 in forming a consistent, distinctive morphological pattern composed of the following characters:

- strong hollowing of the lateral corpus beneath dp3/P3
- prominent dc/dp3 or C/P3 jugum forming a distinct anterolateral "corner" of the mandibular arch in occlusal view
- anteriorly to anterosuperiorly directed mental foramen
- modest to moderate development of the lateral prominence

- superiorly placed root of the ramus, with narrow extramolar sulcus, when distinct
- basally set inferior transverse torus
- anteroposteriorly broad coronoid process with confined mandibular notch.

Juvenile mandibles of other australopith species, even with permanent molars still unerupted, lack most of this morphology. The mandible of the Taung type-specimen of *A. africanus*, for example, lacks lateral corpus hollowing and a strong dc/dp3 jugum, resulting in a smooth, convex transition between the lateral and the anterior components of the mandibular arch (Fig. 10.4). Mandibles of juvenile robust *Australopithecus*, such as KNM-ER 1466 and KNM-ER 1820, both attributed to *A. boisei* (White 1977; Wood 1991), show a highly inflated corpus, an indistinct dc/dp3 jugum, a massive lateral prominence continuous anteriorly with a pronounced lateral torus (especially in KNM-ER 1466) in place of the lateral hollow, and an elevated inferior transverse torus. The morphology of the more extensive sample of juvenile *A. robustus* mandibles from Swartkrans (SK 61, SK 62, SK 63, SK 64) differs from that of Hadar *A. afarensis* in similar ways.

A different pattern of affinity emerges when the ascending rami are compared. Rak et al. (2007) described an unusual feature of the adult *A. afarensis* ascending ramus, namely, its tall, anteroposteriorly broad coronoid process with a posteriorly extended tip and a deep, narrow mandibular notch. As Rak et al. noted, the same morphology appears in *A. robustus* (SK 23 and SK 34; the juvenile SK 63) and *A. africanus* (Sts 7). Among extant hominoids, it is present in gorillas but not in chimpanzees, bonobos, orangutans, or

humans, which have narrower, lower coronoid processes and broader, shallower notches. Two of the Hadar juvenile mandibles (A.L. 333-43, A.L. 333n-1) preserve this region of the ramus, and although they resemble *A. afarensis* adults, they differ from one another in detail. When oriented on the alveolar plane, the coronoid process of A.L. 333-43 (Fig. 10.4) towers over the (partly eroded) condyle, and the notch is very deep and narrow, with a near-vertical anterior edge (the posterior margin of the asymmetric coronoid process). In A.L. 333n-1, there is less disparity between the coronoid and condylar heights, the mandibular notch is shallower, and the coronoid process more symmetric (Fig. 10.3).

Australopithecus afarensis differs diagnostically in anterior mandibular morphology from its likely ancestor, *A. anamensis* (Leakey et al. 1995; Ward et al. 2001). The *A. anamensis* sample does not currently include a juvenile mandible, but the fact that the character list distinguishing the juvenile mandibles of *A. afarensis* from those of other hominin species is so similar to the one distinguishing the adult mandibles (White et al. 1981; Kimbel and Delezene 2009) means that it is reasonable to predict what the juvenile mandible of an ancestor of *A. afarensis* looked like. Indeed, the fossil record holds clues, in the form of the small sample of mandibles from the 3.8–3.6 Ma Laetoli site, which contains the type-specimen of *A. afarensis* (LH-4). As described elsewhere (Kimbel et al. 2006), LH-4 resembles mandibles of *A. anamensis* in its strongly receding, externally convex symphyseal region and inferomedially inflected lateral corpus walls beneath the canine and premolars. The juvenile Laetoli mandible LH-2 (M1 just emerged) reiterates this

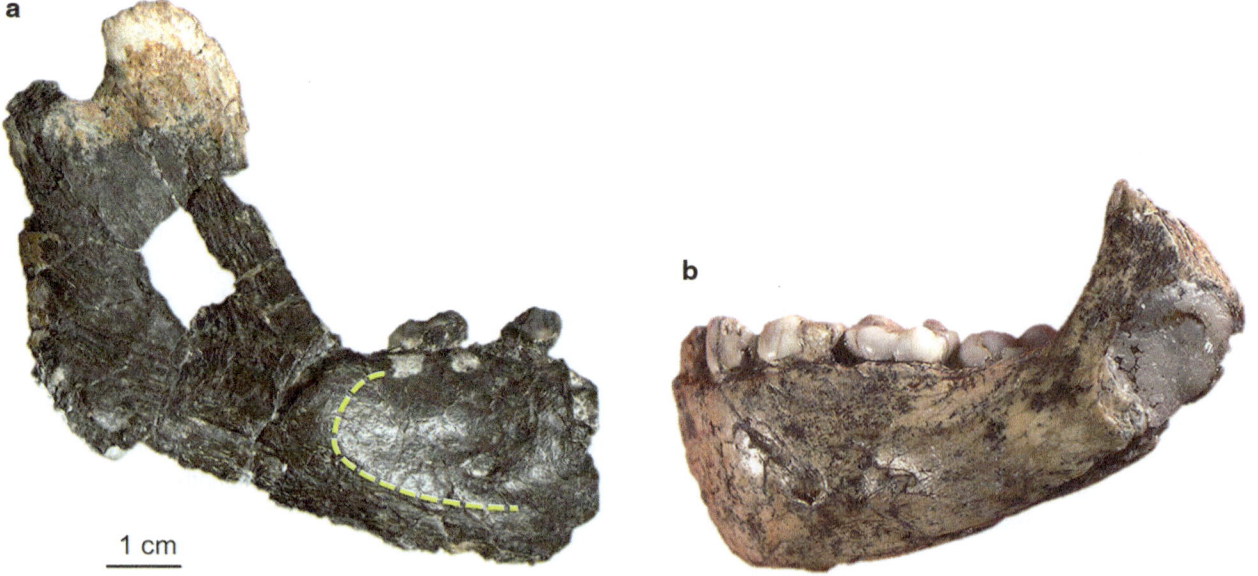

Fig. 10.4 a A.L. 333-43a (*A. afarensis*); **b** Taung (A. africanus). Morphological differences between these two specimens mirror distinctions of their adult counterparts, including lateral corpus hollowing in *A. afarensis* (*anterior to dashed curved line in **a***). See text for discussion

pattern and differs markedly from Hadar juveniles of broadly comparable emergence stage (A.L. 333-43, A.L. 333n-1, A. L. 1030-1). These mandibles feature flatter, less inclined symphyseal profiles and nearly vertical lateral corpus walls, as in adult counterparts. Accordingly, we would expect the juvenile mandible of *A. anamensis* to resemble LH-2 more than the juvenile jaws from Hadar.

The distinctive morphology of the juvenile mandibles of *A. afarensis* from the Hadar site is consistent with an onto-genetically early manifestation of species-specific patterns of mandibular growth across hominoids (e.g., Daegling 1996; Boughner and Dean 2008; Terhune et al. 2014). The fact that the corpus and ramus of the juvenile specimens yield different signals of affinity likely reflects the discrete functional influences on these components of the mandible during growth (Moss and Rankow 1968; Daegling 1996). Among the juvenile mandibles of early hominins, interspecific differences in corpus size and shape appear to track differences in the size of the erupting and erupted dentitions (see more on this point in Discussion). Thus, the more "filled out," relatively thick juvenile mandibular corpora of later *Australopithecus* (*A. africanus*, *A. robustus*, and *A. boisei*) can be related to the dramatically increased size of the developing permanent postcanine teeth in these species as compared to *A. afarensis*. On the other hand, interspecific variation in the morphology of the mandibular ramus is most likely related to masticatory muscle attachments (in the case of the height and shape of the coronoid process, the orientation of the temporalis muscle vector; Ritzman and Spencer 2009). The fact that species-specific mandibular morphology is present in very young individuals makes it unlikely that these differences arise simply as plastic responses to the onset of adult feeding behaviors.

Patterns of Ontogenetic Size and Shape Change in the *A. afarensis* Mandible

In the following section, we metrically describe the ontogeny of the *A. afarensis* mandibular corpus and discuss how it compares to that of modern humans and chimpanzees. As we noted above, an important potential influence on the growth and adult form of the mandible is the developmental trajectory of the permanent dentition. In particular, patterns of canine crown and root growth, which distinguish apes from modern humans and fossil hominins, are expected to differentially affect mandibular corpus growth. Accordingly, here we analyze mandibular corpus growth in terms of the percentage of adult values attained at successive dental emergence stages.

The permanent canines are among the last teeth to emerge in the chimpanzee dental emergence sequence (Table 10.2). If the growth of the canine influences mandibular size during ontogeny, then the relatively late emergence of the canine, its lengthy formation time (Table 10.1), and its large size in the chimpanzee should be reflected in a longer period of corpus size increase during growth. As such, we would expect chimpanzee mandibular size to reach adult values at later dental emergence stages relative to humans and *A. afarensis*. We would expect that corpus height and breadth will be affected by the growth of the canine as the tooth as well as its crypt rotates during ontogeny.

In humans, the permanent canine emerges relatively early in the sequence, prior to the emergence of M2 (Table 10.2). This relatively early emergence, coupled with the small size of the permanent canine and its fast crown formation time (Table 10.1), should result in the canine having a diminished effect on mandibular corpus morphology throughout growth.

Up until now, determining the position of the permanent canine in the dental emergence sequence of *A. afarensis* has been impossible because specimens of most nonadult individuals do not preserve this region of the mandible at the relevant growth stages. Based on the size of the A.L. 1920-1 canine alveolus, we determined that the permanent canine was likely erupted in this individual, indicating that the canine emerged before M2, a sequence that is more human-like than chimpanzee-like (Table 10.2). Data on dental emergence sequence are available for *A. africanus*, which, in at least two specimens (MLD 2, Sts 52), is more similar to chimpanzees in having a relatively late canine emergence, between M2 and M3 (Smith 1994; see Table 10.2). Smith (1994) found that the position of the canine in the emergence sequence of extant primates varies intraspecifically, so, given the small fossil samples, it is unclear if the difference in canine emergence between *A. africanus* and A.L. 1920-1 represents a biologically meaningful distinction or simply reflects sampling of normal

Table 10.2 Sequence of mandibular tooth emergence in chimpanzees, humans, and *A. africanus*

Taxon	Sequence
P. troglodytes	M1 I1 I2 [M2 P4 = P3] C M3
H. sapiens	[M1 = I1] I2 [C P3] [P4 M2] M3*
A. africanus	M1 I1 I2 M2 P3 P4 C M3

All data from Smith (1994)
*Sequence for white females

intraspecific variation. We therefore cannot reliably use the position of the canine in the dental emergence sequence of *Australopithecus* to guide our predictions of mandibular ontogeny in *A. afarensis*. Because canine-crown formation time in *Australopithecus* was similar to that of modern humans (see Table 10.1), however, we should expect that the reduced adult crown and root size in *A. afarensis* will have a diminished effect on corpus proportions during growth.

Methods

We collected metric data on corpus height and breadth from three infant (all deciduous dentition emerged: A.L. 333-43, 333n-1, 1030-1), one juvenile (M1 emerged: A.L. 1920-1), two subadult (M2 emerged: A.L. 128-23, 145-35), and nine adult (M3 erupted: A.L. 198-1, 198-22, 207-13, 225-8, 228-2, 288-1i, 315-22, 330-5, 333w-12) *A. afarensis* mandibles. Although the three infant *A. afarensis* mandibles could not be sexed a priori, the A.L. 1920-1 juvenile (M1 emerged, M2 erupting) is among the smallest mandibles in the *A. afarensis* sample; we therefore assumed that it would have grown into a small adult and so constructed its subadult and adult "target" sample around the 11 smallest (presumptive female) mandibles in the Hadar sample (i.e., the subadults and adults listed above). The Hadar sample construction is important to keep in mind when interpreting our results (see Discussion).

We collected comparative metrical data on cross-sectional ontogenetic skeletal samples of chimpanzees and modern humans. For all three species, specimens were sorted into four dental emergence stages (dp4 emerged, M1 emerged, M2 emerged, M3 emerged). Data on wild-shot *Pan troglodytes* were collected at the Cleveland Museum of Natural History (n = 10, 16, 16, 20 for dental emergence stage dp4, M1, M2, and M3, respectively). A Nubian skeletal collection housed at Arizona State University was used as the modern human sample in this study (n = 9, 12, 2, 10 for dental emergence stage dp4, M1, M2, and M3, respectively). For humans and chimpanzees, the skeletal samples comprised a mix of male and female specimens.[1]

Chimpanzees are strongly sexually dimorphic in canine size. The pattern of mandibular growth exhibited by males and females, especially at the canine position, could therefore differ between the sexes. We performed the same analysis described below without males, and compared the

results between the sample that consisted of both males and females and the sample that contained only females. The pattern of growth was the same, regardless of the sample's sex composition. Chimpanzee males were therefore retained in the analysis.

Using standard calipers and methods described elsewhere (White and Johanson 1982; Kimbel et al. 2006), we measured minimum corpus breadth and perpendicular corpus height at five positions along the tooth row: dc/C, dp3/P3, dp4/P4, M1, and M2. These data were used to calculate, for corpus height and breadth, the percentage of adult sample value attained at each of the four dental emergence stages. To calculate the percentage of adult value attained, we set the mean adult values as the "adult value." We used absolute corpus width and height values to calculate corpus shape (corpus width/corpus height) and examined change in this ratio across tooth-emergence stages. Due to factors of preservation, corpus height at the canine could not be measured for any adult *A. afarensis* specimen in our sample; therefore, percentage of adult height and adult corpus shape at the canine is not reported for *A. afarensis*.

Results

Statistics on absolute size increases during mandibular growth in breadth and height are given in Tables 10.3 and 10.4; these data are translated into percentage increases to mean adult size in Tables 10.5 and 10.6. Humans and *A. afarensis* attain a greater percentage of final adult values at earlier dental emergence stages than chimpanzees. At all measured positions except M1, chimpanzees undergo larger percentage increases in mandibular corpus breadth with the emergence of dp4 and the first two adult molars than either humans or *A. afarensis* (Table 10.5; Fig. 10.5). With emergence of dp4 (and thus the completion of the deciduous dentition), the chimpanzees in our sample have attained between 57–65% of full-adult values, whereas in our human sample these figures are 88–100% and in *A. afarensis* 77–90%. As shown in Table 10.5 and Fig. 10.5, chimpanzees continue to lag behind the two hominin species through M2 emergence. Between dp4 and M1 emergence, the *A. afarensis* size increase is similar to that of chimpanzees measured at dc/C and dp3/P3, but, with eruption of the permanent molars, the *A. afarensis* and modern human growth trajectories flatten. In contrast, mandibular corpus breadth continues to increase in chimpanzees throughout permanent molar emergence, consistent with expectation (see above).

The corpus height data present a strikingly different and more complex pattern. Height at all measured points along the corpus increases in all taxa throughout ontogeny (Table 10.4). Similarly, with the eruption, successively, of

---

[1] Data on sex were available for chimpanzee individuals in the two oldest dental emergence stages. Sex contribution to both categories was as follows: M2 emerged n = 8 males, n = 8 females; M3 emerged, n = 10 males, n = 10 females.

Table 10.3 Minimum corpus breadth (mm) measured at 5 points along the corpus at each dental emergence stage

Dental emergence stage	Species		dc/C	dp3/P3	dp4/P4	M1	M2
dp4 emerged	P. troglodytes	Mean	11.45	9.59	10.05	–	–
		SD	1.19	0.82	0.74	–	–
		n	10	10	10	–	–
	H. sapiens	Mean	10.77	11.22	12.24	–	–
		SD	0.43	1.20	0.84	–	–
		n	9	9	9	–	–
	A. afarensis	Mean	13.90	13.75	15.27	–	–
		SD	–	0.78	0.5	–	–
		n	1	2	3	–	–
M1 emerged	P. troglodytes	Mean	13.91	12.47	12.12	13.28	–
		SD	0.90	1.10	0.83	0.76	–
		n	16	17	17	17	–
	H. sapiens	Mean	11.35	11.61	12.25	13.82	–
		SD	0.94	1.18	1.09	1.23	–
		n	12	12	12	12	–
	A. afarensis	Mean	17.20	17.70	15.70	17.70	–
		SD	–	–	–	–	–
		n	1	1	1	1	–
M2 emerged	P. troglodytes	Mean	16.32	14.51	13.66	13.85	15.64
		SD	1.52	1.37	1.33	1.06	0.92
		n	15	16	16	16	16
	H. sapiens	Mean	11.75	11.70	12.95	14.80	16.10
		SD	1.91	2.40	2.19	2.12	1.13
		n	2	2	2	2	2
	A. afarensis	Mean	18.20	18.95	17.75	19.55	23.85
		SD	–	3.04	1.63	2.19	1.34
		n	1	2	2	2	2
M3 emerged	P. troglodytes	Mean	19.11	16.76	15.57	15.35	16.5
		SD	2.40	2.00	1.45	1.37	1.56
		n	20	20	20	20	20
	H. sapiens	Mean	12.20	12.04	12.28	12.99	14.42
		SD	1.52	1.76	1.92	1.46	1.34
		n	10	10	10	10	10
	A. afarensis	Mean	18.06	17.78	16.91	18.31	19.72
		SD	0.78	1.02	0.92	2.13	1.32
		n	5	5	7	8	6

dp4, M1, and M2, all taxa show stepwise increases in the percentage of adult height attained at each point along the corpus (Table 10.6; Fig. 10.6). Although the increase in corpus height at each dental emergence stage is similar in *A. afarensis* and chimpanzees, in the premolar region, chimpanzees tend to lag behind *A. afarensis* and humans in the percentage of adult values attained up to the emergence of M1. Whereas the trajectory of human corpus height growth flattens in the premolar region after M1 emergence, *A. afarensis* and chimpanzee corpora continue to attain greater percentages of adult values up through M2 emergence.

The divergent growth trajectories in mandibular corpus height and breadth produce differences in the development of adult corpus cross-sectional shape. Thus, because both corpus breadth and height increase during chimpanzee ontogeny, the chimpanzee corpus-shape index does not change appreciably up through M3 emergence (Table 10.7; Fig. 10.7a–c). In contrast, corpus shape in humans and *A. afarensis* becomes more gracile (i.e., the corpus becomes taller in relation to breadth) from stage to stage at most measured points along the corpus (Table 10.7; Fig. 10.7). Although the pattern of shape change is similar in the two hominin species, *A. afarensis* possesses a broader (in relation to height) mandibular corpus, at the dc/C, dp3/P3, and dp4/P4 positions up through M1 emergence, than either humans or chimpanzees, a morphology that is well known in adult *A. afarensis* mandibular material (Kimbel et al. 2004).

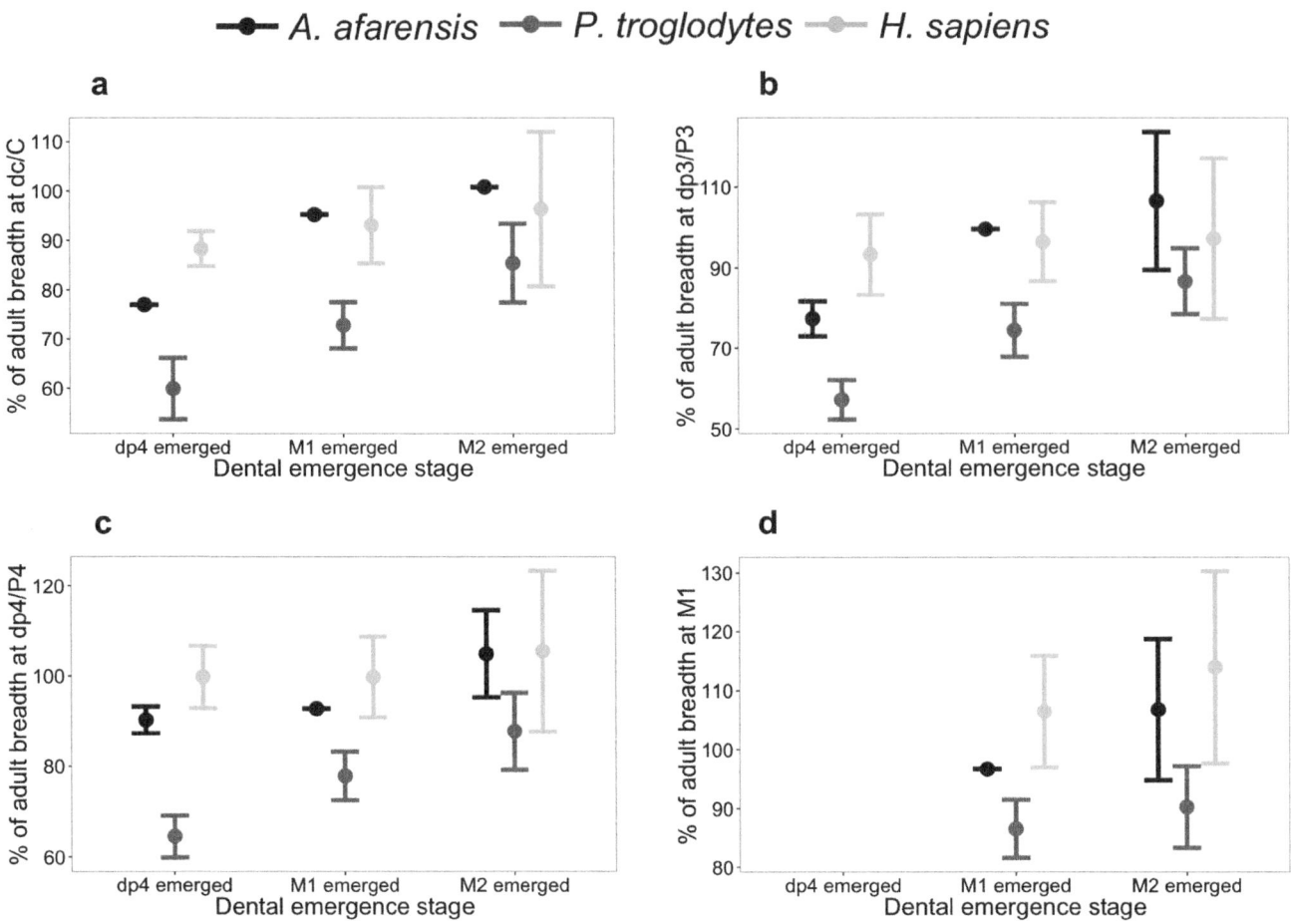

Fig. 10.5 Changes in the percentage of mandibular corpus breadth attained during ontogeny at the positions of dc/C (**a**), dp3/P3 (**b**), dp4/P4 (**c**), and M1 (**d**)

Discussion and Conclusions

The recently expanded Hadar sample of infant and juvenile mandibular specimens attributed to *A. afarensis* confirms that diagnostic morphological features of the species are present at all known ontogenetic stages. This pattern of similarity allows taxonomic identification of new specimens (e.g., Alemseged et al. 2006) and underwrites predictions of subadult morphology for taxa in which nonadults are as yet unknown, such as *A. anamensis*.

Our metrical analysis found that *A. afarensis* is neither exactly human-like nor chimpanzee-like in its pattern of mandibular corpus growth. One important caveat stems from our decision to model mandibular ontogeny of *A. afarensis* using the smaller half of the Hadar adult sample, based on the tiny A.L. 1920-1 juvenile, as noted in the methods section. It is possible that our choice to model the growth of a small mandible precludes us from making inferences regarding the growth of larger (i.e., male) mandibles. There are two reasons why we do not believe this to be the case,

however. First, as Taylor (2002) showed, mandibular growth trajectories do not differ between the sexes in chimpanzees. Our data support this finding: when we ran comparisons of the percentage of adult corpus size attained at each dental emergence stage without male chimpanzees, the results did not differ from those when using the combined-sex sample.[2] As chimpanzee adult corporal and symphyseal metrics are not strongly dimorphic (Taylor 2006), this result should not be surprising (in our adult chimpanzee sample, the average adult M/F size ratio for all breadth and height metrics is 1.056). Second, leaving aside the large, geologically late Hadar mandibles from the KH-2 submember of the Hadar Formation, which do not figure in our study, adult mandibular corpus size variation (as measured at M1

---

[2]When comparing male-only and female-only chimpanzee samples, males exhibit a slightly higher % change in absolute corpus breadth between the M2 and M3 emergence stages, most notably at the canine position. No such difference is apparent when we use corpus height. The very small sample sizes for these sex-specific subsets negate high confidence in this comparison, however.

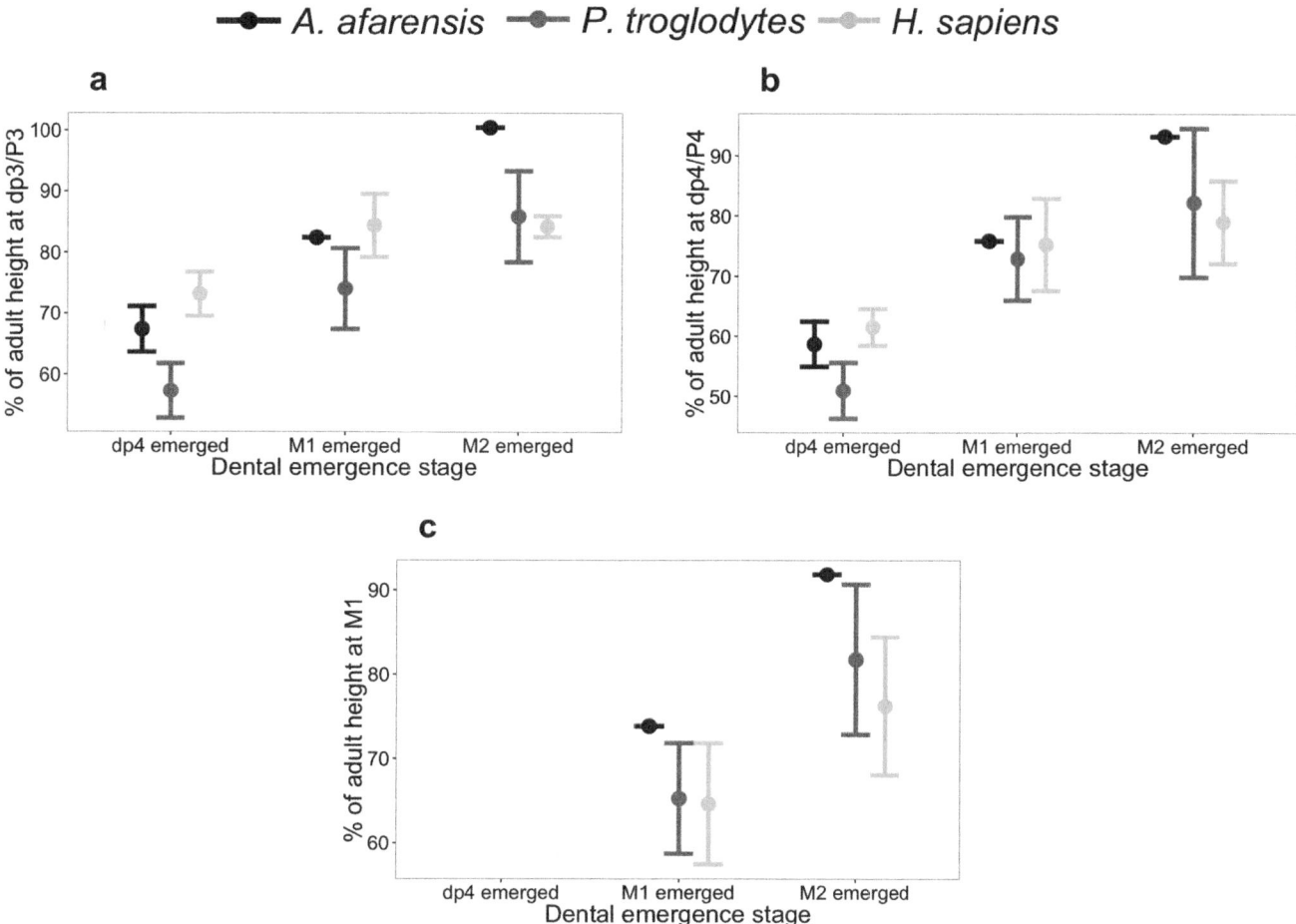

Fig. 10.6 Changes in the percentage of mandibular corpus height attained during ontogeny at the positions of dp3/P3 (**a**), dp4/P4 (**b**), and M1 (**c**)

position, at least) in the Hadar sample is not significantly different from that of chimpanzees (Lockwood et al. 2000). Therefore, we are confident that our results reveal genuine patterns of *A. afarensis* mandibular corpus growth.

Broadly speaking, *A. afarensis* resembles humans rather than chimpanzees in the percentage of adult mandibular corpus breadth and height values attained throughout growth, particularly at anterior dental positions and at earlier dental emergence stages. The similarity is consistent with our expectations based on the reduced, faster growing canine in hominins compared to chimpanzees.

In chimpanzees, as we have noted, the permanent mandibular canine is large and grows for a relatively and absolutely long time (Smith 1994; Schwartz and Dean 2001). The size of the permanent canine influences the shape of the adult mandibular corpus (Plavcan and Daegling 2006) and here we showed the mechanism by which this relationship is achieved during ontogeny. We suggest that besides the considerable size of the canine crown and root, the time period over which these large teeth develop also affects corpus shape. If canine formation proceeded at a faster pace, then adult corpus breadth and height would be reached relatively sooner and the growth trajectory of chimpanzee corpus shape would be more similar to that of humans and *A. afarensis*. Because canine formation is slow in chimpanzees, resulting in the late emergence of the canine in relation to the other permanent teeth, however, the growth of the permanent canine crown and root affects the size and shape of the anterior corpus throughout most of mandibular ontogeny (i.e., final adult size is reached later).

Indeed, at the dental emergence stages covered here, chimpanzees have achieved less of their adult values in corpus breadth – and so undergo more dramatic dimensional changes between emergence events – than humans or *A. afarensis*. In corpus height, this lag holds for the anterior portion of the chimpanzee corpus up to or through the time of M1 emergence. Whereas the height of the corpus of all three species increases throughout the better part of the growth period, *A. afarensis* is similar to humans but not to chimpanzees in that corpus breadth remains more or less stable at the canine and premolar positions from M1 through M2 emergence stages.

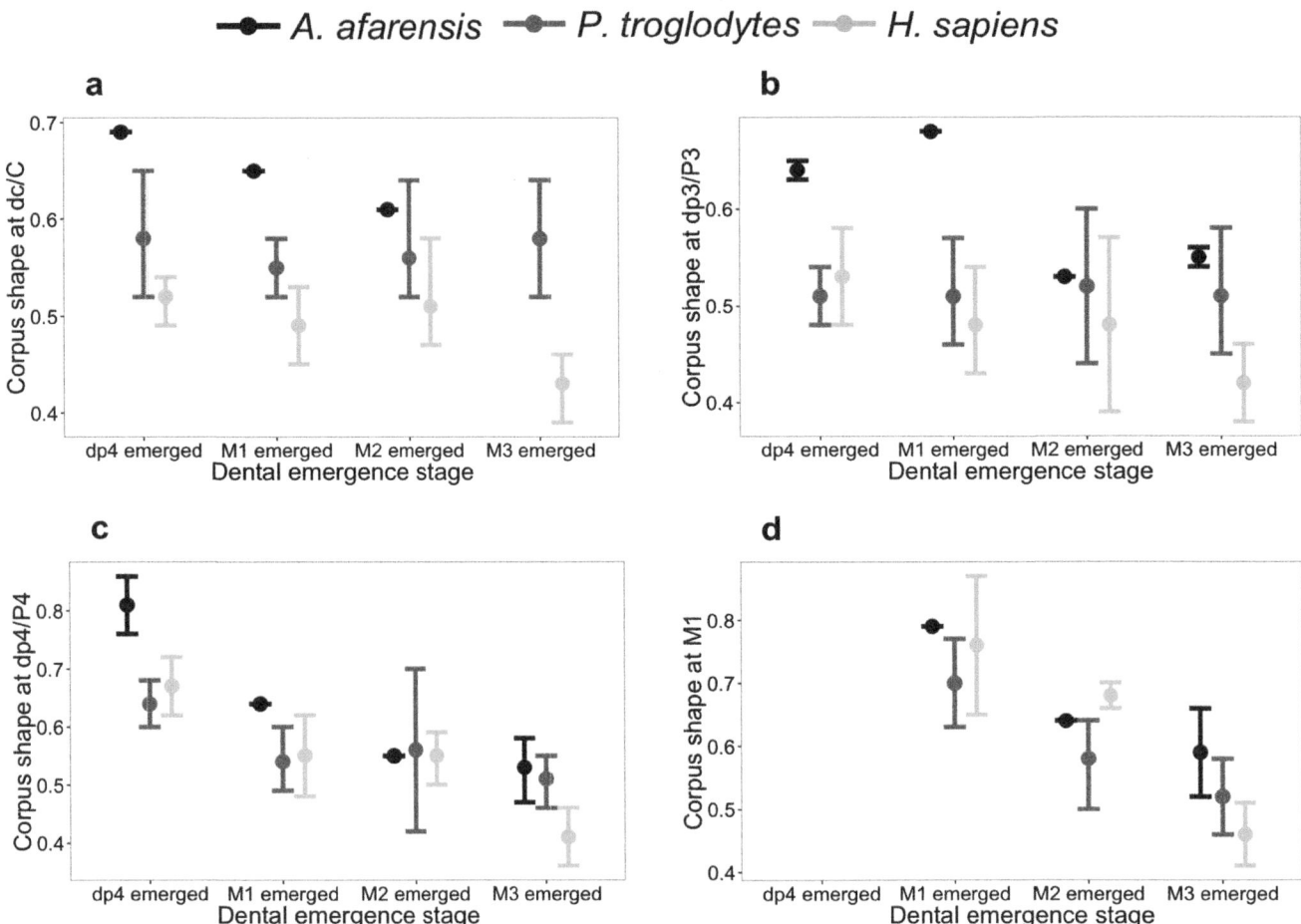

Fig. 10.7 Ontogenetic changes in mandibular corpus shape (breadth/height) at the positions of dc/C (**a**), dp3/P3 (**b**), dp4/P4 (**c**), and M1 (**d**)

The similarities between humans and *A. afarensis* are most apparent in the anterior part of the corpus (the canine and premolar regions) and at earlier dental emergence stages. At later emergence stages (i.e., M1 and especially M2) and further posterior along the corpus, the pattern of change is less distinctive. These observations make sense if both the large size of the growing permanent canine crown and root as well as the relatively long period of canine growth are important determinants of adult chimpanzee mandibular corpus morphology compared to hominins.

Growth of the chimpanzee canine, however, does not appear to influence corpus morphology throughout all of mandibular ontogeny or in all parts of the mandibular corpus. As noted above, at later dental emergence stages (especially "M2 emerged"), differences between chimpanzees, humans, and *A. afarensis* are much more subtle. The anterior corpus is influenced more by growth of the canine than the posterior corpus, which is not unexpected given the anterior position of the canine. Thus, our data imply that canine development is not the only factor molding hominid mandibular morphology.

Both alveolar and basal components of mammalian mandibular bone are deformed under masticatory loads (Daegling and Hylander 1997). While it is far from clear how closely the external morphology of the mandible tracks standard dietary categories or food material properties (Ross et al. 2012), such associations may be obscured by differences in the size and development of the teeth of species that consume foods of similar material properties. This inference accords with the spatial model, which predicts that the morphology of the mandibular corpus reflects the size of the tooth crown and roots that it houses (or the dimensions of the tooth germs that developed within it) (Daegling and Grine 1991; Cobb and Panagiotopoulou 2011). Plavcan and Daegling (2006) found that other than for the canine, tooth size is not closely correlated with corpus size and shape across extant anthropoid primates, suggesting that the mandibular corpus is not influenced by the size of the teeth that it houses. Static adult comparisons may, however, obscure tooth size-jaw size correlations that are strong early in ontogeny, when the tooth crowns and roots are growing, but decline after dental emergence. Indeed, this is what Cobb

Table 10.4 Minimum corpus height (mm) measured at 5 points along the corpus at each dental emergence stage

Dental emergence stage	Species		dc/C	dp3/P3	dp4/P4	M1	M2
dp4 emerged	*P. troglodytes*	Mean	19.69	18.91	15.69	–	–
		SD	1.69	1.47	1.43	–	–
		n	10	10	10	–	–
	H. sapiens	Mean	20.83	21.16	18.41	–	–
		SD	1.27	1.05	0.91	–	–
		n	9	9	9	–	–
	A. afarensis	Mean	20.20	21.33	18.87	–	–
		SD	–	1.18	1.21	–	–
		n	1	3	3	–	–
M1 emerged	*P. troglodytes*	Mean	25.35	24.43	22.46	19.21	–
		SD	2.16	2.19	2.13	1.94	–
		n	16	16	16	17	–
	H. sapiens	Mean	23.40	24.40	22.54	18.43	–
		SD	1.93	1.49	2.30	2.06	–
		n	11	11	11	12	–
	A. afarensis	Mean	26.50	26.10	24.40	22.50	–
		SD	–	–	–	–	–
		n	1	1	1	1	–
M2 emerged	*P. troglodytes*	Mean	29.14	28.32	25.34	24.08	22.74
		SD	2.71	2.45	3.81	2.62	2.73
		n	12	16	15	16	16
	H. sapiens	Mean	23.20	24.35	23.65	21.75	19.90
		SD	0.28	0.49	2.05	2.33	2.40
		n	2	2	2	2	2
	A. afarensis	Mean	30.00	31.80	30.00	28.00	27.80
		SD	–	–	–	–	–
		n	1	1	1	1	1
M3 emerged	*P. troglodytes*	Mean	32.79	33.00	30.82	29.45	27.24
		SD	3.34	3.33	3.04	2.64	2.19
		n	20	20	20	20	20
	H. sapiens	Mean	28.50	28.93	29.96	28.53	26.09
		SD	3.14	2.82	2.02	1.99	1.99
		n	10	10	10	10	10
	A. afarensis	Mean	–	31.67	32.17	30.48	28.87
		SD	–	0.58	2.25	1.07	2.77
		n	–	4	6	8	7

and Panagiotopoulou (2011) inferred from the results of their study of the mandibular symphysis and incisor formation in *Macaca*. The size and shape of the growing mandibular corpus is thus likely to be influenced by the tooth crowns that are developing within it. We would therefore expect that a closer association between corpus dimensions and tooth size would be found if the two were measured during ontogeny rather than in adult forms.

An ontogenetic perspective suggests that interpretations of the process by which adult mandibular morphology is realized must make allowance for the spatial requirements of housing the developing permanent dentition, especially the large canine crown and root in the great apes and, by extension, the expanded postcanine teeth in the geologically late australopiths. Even among the earliest australopiths,

differences in tooth crown and root size arguably influence mandibular form. Thus, *A. anamensis* has been shown to have (on average) more voluminous lower canine roots than *A. afarensis* – even as their canine-crown sizes do not differ appreciably (Ward et al. 2013). A large, fast-growing canine root would be expected to play a role in shaping the more apelike anterior mandibular corpus in *A. anamensis*. This does not necessarily discount the influence of hypothesized differences in feeding behavior on symphyseal form in these species (Ward et al. 2013). It does suggest, however, that analyses of fossil hominin mandibular form need to consider the mandible as something more than simply a lever in the mechanics of the masticatory system. Here we showed that canine reduction was likely an important influence on the growth trajectory of the *A. afarensis* mandibular corpus such

Table 10.5 Percentage of adult minimum corpus breadth attained at 5 points along the corpus at each molar emergence stage

Dental emergence stage	Species		dc/C	dp3/P3	dp4/P4	M1	M2
dp4 emerged	P. troglodytes	Mean	59.89	57.22	64.53	–	–
		SD	6.24	4.90	4.74	–	–
		n	10	10	10	–	–
	H. sapiens	Mean	88.25	93.21	99.71	–	–
		SD	3.53	9.96	6.86	–	–
		n	9	9	9	–	–
	A. afarensis	Mean	76.97	77.33	90.26	–	–
		SD	–	4.37	2.98	–	–
		n	1	2	3	–	–
M1 emerged	P. troglodytes	Mean	72.77	74.43	77.89	86.50	–
		SD	4.72	6.57	5.36	4.95	–
		n	16	17	17	17	–
	H. sapiens	Mean	93.03	96.41	99.76	106.36	–
		SD	7.72	9.78	8.90	9.47	–
		n	12	12	12	12	–
	A. afarensis	Mean	95.24	99.55	92.82	96.66	–
		SD	–	–	–	–	–
		n	1	1	1	1	–
M2 emerged	P. troglodytes	Mean	85.39	86.62	87.77	90.23	94.69
		SD	7.96	8.20	8.56	6.91	5.73
		n	15	16	16	16	15
	H. sapiens	Mean	96.31	97.18	105.46	113.93	111.65
		SD	15.65	19.97	17.85	16.33	7.85
		n	2	2	2	2	2
	A. afarensis	Mean	100.78	106.58	104.94	106.76	120.96
		SD	–	17.10	9.62	11.97	6.81
		n	1	2	2	2	2

Table 10.6 Percentage of adult minimum corpus height attained at 5 points along the corpus at each dental emergence stage

Dental emergence stage	Species		dc/C	dp3/P3	dp4/P4	M1	M2
dp4 emerged	P. troglodytes	Mean	60.07	57.30	50.91	–	–
		SD	5.16	4.45	4.63	–	–
		n	10	10	10	–	–
	H. sapiens	Mean	73.10	73.13	61.45	–	–
		SD	4.45	3.65	3.05	–	–
		n	9	9	9	–	–
	A. afarensis	Mean	–	67.37	58.65	–	–
		SD	–	3.74	3.75	–	–
		n	–	3	3	–	–
M1 emerged	P. troglodytes	Mean	77.31	74.03	72.89	65.25	–
		SD	6.58	6.64	6.90	6.58	–
		n	16	16	16	17	–
	H. sapiens	Mean	82.11	84.34	75.22	64.58	–
		SD	6.77	5.16	7.67	7.21	–
		n	11	11	11	12	–
	A. afarensis	Mean	–	82.42	75.85	73.83	–
		SD	–	–	–	–	–
		n	–	1	1	1	–
M2 emerged	P. troglodytes	Mean	82.11	85.82	82.24	81.76	83.49
		SD	6.77	7.43	12.36	8.90	10.02
		n	12	16	15	16	16
	H. sapiens	Mean	81.40	84.17	78.94	76.24	76.27
		SD	0.99	1.71	6.84	8.18	9.21
		n	2	2	2	2	2
	A. afarensis	Mean	–	100.42	93.26	91.88	96.29
		SD	–	–	–	–	–
		n	–	1	1	1	1

Table 10.7 Corpus shape at 5 points along the corpus for each dental emergence stage

Dental emergence stage	Species		dc/C	dp3/P3	dp4/P4	M1	M2
dp4 emerged	*P. troglodytes*	Mean	0.58	0.51	0.64	–	–
		SD	0.07	0.03	0.04	–	–
		n	10	10	10	–	–
	H. sapiens	Mean	0.52	0.53	0.67	–	–
		SD	0.03	0.05	0.05	–	–
		n	9	9	9	–	–
	A. afarensis	Mean	0.69	0.64	0.81	–	–
		SD	–	0.01	0.05	–	–
		n	1	2	3	–	–
M1 emerged	*P. troglodytes*	Mean	0.55	0.51	0.54	0.70	–
		SD	0.03	0.05	0.06	0.07	–
		n	16	16	16	17	–
	H. sapiens	Mean	0.49	0.48	0.55	0.76	–
		SD	0.04	0.05	0.07	0.11	–
		n	11	11	11	12	–
	A. afarensis	Mean	0.65	0.68	0.64	0.79	–
		SD	–	–	–	–	–
		n	1	1	1	1	–
M2 emerged	*P. troglodytes*	Mean	0.56	0.52	0.56	0.58	0.70
		SD	0.08	0.08	0.14	0.06	0.08
		n	12	16	15	16	16
	H. sapiens	Mean	0.51	0.48	0.55	0.68	0.81
		SD	0.08	0.09	0.05	0.02	0.04
		n	2	2	2	2	2
	A. afarensis	Mean	0.61	0.53	0.55	0.64	0.82
		SD	–	–	–	–	–
		n	1	1	1	1	1
M3 emerged	*P. troglodytes*	Mean	0.58	0.51	0.51	0.52	0.61
		SD	0.06	0.07	0.05	0.05	0.07
		n	20	20	20	20	20
	H. sapiens	Mean	0.43	0.42	0.41	0.46	0.55
		SD	0.04	0.04	0.05	0.05	0.04
		n	10	10	10	10	10
	A. afarensis	Mean	–	0.55	0.53	0.59	0.64
		SD	–	0.01	0.05	0.07	0.04
		n	–	3	6	7	4

that, as in humans, it achieved adult values relatively early. These results underscore the need to consider the influence that the developing dentition has on both juvenile and adult mandibular morphology.

Acknowledgments We thank Gary Schwartz and two anonymous reviewers for comments on the manuscript that improved the final product. We are grateful to Terrence Ritzman for help with the comparative sample and Lyman Jellema and Arelyn Simon for access to skeletal collections at the Cleveland Museum of Natural History and the Archaeological Research Institute at Arizona State University, respectively. This study was funded by the Institute of Human Origins and a Sigma Xi Grant-In-Aid of Research to HG.

References

Alemseged, Z., Spoor, F., Kimbel, W. H., Bobe, R., Geraads, D., Reed, D., et al. (2006). A juvenile early hominin skeleton from Dikika, Ethiopia. *Nature, 443*, 296–301.

Björk, A. (1963). Variations in the growth pattern of the human mandible: Longitudinal radiographic study by the implant method. *Journal of Dental Research, 42*, 400–411.

Boughner, J. C., & Dean, M. C. (2008). Mandibular shape ontogeny and dental development in bonobos (*Pan paniscus*) and chimpanzees (*Pan troglodytes*). *Evolutionary Biology, 35*, 296–308.

Bouvier, M., & Hylander, W. L. (1981). The effect of dietary consistency on the morphology of the mandibular condylar cartilage in macaques. *American Journal of Physical Anthropology, 54*, 203–204.

Brunet, M., Guy, F., Pilbeam, D., Mackaye, H. T., Likius, A., Ahounta, D., et al. (2002). A new hominid from the Upper Miocene of Chad, Central Africa. *Nature, 418,* 145–151.

Cobb, S. N., & Panagiotopoulou, O. (2011). Balancing the spatial demands of the developing dentition with the mechanical demands of the catarrhine mandibular symphysis. *Journal of Anatomy, 218,* 96–111.

Cofran, Z. (2014). Mandibular development in *Australopithecus robustus. American Journal of Physical Anthropology, 154,* 436–446.

Coquerelle, M., Bookstein, F. L., Braga, J., Halazonetis, D. J., Weber, G. W., & Mitteroecker, P. (2011). Sexual dimorphism of the human mandible and its association with dental development. *American Journal of Physical Anthropology, 145,* 192–202.

Daegling, D. J. (1996). Growth in the mandibles of African apes. *Journal of Human Evolution, 30,* 315–341.

Daegling, D. J., & Grine, F. E. (1991). Compact bone distribution and biomechanics of early hominid mandibles. *American Journal of Physical Anthropology, 86,* 321–339.

Daegling, D. J., & Hylander, W. L. (1997). Occlusal forces and mandibular bone strain: Is the primate jaw "overdesigned"? *Journal of Human Evolution, 33,* 705–717.

Dean, M. C., & Beynon, A. D. (1991). Tooth crown heights, tooth wear, sexual dimorphism and jaw growth in hominids. *Zeitschrift für Morphologie und Anthropologie, 78,* 425–440.

Dean, M. C., Beynon, A. D., Thackeray, J. F., & Macho, G. A. (1993). Histological reconstruction of dental development and age at death of a juvenile *Paranthropus robustus* specimen, SK 63, from Swartkrans, South Africa. *American Journal of Physical Anthropology, 91,* 401–419.

Dean, C., Leakey, M. G., Reid, D., Schrenk, F., Schwartz, G. T., Stringer, C., et al. (2001). Growth processes in teeth distinguish modern humans from *Homo erectus* and earlier hominins. *Nature, 414,* 628–631.

Haile-Selassie, Y. (2001). Late Miocene hominids from the Middle Awash, Ethiopia. *Nature, 412,* 187–191.

Haile-Selassie, Y., Suwa, G., & White, T. D. (2004). Late Miocene teeth from Middle Awash, Ethiopia, and early hominid dental evolution. *Science, 303,* 1503–1505.

Humphrey, L. T. (1999). Relative mandibular growth in humans, gorillas and chimpanzees. In R. Hoopa & C. Fitzgerald (Eds.), *Human growth in the past* (pp. 65–97). Cambridge: Cambridge University Press.

Hylander, W. L. (1979). Mandibular function in *Galago crassicaudatus* and *Macaca fascicularis*: An in vivo approach to stress analysis of the mandible. *Journal of Morphology, 159,* 253–296.

Kimbel, W. H., & Delezene, L. K. (2009). "Lucy" redux: A review of research on *Australopithecus afarensis. Yearbook of Physical Anthropology, 52,* 2–48.

Kimbel, W. H., Rak, Y., & Johanson, D. C. (2004). *The skull of* Australopithecus afarensis. New York: Oxford University Press.

Kimbel, W. H., Lockwood, C. A., Ward, C. V., Leakey, M. G., Rak, Y., & Johanson, D. C. (2006). Was *Australopithecus anamensis* ancestral to *A. afarensis*? A case of anagenesis in the hominin fossil record. *Journal of Human Evolution, 51,* 134–152.

Leakey, M. G., Feibel, C. S., McDougall, I., & Walker, A. (1995). New four-million-year-old hominid species from Kanapoi and Allia Bay, Kenya. *Nature, 376,* 565–571.

Lockwood, C. A., Kimbel, W. H., & Johanson, D. C. (2000). Temporal trends and metric variation in the mandibles and dentition of *Australopithecus afarensis. Journal of Human Evolution, 39,* 23–55.

Moss, M. L., & Rankow, R. M. (1968). The role of the functional matrix in mandibular growth. *The Angle Orthodontist, 38,* 95–103.

Plavcan, J. M., & Daegling, D. J. (2006). Interspecific and intraspecific relationships between tooth size and jaw size in primates. *Journal of Human Evolution, 51,* 171–184.

Rak, Y., Ginzburg, A., & Geffen, E. (2007). Gorilla-like anatomy on *Australopithecus afarensis* mandibles suggests *Au. afarensis* link to robust australopiths. *Proceedings of the Natural Academy of Sciences USA, 104,* 6568–6572.

Ravosa, M. J. (2000). Size and scaling in the mandible of living and extinct apes. *Folia Primatologica, 71,* 305–322.

Ritzman, T. B., & Spencer, M. A. (2009). Coronoid process morphology and function in anthropoid primates. *American Journal of Physical Anthropology, 104,* 221.

Ross, C. F., Berthaume, M. A., Dechow, P. C., Iriarte-Diaz, J., Porro, L. B., Richmond, B. G., et al. (2011). *In vivo* bone strain and finite-element modeling of the craniofacial haft in catarrhine primates. *Journal of Anatomy, 218,* 112–148.

Ross, C. F., Iriarte-Diaz, J., & Nunn, C. L. (2012). Innovative approaches to the relationship between diet and mandibular morphology in primates. *International Journal of Primatology, 33,* 632–660.

Schwartz, G. T., & Dean, C. (2001). Ontogeny of canine dimorphism in extant hominoids. *American Journal of Physical Anthropology, 115,* 269–283.

Scott, J. E., McAbee, K. R., Eastman, M. M., & Ravosa, M. J. (2014). Experimental perspective on fallback foods and dietary adaptations in early hominins. *Biological Letters, 10,* 20130789.

Senut, B., Pickford, M., Gommery, D., Mein, P., Cheboi, K., & Coppens, Y. (2001). First hominid from the Miocene (Lukeino Formation, Kenya). *Comptes Rendus de l'Académie des Sciences—Series IIA—Earth and Planetary Science, 332,* 137–144.

Singh, N. (2014). Ontogenetic study of allometric variation in *Homo* and *Pan* mandibles. *The Anatomical Record, 297,* 261–272.

Smith, B. H. (1994). Sequence of emergence of the permanent teeth in *Macaca, Pan, Homo*, and *Australopithecus*: Its evolutionary significance. *American Journal of Physical Anthropology, 6,* 61–76.

Suwa, G., Kono, R. T., Simpson, S. W., Asfaw, B., Lovejoy, C. O., & White, T. M. (2009). Paleobiological implications of the *Ardipithecus ramidus* dentition. *Science, 326,* 94–99.

Terhune, C., Robinson, C. A., & Ritzman, T. B. (2014). Ontogenetic variation in the mandibular ramus of great apes and humans. *Journal of Morphology, 275,* 661–677.

Taylor, A. B. (2002). Masticatory form and function in the African apes. *American Journal of Physical Anthropology, 117,* 133–156.

Taylor, A. B. (2006). Size and shape dimorphism in great ape mandibles and implications for fossil species recognition. *American Journal of Physical Anthropology, 120,* 82–98.

Ward, C. V., Leakey, M., & Walker, A. (2001). Morphology of *Australopithecus anamensis* from Kanapoi and Allia Bay, Kenya. *Journal of Human Evolution, 41,* 255–368.

Ward, C. V., Manthi, F. K., & Plavcan, J. M. (2013). New fossils of *Australopithecus anamensis* from Kanapoi, West Turkana, Kenya (2003–2008). *Journal of Human Evolution, 65,* 501–524.

Weidenreich, F. (1936). *The mandibles of* Sinanthropus pekinensis: *A comparative study*. Beijing: Palaeontologica Sinnica D 4.

White, T. D. (1977). The anterior mandibular corpus of early African Hominidae: Functional significance of shape and size. Ph.D. Dissertation, University of Michigan, Ann Arbor.

White, T. D., & Johanson, D. C. (1982). Pliocene hominid mandibles from the Hadar Formation, Ethiopia (1974–1977 collections). *American Journal of Physical Anthropology, 57,* 501–544.

White, T. D., Johanson, D. C., & Kimbel, W. H. (1981). *Australopithecus africanus*: Its phyletic position reconsidered. *South African Journal Sciences, 77,* 445–470.

White, T. D., Suwa, G., & Asfaw, B. (1994). *Australopithecus ramidus*, a new species of early hominid from Aramis, Ethiopia. *Nature, 371,* 306–312.

Wood, B. A. (1991). *Koobi Fora research project* (Vol. 4). *Hominid cranial remains*. Oxford: Clarendon Press.

Chapter 11
Middle Pleistocene *Homo* Crania from Broken Hill and Petralona: Morphology, Metric Comparisons, and Evolutionary Relationships

G. Philip Rightmire

Abstract A fossilized human cranium was discovered by miners quarrying at Broken Hill (now Kabwe) in 1921. Broken Hill is one of the best preserved hominins ever recovered from a later Middle Pleistocene locality. Remarkably, no comprehensive descriptive or comparative account has been published since 1928. Overall, Broken Hill resembles *Homo erectus*. The frontal is flattened with midline keeling, the vault is low, and the massive face is "hafted" to the braincase in such a way as to accentuate facial projection. At the same time, there are apomorphic features shared with later humans. Brain size is 1280 cm^3, the temporal squama is arch-shaped, and the upper scale of the occipital is expanded relative to its lower nuchal portion. Specialized characters of the temporomandibular joint region include a raised articular tubercle and a sphenoid spine. Reorientation of the nasal aperture and placement of the incisive canal suggest that the face may be more nearly vertical than in *H. erectus*. It is apparent that Broken Hill is similar to other African crania from Bodo, Ndutu, and Elandsfontein as well as European fossils including Arago and Petralona. However, the systematic position of these hominins remains controversial. The material has been grouped into a series of grades within a broad *H. sapiens* category. A very different reading of the record recognizes multiple, distinct taxa and suggests that speciation must have occurred repeatedly throughout the Pleistocene. Still another perspective holds that differences among the African and European specimens are minor and can be attributed to geography and intragroup variation. It is argued that many of the fossils belong together in one widely dispersed taxon. If the Mauer mandible is included within this hypodigm, then the appropriate name is *H. heidelbergensis*. Treated in a broad sense, *H. heidelbergensis* is ancestral to both *H. neanderthalensis* and *H. sapiens*. This study will provide a detailed account of the morphology of Broken Hill and its similarities to other Middle Pleistocene hominins from Africa. Comparisons will include Arago, Petralona, and assemblages such as Sima de los Huesos. My approach will address the taxonomic utility of characters of the vault, cranial base and face, species-level systematics, and evolutionary relationships.

Keywords Craniofacial anatomy • Encephalization • *Homo erectus* • *Homo heidelbergensis* • *Homo rhodesiensis* • *Homo sapiens* • Neanderthals • Phylogeny • Speciation • Intraspecific variation

Introduction

The cranium from Broken Hill (now Kabwe) remains one of the treasures of prehistory. It was found in 1921, when miners quarrying for lead ore broke into the lower part of an extensive cavern containing quantities of mineralized bones and stone artifacts. Accounts of the circumstances surrounding this discovery are contradictory (Hrdlička 1930). Several additional human fossils, along with animal bones, were collected from the cave fill, but claims for the association of any of these elements with the original cranium remain incompletely documented. Comparative studies of the fauna have demonstrated similarities with the large assemblage from Elandsfontein in South Africa, indicating an early Middle Pleistocene age (Klein 1994; Klein et al. 2006). However, more recent efforts to date individual bones directly using Electron Spin Resonance (ESR) suggest that the Broken Hill material may be of late Middle Pleistocene antiquity (Stringer 2011).

The cranium was described initially by Woodward (1921), who saw resemblances to the Neanderthals then known from Europe but attributed the find to a new species ('*Homo rhodesiensis*'). More comprehensive studies were

G.P. Rightmire (✉)
Department of Human Evolutionary Biology, Harvard University, Peabody Museum, Cambridge, MA 02138, USA
e-mail: gprightm@fas.harvard.edu

© Springer International Publishing AG 2017
Assaf Marom and Erella Hovers (eds.), *Human Paleontology and Prehistory*,
Vertebrate Paleobiology and Paleoanthropology, DOI: 10.1007/978-3-319-46646-0_11

published several years later by Pycraft (1928) and Mourant (1928). While pointing to differences in certain features, Mourant (1928) again argued for a close relationship between Broken Hill and Late Pleistocene Neanderthals. It is now recognized that this comparison was inappropriate. Broken Hill lacks the specialized characters of Neanderthals but resembles other crania from Elandsfontein in South Africa and Bodo from the Middle Awash of Ethiopia. As a group, these African fossils are broadly similar to hominins from Middle Pleistocene localities in Europe including the Sima de los Huesos in Spain, Arago Cave in France, and Petralona in Greece.

Interpreting this record has been problematic. The number of taxa represented is disputed, and phylogenetic relationships remain to be clarified. In one view, African and European mid-Pleistocene populations can be grouped with later humans within a broad *Homo sapiens* category. Archaic and modern grades are defined by advances in brain size and skull form. Although changes to the vault and face accumulate in a mosaic pattern, early and late groups are said to follow one another seamlessly, as segments of a single evolving lineage (Bräuer 2007, 2008). A very different reading of the record recognizes multiple, distinct taxa as evidence for speciation occurring repeatedly throughout the Pleistocene (Tattersall and Schwartz 2008; Schwartz and Tattersall 2010). At least two lineages are identified, in addition to *Homo erectus* and recent humans. A European branch can be traced back via Petralona, Arago, and Sima de los Huesos, deeply into the Middle Pleistocene. Proponents of this view (Arsuaga et al. 1989, 1997; see also Hublin 2009) claim that even the oldest European hominins share apomorphies with *Homo neanderthalensis* and can reasonably be attributed to this species. A variation on this phylogenetic scheme has been proposed by Martinón-Torres et al. (2012), who find that the Sima de los Huesos teeth are "more Neanderthal" in form than the Mauer or Arago dentitions. Given this result, Martinón-Torres et al. (2012) suggest that along with an ancient lineage linking the Sima hominins directly with Neanderthals, a second population including Mauer and Arago was present in Europe. This second species must be called *Homo heidelbergensis*.

A key question is how these European lineages are related to the hominins in Africa. If all of the European fossils are subsumed within *Homo neanderthalensis*, then the species represented by Broken Hill, Elandsfontein, and Bodo can be called *Homo rhodesiensis*, following the nomenclature proposed by Woodward (1921). Some (chronologically late) members of this group exhibit morphology that is archaic, coupled with characters suggestive of a link to anatomically modern humans. Still another perspective holds that morphological differences among the most ancient European and African specimens are minor and can be attributed to

geography and intragroup variation (Stringer 1983, 1993; Rightmire 1990, 1996, 1998, 2008; Mounier et al. 2009). It can be argued that many of the fossils belong together in one geographically dispersed taxon. If the Mauer mandible is included within the hypodigm, then the appropriate name for this species is *Homo heidelbergensis*. Treated in this broad sense, *Homo heidelbergensis* must be ancestral to both *Homo neanderthalensis* and *Homo sapiens*.

While many of the Middle Pleistocene fossils are incomplete, Broken Hill is clearly one of the most informative specimens. Another is the cranium from Petralona. It is possible to document the extent to which these African and European fossils differ in their craniofacial morphology. Over the course of nearly a century, Broken Hill has been treated in numerous comparisons involving modern humans, Neanderthals, and earlier *Homo*. Following Mourant (1928), many of these studies have been based on measurements or, more recently, cranial landmarks used in morphometric analysis (Friess 2010; Harvati et al. 2010, 2011). Another approach has emphasized anatomical description, with attention to the relevance of individual characters. Since it was discovered in 1959, the Petralona cranium has also been studied in detail (Stringer et al. 1979, 1983). This research has produced many useful data, but there is (still) no firm consensus as to the evolutionary significance of either specimen. In this review, I introduce further evidence from measurements and comparative anatomy. My goal is to clarify the relationship of Broken Hill to Petralona, with the goal of testing the null hypothesis that these individuals can be grouped together in one taxon.

The Broken Hill Cranium

The Broken Hill frontal and sphenoid are intact, as is the left parietotemporal region (Fig. 11.1). On the right, the temporal bone is missing, and there is a large gap in the occipital. Damage extends along the junction of the upper and lower scales over an area that would include the center of the transverse torus. Fortunately, the braincase is not distorted, and it is possible to reconstruct by mirror-imaging almost all of the missing morphology. The facial skeleton is quite complete. The posterior aspect of the right zygomatic arch has been sheared away, and in the subnasal region, a strip of cortex has been broken out to expose the root of the right central incisor and the alveolus for the (missing) right lateral incisor tooth. CT imaging reveals that bone has been resorbed around the apices of the left molar roots, suggesting periapical infection (Zonneveld and Wind 1985). The incisors have been reduced to stumps. Other teeth are also heavily worn, and many show evidence of severe caries.

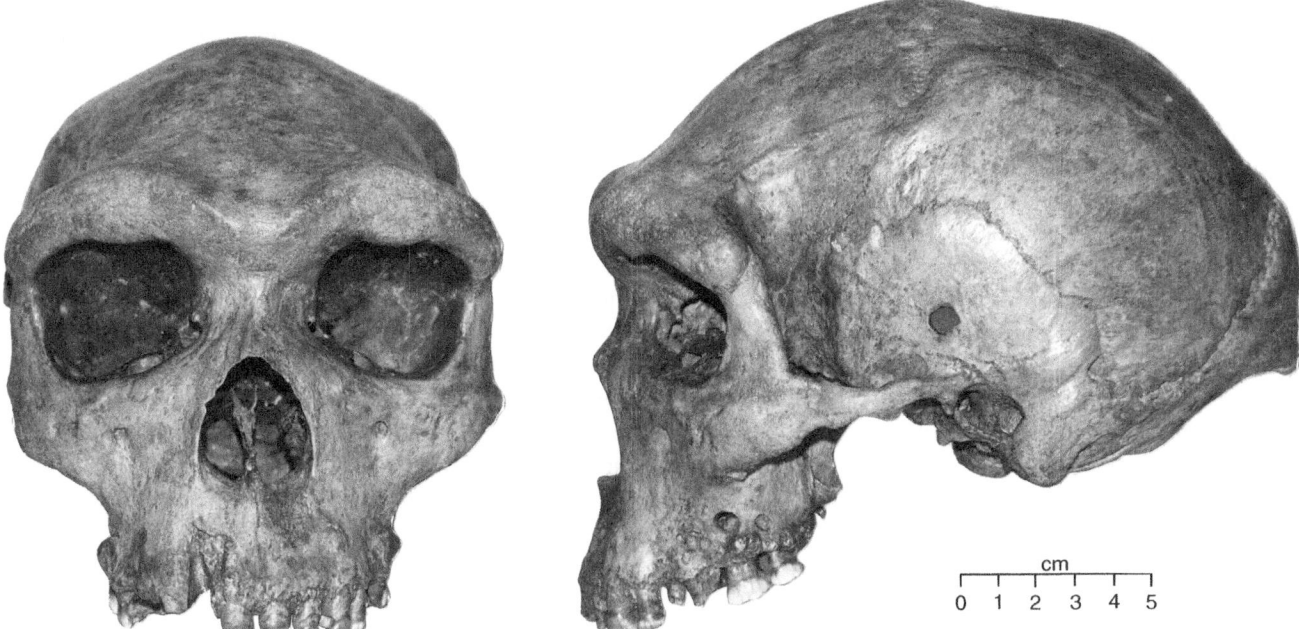

Fig. 11.1 Facial and lateral views of the cranium from Broken Hill. The face is massive, with some of the heaviest brows on record. The frontal is flattened with slight midline keeling, and the vault is low in profile. In its overall morphology, Broken Hill resembles *Homo erectus*, but there are apomorphic features shared with later humans. The temporal squama is high and arch-shaped, and the upper scale of the occipital is expanded relative to its lower nuchal portion. Several discrete characters of the temporomandibular joint region are specialized. More changes are apparent in the face, where the lateral border of the nasal aperture is set vertically, and the palatal anatomy is like that of later people

Anatomy of the Braincase

Endocranial capacity is 1280 cm³. In superior view, the braincase is elliptical in outline. The frontal is somewhat elongated anteriorly, where there is marked postorbital constriction. A striking attribute of Broken Hill is the very formidable supraorbital development. As measured at their most lateral points, the tori span a width of 139 mm. Together with the large orbits, these structures dominate the face. From the side, the vault appears low and angular in profile, rather than globe-like as in modern humans. While Broken Hill has been described as "archaic" in many aspects of its morphology, the parietals are expanded, and the occipital is less strongly flexed than is the case even for late *Homo erectus* from Sangiran, Ngandong, and Zhoukoudian.

The glabellar region is inflated and positioned well forward of the nasal root. The (chord) distance between glabella and nasion is about 8 mm, and there is no trace of a metopic suture. This central section of the brow is the most projecting. From glabella, the tori arch upward and laterally, to become increasingly thickened over the center of each orbit. Here, a distinct bulge is limited medially by the supraorbital notch and laterally by a shallow channel which courses obliquely upward. This channel is the only division between the massive superciliary eminence and the bar-like lateral portion of

the torus. The outer part of the brow curves downward but is still very thick as it nears the zygomatico-frontal suture.

The vermiculate bone of the brow is delimited from the smoother supratoral region by a roughened line. On each side, this line curves posteriorly, so that the facial aspect of the browridge recedes above the orbital opening. The supratoral surface is hollowed laterally. Centrally the frontal forms a shelf behind glabella. This contour rises gradually, and about a third of the way from nasion to bregma, a faint midline keel is developed. The origins of the (inferior) lines are strongly crested, and here there is marked postorbital narrowing. Coupled with the heavy brows, this constriction gives the Broken Hill frontal an archaic appearance.

CT scanning reveals that the frontal sinus is voluminous and complex (Seidler et al. 1997). Several compartments are present, separated by bony partitions. On the left, one chamber reaches almost to the wall of the temporal fossa. As noted by Seidler et al. (1997), these spaces fill the bone directly overlying the orbits to a much greater extent than is characteristic of recent humans. More medially, the sinus invades almost all of the glabellar region and the brow. Finger-like extensions also pass upward from glabella, into the space between the internal and external tables of the bone.

As with other parts of the cranium, the sphenoid of Broken Hill is noteworthy for its robusticity and degree of pneumatization. The lateral face of the greater wing is deeply concave, accentuating the postorbital constriction present in the frontal bone. The effect of this curvature is to increase the volume of the temporal fossa, suggesting a large cross-section for the temporalis muscle and its tendon. At the base of this channel, the infratemporal crest is raised, especially anteriorly where it encroaches upon the pterygomaxillary fissure. Foramen ovale is situated very close to the margin of the greater wing. Only a thin spicule of bone intervenes between the foramen and the petrous apex. Foramen spinosum lies in about the usual position, between ovale and the sphenoid spine. However, the smaller foramen is separated from the larger by the bony spicule, and it therefore penetrates the crevice between the sphenoid and the petrous temporal. Pycraft (1928) indicates that the foramen spinosum is missing, but this is not the case.

The sphenoid spine projects downward. It does not contribute directly to the medial wall of the glenoid fossa, and it is not particularly large. Such a spine is not present in *Homo erectus*, but in Broken Hill, it is oriented in about the same way as in modern humans. Its medial border appears to be flattened to form with the adjacent temporal a narrow groove for the cartilaginous part of the auditory tube.

The pterygoid processes are exceptional in their length and thickness. At their roots, these structures are expanded and heavily pneumatized. Sinus cavities reach not only into the pterygoids but into the sphenoid greater wings as well. The medial plate is relatively long, and the hamulus is preserved. The lateral plate passes obliquely forward, where its area of contact with the posterior surface of the maxilla is extensive. The pterygoid fossa is deeply excavated.

Midline keeling does not extend onto the parietal vault. There is no localized eminence at bregma, and the serrations making up the sagittal suture are in fact slightly depressed, relative to the adjacent bone surfaces. Posteriorly, there is an area of flattening, not only near the suture but also across the entire occipital angle. This flattening is present on both sides and contributes to the "stepped" or angulated profile which the vault displays in side view. Further from the midline, the contour of the parietal is rounded, but no tuber is expressed. The maximum (biparietal) width falls low on the parietals, about midway along the arc of the squamosal suture.

At the coronal suture, the superior and inferior temporal lines are separated, and both turn upward. Their arc-like path here deviates sharply from the more horizontal course established on the frontal. At its closest approach to the midline, the superior line is approximately 87 mm from its homologue on the opposite side. This line reaches posteriorly to a point well behind asterion before curving forward onto the mastoid angle, where there is a moderately prominent angular torus. The inferior line also passes behind asterion as it turns toward the supramastoid crest. Between the two lines and just posterior to the parietal incisure, there is a faint but palapable angular sulcus.

The left temporal bone displays pathology. These lesions do not obscure most aspects of the anatomy of the specimen. The temporal squama is similar in its proportions to that of recent *Homo sapiens*. Its superior border is arched, rather than relatively straight as in *Homo erectus*. Height of the squama measured from porion is 50 mm, while length taken from the parietal incisure to the most anterior point on the sphenotemporal boundary is 72 mm. Martínez and Arsuaga (1997) have devised an angle registering the inclination of the posterior segment of the squamosal suture. This angle is large (>50°) for Broken Hill, as is also the case for several of the Sima de los Huesos individuals, other Middle Pleistocene hominins, and modern humans.

In side view, the upper part of the occipital is nearly vertical. The shelf-like occipital torus is bounded above by a transverse depression. At its closest approach to the (missing) midline, the torus produces a blunt tubercle at its lower margin. This morphology suggests that structures located still more centrally would have been massively developed, but neither the linear tubercle itself nor the surface adjacent to it can be studied. In the region that is preserved, the superior nuchal line is impressed into the base of the torus, which overhangs the nuchal plane below. The line curves toward asterion and can be traced from the occipital onto the lateral aspect of the mastoid process, where it merges with the elevated mastoid crest.

The supramastoid crest is mound-like. Where it extends forward to overhang the auditory opening, this shelf could not have been prominent, and the porus itself is not more deeply recessed than is characteristic for (some) recent crania. The root of the zygomatic process is robust. On its superior surface, a channel accommodating the posterior fibers of the temporalis muscle is clearly outlined. Width of the temporal gutter measured at its most anterior extent (following Wood 1991) is 17 mm.

The mastoid process is elongated rather than conical in form, with a blunt tip. This shape is not an artifact caused by damage that has exposed air cells near the apex but is a consequence of flattening of the posterolateral face. This aspect of the process, which is scarred by muscle attachment, is continuous with the nuchal plane of the occipital. Such morphology is characteristic of *Homo erectus*. However, in Broken Hill, the mastoid region is not so laterally expanded with respect to the parietal walls above. Also, the process descends vertically, with only a slight medial tilt. This orientation differs from that in more archaic specimens, where the long axis of the mastoid is inclined inward. In its size, the process is not exceptional. Length measured from the

Frankfurt Horizontal is within the range observed for *Homo erectus* and close to the averages obtained for modern humans.

The fossa for the (posterior) belly of the digastric muscle is elongated and broad posteriorly. It reaches almost to the mastoid foramen, which is situated just forward of asterion. The fossa deepens as it passes alongside the mastoid process. Here it becomes a groove, covering a distance of 39 mm before ending at the stylomastoid foramen. Near the foramen, the channel is less sharply incised, but it is never bridged over or obliterated. Anterior obliteration of the incisure is recorded for the Zhoukoudian hominins (Weidenreich 1943) and for the Neanderthals. In Broken hill, the digastric fossa, stylomastoid foramen, and the styloid pit are collinear, as is the case for recent humans. This condition differs from that in Asian *Homo erectus*. In the Zhoukoudian specimens, the stylomastoid foramen lies "outside" the line joining the incisure and the styloid pit (Weidenreich 1943). Medially, a parallel segment of the occipitomastoid suture is deeply incised, and this narrow channel may mark the passage of the occipital artery. There is some heaping up of bone alongside the suture, but an occipitomastoid crest in the sense of Weidenreich (1943) is not developed. Protrusion of the entire medial margin of the digastric incisure is not nearly so extreme as in Neanderthals and instead resembles the eminence seen in many *Homo sapiens* crania.

The glenoid fossa is wide. The entoglenoid to ectoglenoid chord is 31 mm. Because the ectoglenoid process peaks at the lateral-most margin of the articular surface, this measurement is equivalent to mandibular fossa breadth as defined by Wood (1991). The articular tubercle is carried well out onto the massive zygomatic root. As a result, the entire fossa is expanded laterally, so as to project beyond the lateral wall of the braincase. In this respect, Broken Hill approaches but does not quite match the condition seen in *Homo erectus*. The articular tubercle is irregular in form. Its lateral part is hollowed and somewhat eroded in appearance. The medial section is convex in the same plane and also prominent anteroposteriorly. Here the tubercle stands out in clear relief against the preglenoid planum. Still more medially, the entoglenoid process is reduced to a thin lip of bone applied to the adjacent sphenoid. Broken Hill thus differs from *Homo erectus*, where no bar-like articular eminence is usually developed, and the entoglenoid pyramid is more robust.

The inner wall of the glenoid fossa is rounded rather than constricted, and there is no crevice-like extension of the cavity between the entoglenoid process and the tympanic plate. The back of the fossa is formed from both squamous bone and the tympanic. The tympanic part is nearly vertical in orientation. Lateral to it, the postglenoid process is very well developed. This structure is elongated in the coronal plane and also deep. Maximum depth of the process as

measured following Wood's (1991) procedure is 11 mm. Its inferior aspect is rugged and carries several small tubercles.

The tympanic bone is damaged. The part which should be applied to the rear of the postglenoid process is broken away. Nevertheless, the bone is complete enough to provide important information. Much of the plate is vertical, but its inferior aspect is inclined posteriorly. Whether this (lower) margin was thin and crested or relatively blunt can no longer be ascertained, but a trace of the tympanomastoid fissure is preserved. The styloid sheath seems to be quite thickened. Within the pit surrounded by this sheath, only the base of the styloid process may be present. Further medially and just anterior to the carotid canal, the tympanic bone produces an irregular tubercle, clearly associated on one side with the spine of the sphenoid and fused on the other to the petrous temporal. The petrous bone itself is unremarkable. Its apex is eroded in appearance as in modern populations, rather than more compact as in *Homo erectus*. The foramen lacerum is spacious, especially between the petrous apex and the basilar part of the occipital.

The Facial Skeleton

The upper face is very broad in relation to both the zygomatic arches and the maxilloalveolar region. From nasion, the frontonasal and frontomaxillary sutures trend downward toward the anterior lacrimal crests. The distance between these crests is 28 mm, while the full width (dacryon-dacryon) of the interorbital pillar is somewhat greater. The superior margins of the orbits are approximately horizontal. These openings are deeper laterally than medially, so that their lower borders slope downward toward the angles of the cheeks. The plane of each orbit is inclined posteroinferiorly.

Individually, the nasal bones and maxillary frontal processes are only slightly convex, and together they form a tented bridge. Only the upper portion of this saddle exhibits faint keeling in the midline. The combined width of the nasal bones superiorly is 18 mm. The bones are narrowed centrally (least breadth is 12 mm) but expand below to produce an hourglass-like shape. In side view, the nasal profile is evenly concave. Rhinion is preserved, and it is apparent that the piriform margin curves backward from this point before dropping vertically toward the nasal floor. This gives the aperture an erect orientation and accentuates the overall orthognathism of the midface. Here Broken Hill resembles recent humans rather than the condition found in *Homo erectus*, where the border of the aperture slopes anteriorly (Weidenreich 1943).

The pear-shaped nasal aperture seems small in comparison to the orbits, infraorbital region, and upper jaw. Its lateral margins are quite thin but not everted. Inferiorly, these margins are slightly thickened and merge smoothly with the maxillary wall. There are no lateral crests. The nasal sill is dominated by the prominent anterior spine, marking attachment of the nasal septum. As in many recent human crania, this spine is bifid, and both tips reach forward to overhang the clivus. From each of these projections, a well defined spinal crest extends posterolaterally to separate the clivus from the nasal floor. These crests pass approximately 5 mm to the rear of the lateral margins of the aperture and then subside into the internal contours of the nasal cavity. Here the anatomy is well enough preserved to demonstrate that there are no large "medial projections" of the sort identified by Schwartz and Tattersall (1996) as characteristic of Neanderthals. The floor of the cavity is inclined gently downward as it recedes from the nasal sill. This topography can be scored as "continuous-smooth" in the terminology of McCollum (2000).

The surface of the nasoalveolar clivus is flattened in the transverse plane. Its midsagittal contour is slightly convex and projects beyond the outline of the canine jugum. At the same time, the entire subnasal portion of the face is angled forward, and the degree of alveolar prognathism is substantial. Length of the nasoalveolar clivus measured from subspinale to prosthion is 29 mm, while the total distance from the tip of the anterior spine to the lowest point between the central incisors is 36 mm. This part of the facial skeleton is exceptionally robust. Anteriorly, the alveolar process presents vertical corrugations associated with the incisor roots. Although the canine clearly has a long, stout root, its jugum is not noticeably more prominent than any associated with the incisors. These structures are confined to the lower half of the clivus and do not encroach upon the nasal margin. Posteriorly, on the left side, there is a prominent (buccal) nodular exostosis at the position of M^2. On the right, a larger and more horizontally elongated bony growth is situated similarly. Such maxillary exostoses are common in Chinese *Homo erectus* (Weidenreich 1943).

The wall of the maxilla adjacent to the piriform aperture is generally smooth. This region is perhaps less "inflated" in Broken Hill than in many of the Neanderthals. Behind the canine/P^3 jugal prominence, there is some localized hollowing, but a canine fossa comparable to that sculpted into modern faces is not present. Infraorbital foramina are centered about 14 mm (left) and 15 mm (right) below the orbital margins. From each opening, a shallow depression extends downward for only a short distance. This morphology differs from that in *Homo erectus*, where there is a deeper and elongated vertical groove (the "maxillary sulcus" of Weidenreich 1943) lying alongside a more expanded canine jugum.

Where it springs from the maxillary body above the position of M^2, the zygomaticoalveolar root is thickened. CT imaging demonstrates that sinus pockets extend from the highly pneumatized maxilla into the zygomatic arch. The incisure is strongly flexed. However, there is no actual notch as would be created by an inferior bend at the lateral extremity of the pillar. Malar (cheek) height is 29 mm, but maximum height of the zygomatic bone is ca. 55 mm, and the cheek is massively constructed. Its anterior face slopes posteroinferiorly. An irregular ridge can be traced along the zygomaticomaxillary suture as it passes obliquely upward toward the inferior orbital margin. The orbital rim itself is blunt, and there is little drop in elevation as this surface turns inward to form the orbital floor.

Laterally, the surface of the cheek is mostly flattened. Lower on the zygomatic bone, there is a distinct protuberance, centered below the zygomaticofacial foramen and reaching anteriorly toward the zygomaticomaxillary border. This swelling is associated neither with the suture nor with the area of masseter origin, and it cannot be likened to the more inferiorly placed malar "tuber" identified by Weidenreich (1943) in the Zhoukoudian hominins. The masseter scar itself is extensive, deeply pitted, and displays several small bony projections at the lower end of the zygomaticomaxillary suture. On the left side, two zygomaticofacial foramina are spaced some 13 mm apart. A third (larger) foramen is situated more anteriorly. These openings are set in an arc that parallels the contour of the orbit.

The Broken Hill palate and dental arcade are parabolic in contour. The palatal surface is rugged and marked by numerous impressions. There are traces of a median torus, expressed to either side of the intermaxillary suture. This torus extends posteriorly almost to the junction with the palatine bones. At the front of the arcade, the palatal roof is inclined steeply upward. Because of this topography, the incisive fossa is situated well above the alveolar margin but only a short distance behind the incisor roots. A channel leading into the incisive canal is formed almost directly behind the septum separating the central incisor teeth. This configuration is comparable to the modern condition.

Petralona

The Petralona braincase is remarkably complete (Fig. 11.2). The frontal bone is undistorted, but the rear of the vault shows slight deformation. This causes the right parietal to bulge a little more than the left, and the right temporal squama is displaced laterally. Both mastoid processes are broken. The long axis of the palate is set at an angle of several degrees to the sagittal plane of the braincase, reflecting a slight twisting of the entire facial skeleton to the right side. The proportions of the face itself are not affected. The supraorbital structures and the front of the interorbital pillar are intact, although bone is missing from the medial

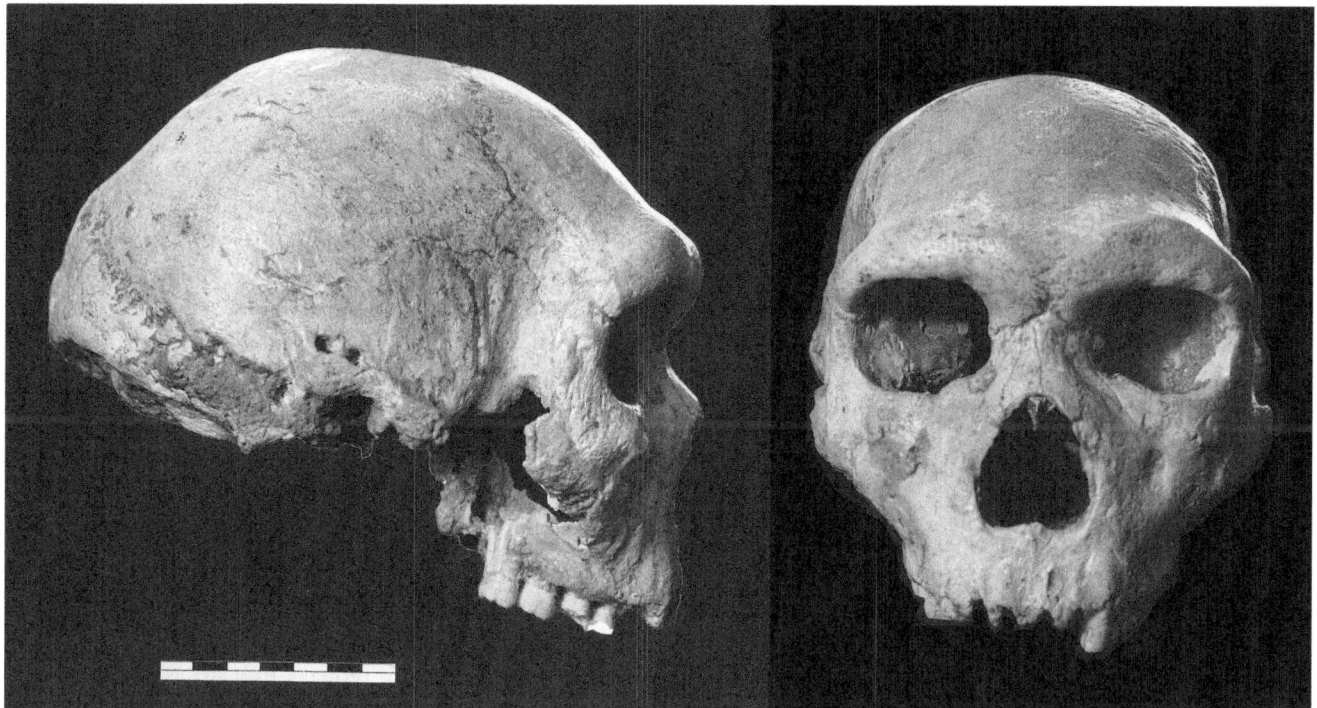

Fig. 11.2 Lateral and facial views of the cranium from Petralona. This specimen from Greece is remarkably complete. It differs from Broken Hill in having a broader base but resembles the African fossil in midfacial morphology, vault proportions, and anatomy of the basicranium

wall and floor of each orbit. There is minor damage to the posterior part of one of the zygomatic bones, but otherwise the cheeks, nasal saddle, and maxillary surfaces are well preserved. The floor of the nasal cavity is crusted with a thin coat of matrix. There is damage to the sphenoid and the body of the right maxilla. The alveolar process is largely complete. All of the incisors are missing, but C to M^3 are present on the left. Portions of the crown/root of C and P^4 to M^3 are preserved on the right.

The Braincase

Capacity of the thick-walled Petralona braincase is estimated at 1230 cm^3. The frontal is relatively short and broad in comparison to that of Broken Hill. The glabellar region is massive and projecting, if slightly indented just at the midline where there are traces of a metopic root. To either side, the superciliary segment is greatly thickened. This portion of the brow curves upward, as in Broken Hill and Bodo, but the arch is not so pronounced as in the later Neanderthals. Vertical thickening is slightly reduced at the center of the orbit. On the left side, there is no noticeable separation between the medial and lateral elements of the torus. On the right, where the bone is completely free of matrix, there is a faint but palpable depression. This is not a clear groove but serves nonetheless as a division between the heavy superciliary eminence and

the attenuated lateral brow. CT scans show that the frontal sinuses are greatly expanded, as in Broken Hill. These air cavities extend posteriorly toward bregma and also laterally, where they are separated from the sphenoid sinuses by only thin bony partitions (Seidler et al. 1997).

Supratoral flattening is marked centrally, above glabella. There is little/no development of any sulcus. Posteriorly, the frontal surface is gently rounded in the coronal plane. There is neither midline keeling nor any eminence at bregma. Temporal cresting is preserved on both sides, and separation of the superior and inferior lines is apparent at the coronal boundary. These lines are abraded on the left but can be followed onto the parietal on the right side. The superior temporal line produces a low, rather elongated angular torus at the mastoid angle of the parietal bone. The inferior line, slightly raised throughout its course, is associated with a massive supramastoid crest. The latter is a striking feature of the temporal bone, where it curves upward from the zygomatic root to form a wedge-shaped bulge behind the ear. This bulge is continued for a short distance onto the parietal. The angular sulcus inferior to it is shallow but extensive. The center of the parietal is rounded, but there is no appreciable boss. Breadth taken at this level on the vault is far less than that measured at the supramastoid crests. Superiorly, near the vertex, there is some heaping up of bone along the sagittal suture. This blunt keeling is very limited and does not reach into the lambdoid region, which is flattened.

The temporal squama is damaged on the right side. Fine parallel striations radiating from its superior margin suggest that originally, there was substantial overlap with the parietal. On the left, most of the squama is complete, and it forms a high arch, peaking near pterion. Posteriorly, the squamosal suture slopes downward toward the supramastoid crest, behind which it then angles sharply in the direction of the (broken) mastoid process. This region is heavily coated with beads of stalagmite on both sides.

The upper scale of the occipital is approximately vertical. The transverse torus does not stand out in high relief and instead presents a mound-like appearance, rather different from that of Broken Hill. Above it, there is a shallow depression extending for some distance from the midline. Neither this faint sulcus nor the torus itself reaches as far laterally as asterion. There is no true external protuberance. The superior nuchal lines meet centrally at a roughened linear tubercle, which is continued forward to join with an external occipital crest. This crest can be followed to the rim of the foramen magnum. Laterally, the superior lines terminate in small, poorly defined retromastoid processes. The nuchal plane is moderately impressed by muscle markings, and bilateral mounds cover the areas filled internally by the cerebellar hemispheres. Because of damage, details concerning the form of the digastric incisure and occipitomastoid junction are mostly lost. A little of the digastric fossa seems to be preserved on the right, and to its medial side there is a trace of raised bone that may represent a juxtamastoid eminence.

The glenoid fossa is relatively wide and deep. Both cavities are coated with a thin layer of stalagmite, but on the left, enough of the bone is exposed to show that there is an articular tubercle. The preglenoid planum is very restricted in its anterposterior dimension. Stringer et al. (1979) describe the postglenoid tubercle as "prominent" and lying directly anterior to the (partially obscured) "thin and relatively vertical" tympanic plate. There is definitely a medially placed, thickened, and projecting styloid sheath. Stringer et al. (1979) comment that the tympanic plates and petrous bones "are apparently only slightly angled in relation to each other."

Facial Morphology

The Petralona face is massively constructed. As with Broken Hill, there is no reduction in overall face size in comparison to *Homo erectus*, and the facial skeleton seems to be "hafted" to the braincase in such a way as to accentuate facial projection. From nasion, the nasofrontal sutures trend downward but become almost horizontal as the frontomaxillary boundaries. Width of the nasal bridge taken at the position of the anterior lacrimal crests is 32 mm. The nasofrontal contact itself is 15 mm wide. Superiorly, the internasal suture is offset from the midline, and here blunt keeling is apparent. The nasal bones fan out below to reach a width of 26 mm. Their ends are broken, but at this level the transverse profile is more rounded.

From the side, the nasal profile is concave, and there is (now) no evidence of any dip toward rhinion. This landmark is no longer present, but most of the lateral margin of the nose is well defined, and this passes posteriorly before dropping steeply toward the nasal floor. In this respect, the aperture differs from the condition in *Homo erectus* and is oriented like that of later humans. The anterior nasal spine is also visible in side view, in a position directly underneath the damaged nasal tips. Below it, the nasoalveolar clivus is flattened. This surface incorporates corrugations associated with the incisor sockets, but it is situated almost vertically. The distance from nasospinale to prosthion is only 22 mm. Here there is a clear contrast to Broken Hill, where the clivus is very much deeper and more prognathic. Canine juga are hardly more prominent than the incisor swellings and do not reach upward quite to the level of the nasal opening.

The anterior nasal spine is slightly eroded. But there is no doubt that this structure projects forward as well as upward from the sill, to form the most anterior point of attachment of the nasal septum. Associated with it, there are several crests. A blunt spinal crest passes laterally and seems to produce two branches. One branch merges directly with the edge of the aperture, which is sharp (not rounded). The other trends posterolaterally, where it is obscured in an area roughened by matrix. The nasal margin itself is continued downward and medially onto the sill, and between it and the first of the spinal crests there is a shallow, crescent-shaped depression. This hollow curves upward at the midline and serves to accentuate protrusion of the nasal spine. The floor of the cavity, partly hidden by bits of stalagmite deposited adjacent to the intermaxillary crest, must be scored as smooth. At the junction of the sill with the nasal floor, there is some change in elevation, but this downward slope is most apparent near the midline. Laterally, no step is present. Both sets of maxillary sinuses are exposed, and it is evident that these cavities extend inferiorly, in relation to the level of the nose.

The midfacial skeleton is relatively and absolutely broad. Bimaxillary breadth is 120 mm as compared to 107 mm for Broken Hill (but >130 mm for Bodo). As noted by other workers (e.g., Stringer 1983), the nasal cavity is large. The walls of the maxillae appear to be inflated. Infraorbital foramina are set 17 to 19 mm below the orbital rims and are not associated with any grooves or furrows. Neither here nor elsewhere in the cheek region is there much indication of hollowing. This the case also for the Bodo face, although the latter individual shows traces of a furrow, inferior to the foramen. Petralona may be contrasted with Broken Hill, where there is at least some localized depression of the surface of the cheek, even if no extensive canine fossa is developed.

The zygomatic root is thickened medially to produce a massive pillar, centered over M^1/M^2. This pillar takes its origin low on the body of the maxilla, just a few mm above the row of bony exostoses that coalesce (on the left) to form an external alveolar torus. From the front, this contour is relatively straight, and there is little development of an incisure. Cheek height measured as a minimum on the maxilla is 38 mm. Total vertical height of the zygomatic bone is at least 56 mm. Below the margin of the (left) orbit, the surface of the cheek is slightly damaged. Small sections of cortex have been separated and raised, so as to give extra relief to a tubercle situated along and lateral to the zygomaticomaxillary suture, at about the level of the infraorbital foramen. On the right, there is a similar prominence, partly coated by stalagmite. Otherwise, the face of the zygomatic bone is flattened. Its inferior margin, irregular because of the masseter attachment, is also partially eroded and/or blunted by preparation. Nowhere in this region is there any expression of a malar tubercle comparable to the structure described by Weidenreich (1943) for Zhoukoudian *Homo erectus*.

The palate is U-shaped, broad, and not particularly deep. The anterior (incisor) portion of the arcade is almost straight, while the tooth rows diverge posteriorly. The palatal surface is marked by low mounds paralleling the midline and by a rough ridge along the alveolar margin on the right. The incisive fossa opens anteriorly, just behind the alveolar process. It encroaches upon the space between the central incisor sockets and is thus situated further forward than in Bodo or even Broken Hill. The canal has been only partly cleared of matrix but must follow a nearly vertical trajectory upward toward the nasal floor.

Metric Comparisons

Measurements for Broken Hill, Bodo, Petralona, and two of the more complete Sima de los Huesos crania are provided in Tables 11.1 and 11.2. Along with standard linear dimensions, endocranial volumes are given, as are angles measuring sagittal curvature of the vault and forward projection of the facial midline. As a guide to overall size, geometric means (GMN) are calculated separately for the braincase and the face. Indices are listed in Table 11.3. Simple ratios of linear dimensions are designed to quantify aspects of geometric shape that are invariant for a particular measure of size (Mosimann 1970). Cranial globularity is calculated as a function of three variables.

Broken Hill and Petralona have cranial capacities that are similar to Bodo and intermediate between SH 5 and SH 4. Brain volume (converted to its cube root) can be compared to cranial base (biauricular) breadth, as the vol/aub ratio. In this ratio, Broken Hill resembles the Sima de los Huesos specimens. Petralona has a substantially broader base, and the index is reduced to 0.071. This vol/aub ratio is the lowest recorded for a sample (N = 8) of African and Eurasian mid-Pleistocene hominins including Omo 2, Steinheim, and Dali (range 0.071–0.094). In its volume-to-base breadth proportions, Petralona is comparable to crania of *Homo erectus* from East Africa (N = 4, range 0.071–0.075) and Indonesia (N = 15, range 0.069–0.081).

The Broken Hill and Petralona braincases are almost identical in length. Petralona is broader across the frontal and parietals and slightly higher at vertex. These differences result in only marginal increases to both the height/length and breadth/height indices (Table 11.3). Globularity is also slightly greater in Petralona, but both vaults are relatively low and/or elongated in comparison to the "rounder" skulls from Sima de los Huesos. Indeed, Broken Hill and Petralona approach the form usual in *Homo erectus* and differ markedly from representatives of recent *Homo sapiens*.

Frontal narrowing can be measured using different landmarks. One method compares minimum frontal breadth (frontotemporale-frontotemporale) to the biorbital chord, so as to quantify postorbital constriction in relation to the brow ridges. Alternatively, postorbital breadth measured lower in the temporal fossae can be paired with maximum frontal breadth, or with maximum cranial width (the frontoparietal index). The index of frontal narrowing (Table 11.3) compares postorbital breadth with maximum biparietal width, in an effort to capture form of the anterior braincase without including the (supra)-mastoid cresting that is common in archaic individuals. This index is essentially the same for Broken Hill and Petralona. It is apparent that a similar degree of frontal narrowing is present in specimens from the Sima de los Huesos.

Although the Broken Hill occipital is damaged, width of the squama can be estimated by doubling to the midline. The resulting biasterionic breadth (129 mm) is comparable to that of the larger *Homo erectus* crania from Sangiran and Ngandong. For Petralona, biasterionic breadth is 120 mm. This dimension would be increased by 8 to 10 mm if the measurement were taken to the centers of bilateral asterionic ossicles. For Broken Hill, inion is not preserved. However, the missing central portion of the squama can be reconstructed by following the contour of the transverse torus where it curves from the left side toward the midline. By this procedure, estimates of 60 mm for the lambda-inion chord and 54 mm for the inion-opisthion chord can be obtained. Inion is here taken to lie at the center of the (reconstructed) linear tubercle. Even if this point were located higher on the most projecting part of the torus, the upper scale would be shortened only marginally, to about the length of the nuchal plane. In these proportions, the Broken Hill occipital differs from most *Homo erectus* but resembles Petralona and the Sima crania, where the lambda-inion chord substantially exceeds the inion-opisthion length.

Table 11.1 Measurements of the vault and cranial base[a]

	Broken Hill	Bodo	Petralona	SH 4[b]	SH 5[b]
Entire neurocranium					
1. Volume	1280	1250	1230	1390	1125
2. Glabella-occipital length (**GOL**)	209?	–	208	201?	185
3. Basion-nasion length (**BNL**)	110	107	110	109	109
4. Basion-bregma ht (**BBH**)	129	131	126	131	125
5. Porion-vertex ht	103	114?	105.6	114?	98?
6. Max br	145?	148	165	164	146
7. Max biparietal br	145	148	151	164	145?
8. Biauricular br	138?	–	150	147?	135?
Frontal bone					
9. Supraorbital torus thickness	22	16	21	11	14
10. Min frontal br (ft-ft)	98	105	110	117	106
11. Postorbital br	104	110	108	117	106
12. Max frontal br (**XFB**)	118	119	120	126	118?
13. Nasion-bregma chd (**FRC**)	120	125	110	115	106
14. Nasion-bregma arc	139	144	129	126?	114
15. Glabella-bregma arc	127	128	112	123?	102?
16. Frontal angle (**FRA**)	140	139	140	140	145
Parietotemporal region					
17. Bregma-lambda chd (**PAC**)	112	–	106	111	105
18. Bregma-lambda arc	120	–	114	118	112
19. Bregma-asterion chd	138	–	137?	135	123
20. Bregma-asterion arc	158	–	162.5?	164	150
21. Lambda-asterion chd	91	–	88	95	85
22. Lambda-asterion arc	100	–	99	–	–
23. Temporal squama ht	50	–	49.4	52	44
Occipital bone					
24. Biasterionic br (**ASB**)	129	–	120	132	116
25. Lambda-opisthion chd (**OCC**)	89	–	92	94	92
26. Lambda-opisthion arc	–	–	128	125	114
27. Occipital angle (**OCA**)	–	–	97	106	114
28. Lambda-inion chd	60	–	65	67	61
29. Inion-opisthion chd	54	–	55	46	49
30. Foramen magnum br	–	–	32.6	30	28
31. Foramen magnum length	41?	–	41.5	42	38
Overall size					
32. Geometric mean[c]	104.2	–	105.3	102.9	96.5

[a]Abbreviations in bold specify measurements defined by Howells (1973)
[b]Measurements are from Arsuaga et al. (1997)
[c]GMN is calculated from 15 linear variables (2–3, 5–9, 11–12, 15, 18, 20, 24, 28–29)

In overall size, the Broken Hill facial skeleton is somewhat smaller than that of either the massive Bodo specimen or Petralona. Further comparisons based on GMN show that the Broken Hill face is about 57.9% as large as the braincase. For Petralona, this figure is somewhat greater (60.7%), and for SH 5, the face is still larger in relation to the vault. Broken Hill and Petralona have similar midorbital (zygoorbitale-zygoorbitale) breadths. Upper facial widths are also comparable in the two specimens, but the bimaxillary chord is greater in Petralona. The facial breadth index (Table 11.3) confirms that in Broken Hill, the upper face is relatively broad, while in Petralona, the cheek region is expanded (resulting in a lower index value). Relatively great bimaxillary widths also characterize the faces of SH 5 and Bodo. How much significance to attach to these differences is uncertain. Within species of hominoids, dimensions of the facial skeleton tend to be more variable than those of the neurocranium or the skull base (Wood and Lieberman 2001). This finding is generally supported by Rightmire (2008), who finds that values of CV (coefficient of variation) or V* (CV adjusted for small sample size) for the biorbital, midorbital, and bimaxillary chords are elevated in relation to coefficients for the vault in samples of both *Homo erectus* and Middle Pleistocene hominins. Increased variability within species may indicate that facial breadths have lower taxonomic valence than other measurements.

Table 11.2 Measurements of the facial skeleton[a]

	Broken Hill	Bodo	Petralona	SH 5[b]
1. Nasion-prosthion ht (**NPH**)	90	88	90	85
2. Basion-prosthion length (**BPL**)	117	121?	116	121
3. Biorbital chd (**FMB**)	124	130	126	112
4. Nasion subtense (**NAS**)	26	22	23.5	22
5. Nasion angle (**NFA**)	134	142	140	137
6. Interorbital br (**DKB**)	30	35?	36?	33
7. Nasal bridge ht	12	–	12	–
8. Nasal bridge angle	103	–	113	–
9. Orbit br (**OBB**)	48	47.5	45	43
10. Orbit ht (**OBH**)	39	39	34	33
11. Midorbital chd	76	76?	75	–
12. Naso-orbital subtense	20	17?	17	–
13. Naso-orbital angle	124	132?	131	–
14. Nasal br (**NLB**)	30	43?	37	38
15. Nasal ht (**NLH**)	57	62	68	57
16. Bimaxillary chd (**ZMB**)	107	134?	120	118
17. Subspinale subtense (**SSS**)	33	28?	36	–
18. Subspinale angle (**SSA**)	116	134?	118	111
19. Prosthion subtense	53	43?	–	–
20. Prosthion angle	90	114.6?	–	–
21. Cheek ht (**WMH**)	29	33.5	38	33
22. Max malar ht	55	>60	56	–
23. Nasoalveolar clivus length	29	–	22	–
24. Palate br (internal)	50	48?	50	44
25. Palate length (internal)	58	–	51	55
26. Palate depth at M^2	20	–	–	–
Geometric mean[c]	60.4	65.6	64.0	60.0

[a]Abbreviations in bold specify measurements defined by Howells (1973)
[b]Measurements are from Arsuaga et al. (1997)
[c]GMN is calculated from 10 linear variables (1–3, 9–10, 14–16, 21, 24)

Table 11.3 Indices relating to brain size, the cranial base, vault proportions, and facial shape

	Broken Hill	Bodo	Petralona	SH 4	SH 5
Volume/base br[a]	0.078	–	0.071	0.076	0.077
Cranial ht/length[b]	49.3	–	50.7	56.7	52.9
Cranial br/ht[b]	141.0	130.0	143.0	144.0	148.0
Globularity[b]	34.2	–	36.8	46.3	41.5
Frontal narrowing[c]	71.7	74.3	71.5	71.3	73.1
Occipital scale[d]	90.0	–	84.6	68.6	80.3
Facial br[e]	116.0	97.0	105.0	–	94.9
Orbital[f]	81.2	82.1	75.5	–	76.7
Palatal[g]	86.2	–	98.0	–	80.0

[a]Cube root of brain volume/biauricular breadth
[b]Indices describing vault proportions are calculated from glabella-occipital length, maximum biparietal breadth, and porion-vertex height. Globularity = (breadth × height/length^2)
[c]Frontal narrowing index = postorbital breadth/maximum biparietal breadth
[d]Occipital scale index = inion-opisthion chord/lambda-inion chord
[e]Facial breadth index = biorbital chord/bimaxillary chord
[f]Orbital index = orbit height/orbit breadth
[g]Palatal index = internal palate breadth/internal palate length

Topography of the face can be described with reference to angles registering midline projection (Table 11.2). The nasion angle measures prominence of the nasal root relative to the biorbital chord. Here Broken Hill differs from Petralona and Bodo, where nasion is somewhat more recessed. The naso-orbital angle reflects elevation of the nasal saddle in relation to the lower orbital margins. For Broken Hill, this angle is lower (indicating greater elevation) than in Petralona. In facial forwardness at subspinale, the two specimens are similar. Both have subspinale angles (the zygomaxillary angle of Howells 1973) that are reduced in comparison to that of the relatively flat-faced Bodo cranium. SH 5 exhibits a lower subspinale angle and hence a more prominent subnasal profile, as does the face from Arago in France. This latter condition is Neanderthal-like.

Broken Hill has a relatively high orbital index, as does Bodo. In Petralona and SH 5, the orbits are lower and relatively broad. It may be noted that the partial cranium from Arago differs still more strongly from Broken Hill in orbit height (Rightmire 2001), and there is some evidence for a pattern in which European specimens have lower orbits than do mid-Pleistocene Africans. However, the shape of the orbital cavity, like the nasal aperture, is subject to substantial within group variation. This is the case for both recent and archaic humans, where values of CV or V* tend to be particularly high for orbit height and nasal width, as well as cheek height.

In Broken Hill, the nasoalveolar clivus is quite projecting, while in Petralona this surface is flattened and situated almost vertically. Broken Hill also differs from Petralona in palate shape. The palate is relatively long in Broken Hill, but it is almost "square" in Petralona. In part, such differences in proportions may reflect variation associated with masticatory function, known to generate mechanical loading in the face. Especially in archaic hominins capable of producing strong bite forces, the palate, like the mandible, will be subject to high twisting and shearing stresses concentrated near the occlusal plane (Lieberman 2011). Both upper and lower jaws demonstrate much phenotypic plasticity (Wood and Lieberman 2001).

Discussion

Hominin fossils are known from numerous Middle Pleistocene localities. It is recognized that these individuals display traits that are derived in comparison to *Homo erectus*. At the same time, the skulls retain numerous primitive features that set them apart from modern humans. How these diverse mid-Pleistocene assemblages should be classified, and how they fit into the "tree" of human evolution, are important questions. The crania from Broken Hill and Petralona are key specimens, from which inferences concerning the morphology of larger African and European regional populations can be drawn. Of course, all biological populations display variation, and the extent of this variation cannot be gauged adequately from small samples. Particularly for Africa, few complete skulls are available. Nevertheless, the detailed anatomical and metric comparisons

conducted here provide information that is useful in evaluating the null hypothesis that Broken Hill and Petralona represent paleodemes of one species.

The two crania are similar in many aspects of form. Both are long with relatively low vertices, and both display massive and projecting supraorbital tori, flattened frontals heavily invaded by complex air sinuses, postorbital narrowing, and occipitals that are flexed relative to those of modern humans. Petralona differs from Broken Hill in having a wider cranial base, a reduced vol/aub ratio, massive supramastoid crests, and a less prominent torus crossing the occipital bone. The well preserved Broken Hill basicranium presents derived (*sapiens*-like) features including an increased petrotympanic angle associated with (coronal) alignment of the petrous and tympanic axes, "erosion" of the pyramid apex leading to enlargement of the foramen lacerum, a projecting sphenoid spine, and clear definition of an articular tubercle at the anterior margin of the mandibular fossa (Rightmire 1990, 2001, 2008). Insofar as the (damaged and partly obscured) cranial base of Petralona can be evaluated, its morphology resembles that of Broken Hill.

In forward placement of the facial skeleton relative to the anterior cranial fossa, Broken Hill and Petralona are comparable to *Homo erectus*. At the same time, the lateral margin of the nose is vertical, rather than forward sloping as in *Homo erectus*. The lower terminus of this border is set back below the overhanging nasal saddle. This reorientation suggests that the facial profile is less prognathic than in *Homo erectus*. In facial forwardness at subspinale, Broken Hill and Petralona are similar, and the angle at subspinale is reduced in relation to that of the flat-faced Bodo cranium. Both Broken Hill and Petralona possess prominent anterior nasal spines, coupled with spinal crests separating the sill from the subnasal portion of the maxilla.

Elsewhere in the facial skeleton, there is more variation. The Petralona face is broader at the zygomatic arches and exhibits a more robust cheek region than does Broken Hill. The orbital cavities are low and relatively broad. The walls of the maxillae appear to be inflated. Infraorbital foramina are not associated with any grooves or furrows. Neither here nor elsewhere in the cheek region is there much indication of hollowing. Petralona thus stands in some contrast to Broken Hill, where there are localized depressions of the infraorbital surface, even if no canine fossa is developed. These observations have been taken to indicate that not only Petralona but also Arago and other European mid-Pleistocene hominins anticipate the distinctive midfacial morphology of later Neanderthals (Hublin 1998, 2009). As described by Arsuaga et al. (1997), the cheek region of SH 5 is not inflated in the extreme manner of Neanderthals, but it can be interpreted as intermediate in form. How such facial features are evaluated (whether any of them can be judged to be true Neanderthal apomorphies) is critical to determining how the Petralona,

Arago, and Sima de los Huesos individuals are related to populations outside of Europe and how the fossils should be treated in phylogenetic schemes.

Information relevant to these questions is advanced by Harvati et al. (2010), who have carried out a geometric morphometric study designed to quantify craniofacial shape in Middle Pleistocene hominins, Neanderthals, and modern humans. This analysis is based (in part) on landmarks situated on the supraorbital torus, the orbits and nasal aperture, the zygomatic bone, and the maxilla. After superimposition with generalized Procrustes analysis (GPA), mean configurations of the groups are assessed visually. When viewed in the transverse plane, configurations confirm that "classic" Neanderthals have "a more convex maxilla" and a more receding infraorbital profile than do recent populations. Importantly, Petralona, Arago, and SH 5 are essentially indistinguishable from Broken Hill and Bodo. Both European and African groups are said to approach (but not match) the Neanderthal condition. Midfacial prognathism is explored by comparing orientation of midsagittal landmarks in relation to lateral portions of the face. The Middle Pleistocene crania appear to have "less anteriorly placed" faces than Neanderthals and to be "nearly identical" to one another in (mean) shape. These findings suggest that there is little basis for claiming that the European mid-Pleistocene hominins are more similar to Neanderthals than their African counterparts. Consequently, Harvati et al. (2010) hypothesize that some facial attributes of Petralona and Arago commonly regarded as "incipient" Neanderthal features may instead be plesiomorphic states.

A different interpretation invokes the effects of scaling. Size has long been recognized as contributing to variation in craniofacial shape (Lahr and Wright 1996). Maddux and Franciscus (2009) have used a geometric morphometric approach to explore the influence of allometry on the infraorbital region in Middle Pleistocene, Late Pleistocene, and recent populations of *Homo*. The authors project a grid onto each specimen, fitting it to the boundaries of the infraorbital plate. Landmarks are digitized as the intersections of grid lines and are intended to capture the topography of the underlying curvilinear surface. GPA serves to superimpose the landmarks of all specimens, aligning them to the mean configuration and allowing quantification of size and shape. Principal components analysis suggests that Neanderthals share with European and African mid-Pleistocene crania (and some Upper Paleolithic anatomically modern individuals) a relatively flat infraorbital surface topography. Most recent human skulls exhibit relatively depressed infraorbital plates. It can be established that the degree of infraorbital depression is clearly correlated with cheek size. There is thus "a growing body of evidence" that changes in facial shape are, at least in part, secondary allometric consequences of reduction in overall size during the evolution of *Homo*. Maddux and Franciscus (2009) caution against

treating features such as an inflated maxilla or a canine fossa as discrete phylogenetic traits. A "puffy" maxilla may not be a Neanderthal apomorphy, and the "canine fossa" of later humans may be a result of decreasing facial size.

Conclusions

If the Broken Hill and Petralona midfacial contours are indeed nearly coincident, and if differences in orbit shape, nasal aperture size, and palatal proportions are taken as indications of the variation to be expected within (all) hominoid populations, then there is little basis in facial form for distinguishing these mid-Pleistocene individuals. Both the African and the European crania seem to approach Neanderthals in flatness of the infraorbital profile and shape of the maxilla, but neither conforms fully to the Neanderthal condition. Harvati et al. (2010) are inclined to view the Middle Pleistocene morphology as plesiomorphic. But size and scaling must also be considered. It is probable that Broken Hill and Petralona share with Neanderthals (and some Upper Paleolithic humans) a relatively flat infraorbital topography because they have larger faces than recent *Homo sapiens* (Maddux and Franciscus 2009).

These findings can be read to show that Petralona does not evince true Neanderthal apomorphies in the midface. Neither Petralona nor Arago can be linked more closely to later European populations than can the African mid-Pleistocene hominins. At the same time, Petralona and Broken Hill share many aspects of facial form, vault proportions, and discrete anatomy. The same conclusion can be drawn from studies of the skull base. It follows that the initial null hypothesis cannot be rejected. The fossils represent paleodemes of a single evolutionary lineage widely dispersed across Africa and Europe. Just how this lineage is related to the Neanderthals, and when the latter emerged as a distinct species, are key questions that remain unresolved. But it is likely that European and African populations of *Homo heidelbergensis* did not separate until relatively late in the Middle Pleistocene.

Acknowledgments For access to the fossils in their care, I thank my friends and colleagues at the National Museum of Ethiopia (Addis Ababa), the Iziko South African Museum (Cape Town), the Natural History Museum (London), the Institut de Paléontologie Humaine (Paris), and the University of Thessaloniki. Ian Tattersall allowed me to study materials held in the American Museum of Natural History (New York). Graham Avery, George Koufos, Robert Kruszynski, and Chris Stringer helped me with the process of gathering data central to this research. My studies of Middle Pleistocene *Homo* have been funded by the National Science Foundation, the Leakey Foundation, the Eckler Fund of Binghamton University, and the American School for Prehistoric Research at Harvard University. In the course of these investigations, I have benefited from the anatomical expertise, sound science, and good humor of Yoel Rak.

References

Arsuaga, J. L., Gracia, A., Martínez, I., Bermúdez de Castro, J.M., Rosas, A., Villaverde, V. et al. (1989). The human remains from Cova Negra (Valencia, Spain) and their place in European Pleistocene human evolution. *Journal of Human Evolution, 18*, 55–92

Arsuaga, J. L., Martínez, I., Gracia, A., & Lorenzo, C. (1997). The Sima de los Huesos crania (Sierra de Atapuerca, Spain). A comparative study. *Journal of Human Evolution, 33*, 219–281.

Bräuer, G. (2007). Origin of modern humans. In W. Henke & I. Tattersall (Eds.), *Handbook of paleoanthropology, phylogeny of hominids* (Vol. 3, pp. 1749–1779). Heidelberg: Springer.

Bräuer, G. (2008). The origin of modern anatomy: By speciation or intraspecific evolution? *Evolutionary Anthropology, 17*, 22–37.

Friess, M. (2010). Calvarial shape variation among Middle Pleistocene hominins: An application of surface scanning in paleoanthropology. *Comptes Rendus Palevol, 9*, 435–443.

Harvati, K., Hublin, J.-J., & Gunz, P. (2010). Evolution of middle-late Pleistocene human craniofacial form: A 3-D approach. *Journal of Human Evolution, 59*, 445–464.

Harvati, K., Hublin, J.-J., & Gunz, P. (2011). Three dimensional evaluation of Neanderthal craniofacial features in the European and African Middle Pleistocene human fossil record (abstract). *American Journal of Physical Anthropology, Supplement, 52*, 157.

Howells, W. W. (1973). Cranial variation in man: A study by multivariate analysis of patterns of difference among recent human populations. *Papers of the Peabody Museum, 67*, 1–259.

Hrdlička, A. (1930). *The skeletal remains of early man*. Smithsonian Miscellaneous Collection 83. Washington: Smithsonian Institution.

Hublin, J.-J. (1998). Climatic changes, paleogeography and the evolution of Neanderthals. In T. Akazawa, K. Aoki, & O. Bar-Yosef (Eds.), *Neanderthals and modern humans in Western Asia* (pp. 295–310). New York: Plenum Press.

Hublin, J.-J. (2009). The origin of Neandertals. *Proceedings of the National Academy of Sciences USA, 106*, 16022–16027.

Klein, R. G. (1994). Southern Africa before the iron age. In R. S. Corruccini & R. L. Ciochon (Eds.), *Integrative paths to the past. Paleoanthropological advances in honor of F. Clark Howell* (pp. 471–519). Englewood Cliffs: Prentice-Hall.

Klein, R. G., Avery, G., Cruz-Uribe, K., & Steele, T. E. (2006). The mammalian fauna associated with an archaic hominin skullcap and later Acheulean artifacts at Elandsfontein, Western Cape Province, South Africa. *Journal of Human Evolution, 52*, 164–186.

Lahr, M. M., & Wright, R. V. S. (1996). The question of robusticity and the relationship between cranial size and shape in *Homo sapiens*. *Journal of Human Evolution, 31*, 157–191.

Lieberman, D. E. (2011). *The evolution of the human head*. Cambridge: Belknap Press.

Maddux, S. D., & Franciscus, R. G. (2009). Allometric scaling of infraorbital surface topography in *Homo*. *Journal of Human Evolution, 56*, 161–174.

Martínez, I., & Arsuaga, J.-L. (1997). The temporal bones from Sima de los Huesos Middle Pleistocene site (Sierra de Atapuerca, Spain). A phylogenetic approach. *Journal of Human Evolution, 33*, 283–318.

Martinón-Torres, M., Bermúdez de Castro, J. M., Gómez-Robles, A., Prado-Simón, L., & Arsuaga, J.-L. (2012). Morphological description and comparison of the dental remains from Atapuerca-Sima de los Huesos site (Spain). *Journal of Human Evolution, 62*, 7–58.

McCollum, M. A. (2000). Subnasal morphological variation in fossil hominids: A reassessment based on new observations and recent developmental findings. *American Journal of Physical Anthropology, 112*, 275–283.

Mosimann, J. E. (1970). Size allometry: Size and shape variables with characterizations of the lognormal and generalized gamma distributions. *Journal of the American Statistical Association, 65*, 930–945.

Mounier, A., Marchal, F., & Condemi, S. (2009). Is *Homo heidelbergensis* a distinct species? New insights on the Mauer mandible. *Journal of Human Evolution, 56*, 219–246.

Mourant, G. M. (1928). Studies of Palaeolithic man. III. The Rhodesian skull and its relationships to Neanderthaloid and modern types. *Annals of Eugenics, 3*, 337–360.

Pycraft, W. P. (1928). Description of the human remains. In W. P. Pycraft, G. E. Smith, M. Yearsley, J. T. Carter, R. A. Smith, A. T. Hopwood, D. M. A. Bate, W. E. Swinton, & F. A. Bather (Eds.), *Rhodesian man and associated remains* (pp. 1–51). London: British Museum.

Rightmire, G. P. (1990). *The evolution of* Homo erectus. *Comparative anatomical studies of an extinct human species*. Cambridge: Cambridge University Press.

Rightmire, G. P. (1996). The human cranium from Bodo, Ethiopia: Evidence for speciation in the Middle Pleistocene? *Journal of Human Evolution, 31*, 21–39.

Rightmire, G. P. (1998). Human evolution in the Middle Pleistocene: The role of *Homo heidelbergensis*. *Evolutionary Anthropology, 6*, 218–227.

Rightmire, G. P. (2001). Comparison of Middle Pleistocene hominids from Africa and Asia. In L. Barham & K. Robson-Brown (Eds.), *Human roots. Africa and Asia in the Middle Pleistocene* (pp. 123–133). Bristol: Western Academic and Specialist Press.

Rightmire, G. P. (2008). *Homo* in the Middle Pleistocene: Hypodigms, variation, and species recognition. *Evolutionary Anthropology, 17*, 8–21.

Schwartz, J. H., & Tattersall, I. (1996). Significance of some previously unrecognized apomorphies in the nasal region of *Homo neanderthalensis*. *Proceedings of the National Academy of Sciences USA, 93*, 10852–10854.

Schwartz, J. H., & Tattersall, I. (2010). Fossil evidence for the origin of *Homo sapiens*. *Yearbook of Physical Anthropology, 153*, 94–121.

Seidler, H., Falk, D., Stringer, C., Wilfing, H., Muller, G. B., zur Nedden, D., et al. (1997). A comparative study of stereolithographically modeled skulls of Petralona and Broken Hill: Implications for further studies of Middle Pleistocene hominid evolution. *Journal of Human Evolution, 33*, 691–703.

Stringer, C. B. (1983). Some further notes on the morphology and dating of the Petralona hominid. *Journal of Human Evolution, 12*, 731–742.

Stringer, C. B. (1993). New views on modern human origins. In D. T. Rasmussen (Ed.), *The origin and evolution of humans and humanness* (pp. 75–94). Boston: Jones and Bartlett.

Stringer, C. B. (2011). The chronological and evolutionary position of the Broken Hill cranium. *American Journal of Physical Anthropology, Supplement, 52*, 287.

Stringer, C. B., Howell, F. C., & Melentis, J. K. (1979). The significance of the fossil hominid skull from Petralona, Greece. *Journal of Archaeological Science, 6*, 235–253.

Tattersall, I., & Schwartz, J. H. (2008). The morphological distinctiveness of *Homo sapiens* and its recognition in the fossil record: Clarifying the problem. *Evolutionary Anthropology, 17*, 49–54.

Weidenreich, F. (1943). The skull of *Sinanthropus pekinensis*: A comparative study of a primitive hominid skull. *Palaeontologia Sinnica, new series, D10*, 1–484.

Wood, B. (1991). *Koobi Fora research project. Hominid cranial remains* (Vol. 4). Oxford: Clarendon Press.

Wood, B., & Lieberman, D. E. (2001). Craniodental variation in *Paranthropus boisei*: A developmental and functional perspective. *American Journal of Physical Anthropology, 116*, 13–25.

Woodward, A. S. (1921). A new cave man from Rhodesia, South Africa. *Nature, 108*, 371–372.

Zonneveld, F. W., & Wind, J. (1985). High-resolution computed tomography of fossil hominid skulls: A new method and some results. In P. V. Tobias (Ed.), *Hominid evolution: Past, present and future* (pp. 427–436). New York: Alan R. Liss.

Chapter 12
Thermoregulation in *Homo erectus* and the Neanderthals: A Reassessment Using a Segmented Model

Mark Collard and Alan Cross

Abstract Thermoregulation is widely believed to have influenced body size and shape in the two best-known extinct members of genus *Homo*, *Homo erectus* and *Homo neanderthalensis*, and to have done so in contrasting ways. *H. erectus* is thought to have been warm adapted, while *H. neanderthalensis* is widely held to have been cold adapted. However, the methods that have been used to arrive at these conclusions ignore differences among body segments in a number of thermoregulation-related variables. We carried out a study designed to determine whether the current consensus regarding the thermoregulatory implications of the size and shape of the bodies of *H. erectus* and *H. neanderthalensis* is supported when body segment differences in surface area, skin temperature, and rate of movement are taken into account.

The study involved estimating heat loss for a number of Holocene modern human skeletal samples and several fossil hominin specimens, including five Pleistocene *H. sapiens*, the well-known *H. erectus* partial skeleton KNM-WT 15000, a *H. erectus* specimen from Dmanisi, Georgia, and three Neanderthals. The resulting heat loss estimates were then used in two sets of comparative analyses. In the first, we focused on whole-body heat loss and tested predictions concerning heat loss in KNM-WT 15000 and European Neanderthals relative to modern humans, and within *H. erectus* and *H. neanderthalensis*. In the second set of analyses we again tested predictions concerning heat loss in *H. erectus* and *H. neanderthalensis* relative to modern humans, and within *H. erectus* and *H. neanderthalensis*, but this time we focused on the contribution of their limbs to heat loss.

The results of the study do not fully support the current consensus regarding the thermoregulatory adaptations of *Homo erectus* and *Homo neanderthalensis*. The whole-body heat loss estimates were consistent with the idea that KNM-WT 15000 was warm adapted and that European Neanderthals were cold adapted, and with the notion that there are thermoregulation-related differences in body size and shape within *H erectus* and *H. neanderthalensis*. The whole-limb estimates told a similar story. In contrast, the results of our analysis of limb segment-specific heat loss were not consistent with the current consensus regarding the thermoregulatory significance of distal limb length in *H. erectus* and *H. neanderthalensis*. Contrary to expectation, differences between the proximal and distal limb segments did not follow any particular trend.

The obvious implication of these results is that, while we can be more confident about the basic idea that thermoregulation influenced the evolution of body size and shape in *H. erectus* and *H. neanderthalensis*, we need to be more cautious in attributing differences in limb segment size to thermoregulation. Based on our results, it is possible that other factors influenced limb segment size in these species more than thermoregulation. Identifying these factors will require further research.

Keywords Body size • Hominin evolution • Thermoregulation • *Homo neanderthalensis* • Limb proportions

M. Collard (✉) · A. Cross
Human Evolutionary Studies Program and Department of Archaeology, Simon Fraser University, 8888 University Drive, Burnaby, British Columbia V5A 1S6, Canada
e-mail: mcollard@sfu.ca

A. Cross
e-mail: fishing4pike@hotmail.com

M. Collard
Department of Archaeology, University of Aberdeen St. Mary's Building, Elphinstone Road, Aberdeen, AB24 3UF, UK

Introduction

Thermoregulation is widely believed to have influenced body size and shape in the two best-known extinct members of genus *Homo*, *Homo erectus* and *Homo neanderthalensis*, and to have done so in contrasting ways. KNM-WT 15000, the famous nearly-complete juvenile male *H. erectus*

Assaf Marom and Erella Hovers (eds.), *Human Paleontology and Prehistory*,
Vertebrate Paleobiology and Paleoanthropology, DOI: 10.1007/978-3-319-46646-0_12

skeleton from 1.5 million year old deposits in West Turkana, Kenya, is reconstructed as relatively narrow bodied and long limbed, and these characteristics are usually interpreted as adaptations to hot conditions (Ruff and Walker 1993; Ruff 1994). In contrast, the Neanderthals are reconstructed as having stocky bodies and relatively short forearms and lower legs. These traits are generally accepted to be adaptations to cold conditions—so much so that the shape of the Neanderthal body is often described as "hyperpolar" (Holliday 1997; Weaver 2003; Tilkens et al. 2007).

The rationale for both these hypotheses is that altering the breadth of the trunk and the length of the distal limb segments affects the ratio of surface area to body mass (SA: BM), and this in turn affects heat loss (Trinkaus 1981; Ruff 1991). The reason for this is that more heat is lost when SA: BM is large than when SA:BM is small (Trinkaus 1981; Ruff 1991). Reducing trunk breadth and lengthening the distal limb segments should increase SA:BM and therefore increase heat loss, whereas broadening the trunk and shortening the distal limb segments should decrease SA:BM and therefore decrease heat loss (Trinkaus 1981; Holliday and Ruff 2001). Thus, the relatively narrow trunk and relatively long distal limb segments of KNM-WT 15000 would have given him an advantage in high ambient temperatures, while the broad trunks and relatively short distal limb segments of the Neanderthals would have given them an advantage in low ambient temperatures.

While changing the ratio of surface area to body mass undoubtedly has the potential to impact heat loss, there are reasons for questioning the consensus that KNM-WT 15000 was hot climate adapted and the Neanderthals were cold climate adapted. One is that the hypotheses do not take into account the fact that the segments of the body move at different speeds during locomotion and therefore experience different wind speeds. Because wind speed influences heat loss, it is possible that the relationship between trunk breadth and limb length on the one hand and heat loss on the other is more complicated than the thermoregulatory interpretation of body size and shape in KNM-WT 15000 and the Neanderthals assumes. Another reason for questioning the consensus view of these hominins is that in living humans skin temperature varies among body segments (e.g., Houdas and Ring 1982). This too suggests SA:BM may be too simple to adequately represent the thermoregulatory abilities of KNM-WT 15000 and the Neanderthals. Lastly, while the impact of differences in whole-body SA:BM on thermoregulation have been quantified in various ways (e.g., Wheeler 1993; Ruff 1993, 1994), no study has attempted to quantify the specific contribution of the limbs to thermoregulation in fossil hominins. Consequently, it has not been demonstrated that the limb proportion differences between KNM-WT 15000 and modern humans, or between

the latter and Neanderthals, actually translate into significant heat loss differences.

With the foregoing in mind, we carried out a study designed to determine whether the current consensus regarding the thermoregulatory implications of the size and shape of the bodies of *H. erectus* and *H. neanderthalensis* is supported when body segment differences in surface area, skin temperature, and rate of movement are considered. The study involved estimating heat loss for a number of modern human skeletal samples, and for fossil specimens that have been assigned to *H. erectus* and *H. neanderthalensis*. The resulting heat loss estimates were then used in two sets of comparative analyses. In the first, we focused on whole-body heat loss and tested predictions concerning heat loss in KNM-WT 15000 and European Neanderthals relative to modern humans. We also tested predictions concerning heat loss within *H. erectus* and *H. neanderthalensis*. In the second set of analyses, we again tested predictions concerning heat loss in *H. erectus* and *H. neanderthalensis* relative to modern humans, and within *H. erectus* and *H. neanderthalensis*, but this time we focused on the contribution of their limbs to heat loss. The results of the study suggest that the current consensus requires some modification.

Materials and Methods

The limb bone data used in the study are presented in Table 12.1. The humerus, femur, and tibia data for the Holocene modern human samples are the male means provided by Trinkaus (1981). The ulna values for the Holocene samples were estimated by adding 5% to the length of the radius values given by Trinkaus (1981), as per Haeusler and McHenry (2004). The humerus, femur, and tibia data for the five Pleistocene human specimens (Skhul IV, Skhul V, Predmosti 3, Predmosti 14, Caviglione 1) and the three Neanderthal specimens (La Ferrassie 1, La Chapelle 1, and Shanidar 4) are also from Trinkaus (1981). As with the Holocene modern human samples, the ulna values for these specimens were estimated by adding 5% to the length of the radius. The long bone lengths for the Dmanisi individual are for the large adult from the site. They were taken from Lordkipanidze et al. (2007), with the exception of ulna length, which was estimated from the length of the humerus using the equation provided by Haeuseler (2001). The lengths of KNM-WT 15000's long bones were taken from Ruff and Walker (1993). They are the lengths at the time of death rather than the lengths that have been estimated for KNM-WT 15000 as an adult.

Table 12.2 lists the stature and body mass estimates used in the study. Some estimates were taken directly from the

Table 12.1 Limb bone lengths (mm) for the samples used in this study. Values in square brackets are estimates

Sample	Taxon	Humerus	Ulna	Femur	Tibia	Notes
Inuit	Holocene *H. sapiens*	30.4	[24.0]	40.8	33.1	Data from Trinkaus (1981). Eskimo male mean. Ulna value estimated by adding 5% to the length of radius, as per Haeusler and McHenry (2004).
Yugoslavians	Holocene *H. sapiens*	33.0	[25.8]	45.5	38.1	Data from Trinkaus (1981). Male mean. Ulna value estimated by adding 5% to the length of radius, as per Haeusler and McHenry (2004).
Lapps	Holocene *H. sapiens*	30.6	[23.8]	41.0	32.5	Data from Trinkaus (1981). Male mean. Ulna value estimated by adding 5% to the length of radius, as per Haeusler and McHenry (2004).
Amerinds	Holocene *H. sapiens*	30.8	[25.1]	42.3	35.9	Data from Trinkaus (1981). New Mexico Amerindian male mean. Ulna value estimated by adding 5% to the length of radius, as per Haeusler and McHenry (2004).
Melanesians	Holocene *H. sapiens*	31.7	[26.0]	43.6	37.1	Data from Trinkaus (1981). Male mean. Ulna value estimated by adding 5% to the length of radius, as per Haeusler and McHenry (2004).
Egyptians	Holocene *H. sapiens*	32.5	[26.8]	45.3	38.7	Data from Trinkaus (1981). Male mean. Ulna value estimated by adding 5% to the length of radius, as per Haeusler and McHenry (2004).
Skhul IV	Pleistocene *H. sapiens*	33.7	[28.8]	49.0	43.4	Data from Trinkaus (1981). Ulna value estimated by adding 5% to the length of radius, as per Haeusler and McHenry (2004).
Skhul V	Pleistocene *H. sapiens*	38.0	[28.1]	51.5	41.2	Data from Trinkaus (1981). Ulna value estimated by adding 5% to the length of radius, as per Haeusler and McHenry (2004).
Predmosti 3	Pleistocene *H. sapiens*	35.7	[29.3]	48.7	42.1	Data from Trinkaus (1981). Ulna value estimated by adding 5% to the length of radius, as per Haeusler and McHenry (2004).
Predmosti 14	Pleistocene *H. sapiens*	33.6	[27.8]	45.2	39.5	Data from Trinkaus (1981). Ulna value estimated by adding 5% to the length of radius, as per Haeusler and McHenry (2004).
Caviglione 1	Pleistocene *H. sapiens*	34.2	[27.6]	47.0	41.2	Data from Trinkaus (1981). Ulna value estimated by adding 5% to the length of radius, as per Haeusler and McHenry (2004).
Shanidar 4	*H. neanderthalensis*	30.5	[24.7]	42.2	33.4	Data from Trinkaus (1981). Ulna value estimated by adding 5% to the length of radius, as per Haeusler and McHenry (2004).
La Chapelle 1	*H. neanderthalensis*	31.2	[23.8]	43.0	34.0	Data from Trinkaus (1981). Ulna value estimated by adding 5% to the length of radius, as per Haeusler and McHenry (2004).
La Ferrassie 1	*H. neanderthalensis*	33.7	[25.6]	45.8	37.0	Data from Trinkaus (1981). Ulna value estimated by adding 5% to the length of radius, as per Haeusler and McHenry (2004).
Dmanisi	*H. erectus*	29.5	[24.3]	38.6	30.6	Data from Lordkipandze et al. (2007). Ulna length estimated from humerus length using the equation provided by Haeusler (2001).
KNM-WT 15000	*H. erectus*	31.9	27.0	42.9	38.0	Ruff and Walker (1993); juvenile values.

literature; others were obtained with the aid of published equations for estimating body mass and stature. In the latter cases, equations derived from geographically appropriate reference samples were employed as far as possible.

All the statures and body masses of the Holocene human samples were estimated with published equations. The stature estimate for the Inuit sample was obtained from femur length with Feldesman and Fountain's (1996) equation; encouragingly, it is the same as the Eskimo/Inuit estimate used by Ruff (1994). Raxter et al.'s (2008) femur-based stature equation was used for the Egyptian sample because it is specific to Egyptians. Yugoslav, Lapp, and Amerindian statures were calculated from femur length using Trotter and Gleser's (1958) equation for whites, while the stature of the Melanesian sample was estimated from femur length using Trotter and Gleser's (1958) equation for blacks. The body masses of most of the samples were estimated from stature with Ruff and Walker's (1993) male equation. While this

Table 12.2 Stature (cm) and body mass (kg) estimates used in this study. See Materials and Methods section for details

Sample	Taxon	Stature	Body mass
Inuit	Holocene *H. sapiens*	159	67
Yugoslavians	Holocene *H. sapiens*	171	65
Egyptians	Holocene *H. sapiens*	167	61
Lapps	Holocene *H. sapiens*	161	56
Amerinds	Holocene *H. sapiens*	164	59
Melanesians	Holocene *H. sapiens*	164	59
Skhul IV	Pleistocene *H. sapiens*	179	66
Skhul V	Pleistocene *H. sapiens*	185	70
Predmosti 3	Pleistocene *H. sapiens*	179	71
Predmosti 14	Pleistocene *H. sapiens*	170	66
Caviglione 1	Pleistocene *H. sapiens*	175	65
Shanidar 4	*H. neanderthalensis*	162	71
La Chapelle 1	*H. neanderthalensis*	164	76
La Ferrassie 1	*H. neanderthalensis*	171	85
Dmanisi	*H. erectus*	153	50
KNM-WT 15000	*H. erectus*	160	48

equation does not account for variation in body breadth, the latter variable was not available for the samples in question. We considered using body breadths from other sources but decided that the additional error introduced by this procedure outweighed the benefits. The only human sample for which we used both stature and body breadth to estimate body mass was the Inuit one. The difference between a stature-based estimate for this sample and published estimates (e.g., Ruff 1994) was sufficiently large that using stature and body breadth method seemed warranted. The Inuit sample's body mass was estimated from stature and bi-iliac breadth with Ruff et al.'s (2005) equation for males; we used the mean bi-iliac breath for Eskimo/Inuit presented in Ruff (1994).

Turning now to the fossil specimens, the stature and body mass estimates for La Ferrassie 1 and KNM-WT 15000 were obtained directly from the literature (Ruff et al. 1997, 2005; Ruff and Walker 1993). Ruff et al. (2005) give a stature estimate of 162 cm for La Chappelle 1. Using the same femur length and formula (Trotter and Gleser' (1952) equation for whites) we obtained an estimated stature of 164 cm. We opted to use the latter value. The stature estimate for the Dmanisi individual was taken from Ruff (2010). We used Lordkipanidze et al. (2007)'s femoral head-derived body mass estimate for the Dmanisi specimen rather than their average value because the latter involves variables whose connection with body mass is unclear. The stature estimates for Skhul IV, Skhul V, Predmosti 3, Predmosti 14, Caviglione 1, and Shanidar 4 were obtained using Trotter and Gleser's (1958) femur-length based equation for whites. It has been argued that this equation is less accurate for early modern humans than Trotter and Gleser's (1958) formula for blacks or taking an average of the estimates yielded by the

two formulae (Holliday 1997; Ruff et al. 1997). However, we found that the latter course of action produced estimates that fell within the standard error for the white formula (SE = 3.94). The body mass estimates for Skhul IV, Skhul V, Predmosti 3, Predmosti 14, Caviglione 1, and Shanidar 4 were taken from Froehle and Churchill (2009).

Having compiled the limb, stature, and body mass data, we estimated the surface areas of each taxon's body segments. The approach we used is rooted in the segmented method of estimating surface area employed by Haycock et al. (1978), Cross et al. (2008), and Cross and Collard (2011). For the limb segments, long bone lengths were combined with surface area per unit of length values derived from Cross et al.'s (2008) data. Cross et al. (2008) estimated that approximately 27% of the femur is situated within the trunk segment. They based this value on the observation that crotch height marks the lower boundary of the trunk segment and that palpation of the greater trochanter indicated that 27% of the femur was above the crotch. In an analysis of Cross et al.'s (2008) segment displacement data, we observed no difference between the displacement of markers placed on the greater trochanters and markers placed on the trunk, which supports the inclusion of the upper portion of the femur in the trunk segment. Accordingly, 27% was subtracted from the femora before the surface area of the upper leg was estimated. The surface areas of the non-limb segments were estimated by summing the limb segment surface areas, dividing the resulting figure by the percentage of total body surface area that the limbs represent in Cross et al.'s (2008) sample, and then multiplying the quotient by the percentage of surface area that the non-limb segments represent in Cross et al.'s (2008) sample. Total surface area is the sum of all segment surface areas. For comparative purposes, the total surface area for each sample/specimen was also estimated using the standard Du Bois and Du Bois (1916) equation: Surface area (cm^2) = 0.007184 * H$^{0.725}$ * W$^{0.425}$. Segment and total surface area estimates are listed in Table 12.3.

After obtaining the surface areas, we estimated displacement distances for the segments and walking cycle durations (Table 12.4). We accomplished this with the aid of Cross et al.'s (2008) 3D motion capture data. First, we estimated total arm length. This was necessary because the

skeletal samples and fossil hominin specimens lacked data on hand length. We found that, on average, hand length was 75% of lower arm length in Cross et al.'s (2008) dataset, and we assumed this to be the case for our sample. Next, we estimated displacement distances for the trunk and head/neck from total arm length. We used this approach because we found that the displacement distances of the trunk and head/neck were most strongly correlated with total arm length in Cross et al.'s (2008) data (r^{2}s > 0.96). Subsequently, we estimated upper arm displacement distances

Table 12.3 Segment surface area estimates (cm^2) for the samples used in this study. UA = upper arms; LA = lower arms; UL = upper legs; LL = lower legs; HN = head and neck; Total = sum of segment surface areas; Standard = Estimate of total surface area obtained with the standard, Dubois and Dubois method

Sample	UA	LA	UL	LL	HN	Trunk	Hands	Feet	Total	Standard
Inuit	1782.0	1032.5	2897.4	2152.2	1077.7	5395.0	648.6	1269.5	16255	15440
Yugoslavians	1934.5	1109.9	3231.2	2477.3	1199.5	6004.7	721.9	1413.0	18092	17495
Egyptians	1905.2	1152.9	3217.0	2516.3	1204.8	6031.1	725.0	1419.2	18171	16850
Lapps	1858.3	1118.5	3096.2	2412.2	1162.8	5821.1	699.8	1369.8	17539	15823
Amerinds	1805.5	1079.8	3003.9	2334.2	1126.9	5641.5	678.2	1327.5	16998	16396
Melanesian	1793.8	1023.9	2911.6	2113.2	1074.7	5380.1	646.8	1266.0	16210	16396
Skhul IV	1975.5	1239.0	3479.7	2821.9	1304.1	6528.3	784.8	1536.2	19669	18323
Skhul V	2227.6	1208.9	3657.2	2678.8	1339.2	6704.2	806.0	1577.6	20199	19241
Predmosti 3	2092.7	1260.5	3458.4	2737.3	1308.6	6550.8	787.5	1541.5	19737	18900
Predmosti 14	1969.6	1196.0	3209.9	2568.3	1225.6	6135.6	737.6	1443.8	18486	17650
Caviglione 1	2004.8	1187.4	3337.7	2678.8	1262.0	6317.4	759.5	1486.6	19034	17908
La Ferrassie 1	1975.5	1101.3	3252.5	2405.7	1197.0	5992.5	720.4	1410.1	18055	19737
La Chapelle 1	1828.9	1023.9	3053.6	2210.7	1112.4	5568.6	669.4	1310.3	16778	18259
Shanidar 4	1787.9	1062.6	2996.8	2171.7	1098.9	5501.2	661.3	1294.5	16575	17581
Dmanisi	1729.3	1045.4	2741.2	1989.6	1028.5	5148.9	619.0	1211.6	15514	14532
KNM-WT 15000	1870.0	1161.5	3067.8	2470.8	1174.4	5879.3	706.8	1383.5	17670	14753

Table 12.4 Segment displacement estimates (per cycle) for the samples used in this study. UA = upper arm; LA = lower arm; UL = upper leg; LL = lower leg. HN = head and neck. Cycle duration = heel strike to the next heel strike of the same foot

Sample	UA	LA	UL	LL	HN	Trunk	Hand	Foot	Cycle duration
Inuit	145.33	173.64	158.26	157.01	136.04	138.50	207.34	155.22	1.05
Yugoslavians	160.81	192.14	164.44	163.14	148.01	149.36	229.43	161.28	1.10
Egyptians	164.18	196.16	164.77	163.47	150.61	151.73	234.24	161.61	1.10
Lapps	158.26	189.08	162.69	161.40	146.03	147.57	225.78	159.57	1.08
Amerinds	151.59	181.12	161.11	159.84	140.88	142.89	216.28	158.02	1.07
Melanesians	144.93	173.16	157.93	156.68	135.73	138.21	206.77	154.89	1.05
Skhul IV	176.84	211.29	170.20	168.85	160.40	160.61	252.30	166.93	1.14
Skhul V	185.12	221.18	169.92	168.58	166.80	166.42	264.11	166.66	1.13
Predmosti 3	184.58	220.54	169.08	167.75	166.38	166.04	263.34	165.84	1.13
Predmosti 14	171.86	205.33	165.30	164.00	156.55	157.11	245.19	162.13	1.10
Caviglione 1	172.53	206.14	167.51	166.19	157.07	157.59	246.15	164.30	1.12
La Ferrassie 1	161.76	193.27	163.79	162.50	148.74	150.03	230.78	160.65	1.09
La Chapelle 1	146.54	175.09	160.10	158.83	136.98	139.35	209.07	157.02	1.07
Shanidar 4	148.90	177.90	159.23	157.97	138.80	141.00	212.43	156.17	1.06
Dmanisi	144.32	172.43	155.25	154.02	135.26	137.79	205.90	152.27	1.03
KNM-WT 15000	163.51	195.36	162.97	161.69	150.09	151.25	233.28	159.85	1.09

from total arm length with a regression equation developed on the basis of Cross et al.'s (2008) data. Estimating displacement distances for the lower arm and hand is complicated by the fact that the strength of the correlation between segment length and displacement varies within limbs, because the displacement of a segment is related not only to its length but also to the properties of the segments with which it articulates. As a consequence, simply summing segment displacement distances for segments without taking into account their interactions produces an unrealistic arm-swing pattern. We dealt with this problem by calculating the percentage of upper arm displacement that lower arm and hand displacement represent in Cross et al.'s (2008) sample. Lower arm displacement was found to be 119.48% of upper arm displacement, and hand displacement was found to be 142.67% of upper arm displacement. These values were then used to estimate lower arm and hand displacement in the skeletal samples and fossil hominin specimens. Thereafter, we estimated displacement distances for the legs. The approach we used was similar to the one we employed for the arms: The displacement distance of the upper leg was estimated from total leg length, and the displacement distances of the lower leg and foot were calculated from upper leg displacement using percentages derived from Cross et al.'s (2008) data (+99.21% and +98.08%, respectively). Lastly, we estimated walking cycle duration. To do so, we used Cross et al.'s (2008) data to generate a regression equation that allowed walking cycle duration to be estimated from upper leg displacement.

Having estimated the surface areas and displacement rates of the body segments, we then modeled each sample as walking bipedally at 1.2 m/s and used Cross et al.'s (2008) methods to calculate individual heat production (Table 12.5), convective heat loss, radiant heat loss, and heat balance. One-point-two meters per second is widely accepted to be the average human walking speed (Hinrichs and Cavanagh 1981; Langlois et al. 1997; Orendurff et al. 2004; Neptune et al. 2008), and has been used in many studies of this type (e.g., Hinrichs and Cavanagh 1981; Orendurff et al. 2004; Neptune et al. 2008). In addition, it is employed in such tasks as setting crossing signals (Langlois et al. 1997). Cross et al.'s (2008) method involves three steps. First, the target individual's heat production is calculated with the following equation:

$$\text{Heat production} = w * v * a \qquad (1)$$

where w the individual's total body weight in kilograms, v is their walking speed (1.2 m/s), and a is a constant pertaining to the production of heat by metabolism and work and is equal to 2. Convective and radiant heat loss are then estimated for each body segment with the following equations:

Convective heat loss (in Watts)
$$= (STsk - Ta) * \sqrt{c} * SSA * 8.3 \qquad (2)$$

Radiant heat loss (in Watts) $= (STsk - Tr) * SSA * 5.2$
$$\qquad (3)$$

where Ta is ambient temperature in degrees centigrade, $STsk$ is segment-specific skin temperature in degrees centigrade in Ta, c is the segment-specific displacement rate in meters per second (i.e., the square root of total displacement divided by cycle duration), SAA is segment-specific surface areas, Tr is radiant temperature, and 8.3 and 5.2 are heat transfer coefficients. The last step of Cross et al.'s (2008) method is to sum the segment-specific estimates for convective and radiant heat loss, and then divide this value by the estimate for heat production. The resulting values represent the individuals' whole-body relative heat loss. We made estimates for each sample/specimen in ambient temperatures of 20°C, 25°C, 30°C, and 35°C (Table 12.6).

Relative heat loss for each limb segment was estimated by summing the segment-specific convective and radiant heat loss values and dividing the resulting figure by the estimate for total body heat production. Estimates were again made for each sample/specimen in ambient temperatures of 20°C, 25°C, 30°C, and 35°C (Table 12.7). Whole-limb values (i.e., the sum of heat loss estimates for the proximal and distal segments of each limb) were also calculated to assess the responses of entire limbs.

The task of estimating the thermal responses of extinct, culture-using hominins has the potential to be extremely complex. The model employed in this study was kept simple

Table 12.5 Heat production estimates in Watts for the samples used in this study. See main text for details of how heat production was estimated

Sample	Heat production
Inuit	160.8
Yugoslavians	156.0
Egyptians	146.4
Lapps	134.4
Amerinds	141.6
Melanesians	141.6
Skhul IV	158.4
Skhul V	168.0
Predmosti 3	170.4
Predmosti 14	158.4
Caviglione 1	156.0
Shanidar 4	170.4
La Chapelle 1	182.4
La Ferrassie 1	204.0
Dmanisi	120.0
KNM-WT 15000	115.2

in order to establish, all else being equal, what the thermal implications of observed proportional differences would have been. No attempt was made to account for possible inter-population variability in adipose characteristics, vaso-constriction or vasodilation, sweat gland distribution and production, the amount or density of body hair, or the thermal properties of clothing. Individuals were modeled as if they were hairless, naked bipeds employing a modern human striding bipedal gait. The same segment skin temperatures were used for both modern and fossil individuals. Skin temperatures were taken from Houdas and Ring (1982). These values were derived from motionless adult humans in each of the ambient temperatures considered in this study. Following Cross et al. (2008), Tr was treated as equal to Ta. Research employing thermal mannequins has shown that convective and radiant heat transfer coefficients vary somewhat from one segment to the next (e.g., Quintela et al. 2004; Oliveira et al. 2011). However, attempts to use mannequins to model the thermal properties of body segments during walking (e.g., Oliveira et al. 2011) have not yet included sufficiently realistic segment kinematics to believe that their segment-specific heat transfer coefficients would provide more accurate estimates of thermal response during locomotion than the coefficients employed here.

Once the relative heat loss values had been calculated, we carried out two sets of analyses. The first focused on whole-body heat loss. Initially, we compared the whole-body heat loss estimates for KNM-WT 15000 and the two European Neanderthals, La Ferrassie 1 and La Chapelle 1, with the whole-body heat loss estimates for the modern human samples. Because arguments in the literature have focused on the relationship between total body surface area,

limb proportions, and mean annual temperature or latitude (the assumption being that total body surface area and limb proportions reflect adaptation to thermal stress) we used comparable ratios that incorporate the contributions of segment-specific data. The first ratio we employed is the ratio of the sum of segment heat loss to heat production (SSHL:HP). In this ratio, SSHL represents the variable for which SA is assumed to be a proxy, and HP represents the amount of heat generated by a walking hominin of a given weight. We predicted that, if the current consensus regarding the thermoregulatory adaptations of KNM-WT 15000 and the Neanderthals is correct, then KNM-WT 15000 should consistently have a higher SSHL:HP (i.e., dissipate relatively more heat) than the modern humans in our sample, and that the European Neanderthals should consistently have lower SSHL:HP (i.e., retain relatively more heat) than our modern human sample.

Having compared whole-body heat loss across the species, we examined whole-body heat loss within the *H. erectus* and *H. neanderthalensis* samples. There is reason to think that the mean annual temperature at Dmanisi would have been cooler at 1.7 Ma than the mean annual temperature at West Turkana at 1.6 Ma (Lordkipanidze et al. 2007), and that the mean annual temperature at La Ferrassie and La Chapelle at the time they were occupied by Neanderthals would have been cooler than the mean annual temperature at Shanidar when it was occupied by Neanderthals (Froehle and Churchill 2009). Thus, the prediction we tested was that the Dmanisi specimen should exhibit lower SSHL:HP than KNM-WT 15000, and that the two European Neanderthals should exhibit lower SSHL:HP than the Neanderthal from Shanidar.

Table 12.6 Segmented (SEG) and conventional (CON) method estimates of whole-body relative heat loss (Total Heat Loss in Watts/Heat Production in Watts), in ambient temperatures of 20°C, 25°C, 30°C, and 35°C

Sample	20°C		25°C		30°C		35°C	
	SEG	CON	SEG	CON	SEG	CON	SEG	CON
Inuit	1.280	1.266	1.018	0.884	0.589	0.538	0.089	0.087
Yugoslavians	1.506	1.371	1.205	0.957	0.692	0.583	0.105	0.094
Egyptians	1.592	1.385	1.273	0.967	0.732	0.589	0.111	0.095
Lapps	1.528	1.417	1.215	0.989	0.703	0.602	0.106	0.098
Amerinds	1.524	1.393	1.216	0.973	0.701	0.592	0.106	0.096
Melanesians	1.582	1.393	1.263	0.973	0.728	0.592	0.110	0.096
La Ferrassie 1	1.135	1.164	0.907	0.813	0.522	0.495	0.079	0.080
La Chapelle 1	1.149	1.195	0.917	0.835	0.529	0.508	0.080	0.082
Shanidar 4	1.219	1.231	0.970	0.860	0.561	0.524	0.085	0.085
Dmanisi	1.645	1.457	1.300	1.018	0.757	0.620	0.115	0.100
KNM-WT 15000	1.973	1.983	1.572	1.076	0.907	0.655	0.138	0.106
Skhul IV	1.601	1.392	1.289	0.972	0.736	0.592	0.111	0.096
Skhul V	1.570	1.378	1.261	0.962	0.720	0.586	0.110	0.095
Predmosti 3	1.511	1.335	1.211	0.932	0.694	0.568	0.106	0.092
Predmosti 14	1.509	1.341	1.205	0.936	0.694	0.570	0.106	0.092
Caviglione 1	1.573	1.382	1.261	0.965	0.723	0.587	0.110	0.095

Table 12.7 Segment-specific relative heat loss (Segment Heat Loss in Watts/Total Body Heat Production in Watts) in ambient temperatures of 20°C, 25°C, 30°C, and 35°C

Sample	20°C				25°C			
	UA	LA	UL	LL	UA	LA	UL	LL
Inuit	0.132	0.078	0.219	0.119	0.096	0.054	0.185	0.080
Yugoslavians	0.154	0.090	0.255	0.143	0.111	0.062	0.219	0.096
Egyptians	0.160	0.099	0.267	0.153	0.116	0.068	0.229	0.103
Lapps	0.160	0.093	0.263	0.140	0.116	0.064	0.222	0.094
Amerinds	0.154	0.094	0.257	0.147	0.111	0.065	0.219	0.099
Melanesians	0.160	0.098	0.265	0.151	0.116	0.068	0.227	0.102
La Ferrassie 1	0.119	0.068	0.194	0.105	0.086	0.046	0.166	0.070
La Chapelle 1	0.118	0.068	0.201	0.106	0.086	0.046	0.170	0.071
Shanidar 4	0.124	0.076	0.211	0.112	0.090	0.052	0.179	0.075
Dmanisi	0.173	0.107	0.277	0.147	0.125	0.073	0.233	0.099
KNM-WT 15000	0.200	0.127	0.321	0.191	0.145	0.087	0.275	0.128
Skhul IV	0.155	0.099	0.267	0.158	0.113	0.068	0.232	0.106
Skhul V	0.168	0.093	0.264	0.142	0.122	0.064	0.230	0.095
Predmosti 3	0.155	0.096	0.246	0.143	0.113	0.066	0.214	0.096
Predmosti 14	0.155	0.096	0.246	0.144	0.112	0.066	0.212	0.097
Caviglione 1	0.159	0.097	0.260	0.153	0.116	0.066	0.224	0.103
Sample	30°C				35°C			
	UA	LA	UL	LL	UA	LA	UL	LL
Inuit	0.056	0.037	0.094	0.055	0.018	0.008	0.003	0.008
Yugoslavians	0.065	0.042	0.110	0.067	0.021	0.009	0.003	0.010
Egyptians	0.068	0.046	0.115	0.071	0.022	0.010	0.003	0.011
Lapps	0.068	0.043	0.113	0.065	0.022	0.010	0.003	0.010
Amerinds	0.065	0.044	0.111	0.068	0.021	0.010	0.003	0.010
Melanesians	0.068	0.046	0.114	0.070	0.022	0.010	0.003	0.010
La Ferrassie 1	0.050	0.032	0.083	0.049	0.016	0.007	0.002	0.007
La Chapelle 1	0.050	0.032	0.086	0.049	0.016	0.007	0.003	0.007
Shanidar 4	0.053	0.035	0.091	0.052	0.017	0.008	0.003	0.008
Dmanisi	0.074	0.050	0.119	0.069	0.024	0.011	0.004	0.010
KNM-WT 15000	0.085	0.059	0.138	0.089	0.027	0.013	0.004	0.013
Skhul IV	0.066	0.047	0.115	0.074	0.021	0.010	0.003	0.011
Skhul V	0.071	0.043	0.114	0.066	0.023	0.010	0.003	0.010
Predmosti 3	0.066	0.045	0.106	0.066	0.021	0.010	0.003	0.010
Predmosti 14	0.066	0.045	0.106	0.067	0.021	0.010	0.003	0.010

In the second set of analyses, we examined the contribution of the limbs and limb segments to heat loss. In these analyses we focused on limb-specific and limb segment-specific ratios of HL to HP. Here HL represents the amount of convective and radiant heat lost by a given pair of limbs (e.g., both arms) or limb segments (e.g., both forearms), and HP represents the amount of heat generated by the body as a whole for a walking hominin of a given weight. We began by testing the prediction that the limbs of European Neanderthals should have lower segment HL:HP values than those of modern humans, while the limbs of KNM-WT 15000 should have higher segment HL:HP values than those of modern humans. Next, we tested the prediction that that the limbs of KNM-WT 15000 should lose more heat than those of the Dmanisi specimen, while the limbs of the European Neanderthals should lose less heat than those of the Middle Eastern Neanderthal. Subsequently, we investigated the contribution of the upper and lower limb segments to heat loss. Based on the argument of Trinkaus (1981) and Holliday and Ruff (2001) that the distal segments of the limbs are particularly evolutionarily labile with respect to thermoregulation, we predicted that differences in segment-specific relative heat loss between Neanderthals and *H. erectus* should be more pronounced in the distal segments of each limb than in their proximal segments.

Results

The pattern of relative whole-body heat loss (Table 12.6) in our sample is consistent with the current consensus concerning the thermoregulatory implications of the size and shape of the bodies of *H. erectus* and *H. neanderthalensis*. As predicted, KNM-WT 15000 is estimated to have lost more heat than the modern human samples, and the two European Neanderthals are estimated to have lost less heat than the modern human samples.

Table 12.8 lists the mean SSHL:HP for our modern human sample as well as the number of standard deviations above or below these means that the estimates for each of the fossil hominins depart. Of the modern humans samples, the Inuit were estimated to have the lowest SSHL:HP and the Egyptians were estimated to have the highest HL:HP in all four ambient temperatures. The Inuit departed from the human mean by −1.8 to −2.0SD while the Egyptians departed from the human mean by +0.8 to +1.8SD. Results for the European Neanderthals were consistent with the arguments for polar adaptation. The Neanderthals displayed the lowest HL:HP of all the samples including the Eskimo. The two European Neanderthals differed from the human mean by −3.0SD or greater in each ambient temperature, while Shanidar 4 differed from the human mean by −2.4 or greater. Also consistent with the arguments for thermal adaptation, KNM-WT 15000 consistently had the highest SSHL:HP of all of the specimens, departing from the human mean by +4.0–6.7SD.

The results of the intra-species comparisons were also consistent with the current consensus regarding the thermoregulatory implications of the size and shape of the bodies of *H. erectus* and *H. neanderthalensis* (Tables 12.6 and 12.8). As predicted, the European Neanderthals had lower HL:HP values than the Near Eastern Neanderthal from Shanidar, and KNM-WT 15000 had a higher HL:HP value than the Dmanisi *H. erectus* specimen.

Our assessment of the thermal responses of hominin limbs indicated that whole-limb relative heat loss estimates (i.e., the sum of heat loss estimates for the proximal and distal segments of each limb) followed a similar pattern to that for whole-body heat loss (Table 12.8). As with overall SSHL:HP, the relative ranking of specimens and populations remained constant across the four ambient temperatures. Of the modern humans samples, the Inuit were estimated to have the lowest segment HL:HP values, differing from the modern human mean by −0.8 to −2.8SD depending on the limb segment and ambient temperature. The Egyptians were estimated to have the highest segment HL:HP values, differing from the modern human mean by less than +1.3SD for all four limb segments regardless of ambient temperature. As predicted, and consistent with the arguments for polar

adaptation, the limbs of the Neanderthals consistently had the lowest segment HL:HP of the other samples including the Inuit. The limbs of KNM-WT 15000 consistently had the highest segment HL:HP of the other hominins, typically losing between +3.6 and +6.0SD more heat than the modern human mean. Also as predicted, the two European Neanderthal specimens were found to lose relatively less heat from their limbs than the Middle Eastern Neanderthal. The limbs of the two European Neanderthals lost approximately −3SD (−1.5SD to −3.4SD) less heat than the modern human mean while the Shanidar Neanderthal lost around −2SD (−1.1SD to −2.6SD) less heat than the modern human mean. The predicted pattern was also identified in our *H. erectus* sample. The *H. erectus* specimen from Dmanisi lost relatively less heat from its limbs than did the African *H. erectus*, KNM-WT 15000. Thus, the pattern of relative heat loss for the limbs of *H. erectus* and *H. neanderthalensis* is also consistent with the current consensus regarding the thermoregulatory implications of the size and shape of the bodies of *H. erectus* and *H. neanderthalensis*.

In contrast, our findings regarding the contribution of the proximal and distal segments of each limb to heat loss were not consistent with the predictions of the hypothesis that the distal segments of the limbs are particularly evolutionarily labile in relation to thermoregulation. The differences in the number of standard deviations by which the upper and lower limb segments depart from the modern human means were often small in both *H. neanderthalensis* and *H. erectus* (Table 12.8). More importantly, there was no obvious pattern in the differences between average heat loss estimates for the upper and lower limb segments (Table 12.9). At 20° C, for example, there is no difference between the average heat loss for the Neanderthals' upper and lower arm segments. The same holds for the average heat loss values for their upper and lower leg segments. At 35°C, in contrast, the lower arm loses more heat than the upper arm, while the lower leg loses less heat than upper leg. The estimates for *H. erectus* are also not consistent with the predictions of the hypothesis. At 20°C, the lower arm loses more heat than the upper arm, which is the predicted pattern. But the lower leg loses less heat than the upper leg, which is not the predicted pattern. At 35°C, neither set of segments is consistent with the predictions of the hypothesis. The lower arm loses less heat than upper arm, and the lower leg loses less heat than the upper leg.

The change in differences between the upper and lower segments as we move from colder to warmer ambient temperature does not conform to expectation either. Given that Neanderthal arms and legs are supposed to be adapted to cold conditions, we should see a closer fit with the predictions of the hypothesis as temperature declines, yet the differences between the upper and lower segments actually disappear at the lowest temperature, 20°C. The same holds

Table 12.8 Relative heat loss in Watts in TA = 20°C, 25°C, 30°C, and 35°C. Holocene mean and standard deviation (SD) are for the Holocene human skeletal sample (i.e., excluding the fossil modern human samples). Values for other samples represent the number of standard deviations to which they depart from the modern human mean. WB = whole-body heat loss/total heat production; UA = upper arm heat loss/total heat production; LA = lower arm heat loss/total heat production; UL = upper leg heat loss/total heat production; LL = lower leg heat loss/total heat production

Sample	20°C WB	20°C UA	20°C LA	20°C UL	20°C LL	25°C WB	25°C UA	25°C LA	25°C UL	25°C LL
Inuit	−1.94	−1.91	−2.00	−1.94	−1.92	−1.94	−1.87	−1.81	−1.99	−2.00
Yugoslavians	0.03	0.09	−0.29	0.06	0.08	0.08	0.05	−0.20	0.13	0.05
Egyptians	0.79	0.64	1.00	0.72	0.92	0.81	0.60	1.00	0.76	0.84
Lapps	0.11	0.42	0.13	0.50	−0.19	0.04	0.41	−0.54	0.34	−0.25
Amerinds	0.07	−0.12	0.25	0.19	0.38	0.04	−0.13	−0.46	0.15	0.31
Melanesians	0.01	0.41	0.86	0.58	0.75	−0.47	0.44	0.15	−2.05	0.61
Holocene human mean (SD)	1.502 (0.114)	0.153 (0.011)	0.092 (0.007)	0.254 (0.018)	0.142 (0.012)	1.198 (0.093)	0.111 (0.008)	0.063 (0.005)	0.217 (0.016)	0.096 (0.008)
La Ferrassie 1	−3.2	−3.1	−3.4	−3.3	−3.1	−3.1	−3.1	−3.3	−3.2	−3.2
La Chapelle 1	−3.1	−3.2	−3.4	−2.9	−3.0	−3.0	−3.2	−3.3	−2.9	−3.1
Shanidar 4	−2.5	−2.6	−2.3	−2.4	−2.5	−2.4	−2.6	−2.2	−2.4	−2.6
Dmanisi	1.3	1.8	2.1	1.3	0.4	1.1	1.8	2.1	1.0	0.4
KNM-WT 15000	4.1	4.3	5.0	3.8	4.1	4.0	4.2	4.8	3.6	4.0
Skhul IV	0.9	0.2	1.0	0.7	1.3	1.0	0.2	1.1	0.9	1.3
Skhul V	0.6	1.4	0.1	0.6	0.0	0.7	1.3	0.2	0.8	−0.1
Predmosti 3	0.1	0.2	0.6	−0.4	0.1	0.1	0.2	0.6	−0.2	0.0
Predmosti 14	0.1	0.2	0.6	−0.4	0.2	0.1	0.2	0.6	−0.3	0.1
Caviglione 1	0.6	0.5	0.7	0.3	0.9	0.7	0.6	0.7	0.5	0.8

Sample	30°C WB	30°C UA	30°C LA	30°C UL	30°C LL	35°C WB	35°C UA	35°C LA	35°C UL	35°C LL
Inuit	−2.0	−2.2	−2.1	−2.0	−1.8	−1.8	−2.8	−0.9	−0.8	−1.8
Yugoslavians	0.0	0.1	−0.3	0.0	0.1	0.8	0.1	0.4	0.8	−0.1
Egyptians	0.8	0.7	1.1	0.6	0.9	1.8	1.0	1.3	1.3	0.5
Lapps	0.1	0.5	−0.7	0.4	−0.2	0.8	0.7	0.1	1.1	−0.4
Amerinds	0.0	−0.1	−0.6	0.1	0.4	0.7	−0.1	0.2	0.9	0.1
Melanesians	−0.5	0.5	0.1	−2.2	0.6	0.6	0.7	0.6	−0.9	0.4
Holocene human mean (SD)	0.691 (0.052)	0.065 (0.004)	0.043 (0.003)	0.110 (0.008)	0.066 (0.006)	0.100 (0.006)	0.021 (0.001)	0.009 (0.001)	0.003 (0.0003)	0.010 (0.001)
La Ferrassie 1	−3.3	−3.7	−3.8	−3.3	−2.9	−3.5	−4.7	−2.0	−1.8	−2.8
La Chapelle 1	−3.1	−3.7	−3.8	−3.0	−2.8	−3.4	−4.8	−2.0	−1.5	−2.7
Shanidar 4	−2.5	−3.0	−2.6	−2.4	−2.3	−2.6	−3.9	−1.2	−1.1	−2.3
Dmanisi	1.3	2.1	2.3	1.2	0.4	2.5	2.8	2.1	1.7	0.2
KNM-WT 15000	4.2	5.0	5.3	3.5	3.8	6.7	6.0	4.0	3.3	3.0
Skhul IV	0.9	0.2	1.2	0.6	1.3	1.9	0.3	1.3	1.2	0.9
Skhul V	0.6	1.6	0.2	0.5	0.0	1.6	2.1	0.7	1.1	−0.2
Predmosti 3	0.1	0.3	0.6	−0.5	0.1	1.0	0.4	0.9	0.4	−0.2
Predmosti 14	0.0	0.2	0.6	−0.5	0.2	0.9	0.3	1.0	0.4	−0.1
Caviglione 1	0.6	0.7	0.7	0.2	0.8	1.6	0.9	1.0	1.0	0.5

Table 12.9 Comparison of average heat loss estimates for Neanderthals and *H. erectus* upper and lower limb segments. UA = upper arm. LA = lower arm. UL = upper leg. LL = lower leg

		20°C	25°C	30°C	35°C
Neanderthals	UA	−3.0	−3.0	−3.5	−4.5
	LA	−3.0	−2.9	−3.4	−1.7
		No difference	LA loses more heat	LA loses more heat	LA loses more heat
	UL	−2.9	−2.8	−2.9	−1.5
	LL	−2.9	−3.0	−2.7	−2.6
		No difference	LL loses less heat	LL loses more heat	LL loses less heat
H. erectus	UA	3.1	3.0	3.6	4.4
	LA	3.6	3.5	3.8	3.1
		LA loses more heat	LA loses more heat	LA loses more heat	LA loses less heat
	UL	2.6	2.3	2.4	2.5
	LL	2.3	2.2	2.1	1.6
		LL loses less heat	LL loses less heat	LL loses less heat	LL loses less heat

for *H. erectus*. Given that its arms and legs are supposed to be adapted to warmer temperatures, we might expect to see a closer fit with the predictions of the hypothesis as temperature increases, but the differences between heat loss estimates for the arm segments at 35°C are the reverse of what the hypothesis predicts, whereas those at 20°C, 25°C, and 30°C are consistent with the hypothesis.

Looking at the amount of change in limb segment heat loss across the four ambient temperatures does not alter the picture. As we move from 20°C to 35°C, we see that the upper arms of Neanderthals change by 1.5SD while their lower arms change by 1.3SD, which means that the lower arms respond to the change in ambient temperature less than the upper arms. The same is true for the Neanderthals' leg segments: the upper leg segments change by 1.4SD from 20°C to 35°C while the lower leg segments change by 0.3SD. Both of these findings are inconsistent with the predictions of the hypothesis. It is a similar story for *H. erectus*. Moving from 20°C to 35°C, the upper arms change by 1.3SD while the lower arms change by 0.5SD, which means that the lower arms of *H. erectus* also respond to the change in ambient temperature less than its upper arms. Turning to the leg segments of *H. erectus*, the amount of change is greater in the lower leg than in the upper leg as we move from 20°C to 35°C. The upper leg changes by only 0.1SD while the lower leg changes by 0.7SD. But the change in the lower leg is in the opposite direction to the one predicted by the hypothesis.

In sum, then, the limb segments' heat loss estimates do not support the hypothesis that the distal limb segments of *H. erectus* and *H. neanderthalensis* were more affected by heat loss-related selection than their proximal limb segments.

Discussion

The results of the study were mixed. The whole-body heat loss estimates we obtained follow the pattern predicted by the thermoregulation hypothesis. They suggest that the African *H. erectus* specimen in our sample, KNM-WT 15000, would have lost more heat than the humans in our sample, and that the European Neanderthal specimens in our sample would have conserved more heat than the humans in our sample. They also suggest that the hot-climate-dwelling KNM-WT 15000 would have lost less heat than the colder-climate *H. erectus* specimen from Dmanisi, and that the two cold-climate Neanderthal specimens, La Ferrassie 1 and La Chapelle 1, would have conserved more heat than the warmer-climate Middle Eastern Neanderthal specimen, Shanidar 4. The whole-limb heat loss estimates we obtained follow the pattern predicted by the thermoregulation hypothesis too. They suggest that the limbs of KNM-WT 15000 would have lost more heat than those of the humans in our sample, while the limbs of the European Neanderthal specimens in our sample would have conserved more heat than those of the humans in our sample. The whole-limb heat loss estimates also suggest that, as predicted, the limbs of KNM-WT 15000 would have lost less heat than those of the Dmanisi specimen, and that the limbs of La Ferrassie 1 and La Chapelle 1 would have lost less heat than those of Shanidar 4. In contrast, the limb segment heat loss estimates we obtained are not consistent with the predictions of the hypothesis that the distal limb segments of *H. erectus* and *H. neanderthalensis* were more affected by heat loss-related selection than their proximal limb segments. The heat loss differences between the proximal and distal limb segments did not exhibit any obvious pattern. Thus, our results generally support the current consensus regarding the thermoregulatory implications of the size and shape of the bodies of *H. erectus* and *H. neanderthalensis*, but they are not entirety consistent with it.

Because the methods used in this study differ from those employed in previous studies it is important that we ensure that our data are as reliable as if we had used the conventional methods. The equation used in this study for estimating heat production is the same as the one used in other studies (e.g., Dennis and Noakes 1999; Marino et al. 2004), so the variables of interest in this regard are the estimates of total skin surface area and relative heat loss. To assess the reliability of the former, we estimated total skin surface area with both our segmented method and the conventional Du Bois and Du Bois (1916) method. Consistent with the findings of Cross et al. (2008) and Cross and Collard (2011), a paired t-test found these two sets of estimates (N = 16) to be statistically indistinguishable (p = 0.118). We also tested the reliability of our proxy for relative heat loss, HL:HP, to ensure that the additional heat loss variables included in our method (i.e., segment specific skin temperatures, segment specific wind speeds derived from 3D kinematic data, and segment specific surface areas) were tracking the patterns identified with the conventional proxy for relative heat loss, SA:BM. When this analysis was performed SA:BM consistently and significantly correlated with HL:HP in each of the four ambient temperatures with r-values ranging from 0.980 to 0.992. This suggests that the additional variables used to establish the ratios of HL to HP produce a thermoregulatory proxy that is consistent with the SA:BM ratio. As a further check on the reliability of our proxy for relative heat loss, we calculated HL:HP ratios for each of our samples using conventional methods. To do this we estimated skin surface area using the Du Bois and Du Bois (1916) equation and then used these along with a weighted mean skin temperature and a wind speed equal to walking speed to estimate convective and radiant heat loss following the method outlined by Dennis and Noakes (1999). When these conventional method HL:HP values were compared to those derived from our segmented approach, we found the two sets of estimates to be strongly and significantly correlated at all four ambient temperatures (r = 0.950 to 0.965, p = 0.000). Given these results, there is reason to believe that our method is as reliable as the conventional method of estimating heat loss in humans and other hominins.

Another "quality control" issue that needs to be addressed is whether the segment-specific data we employed had any effect on the results. Cross and Collard (2011) found that variation in limb proportions explained most of the difference between the results yielded by the conventional approach to estimating skin surface area and a segmented approach similar to the one we have used here. With this finding in mind, we revisited the HL:HP estimates that we generated with the

segmented and conventional methods, and performed regression analyses in which we investigated how much of the difference between the estimates yielded by the two methods could be explained by the brachial, crural, and intermembral indices. The results indicated that limb segment length differences explained more than 70% of the variation in the differences between methods. When limb segment relative heat loss (i.e., segment HL:HP) values were used as the independent variables in the regression analyses we found that they explained over 98% of the difference between the two sets of HL:HP estimates (r = 0.982–0.998, p = 0.000). It is clear from these results that the segment-specific data did have an effect on the results, as intended.

The main implication of our study is that the current consensus regarding the thermoregulatory implications of the size and shape of the bodies of *H. erectus* and *H. neanderthalensis* may need some revision. The fact that the results of our whole-body and whole-limb analyses suggest that KNM-WT 15000 was warm-adapted and that the European Neanderthals were cold-adapted suggests that the basic idea that thermoregulation affected the evolution of body size and shape in *H. erectus* and *H. neanderthalensis* is correct. The same holds for fact that our whole-body and whole-limb analyses suggest that the Dmanisi *H. erectus* specimen was more cold adapted than KNM-WT 15000, and that the Middle Eastern Neanderthal in our sample was more warm adapted than the two European Neanderthals in our sample. However, the failure of our limb segment analyses to identify consistent differences between the proximal and distal limb segments in terms of heat loss raises the possibility that the idea that selection altered the lengths of the distal limb segments in *H. erectus* and *H. neanderthalensis* to improve thermoregulation is incorrect.

With regard to future research, the most obvious task concerns the hypothesis that selection altered the lengths of the distal limb segments in *H. erectus* and *H. neanderthalensis* to improve thermoregulation. Given the kinematics of walking, and especially the fact that the distal segments of limbs experience greater displacement than the upper segments of limbs, it is somewhat surprising that the distal limb segments do not demonstrate a greater sensitivity to thermoregulation-related selection. The results of modeling exercises like the one reported here are heavily assumption-dependent. So, one possibility is that the hypothesis is correct and that our results did not support its predictions because some of the assumptions we made are wrong. Repeating the exercise with a different set of assumptions will indicate whether such is the case. Unfortunately, this is currently impossible for one important assumption – that the impact of evaporative heat loss (i.e., sweating) can be safely ignored. We know that there are differences among body segments in both the number of sweat glands and their recruitment pattern (e.g., Buono 2000), so it is feasible that taking evaporative heat loss into account would have reduced the number of results that do not fit the predictions of the hypothesis. However, as far as we are aware, segment-specific data for the dynamics of evaporative heat loss during walking do not exist at this time. Collecting such data would be a useful undertaking, needless to say.

While "assumption error" may be the most obvious explanation for the failure of the analyses to support the hypothesis, it is worth considering the possibility that the hypothesis is incorrect and that some other factor or set of factors had a stronger influence on the variation in distal limb segment length within and between *H. erectus* and *H. neanderthalensis* than did temperature. The obvious candidate for the factor affecting the lower legs is locomotion. Is it possible that the lengths of the distal segments of the legs of *H. erectus* and *H. neanderthalensis* have been selected in relation to a locomotion-related variable, such as terrain (e.g., Higgins and Ruff 2011)? As far as the forearms are concerned, one possibility worth investigating is that the within and between species differences are connected with differences in weapon use. Perhaps, for example, long forearms are useful for throwing objects, while short forearms are beneficial when using a thrusting spear. A less obvious factor that could have affected both the lower leg and the forearm is genetic drift. In recent years it has become increasingly clear that drift, in the form of the iterative founder effect, has played an important role in structuring modern human genetic and phenotypic variation (e.g., Weaver et al. 2007). There seems to be no reason why it might not have also played an important role in structuring genetic and phenotypic variation in other fossil hominin species such as *H. erectus* and *H. neanderthalensis*. Lastly, it is also worth considering the possibility that clothing may have reduced the impact of thermoregulation-related selection. To the best of our knowledge, nobody has suggested that *H. erectus* used clothing, but it has been argued that Neanderthals utilized clothing (e.g., Sørenson 2009; Collard et al. in press). If Neanderthals did in fact use clothing, then the nature of thermoregulation-related selection on the limbs could well have been reduced to the extent that other factors became more important influences on the size of the proximal and distal limb segments.

Conclusions

In the study presented here, we employed a novel way of assessing hominin thermoregulatory responses to ambient thermal stress during normal walking. The method we used differs from the conventional approach in that it takes into account the fact that different parts of the body differ in surface area, skin temperature, and 3D kinematics rather than treating the body as an undifferentiated mass. Importantly, this allows for the estimation of differences in thermal response due to differences in both body size *and* proportions.

In the study we used the segmented method to determine whether the current consensus regarding the thermoregulatory implications of the size and shape of the bodies of *H. erectus* and *H. neanderthalensis* is supported when body segment differences in surface area, skin temperature, and rate of movement are taken into account. Based on comparisons with modern humans, we tested the hypothesis that the well known African *H. erectus* specimen KNM-WT 15000 was adapted for warm conditions. We also tested the hypothesis that the European Neanderthals were adapted for cold conditions. In addition, by comparing specimens of conspecifics from locations with markedly different ambient temperatures, we investigated whether there is evidence of adaptation to thermal conditions within *H erectus* and within *H. neanderthalensis*.

The results of our study only partly supported the current consensus. The whole-body heat loss estimates were consistent with the idea that KNM-WT 15000 was warm adapted, and that European Neanderthals were cold adapted. The whole-body heat loss estimates were also consistent with the notion that there are thermoregulation-related differences in body size and shape within *H erectus* and *H. neanderthalensis*. The whole-limb estimates told a similar story. They too followed the predicted pattern. However, the results of our analysis of limb segment-specific heat loss were not consistent with the current consensus regarding the thermoregulatory implications of the size and shape of the bodies of *H. erectus* and *H. neanderthalensis*. Contrary to expectation, differences between the proximal and distal limb segments did not follow any particular trend.

The obvious implication of these results is that, while we can be more confident about the idea that thermoregulation influenced the evolution of body size and shape in *H. erectus* and *H. neanderthalensis*, we need to be more cautious in attributing differences in limb segment size to thermoregulation. Based on our results, the possibility that other factors influenced limb segment size in these species more than thermoregulation should be given serious consideration. Identifying these factors will require further research.

Acknowledgments We would like to thank Yoel Rak not only for his friendship and guidance but also for the inspiration he has provided through his research. In addition, we are grateful to Assaf Maron and Erella Hovers for inviting us to contribute to this volume, and to two anonymous reviewers for their helpful comments. Our work was supported by the Social Sciences and Humanities Research of Canada, the Canada Research Chairs Program, the Canada Foundation for Innovation, the British Columbia Knowledge Development Fund, and Simon Fraser University.

References

Buono, M. J. (2000). Limb vs. trunk sweat gland recruitment patterns during exercise in humans. *Journal of Thermal Biology, 25,* 263–266.

Collard, M., Tarle, L., Sandgathe, D., & Allan, A. (2016). Faunal evidence for a difference in clothing use between Neanderthals and early modern humans in Europe. *Journal of Anthropological Archaeology, 44,* 235–246.

Cross, A., & Collard, M. (2011). Estimating surface area in early hominins. *PLoS ONE, 6,* e16107.

Cross, A., Collard, M., & Nelson, A. (2008). Body segment differences in surface area, skin temperature and 3D displacement and the estimation of heat balance during locomotion in hominins. *PLoS ONE, 3,* e2464.

Dennis, S. C., & Noakes, T. D. (1999). Advantages of smaller body mass in humans when distance-running in warm, humid conditions. *European Journal of Applied Physiology, 79,* 280–284.

Du Bois, D., & Du Bois, E. F. (1916). A formula to estimate the approximate surface area if height and weight be known. *Archives of Internal Medicine, 17,* 863–871.

Feldesman, M. R., & Fountain, R. L. (1996). "Race" specificity and the femur/stature ratio. *American Journal of Physical Anthropology, 100,* 207–224.

Froehle, A., & Churchill, S. E. (2009). Energetic competition between Neandertals and anatomically modern humans. *PaleoAnthropology, 2009,* 96–116.

Haeuseler, M. (2001). New insights into the locomotion of *Australopithecus africanus*: Implications of the partial skeleton of STW 431 (Sterkfontein, South Africa). Ph.D. Dissertation, Universität Zürich.

Haeusler, M., & McHenry, H. M. (2004). Body proportions of *Homo habilis* reviewed. *Journal of Human Evolution, 46,* 433–465.

Haycock, G. B., Chir, B., Schwartz, G. J., & Wisotsky, D. H. (1978). Geometric method for measuring body surface area: A height-weight formula validated in infants, children, and adults. *The Journal of Pediatrics, 93,* 62–66.

Hinrichs, R., & Cavanagh, P. (1981). Upper extremity function in treadmill walking. *Medicine and Science in Sports and Exercise, 13,* 96.

Higgins, R. W., & Ruff, C. B. (2011). The effects of distal limb segment shortening on locomotor efficiency in sloped terrain: Implications for Neandertal locomotor behavior. *American Journal of Physical Anthropology, 146,* 336–345.

Holliday, T. W. (1997). Postcranial evidence of cold adaptation in European Neandertals. *American Journal of Physical Anthropology, 104,* 245–258.

Holliday, T. W., & Ruff, C. B. (2001). Relative variation in human proximal and distal limb segment lengths. *American Journal of Physical Anthropology, 116,* 26–33.

Houdas, Y., & Ring, E. F. J. (1982). *Human body temperature: Its measurement and regulation.* New York: Plenum Press.

Langlois, J. A., Keyl, P. M., Guralnik, J. M., Foley, D. J., Marottoli, R. A., & Wallace, R. B. (1997). Characteristics of older pedestrians who have difficulty crossing the street. *American Journal of Public Health, 87,* 393–397.

Lordkipanidze, D., Jashashvili, T., Vekua, A., Ponce de León, M. S., Zollikofer, C. P. E., Rightmire, G. P., et al. (2007). Postcranial evidence from early *Homo* from Dmanisi, Georgia. *Nature, 449,* 305–310.

Marino, F. E., Lambert, M. I., & Noakes, T. D. (2004). Superior performance of African runners in warm humid but not in cool environmental conditions. *Journal of Applied Physiology, 96,* 124–130.

Neptune, R. R., Sasaki, K., & Kautz, S. A. (2008). The effect of walking speed on muscle function and mechanical energetics. *Gait Posture, 28,* 135–143.

Oliveira, A. V. M., Gaspa, A. R., Francisco, S. C., & Quintela, D. A. (2011). Convective heat transfer from nude body under calm conditions: Assessment of the effect of walking with a thermal manikin. *International Journal of Biometeorology, 56,* 319–332.

Orendurff, M. S., Segal, A. D., Klute, G. K., Berge, J. S., Rohr, E. S., & Kadel, N. J. (2004). The effect of walking speed on center of mass displacement. *Journal of Rehabilitation Research and Development, 41,* 829–834.

Quintela, D., Gaspar, A., & Borges, C. (2004). Analysis of sensible heat exchanges from a thermal manikin. *European Journal of Applied Physiology, 92,* 663–668.

Raxter, M. H., Ruff, C. B., Azab, A., Erfan, M., Soliman, M., & El-Sawaf, A. (2008). Stature estimation in ancient Egyptians: A new technique based on anatomical reconstruction of stature. *American Journal of Physical Anthropology, 136,* 147–155.

Ruff, C. B. (1991). Climate and body shape in hominid evolution. *Journal of Human Evolution, 21,* 81–105.

Ruff, C. B. (1993). Climatic adaptation and hominid evolution: The thermoregulatory imperative. *Evolutionary Anthropology, 2,* 53–60.

Ruff, C. B. (1994). Morphological adaptations to climate in modern and fossil hominids. *Yearbook of Physical Anthropology, 37,* 65–107.

Ruff, C. B. (2010). Body size and body shape in early hominins – implications of the Gona pelvis. *Journal of Human Evolution, 58,* 166–178.

Ruff, C. B., & Walker, A. (1993). Body size and body shape. In A. Walker & R. E. Leakey (Eds.), *The Nariokotome* Homo erectus *skeleton* (pp. 234–265). Cambridge: Harvard University Press.

Ruff, C. B., Trinkaus, E., & Holliday, T. W. (1997). Body mass and encephalization in Pleistocene *Homo. Nature, 387,* 173–176.

Ruff, C. B., Niskanen, M., Junno, J.-A., & Jamison, P. (2005). Body mass prediction from stature and bi-iliac breadth in two high latitude populations, with application to earlier higher latitude humans. *Journal of Human Evolution, 48,* 381–392.

Sørenson, B. (2009). Energy use by Eem Neanderthals. *Journal of Archaeological Science, 36,* 2201–2205.

Tilkens, M. J., Wall-Scheffler, C., Weaver, T., & Steudel-Numbers, K. (2007). The effects of body proportions on thermoregulation: An experimental assessment of Allen's rule. *Journal of Human Evolution, 53,* 286–291.

Trinkaus, E. (1981). Neanderthal limb proportions and cold adaptation. In C. B. Stringer (Ed.), *Aspects of human evolution* (pp. 187–224). London: Taylor and Francis.

Trotter, M. L., & Gleser, G. C. (1952). Estimation of stature from long bones of American whites and negroes. *American Journal of Physical Anthropology, 10,* 463–514.

Trotter, M. L., & Gleser, G. (1958). A re-evaluation of stature based on measurements taken during life and of long bones after death. *American Journal of Physical Anthropology, 16,* 79–123.

Weaver, T. D. (2003). The shape of the Neanderthal femur is primarily the consequence of a hyperpolar body form. *Proceedings of the National Academy of Sciences USA, 100,* 6926–6929.

Weaver, T. D., Roseman, C. C., & Stringer, C. B. (2007). Were Neandertal and modern human cranial differences produced by natural selection or genetic drift? *Journal of Human Evolution, 53,* 135–145.

Wheeler, P. E. (1993). The influence of stature and body form on hominid energy and water budgets; a comparison of *Australopithecus* and early *Homo* physiques. *Journal of Human Evolution, 24,* 13–28.

Chapter 13
Behavioral Differences Between Near Eastern Neanderthals and the Early Modern Humans from Skhul and Qafzeh: An Assessment Based on Comparative Samples of Holocene Humans

Osbjorn M. Pearson and Vitale S. Sparacello

Abstract The differences and similarities between Near Eastern Neanderthals and the early modern humans from Skhul and Qafzeh in Israel have long been a point of study and debate. Conclusions about the magnitude and especially the implications of the differences have served as evidence to support or refute competing hypotheses about their cultural and biological differences. Here we revisit the controversy by assessing the midshaft shapes and robusticity of the femur, tibial, humerus, and radius of these Middle Paleolithic samples in comparison to European Neanderthals, Gravettian modern humans, several modern individuals from other late Pleistocene cultures in Europe and Israel, and a diverse set of Holocene humans from around the globe. The results show that the Near Eastern Neanderthals resemble European Neanderthals as well as a diverse array of modern agriculturalists and intensive foragers. In contrast, the people from Skhul and Qafzeh are much more distinct from recent samples but bear a degree of resemblance to Khoesan and Zulu males and females, Amud 1, and Ohalo 2. Additional insights emerge when the upper and lower limb are considered separately, but the result remains that the early moderns rather than Neanderthals seem to have faced an unusual, or at least uncommon, set of mechanical demands in comparison to most of the more recent groups.

Keywords Cross-sectional geometry • Habitual activity • Limb robusticity

Introduction

The goal of this chapter is to revisit the structural adaptations of the long bones of Near Eastern Neanderthals and early modern humans from Skhul and Qafzeh in order to gain insights into how their habitual physical activities differed and what more recent groups (if any) their activities resembled. The Neanderthals and early modern humans from Israel and elsewhere in Southwest Asia have been the subject of curiosity and controversy since the first skeletons were discovered at Mughuret et-Tabun, Mughuret es-Skhul, and Jebel Qafzeh in the 1930s (McCown and Keith 1939; Vandermeersch 1981). Subsequent excavations at Shanidar in northern Iraq in the 1950s yielded a trove of Neanderthal burials (Solecki 1963; Trinkaus 1983a). The skeleton of an adult male Neanderthal from Amud Cave (Endo and Kimura 1970) and an adult male partial skeleton from Kebara Cave (Arensburg et al. 1985; Rak and Arensburg 1987; Rak 1990; Bar-Yosef and Vandermeersch 1991) added many more fossil specimens and raised new questions about possible differences in physical adaptations and dissimilarities in lifeways between the early moderns and Neanderthals as a whole and also between Southwest Asian and European Neanderthals. Endo and Kimura (1970) included a substantial amount of interpretation of the functional morphology of Amud 1 in their description of the skeleton, but Erik Trinkaus, more than any previous researcher, adopted a strongly functional approach in his analysis and interpretation of the hominin fossils from the Middle Paleolithic. Trinkaus (1983a, b, 1984) described a large number of functional traits that differentiated the Skhul-Qafzeh people from Neanderthals and argued that most of those traits indicated that the Neanderthals were substantially stronger and had better mechanical advantages for many muscle groups than the early moderns. This, in turn, provided support for the hypothesis that the difference between the two groups reflected a major adaptive shift – to which bodies had adapted, whether by evolution, plasticity during the life

O.M. Pearson (✉) · V.S. Sparacello
Department of Anthropology, University of New Mexico, Albuquerque, NM 87121, USA
e-mail: ompear@unm.edu

V.S. Sparacello
e-mail: grifis1979@gmail.com

Assaf Marom and Erella Hovers (eds.), *Human Paleontology and Prehistory*,
Vertebrate Paleobiology and Paleoanthropology, DOI: 10.1007/978-3-319-46646-0_13

course, or both – that produced the more gracile and physically weaker humans of today.

During the 1980 and 1990s, analyses of the cross-sectional geometry of long bones to gain a more precise understanding of the likely mechanical environment to which they were adapted altered some aspects of Trinkaus' conclusions from the early 1980s. Traditional indices of long bone robusticity that were based on external measurements had shown that many Neanderthal long bones were very robust, at or slightly beyond the limit of what could be found among living humans. In contrast, individuals from Skhul and Qafzeh tended to be quite slender. However, when one considered the cross-sectional geometry of these bones, especially when standardized for bone length (the length of the beam being bent) and body mass (one component of the force acting to bend the beam), the tibiae and femora of Neanderthals and Skhul-Qafzeh people were much more similar in size-adjusted strength (Trinkaus and Ruff 1999a, b), but a major difference remained in the shape of the midshaft sections of their femora. The Skhul-Qafzeh individuals tended to have a much more strongly developed femoral pilaster at midshaft and a much higher ratio of maximum to minimum second moment of area (I_{max}/I_{min}) and a higher ratio of the second moment of area in the x and y planes (I_x/I_y) when these planes deviated somewhat from the directions of I_{max} and I_{min}. Trinkaus and his collaborators emphasized differences between the robusticity and shape of bones, arguing that robusticity essentially reflected (and responded to) the gross magnitude of loading that the bone had experienced while its shape recorded information about the regularity or irregularity of the directions from which those loads had been applied (Trinkaus et al. 1991). Thus the large pilasters on femora of the people from Skhul and Qafzeh presumably reflected a large amount of movement, whether walking or running, in a single direction for long periods of time, while the rounder femoral midshafts of Neanderthals presumably reflected a much more irregular pattern of motion with many changes in direction. Recent work comparing the CT scans of the tibiae of collegiate athletes lends support to this interpretation (Shaw and Stock 2013). Subsequent work by Ruff and colleagues (Ruff 1995; Ruff et al. 2005) argued that Neanderthals had medio-laterally reinforced femoral (and perhaps tibial) midshafts because they had wide hips. Across and within human populations, bi-iliac breadth does in fact correlate with the shape of femoral but not tibial midshaft sections; for femora the magnitude of this effect is weak (Pearson et al. 2014; Shaw and Stock 2011).

Another major theme of studies of cross-sectional geometry from the 1980s onward was that human evolution, especially since the end of the Pleistocene, had been accompanied by a gradual, and ultimately substantial, decrease in robusticity, especially in the lower limb (Ruff et al. 1993; Trinkaus and Ruff 2012). This change appeared to correspond to declines in mobility (Holt 2003). More generally, these conclusions also supported arguments that Holocene hunter-gatherers were more robust and mobile than agricultural people (Ruff et al. 1993; Larsen 1995). However, not all studies of the lower limbs of hunter-gatherers and later agriculturalists supported this pattern (Bridges 1989; Collier 1989; Carlson et al. 2007).

Temporal trends for the upper limb proved to be more complicated. Neanderthal males and many groups of modern humans from the Pleistocene have very high levels of bilateral asymmetry in their humeri (Trinkaus et al. 1994; Churchill 1994; Churchill and Formicola 1997), most likely from vigorous activity (quite possibly throwing) done with the right arm but not the left. Some studies suggested that the humerus, and thus presumably the entire upper limb, had also undergone a marked reduction in from the Middle Pleistocene to today (Trinkaus 1983a, b; Smith et al. 1984; Ben-Itzhak et al. 1988), but other studies noted a great deal of variation in the robusticity of recent human's humeri (Collier 1989; Churchill 1994; Churchill and Formicola 1997; Pearson 1997). A further problem arose from the fact that different studies used different standards to attempt to control for body size in assessments of "robusticity," and ecogeographic variation causes human populations from cold climates to have large epiphyses and wide diaphyses relative to length (Pearson 2000). Differences in activity appear to thicken diaphyses even further but have much less effect on the breadth of epiphyses (Trinkaus et al. 1994; Lieberman et al. 2001).

The state of the art for controlling for the effects of body size in cross-sectional geometry is to use section moduli (often abbreviated as Z, and calculated by dividing a second moment of area by the length of the line from the bending axis to the outer-most point of bone in the direction of bending) divided by the product of bone length and body mass (Martin et al. 1998). Many of the earlier studies used other adjustments that do not completely remove the strong signal of body mass for height that is present in data sets human cross-sectional geometry (and external dimension of long bones) and which reflects human ecogeographic variation (Pearson 2000; Ruff 2000).

This chapter presents an evaluation of the size-adjusted strength of the femur, tibia, humerus, and radius of Southwest Asian Neanderthals and the early modern humans from Skhul-Qafzeh in comparison to a broad suite of other groups of Pleistocene fossil hominins and populations of recent (Holocene) humans that differ in lifeways, geographic origin, and ecogeographic adaptations. An earlier study based on the same dataset considered indices based on simple, external measurements (e.g., the classic pilastric index) and concluded that the level of humeral robusticity adjusted for size in both European and Southwest Asian Neanderthals tended to resemble recent hunter-gatherers who practiced intensive forms of subsistence within geographically smaller territories than those employed by hunters of large herbivores, while their femoral robusticity resembled those of mobile people like Italian Epigravettians who lived in topographically rugged areas. In stark contrast, the Skhul-Qafzeh people had notably gracile humeri when adjusted for size and proportionately strong and highly pilastered femora. The closest analogs for the Skhul seemed to be Khoesan herder/foragers and Gravettian hunter-gatherers, both of whom were (presumably) quite mobile. Those results supported the proposal by Lieberman and Shea that archaeological evidence indicated that the lifestyles of Neanderthals and early modern humans in Israel may have differed substantially (Lieberman and Shea 1994; Lieberman 1993, 1998; Shea 1998). Given that the adjustment we used for body size (Pearson et al. 2006) did not attract much interest or use by other researchers, it seems logical to revisit the issue using cross-sectional geometry.

Materials and Methods

For fossil and recent human groups, we have used the same set of data (Pearson et al. 2006), but with riverine and coastal groups of Inuit pooled and all of the European Mesolithic samples pooled. The groups included in the analyses and sample sizes are listed in Table 13.1. Details of age and archaeological or geographic provenience each of the recent samples are presented in Pearson et al. (2006), and elsewhere (Pearson 1997, 2000). The raw data for all of these individuals consists of external dimensions of long bones, including their lengths, midshaft dimensions, and epiphyseal breadths. Measurements were collected from the right side whenever possible; if the right side element was missing or damaged, measurements from the left side were used. All of

Table 13.1 Groups and sample sizes used in the analysis

Group	Females[a]	Males[a]	Notes
SW Asian Neanderthals	3	6	Males: Amud 1, Kebara 2, Shanidar 1, 3, 4, and 5; females: Tabun C1, Shanidar 6 and 8
European Neanderthals	2	8	Males: La Ferrassie 1, La Chapelle-aux-Saints, Le Régourdou, Spy 1 and 2, Fond-de-Forêt, Feldhofer (Neanderthal) 1, and Kiik Koba; females: La Ferrassie 2 and La Quina 5
Skhul-Qafzeh	3	5	Males: Skhul III, IV, V, IX, and Qafzeh 8; females: Skhul II, VII, and Qafzeh 9
Australian Aborigines	8	18	See Pearson (1997, 2000) for details
Buriat	2	1	From the collection of the NMNH[b]
Chinese	0	28	From the collection of the NMNH[b]; Individuals buried at the Karluk cannery, Alaska
Epigravettian Italians	0	7	Primarily from Arene Candide; see Pearson (1997, 2000) for details
Gravettian	5	17	See Pearson (1997, 2000) for details
Inuit	25	62	From the collections of the NMNH and AMNH[b]. See Pearson (1997, 2000) and Pearson et al. (2006) for details
Jebel Sahaba (Sudan)	13	18	See Pearson (1997, 2000) for details
Kebaran	0	1	Ohalo 2
Khoesan	25	37	See Pearson (1997, 2000) for details
Magdalenian	2	6	See Pearson (1997, 2000) for details
Maori	1	1	From the collection of the AMNH[b]
Mesolithic Europeans	13	16	See Pearson (1997, 2000) and Pearson et al. (2006) for details
Pygmy	2	2	From the collections of the IRSNB[b]
Sami	25	34	From the collection of the SMUO[b]; many of these skeletons have been returned subsequently to Sami communities
Tierra del Fuego	9	21	From various institutions in Argentina, Chile, and from the literature. See Pearson and Millones (2005) for details
African American	31	41	See Pearson (1997, 2000) for details
European American	25	25	See Pearson (1997, 2000) for details
Zulu	31	31	From the Dart Collection. See Pearson (1997, 2000) for details
Wadi Kubbaniya	0	1	Plaster casts from the collection of the NMNH[b]

[a]Maximum number; missing data result in smaller sample sizes in many analyses
[b]Abbreviations for collections: AMNH, American Museum of Natural History; IRSNB, Institute Royale des Sciences Naturelles de Belgique; NMNH, National Museum of Natural History (Smithsonian Institution); SMUO, School of Medicine, University of Oslo

the measurements on fossil specimens is presented in Pearson (1997). Due to the way in which second moments of area are calculated, the external dimensions of diaphyses are guaranteed to be strongly correlated with second moments of area (Jungers and Minns 1979; Pearson et al. 2006; Stock and Shaw 2007). Indeed, although variation in the percentage of cortical bone in a section has received a substantial amount of attention in the literature, the external contour of a section is so strongly predictive of second moments of area that non-pathological variation in the percentage of cortical bone present has comparatively little

influence on the resultant second moments of area (Sparacello and Pearson 2010).

Given the close association between external dimensions and second moments of area, we have chosen to predict two cross-sectional properties (torsional second moment of area [J] and the ratio of maximum to minimum second moment of area [I_{max}/I_{min}]) of long bone sections from external dimensions using a set of formulae (Table 13.2) developed by Petersen et al. (in preparation) from a dataset that combines actual measurements of cross-sectional geometry with external dimensions of long bones for humeri, radii, ulnae,

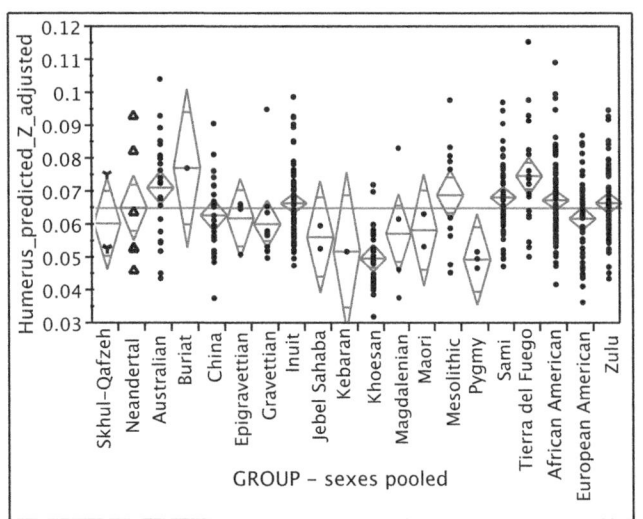 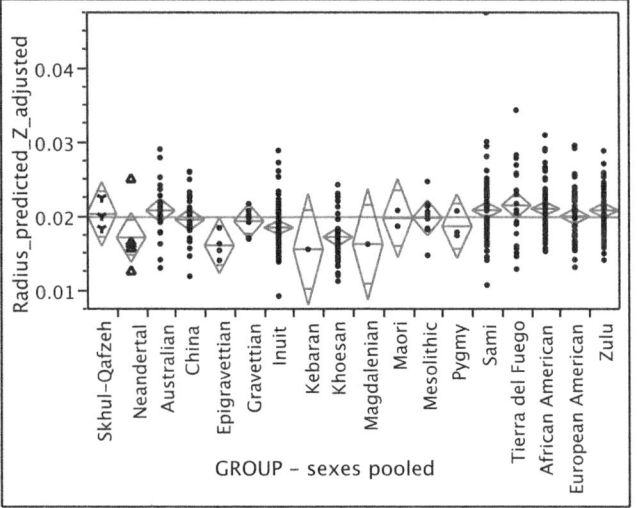

Fig. 13.1 Predicted values for the section modulus (Z) adjusted for body size (mass and beam length) in the midshaft section of the humerus and radius. Note the Neanderthals' relatively weak radii. See text for meaning of the graph symbols

Table 13.2 Prediction equations for selected cross-sectional geometric properties based on external dimensions of long bone midshafts

Property	Bone	Formula	R^2	SEE
J	Humerus	$J = 0.3515 \times MaxD^{2.4145} \times MinD^{1.0992}$	0.92	2104
J	Radius	$J = 0.2617 \times MaxD^{1.3472} \times MinD^{2.2418}$	0.92	364.7
J	Femur[a]	$J = 0.0605 \times MidAP^{1.9274} \times MidML^{2.1571}$	0.91	5736
J	Tibia	$J = 0.2244 \times MidAP^{2.3519} \times MidML^{1.2892}$	0.91	4269
I_{max}/I_{min}	Humerus	$I_{max}/I_{min} = 1.0136 \times MaxD^{1.7550} \times MinD^{-1.7521}$	0.76	0.112
I_{max}/I_{min}	Radius	$I_{max}/I_{min} = 1.1951 \times MaxD^{1.4322} \times MinD^{-1.4960}$	0.63	0.163
I_{max}/I_{min}	Femur	$I_{max}/I_{min} = 0.4830 \times MidAP^{1.4327} \times MidML^{-1.1564}$	0.51	0.184
I_{max}/I_{min}	Tibia	$I_{max}/I_{min} = 0.6744 \times MidAP^{1.9506} \times MidML^{-1.7741}$	0.76	0.231

Abbreviations are as follows: J, torsional second moment of area; I_{max}/I_{min}, ratio of maximum to minimum second moment of area; MaxD, maximum midshaft diameter; MinD, minimum midshaft diameter; MidAP, midshaft antero-pasterior (AP) diameter; MidML, misdshaft medio-lateral (ML) diameter

[a]The wrong equation for femoral J presented in Pearson et al. (2006); the correct formula should have been $J = 10^{-0.5227} \times MaxD^{2.4867} \times MinD^{1.0775}$ which differs from (and is less accurate than) the equation presented here. We thank Dr. Daniel Wescott for alerting us to the problem

femora, and tibiae of African Americans, Zulu, and prehistoric Khoesan. Data on these populations and details of the CT scans have been presented elsewhere (Grine et al. 1995; Churchill et al. 1996; Pearson and Grine 1996, 1997). Inspection of the R^2 values of the prediction formulae in Table 13.2 show a very predictable relationship between external diameters and second moments of area (with R^2 between 0.91 to 0.92), but the predicted ratios of I_{max}/I_{min} fare less well, with R^2 values between 0.51 and 0.76). This degree of predictability for I_{max}/I_{min} means that results obtained from those comparisons are less reliable and could conceivably be overturned by studies based on actual values obtained from CT scans or external contours of the bones.

To conform to the current best practice for size adjustment, we transformed the predicted values of J into section moduli (Z_p) by dividing each by one-half the larger of the two diameters used to predict J and then divided Z_p by the product of the bone's length and the body mass predicted for that individual using the predictive formula published by Grine et al. (1995) (Table 13.3). Auerbach and Ruff (2004) reported that this predictive formula tends to overestimate slightly the mass of very small people (e.g., African pygmies

and perhaps the Khoesan), but the predictions should be reasonable for Middle and Upper Paleolithic humans.

In the results and tables of data that follow, patterns are described for each sex separately using the sex assignments for Middle Paleolithic specimens listed in Table 13.1. However, the sex of many of the Middle Paleolithic individuals are, in fact, debatable. One reviewer of this chapter recommended pooling the sexes for each Middle Paleolithic sample. Figures 13.1–13.4 follow this recommendation, pooling the Southwest Asian and European Neanderthals into a single sample, and also pool all the sexes within each comparative sample. Thus the figures present a simplified and condensed version of the information presented in the text and tables.

All the plots were generated using JMP 6.0 (SAS Institute, Cary, NC), which places a diamond around the mean of each sample. The horizontal line in the center of each diamond indicates the mean of the sample the vertical axis of the diamond indicates the 95% confidence interval for the mean. Samples whose means diamonds do not overlap on the vertical (y) axis have a statistically significant difference ($p < 0.05$) between their means.

Table 13.3 Summary statistics for predicted body mass[a] (in kg) for the samples

Group	Females Mean ± SD (n)	Males Mean ± SD (n)
SW Asian Neanderthals	61 ± – (1)	73.7 ± 4.3 (4)
European Neanderthals	65.7 ± 3.7 (2)	81.0 ± 4.3 (6)
Skhul-Qafzeh	60.4 ± 4.2 (2)	74.8 ± 7.5 (4)
Australian Aborigines	49.7 ± 4.1 (6)	62.1 ± 4.4 (18)
Buriat	60.8 ± 5.5 (2)	66.5 ± – (1)
Chinese	–	64.9 ± 4.7 (28)
Epigravettian Italians	–	72.5 ± 2.6 (5)
Gravettian	66.2 ± 5.3 (4)	71.9 ± 9.4 (10)
Inuit	57.9 ± 3.7 (25)	70.0 ± 4.6 (62)
Jebel Sahaba (Sudan)	52.2 ± 6.7 (4)	69.2 ± 4.8 (8)
Kebaran	–	74.4 ± – (1)
Khoesan	47.8 ± 5.3 (22)	55.9 ± 4.8 (28)
Magdalenian	–	72.1 ± – (1)
Maori	63.5 ± – (1)	77.1 ± – (1)
Mesolithic Europeans	59.0 ± 3.7 (10)	72.3 ± 7.2 (11)
Pygmy	43.7 ± 5.3 (2)	44.2 ± – (1)
Sami	56.8 ± 5.0 (25)	70.1 ± 5.5 (33)
Tierra del Fuego	59.7 ± 0.5 (6)	68.8 ± 5.1 (15)
African American	57.6 ± 5.3 (31)	70.8 ± 6.7 (41)
European American	59.8 ± 6.3 (25)	74.4 ± 6.6 (25)
Zulu	54.7 ± 4.6 (31)	66.2 ± 5.7 (31)
Wadi Kubbaniya	–	–

[a]Estimated from vertical diameter of the femoral head using the prediction equation published by Grine et al. (1995). To maximize the sample sizes for Near Easter Neanderthals and the Skhul Qafzeh individuals, we first estimated the vertical diameter of the femoral head from other, preserved dimensions of epiphyses in the following cases. For Kebara 2, we estimated femoral head diameter from acetabulum diameter and vertical diameter of the humeral head (Pearson et al. 2008), obtaining an estimate of 46.71 mm, which produced an estimate of mass of 69.4 kg. For Qafzeh 9 and Skhul IX, we set the vertical diameter of the femoral head to be equal to its horizontal diameter (44.7 and 49 mm, respectively), producing estimates of mass of 64.9 and 74.6 kg, respectively. For Qafzeh 8, we estimated the vertical diameter of the femoral head from the vertical diameter of the humeral head (56.2 mm), obtaining an estimate of 53.8 mm for the femoral head and an inferred mass of 85.5 kg. For Tabun C1 and Skhul V, we used estimates of mass published by Trinkaus and Ruff (2012)

Results

The results for the humerus and radius (Tables 13.4 and 13.5, Figs. 13.1 and 13.2) show that Neanderthal males have unremarkable – or even slightly low – size-adjusted radial strength. Southwest Asian Neanderthal males, however, have very high values for size-adjusted humeral strength while European Neanderthals males are average for this variable and European Neanderthal females have relatively weak size-adjusted humeri. The Skhul-Qafzeh males have a

Table 13.4 Predicted values for the cross-sectional geometry of the radial and humeral midshafts for males

Group	Humerus		Radius	
	Size-adjusted Z_p	I_{max}/I_{min}	Size-adjusted Z_p	I_{max}/I_{min}
	Mean ± SD (n)	Mean ± SD (n)	Mean ± SD (n)	Mean ± SD (n)
SW Asian Neanderthals	0.0872 ± 0.0045 (2)	1.7682 ± 0.0754 (3)	0.0188 ± 0.0027 (3)	1.5359 ± 0.0624 (4)
European Neanderthals	0.0577 ± 0.0045 (2)	1.8247 ± 0.0584 (5)	0.0164 ± – (1)	1.5933 ± 0.0558 (5)
Skhul-Qafzeh	0.0524 ± 0.0045 (2)	1.3519 ± 0.0754 (3)	0.0203 ± 0.0033 (2)	1.2660 ± 0.0882 (2)
Australian Aborigines	0.0765 ± 0.0028 (17)	1.6476 ± 0.0521 (17)	0.0219 ± 0.0009 (17)	1.6075 ± 0.0518 (17)
Buriat	–	–	–	–
Chinese	0.0624 ± 0.0022 (27)	1.7048 ± 0.0413 (27)	0.0195 ± 0.0007 (28)	1.5268 ± 0.0404 (28)
Epigravettian Italians	0.0615 ± 0.0057 (4)	1.7651 ± 0.0876 (6)	0.01594 ± 0.0019 (4)	1.5884 ± 0.0807 (7)
Gravettian	0.0623 ± 0.0040 (8)	1.5374 ± 0.0647 (11)	0.0190 ± 0.0014 (7)	1.4926 ± 0.0676 (10)
Inuit	0.0699 ± 0.0017 (45)	1.7462 ± 0.0320 (45)	0.0189 ± 0.0006 (44)	1.7168 ± 0.0318 (45)
Jebel Sahaba (Sudan)	0.0523 ± – (1)	1.3500 ± 0.0620 (12)	–	–
Kebaran	0.0514 ± – (1)	1.3916 ± 0.1518 (2)	0.0154 ± – (1)	1.5494 ± 0.1510 (2)
Khoesan	0.0486 ± 0.0028 (16)	1.5528 ± 0.0438 (24)	0.0174 ± 0.0009 (19)	1.4528 ± 0.0427 (25)
Magdalenian	0.0568 ± 0.0057 (4)	1.6664 ± 0.1073 (4)	0.0161 ± – (1)	1.6362 ± 0.1233 (3)
Maori	0.0628 ± – (1)	1.7907 ± – (1)	0.0207 ± – (1)	1.7178 ± – (1)
Mesolithic Europeans	0.0797 ± 0.0057 (4)	1.6222 ± 0.0679 (10)	0.0208 ± 0.0019 (4)	1.5856 ± 0.0755 (8)
Pygmy	0.0463 ± – (1)	1.5193 ± 0.1518 (2)	0.0178 ± – (1)	1.6096 ± 0.1510 (2)
Sami	0.0718 ± 0.0020 (33)	1.6297 ± 0.0368 (34)	0.0206 ± 0.0007 (28)	1.7595 ± 0.0372 (33)
Tierra del Fuego	0.0798 ± 0.0031 (13)	1.7107 ± 0.0521 (17)	0.0232 ± 0.0010 (14)	1.4818 ± 0.0534 (16)
African American	0.0689 ± 0.0018 (41)	1.4800 ± 0.0335 (41)	0.02193 ± 0.0006 (41)	1.5836 ± 0.0334 (41)
European American	0.0662 ± 0.0023 (25)	1.5103 ± 0.0429 (25)	0.0209 ± 0.0007 (25)	1.6746 ± 0.0427 (25)
Zulu	0.0701 ± 0.0020 (31)	1.4524 ± 0.0386 (31)	0.0220 ± 0.0007 (31)	1.4863 ± 0.0384 (31)
Wadi Kubbaniya	–	1.4142 ± – (1)	–	1.2005 ± – (1)

Table 13.5 Predicted values for the cross-sectional geometry of the radial and humeral midshafts for females

Group	Humerus		Radius	
	Size-adjusted Z_p	I_{max}/I_{min}	Size-adjusted Z_p	I_{max}/I_{min}
	Mean ± SD (n)	Mean ± SD (n)	Mean ± SD (n)	Mean ± SD (n)
SW Asian Neanderthals	0.05225 ± – (1)	1.6248 ± – (1)	0.0124 ± – (1)	1.5343 ± 0.0720 (3)
European Neanderthals	0.0456 ± – (1)	1.8187 ± 0.0924 (2)	–	2.0833 ± – (1)
Skhul-Qafzeh	0.0748 ± – (1)	1.5145 ± 0.0754 (3)	0.0199 ± 0.0033 (2)	1.3444 ± 0.0882 (2)
Australian Aborigines	0.0545 ± 0.0046 (6)	1.8875 ± 0.0759 (8)	0.0173 ± 0.0015 (6)	1.6012 ± 0.0807 (7)
Buriat	0.0767 ± – (1)	1.6188 ± – (1)	–	2.0290 ± – (1)
Chinese	–	–	–	–
Epigravettian Italians	–	–	–	–
Gravettian	0.0529 ± 0.0065 (3)	1.5556 ± 0.1073 (4)	0.0199 ± 0.0021 (3)	1.6316 ± 0.1233 (3)
Inuit	0.0591 ± 0.0023 (25)	1.7691 ± 0.0429 (25)	0.0173 ± 0.0008 (23)	1.7180 ± 0.0427 (25)
Jebel Sahaba (Sudan)	0.0593 ± – (1)	1.5319 ± 0.0679 (10)	–	–
Kebaran	–	1.5033 ± – (1)	–	–
Khoesan	0.0499 ± 0.0027 (18)	1.5669 ± 0.0468 (21)	0.0167 ± 0.0009 (17)	1.5516 ± 0.0466 (21)
Magdalenian	–	1.3256 ± – (1)	–	–
Maori	0.0529 ± – (1)	1.3004 ± – (1)	0.0186 ± – (1)	1.9251 ± – (1)
Mesolithic Europeans	0.0611 ± 0.0046 (6)	1.7632 ± 0.0715 (9)	0.0189 ± 0.0015	1.5212 ± 0.0872 (6)
Pygmy	0.0503 ± 0.0080 (2)	2.0259 ± 0.1518 (2)	0.0190 ± 0.0026 (2)	1.4914 ± 0.1510 (2)
Sami	0.0626 ± 0.0023 (25)	1.6878 ± 0.0429 (25)	0.0208 ± 0.0008 (24)	1.8617 ± 0.0427 (25)
Tierra del Fuego	0.0600 ± 0.0051 (5)	1.8820 ± 0.0759 (8)	0.0164 ± 0.0017 (5)	1.6607 ± 0.0755 (8)
African American	0.0644 ± 0.0020 (31)	1.5436 ± 0.0386 (31)	0.0196 ± 0.0007 (31)	1.6956 ± 0.0384
European American	0.0566 ± 0.0023 (25)	1.6606 ± 0.0429 (25)	0.0188 ± 0.0007 (25)	1.7988 ± 0.0427 (25)
Zulu	0.0620 ± 0.0020 (31)	1.5790 ± 0.0386 (31)	0.0194 ± 0.0007 (30)	1.4977 ± 0.0390 (30)

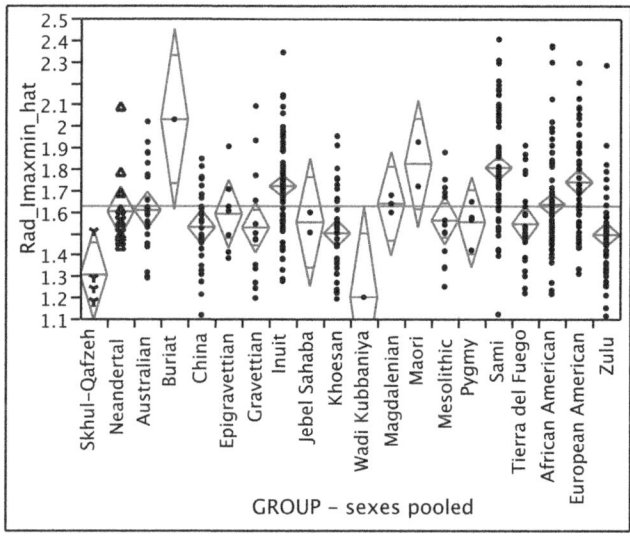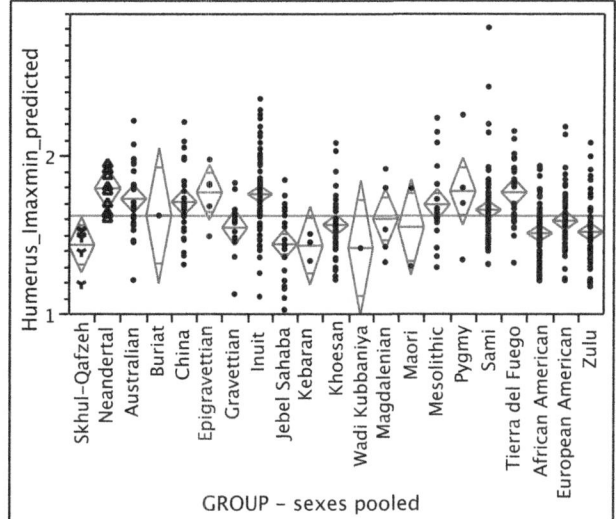

Fig. 13.2 Predicted values for I_{max}/I_{min} in the midshaft of the radius (rad) and humerus. The Skhul-Qafzeh individuals differ from Neanderthals in both measures of shape and tend to have rounder sections. See text for meaning of the graph symbols

low average humeral Z_p, which resembles the sole Kebaran male (Ohalo 2), Magdalenian males, the single sufficiently well preserved male from Jebel Sahaba, and European Neanderthal males. The sole Skhul-Qafzeh female that was complete enough to assess (Qafzeh 9) had high values for size-adjusted humeral strength, and is approached most closely by the sole Buriat female, Sami females, and then by women from a series of industrial or farming populations.

For midshaft radial shape (I_{max}/I_{min}), La Ferrassie 2, the sole European Neanderthal female for whom the index could be calculated, shows a very high value (which indicates considerable development of the interosseous crest), while the other subsamples of Neanderthals have approximately average values (Fig. 13.2). In contrast, the hominins from Skhul and Qafzeh have strikingly low values of radial I_{max}/I_{min}. Other samples with low values for this index include Kebaran males (Ohalo 2), Khoesan males, and Gravettian males. The sole Skhul-Qafzeh female for whom radial I_{max}/I_{min} was estimated has a high value that is approached most closely by the sole Buriat female, the single Maori female, and the mean for Sami females.

In the lower limb (Tables 13.6 and 13.7, Figs. 13.3 and 13.4), Southwest Asian Neanderthal males have a very high mean vale for size-adjusted femoral Z_p and are most closely approximated by the sole Buriat male followed by Epigravettian Italian males as well as Magdalenian and Tierra del Fuego males. European Neanderthal males as well as the Skhul-Qafzeh males are less striking in this regard and fall comfortably within most of the recent male samples. Values for size-adjusted tibial Z_p for European Neanderthals (no Southwest Asian Neanderthal tibiae could be included) and Skhul-Qafzeh males are high but exceeded by Epigravettian

Italian, Tierra del Fuego, and Magdalenian males. For females, the sole Southwest Asian Neanderthal (Tabun C1) for whom size-adjusted femoral Z_p could be calculated has a low value for the feature and falls almost exactly on the mean for European American females. In contrast, La Ferrassie 2, the single European Neanderthal female for whom size-adjusted femoral Z_p could be calculated, has a very high value and Qafzeh 9 (here assumed to be female following Vandermeersch (1981)) is even higher. Among more recent samples, lower but still moderately elevated average values for femoral Z_p occur among Buriat, Gravettian, and Magdalenian females as well as in the sole Maori woman. The values for female Southwest Asian and European Neanderthal females' tibial Z_p are only moderately elevated. No size-adjusted value for tibial Z_p could be inspected for Skhul-Qafzeh females.

Turning to the indices of femoral and tibial I_{max}/I_{min}, as previously observed by Trinkaus and Ruff (1999a, b, 2012), Neanderthals have low values for femoral I_{max}/I_{min}, and the value for the Southwest Asian Neanderthal males is higher. As might be expected from their high pilastric indices, the Skhul-Qafzeh males have a very high ratio of femoral I_{max}/I_{min} that is exceeded only by Ohalo 2 and approached by Khoesan males. In tibial I_{max}/I_{min}, Southwest Asian Neanderthal and Skhul-Qafzeh males have a low value while European Neanderthals have a moderately elevated mean that is exceeded only by Epigravettian Italian, Maori, Magdalenian, Gravettian, and Khoesan males. Among the females, both Neanderthal samples have low values for femoral I_{max}/I_{min} that are most closely approximated by Sami, Mesolithic European, Magdalenian, and Gravettian females. In contrast, the sole Skhul-Qafzeh female

Table 13.6 Predicted values for the cross-sectional geometry of the femoral and tibial midshafts for males

Group	Femur		Tibia	
	Size-adjusted Z_p	I_{max}/I_{min}	Size-adjusted Z_p	I_{max}/I_{min}
	Mean ± SD (n)	Mean ± SD (n)	Mean ± SD (n)	Mean ± SD (n)
SW Asian Neanderthals	0.1488 ± 0.0115 (3)	1.5109 ± 0.1207 (3)	–	1.9168 ± – (1)
European Neanderthals	0.1165 ± 0.0115 (3)	1.3410 ± 0.0935 (5)	0.1122 ± – (1)	2.6279 ± 0.1409 (4)
Skhul-Qafzeh	0.1127 ± 0.0141 (2)	1.7656 ± 0.1045 (4)	0.1151 ± – (1)	1.9107 ± 0.1993 (2)
Australian Aborigines	0.1039 ± 0.0044 (18)	1.4656 ± 0.0390 (18)	0.1054 ± 0.0038 (18)	2.3247 ± 0.0865 (18)
Buriat	0.1524 ± – (1)	1.1486 ± – (1)	0.1015 ± – (1)	2.1734 ± – (1)
Chinese	0.0993 ± 0.0035 (28)	1.3384 ± 0.0313 (28)	0.0939 ± 0.0032 (26)	2.1351 ± 0.0719 (26)
Epigravettian Italians	0.1289 ± 0.0083 (5)	1.5138 ± 0.0740 (5)	0.1264 ± 0.0093 (3)	3.2188 ± 0.1834 (4)
Gravettian	0.1053 ± 0.0065 (8)	1.5044 ± 0.0442 (14)	0.1084 ± 0.0081 (4)	2.8634 ± 0.1059 (12)
Inuit	0.1171 ± 0.0027 (47)	1.4494 ± 0.0241 (47)	0.1030 ± 0.0024 (47)	2.6067 ± 0.0535 (47)
Jebel Sahaba (Sudan)	0.1003 ± – (1)	1.5158 ± 0.0390 (18)	0.0983 ± 0.0114 (2)	2.4354 ± 0.0890 (17)
Kebaran	0.1128 ± – (1)	1.8439 ± 0.1169 (2)	0.1034 ± – (1)	2.5821 ± 0.2594 (2)
Khoesan	0.1021 ± 0.0036 (27)	1.6039 ± 0.0297 (31)	0.1079 ± 0.0037 (19)	2.6994 ± 0.0765 (23)
Magdalenian	0.1225 ± 0.0107 (3)	1.5148 ± 0.0827 (4)	0.1135 ± 0.0093 (3)	3.0064 ± 0.2118 (3)
Maori	0.0975 ± – (1)	1.4838 ± – (1)	0.0967 ± – (1)	3.1000 ± – (1)
Mesolithic Europeans	0.1199 ± 0.0075 (6)	1.3786 ± 0.0523 (10)	0.1187 ± 0.0066 (6)	2.4751 ± 0.1160 (10)
Pygmy	0.0672 ± – (1)	1.2906 ± 0.1169 (2)	0.1400 ± – (1)	2.3491 ± 0.2594 (2)
Sami	0.1032 ± 0.0033 (32)	1.2759 ± 0.0288 (33)	0.0896 ± 0.0029 (31)	2.0752 ± 0.0648 (32)
Tierra del Fuego	0.1211 ± 0.0048 (15)	1.4011 ± 0.0413 (16)	0.1263 ± 0.0051 (10)	2.2863 ± 0.0890 (17)
African American	0.1124 ± 0.0029 (41)	1.3112 ± 0.0258 (41)	0.0981 ± 0.0025 (41)	2.1423 ± 0.0573 (41)
European American	0.1079 ± 0.0037 (25)	1.3339 ± 0.0331 (25)	0.0980 ± 0.0033 (24)	2.1773 ± 0.0749 (24)
Zulu	0.1128 ± 0.0033 (31)	1.3929 ± 0.0297 (31)	0.1112 ± 0.0029 (31)	2.2657 ± 0.0659 (31)

Table 13.7 Predicted values for the cross-sectional geometry of the femoral and tibial midshafts for females

Group	Femur		Tibia	
	Size-adjusted Z_p	I_{max}/I_{min}	Size-adjusted Z_p	I_{max}/I_{min}
	Mean ± SD (n)	Mean ± SD (n)	Mean ± SD (n)	Mean ± SD (n)
SW Asian Neanderthals	0.1077 ± – (1)	1.1419 ± 0.1478 (2)	0.0997 ± – (1)	2.0013 ± – (1)
European Neanderthals	0.1424 ± – (1)	1.1711 ± – (1)	0.1059 ± – (1)	2.0513 ± – (1)
Skhul-Qafzeh	0.1719 ± – (1)	1.5154 ± 0.1478 (2)	–	–
Australian Aborigines	0.0961 ± 0.0075 (6)	1.3242 ± 0.0675 (6)	0.1029 ± 0.0066 (6)	2.1149 ± 0.1497 (6)
Buriat	0.1242 ± 0.0131 (2)	1.3020 ± 0.1169 (2)	0.1128 ± – (1)	1.9894 ± – (1)
Chinese	–	–	–	–
Epigravettian Italians	–	–	–	–
Gravettian	0.1222 ± 0.0107 (3)	1.2180 ± 0.0955 (3)	0.1231 ± – (1)	2.6545 ± 0.1834 (4)
Inuit	0.1107 ± 0.0037 (25)	1.2997 ± 0.0331 (25)	0.0917 ± 0.0032 (25)	2.4276 ± 0.0734 (25)
Jebel Sahaba (Sudan)	0.1121 ± – (1)	1.3196 ± 0.0477 (12)	–	2.1718 ± 0.1059 (12)
Kebaran	–	1.3699 ± – (1)	–	–
Khoesan	0.0997 ± 0.0041 (20)	1.3916 ± 0.0345 (23)	0.0976 ± 0.0037 (19)	2.2941 ± 0.0841 (19)
Magdalenian	0.1225 ± 0.0107 (3)	1.1839 ± – (1)	–	–
Maori	0.1217 ± – (1)	1.4114 ± – (1)	0.1371 ± – (1)	3.0046 ± – (1)
Mesolithic Europeans	0.1083 ± 0.0070 (7)	1.2219 ± 0.0585 (8)	0.0884 ± 0.0081 (4)	2.9192 ± 0.1497 (6)
Pygmy	0.1040 ± – (1)	1.3717 ± 0.1169 (2)	0.1027 ± 0.0114 (2)	2.1486 ± 0.2594 (2)
Sami	0.1008 ± 0.0037 (25)	1.1957 ± 0.0331 (25)	0.0913 ± 0.0033 (24)	1.9837 ± 0.0749 (24)
Tierra del Fuego	0.1169 ± 0.0075 (6)	1.2419 ± 0.0675 (6)	0.1008 ± 0.0066 (6)	2.3886 ± 0.1297 (8)
African American	0.1124 ± 0.0029 (41)	1.3866 ± 0.0297 (31)	0.0966 ± 0.0029 (31)	2.0445 ± 0.0659 (31)
European American	0.1079 ± 0.0037 (25)	1.3303 ± 0.0331 (25)	0.0891 ± 0.0032 (25)	1.9896 ± 0.0734 (25)
Zulu	0.1103 ± 0.0033 (31)	1.4024 ± 0.0297 (31)	0.1074 ± 0.0029 (31)	2.1533 ± 0.0659 (31)

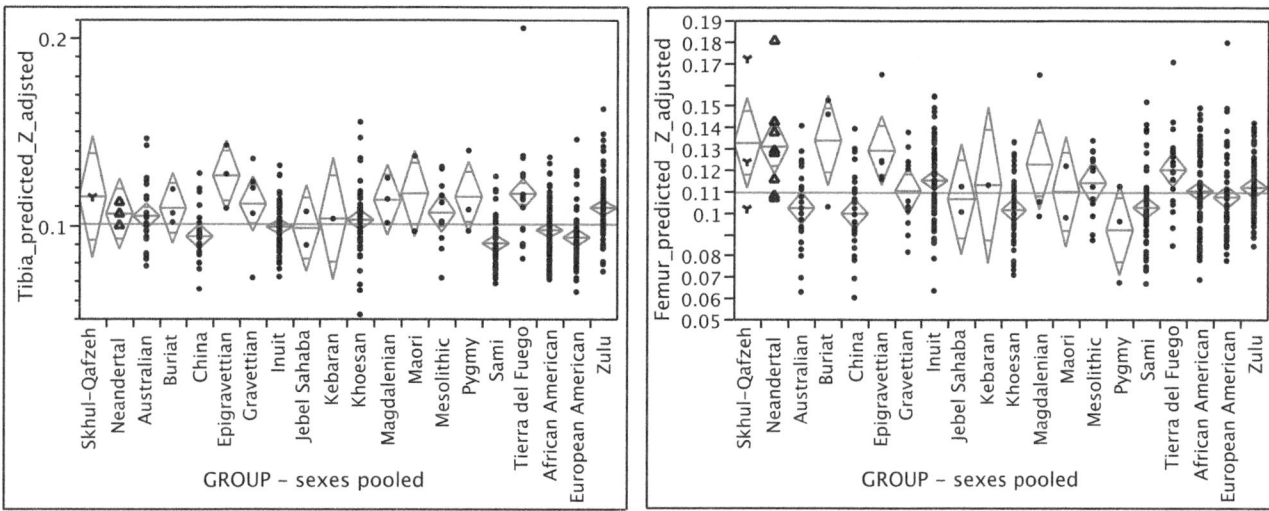

Fig. 13.3 Predicted values for the section modulus (Z) adjusted for body size (mass and beam length) in the midshaft section of the tibia and femur. Neanderthals and the Skhul-Qafzeh individuals share strong lower limbs relative to body size. See text for meaning of the graph symbols

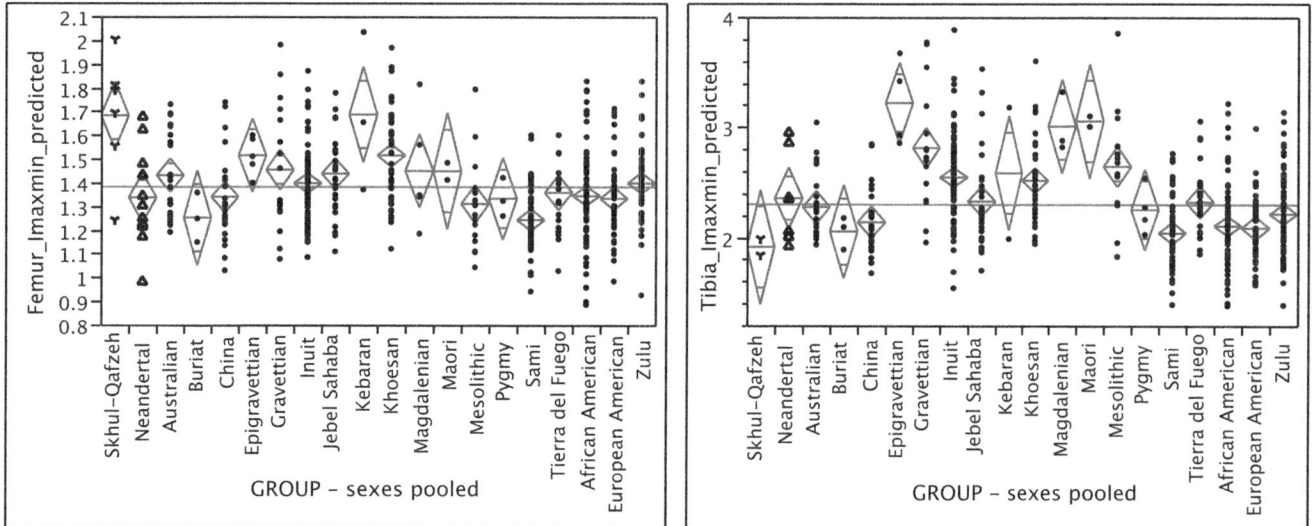

Fig. 13.4 Predicted values for I_{max}/I_{min} in the midshaft section of the femur and tibia. The Skhul-Qafzeh sample has a very high mean for femoral I_{max}/I_{min} but curiously low values for tibial I_{max}/I_{min}. The contrast between Skhul-Qafzeh and Neanderthals in the femur suggests that some of behavioral difference existed, but the exact nature of that difference remains unclear. See text for meaning of the graph symbols

(Qafzeh 9) has a very high value that surpasses all of the other female means but is approached the closest by Zulu, the sole Maori, and Khoesan females. No Skhul-Qafzeh female could be assessed for tibial I_{max}/I_{min}, but both female Neanderthal samples have low values for the index that resemble the averages for European American, African American, Zulu, Sami, Pygmy, Buriat, and Australian Aboriginal females. Some but not all of those groups are supposedly sedentary and women of the "mobile" groups were likely less mobile than their men (Ruff 1987).

Discussion and Conclusions

With the exception of the data for the radius that we report, most of the results for the shape (I_{max}/I_{min}) of the long bones reported here are already well known for Neandertals, as are the general patterns of femoral and tibial size-adjusted strength (Trinkaus and Ruff 2012) and many of the conclusions about the size-adjusted strength of the humerus (Trinkaus and Churchill 1999). The results also largely mirror the findings based on indices of robusticity based on external measurements that we reported in 2006 (Pearson et al. 2006). In the lower limb, European and Southwest Asian Neandertals tend to resemble groups who may have been active in a general sense but were not highly mobile over long distances. The sample of Mesolithic Europeans, which is numerically dominated by individuals from coastal Brittany and coastal Denmark, and the sample from Tierra del Fuego provide some of the best analogs for Neandertals' lower limbs. The Skhul-Qafzeh samples (both males and females) have extraordinarily high values for femoral I_{max}/I_{min}, which are most closely approximated by Kebara, Khoesan, Gravettian, and Southwest Asian Neandertal males. Skhul-Qafzeh individuals also tend to have moderate values for tibial I_{max}/I_{min}. Southwest Asian Neandertals follow the same pattern, but all other samples of modern humans that have high values for femoral I_{max}/I_{min} have moderately to notably high means for tibial I_{max}/I_{min}.

In the upper limb, Neandertals tend to have high values of I_{max}/I_{min} in their humeri and radii coupled with markedly stronger humeri than radii once the influence of size is controlled. The disparity between the size-adjusted strength of the humerus and radius is a puzzle: there is no clear analog for it among recent humans and it may well be a species-level characteristic of Neandertals. The other feature of Neandertals that is remarkable is their strong degree of flattening of the humeral midshaft (i.e., high values of humeral I_{max}/I_{min}). This feature has long been recognized (Trinkaus 1983a) and characterizes a number of other populations as well (Tables 13.4 and 13.5). It may well develop from some kind (or kinds) of intensive physical activity. Recently, Shaw and colleagues (2012) used electromyography to test whether more and higher muscle activation, especially in the dominant (right) arm were higher in spear-thrusting or scraping activities and found that scraping rather than spear-thrusting was more likely to exert a plausible set of stresses generated by muscle contractions that could account for the shape of Neandertal humeri and their strong bilateral asymmetry in strength. This provides one more piece of evidence that helps make sense of the similarities between Neandertals and intensive foragers or others who engage in intensive activities with their upper limb.

The Skhul-Qafzeh hominins' upper limbs resemble most closely those of Khoesan and Kebaran foragers, especially in ratios of humeral and radial I_{max}/I_{min}. This similarity suggests that each of these groups had patterns of physical activity that did not place high or perhaps frequent (or both) mechanical demands on their upper limb. The lifeways of Neandertals and the Skhul-Qafzeh humans thus seem to have contrasted markedly, and other dissimilarities in aspects of their appendicular morphology corroborate this view (Trinkaus 1984, 1992, 1993). One must be cautious in assuming that these differences reflect fixed, species-level

differences given how much variation exists among ethnographically documented foragers (Kelly 1995; Marlowe 2005) and the fact that modern humans in Israel went through their own transition from more mobile lifeways with seasonal mobility in the Upper Paleolithic to a more sedentary by the Natufian period (Lieberman 1993). Hovers (2001, 2006; Hovers and Belfer-Cohen 2013) has argued that variation in lithic production in the Middle Paleolithic of the Levant likely records fluctuations in the intensity of hominin exploitation of the environments around the sites, and that these archaeological patterns form a continuum of responses rather than a dichotomy in behavior between sites with Tabun C industries (which include the levels that contain the early modern humans from Skhul and Qafzeh) and Tabun B industries (which are associated with Neandertal remains (Bar-Yosef 2000)).

Acknowledgments It is a great privilege to participate in this volume in honor of Professor Yoel Rak. Paleoanthropology always feels more exciting, more vivid, and more fun whenever Yoel is involved. His passion for the subject is perhaps only matched by his artist's eye for morphology and his amazing kindness and generosity to colleagues as well as to students just beginning their careers. One of OMP's happiest memories of working on the 1992 field season at Hayonim Cave came when, during our visit to his office in Tel Aviv, Yoel brought out the skeleton of Kebara 2 for a small crowd of excited graduate students. It was OMP's first look at a real Neanderthal skeleton and many of the impressions of the skeleton's morphology that flooded into OMP's mind are emblazoned there permanently, associated with a cozy feeling of happiness and excitement.Funds to collect the data were provided by the National Science Foundation, the Wenner-Gren Foundation, the Boise Fund, John C. Pearson, and the University of New Mexico. Lastly, many curators at a large number of collections as well as the heirs of Max Lohest graciously allowed access to the fossils and recent skeletons in their care. Comments from two anonymous reviewers and Professor Erella Hovers helped to improve the manuscript. We are very grateful to them all.

References

Arensburg, B., Bar-Yosef, O., Chech, M., Goldberg, P., Laville, H., Meignen, L., et al. (1985). Une sepulture Neanderthalienne dans la grotte de Kebara (Israel). *Comptes Rendue de l'Académie des Sciences Paris, 300,* 227–230.

Auerbach, B. M., & Ruff, C. B. (2004). Human body mass estimation: A comparison of morphometric and mechanical methods. *American Journal of Physical Anthropology, 125,* 331–342.

Bar-Yosef, O. (2000). The Middle and early Upper Paleolithic in Southwest Asia and neighboring regions. In O. Bar-Yosef & D. Pilbeam (Eds.), *The geography of Neanderthals and modern humans in Europe and the greater Mediterranean (Peabody Museum Bulletin 8)* (pp. 107–156). Cambridge, MA: Peabody Museum of Archaeology and Ethnology, Harvard University.

Bar-Yosef, O., & Vandermeersch, B. (Eds.). (1991). *Le Squelette Moustérien de Kébara 2 (Cahiers de Paléoanthropologie).* Paris: Éditions du CNRS.

Ben-Itzhak, S., Smith, P., & Bloom, R. A. (1988). Radiographic study of the humerus in Neanderthals and *Homo sapiens sapiens. American Journal of Physical Anthropology, 77,* 231–242.

Bridges, P. S. (1989). Changes in activities with the shift to agriculture in the Southeastern United States. *Current Anthropology, 30*, 385–394.

Carlson, K. J., Grine, F. E., & Pearson, O. M. (2007). Robusticity and sexual dimorphism in the postcranium of modern hunter-gatherers from Australia. *American Journal of Physical Anthropology, 134*, 9–23.

Churchill, S. E. (1994). *Human upper body evolution in the Eurasian later pleistocene.* Ph.D. Dissertation, The University of New Mexico at Albuquerque.

Churchill, S. E., & Formicola, V. (1997). A case of marked bilateral asymmetry in the upper limbs of an Upper Palaeolithic male from Barma Grande (Liguria), Italy. *International Journal of Osteoarchaeology, 7*, 18–38.

Churchill, S. E., Pearson, O. M., Grine, F. E., Trinkaus, E., & Holliday, T. W. (1996). Morphological affinities of the proximal ulna from Klasies River Mouth Main Site: Archaic or modern? *Journal of Human Evolution, 31*, 213–237.

Collier, S. (1989). The influence of economic behaviour and environment upon robusticity of the post-cranial skeleton: A comparison of Australian Aborigines and other populations. *Archaeology in Oceania, 24*, 17–30.

Endo, B., & Kimura, T. (1970). Postcranial skeleton of the Amud Man. In H. Suzuki & F. Takai (Eds.), *The Amud man and his cave site* (pp. 231–406). Tokyo: Academic Press of Japan.

Grine, F. E., Jungers, W. L., Tobias, P. V., & Pearson, O. M. (1995). Fossil *Homo* femur from Berg Aukas, northern Namibia. *American Journal of Physical Anthropology, 97*, 151–185.

Holt, B.M. (2003). Mobility in Upper Paleolithic and Mesolithic Europe: Evidence from the lower limb. *American Journal of Physical Anthropology, 122*, 200–215.

Hovers, E. (2001). Territorial behavior in the Middle Paleolithic of the southern Levant. In N. J. Conard (Ed.), *Settlement dynamics of the Middle Paleolithic and Middle Stone Age* (pp. 123–152). Tübingen: Kerns Verlag.

Hovers, E. (2006). Neanderthals and modern humans in the Middle Paleolithic of the Levant: What kind of interaction? In N. J. Conard (Ed.), *When Neanderthals and modern humans met* (pp. 65–75). Tübingen: Kerns Verlag.

Hovers, E., & Belfer-Cohen, A. (2013). On variability and complexity: Lessons from the Levantine Middle Paleolithic record. *Current Anthropology, 54*, S337–S357.

Jungers, W. L., & Minns, R. J. (1979). Computed tomography and biomechanical analysis of fossil long bones. *American Journal of Physical Anthropology, 50*, 285–290.

Kelly, R. L. (1995). *The foraging spectrum: Diversity in hunter-gatherer lifeways.* Washington DC: Smithsonian Institution Press.

Larsen, C. S. (1995). Biological changes in human populations with agriculture. *Annual Review of Anthropology, 24*, 185–213.

Lieberman, D. E. (1993). The rise and fall of seasonal mobility among hunter-gatherers: The case of the southern Levant. *Current Anthropology, 34*, 599–631.

Lieberman, D. E. (1998). Neanderthal and early modern human mobility patterns: Comparing archaeological and anatomical evidence. In T. Akazawa, K. Aoki, & O. Bar-Yosef (Eds.), *Neanderthals and modern humans in Western Asia* (pp. 263–275). New York: Plenum.

Lieberman, D. E., & Shea, J. J. (1994). Behavioral differences between archaic and modern humans in the Levantine Mousterian. *American Anthropologist, 96*, 300–332.

Lieberman, D. E., Devlin, M. J., & Pearson, O. M. (2001). Articular surface area responses to mechanical loading: Effects of exercise, age and skeletal location. *American Journal of Physical Anthropology, 116*, 266–277.

Marlowe, F. W. (2005). Hunter-gatherers and human evolution. *Evolutionary Anthropology, 14*, 54–67.

Martin, B. D., Burr, D. B., & Sharkey, N. A. (1998). *Skeletal tissue mechanics.* New York: Springer.

McCown, T. D., & Keith, A. (1939). *The Stone Age of Mount Carmel, II: The fossil human remains from the Levalloiso-Mousterian.* Oxford: Clarendon Press.

Pearson, O. M. (1997). Postcranial morphology and the origin of modern humans. Ph.D. Dissertation, Stony Brook University.

Pearson, O. M. (2000). Activity, climate, and postcranial robusticity: Implications for modern human origins and scenarios of adaptive change. *Current Anthropology, 41*, 569–607.

Pearson, O. M., & Grine, F. E. (1996). Morphology of the Border Cave hominid ulna and humerus. *South African Journal of Science, 92*, 231–236.

Pearson, O. M., & Grine, F. E. (1997). Re-analysis of the hominid radii from Cave of Hearths and Klasies River Mouth, South Africa. *Journal of Human Evolution, 32*, 577–592.

Pearson, O. M., & Millones, M. (2005). Rasgos esqueléticos de adaptación al clima y a la actividad entre los habitantes aborígenes de Tierra del Fuego (Skeletal traces of adaptation to climate and activity among the aboriginal inhabitants of Tierra del Fuego). *Magellania, 33*, 37–51.

Pearson, O. M., Cordero, R. M., & Busby, A. M. (2006). How different were Neanderthals' habitual activities? A comparative analysis with diverse groups of recent humans. In K. Harvati & T. Harrison (Eds.), *Neanderthals revisited: New approaches and perspectives* (pp. 89–112). New York: Springer.

Pearson, O.M., Grine, F. E., Fleagle, J. G., & Royer, D. F. (2008). A description of the Omo 1 postcranial skeleton, including newly discovered fossils. *Journal of Human Evolution, 55*, 421–437.

Pearson, O. M., Petersen, T. R., Sparacello, V. S., Daneshvari, S., & Grine, F. E. (2014). Activity, body shape, and cross-sectional geometry of the femur and tibia. In K. Carlson & D. Marchi (Eds.), *Mobility: Interpreting behavior from skeletal adaptations and environmental interactions* (pp. 133–151). New York: Springer.

Petersen, T. R., Pearson, O. M., & Grine, F. E. (in preparation). Prediction of long bone cross-sectional geometric properties from external dimensions.

Rak, Y. (1990). On the differences between two pelvises of Mousterian context from Qafzeh and Kebara Caves, Israel. *American Journal of Physical Anthropology, 81*, 323–332.

Rak, Y., & Arensburg, B. (1987). Kebara 2 Neanderthal pelvis: First look at a complete inlet. *American Journal of Physical Anthropology, 73*, 227–231.

Ruff, C. (1987). Sexual dimorphism in human lower limb bone structure: Relationship to subsistence strategy and sexual division of labor. *Journal of Human Evolution, 16*, 391–416.

Ruff, C. B. (1995). Biomechanics of the hip and birth in early *Homo*. *American Journal of Physical Anthropology, 98*, 527–574.

Ruff, C. B. (2000). Body size, body shape, and long bone strength in modern humans. *Journal of Human Evolution, 38*, 269–290.

Ruff, C. B., Trinkaus, E., Walker, A., & Larsen, C. S. (1993). Postcranial robusticity in *Homo*. I: Temporal trends and mechanical interpretation. *American Journal of Physical Anthropology, 91*, 21–53.

Ruff, C. B., Holt, B. M., Sládek, V., Berner, M., Murphy, Jr. W. A., zur Nedden, D., et al. (2005). Body size, body shape, and mobility of the Tyrolean "Iceman". *Journal of Human Evolution, 51*, 91–101.

Shaw, C. N., & Stock, J. T. (2011). The influence of body proportions on femoral and tibial midshaft shape in hunter-gatherers. *American Journal of Physical Anthropology, 144*, 22–29.

Shaw, C. N., & Stock, J. T. (2013). Extreme mobility in the Late Pleistocene? Comparing limb biomechanics among fossil *Homo*, varsity athletes and Holocene foragers. *Journal of Human Evolution, 64*, 242–249.

Shaw, C. N., Hofmann, C. L., Petraglia, M. D., Stock, J. T., & Gottschall, J. S. (2012). Neanderthal humeri may reflect adaptation to scraping tasks, but not spear thrusting. *PLoS ONE, 7*, e40349.

Shea, J. J. (1998). Neanderthal and early modern human behavioral variability: A regional-scale approach to lithic evidence for hunting in the Levantine Mousterian. *Current Anthropology, 39*, S45–S78.

Smith, P., Bloom, R. A., & Berkowitz, J. (1984). Diachronic trends in humeral cortical thickness of Near Eastern populations. *Journal of Human Evolution, 13*, 603–611.

Solecki, R. S. (1963). Prehistory of Shanidar valley, northern Iraq. *Science, 139*, 179–193.

Sparacello, V. S., & Pearson, O. M. (2010). The importance of accounting for the area of the medullary cavity in cross-sectional geometry: A test based on the femoral midshaft. *American Journal of Physical Anthropology, 143*, 612–624.

Stock, J. T., & Shaw, C. N. (2007). Which measures of diaphyseal robusticity are robust? A comparison of external methods of quantifying the strength of long bone diaphyses to cross-sectional geometric properties. *American Journal of Physical Anthropology, 134*, 412–423.

Trinkaus, E. (1983a). *The Shanidar Neanderthals*. New York: Academic Press.

Trinkaus, E. (1983b). Neanderthal postcrania and the adaptive shift to modern humans. In Trinkaus, E. (Ed.), *The Mousterian legacy: Human biocultural changes in the Upper Pleistocene* (pp. 165–200). (BAR International Series 164). Oxford: BAR.

Trinkaus, E. (1984). Western Asia. In F. H. Smith & F. Spencer (Eds.), *The origins of modern humans: A world survey of the fossil evidence* (pp. 251–293). New York: Alan R. Liss.

Trinkaus, E. (1992). Morphological contrasts between the Near Eastern Qafzeh-Skhul and late archaic human samples: Grounds for a behavioral difference? In T. Akazawa, K. Aoki, & T. Kimura (Eds.), *The evolution and dispersal of modern humans in Asia* (pp. 277–294). Tokyo: Hokusen-Sha Publishing Co.

Trinkaus, E. (1993). Femoral neck-shaft angles of the Qafzeh-Skhul early modern humans, and activity levels among immature Near Eastern Middle Paleolithic hominids. *Journal of Human Evolution, 25*, 393–416.

Trinkaus, E., & Churchill, S. E. (1999). Diaphyseal cross-sectional geometry of Near Eastern Middle Paleolithic humans: The humerus. *Journal of Archaeological Science, 26*, 173–184.

Trinkaus, E., & Ruff, C. B. (1999a). Diaphyseal cross-sectional geometry of Near Eastern Middle Paleolithic humans: The femur. *Journal of Archaeological Science, 26*, 409–424.

Trinkaus, E., & Ruff, C. B. (1999b). Diaphyseal cross-sectional geometry of Near Eastern Middle Palaeolithic humans: The tibia. *Journal of Archaeological Science, 26*, 1289–1300.

Trinkaus, E., & Ruff, C. B. (2012). Femoral and tibial diaphyseal cross-sectional geometry in Pleistocene *Homo. PaleoAnthropology, 2012*, 13–62.

Trinkaus, E., Churchill, S. E., Villemeur, I., Riley, K. G., Heller, J. A., & Ruff, C. B. (1991). Robusticity versus shape: The functional interpretation of Neandertal appendicular morphology. *Journal of the Anthropological Society of Nippon, 99*, 257–278.

Trinkaus, E., Churchill, S. E., & Ruff, C. B. (1994). Postcranial robusticity in *Homo*. II: Bilateral asymmetry and bone plasticity. *American Journal of Physical Anthropology, 93*, 1–34.

Vandermeersch, B. (1981). *Les hommes fossiles de Qafzeh (Israël)*. (Cahiers de Paléontologie). Paris: CNRS.

Chapter 14
The Acheulo-Yabrudian – Early Middle Paleolithic Sequence of Misliya Cave, Mount Carmel, Israel

Mina Weinstein-Evron and Yossi Zaidner

Abstract Misliya Cave, Mount Carmel, Israel was occupied between 250 and 160 ka. During this time the site was inhabited by bearers of the Acheulo-Yabrudian and Early Middle Paleolithic (Mousterian) techno-complexes. The Acheulo-Yabrudian industry is characterized by production of thick and wide flakes and shows no evidence of laminar or Levallois methods. The varied assemblage encompasses true bifaces, artifacts fully worked on one face and only partially on the other, unifaces and scrapers. All these morphological groups were produced using the same flaking and retouching modes. The emergence of the Early Middle Paleolithic is manifested by a technological break, marked by the disappearance of bifaces and thick-flake production technology and the introduction of blade manufacture using laminar and Levallois production methods, and Levallois points and triangular flakes. The mean TL ages of the Acheulo-Yabrudian assemblage indicate production of this cultural complex 257 ± 28 ka – 247 ± 24 ka. The mean TL ages of the Early Middle Paleolithic industries range from 212 ± 27 to 166 ± 23 ka. The pronounced differences in lithic technology together with TL chronology indicate that the transition from the Lower to the Middle Paleolithic in the Levant was rapid and may imply the arrival of a new population around 250 ka.

Keywords Acheulo-Yabrudian • Early Middle Paleolithic • Levant • Lower-Middle Paleolithic transition

M. Weinstein-Evron (✉) · Y. Zaidner
Zinman Institute of Archaeology, University of Haifa, Mount Carmel, 3498838 Haifa, Israel
e-mail: evron@research.haifa.ac.il

Y. Zaidner
e-mail: yzaidner@mail.huji.ac.il

Y. Zaidner
Institute of Archaeology, The Hebrew University of Jerusalem, Mt. Scopus, 91905 Jerusalem, Israel

Introduction

One of the hottest issues in current prehistoric and evolutionary research is that of the emergence of modern humans. While various scenarios have been postulated based on skeletal and cultural material dating to the later part of the MP (Middle Paleolithic), human remains from the Early Middle Paleolithic (EMP) and the very end of the Late Lower Paleolithic (LLP) are still rare and mostly amount to dental finds (e.g., Hershkovitz et al. 2011). Moreover, their taxonomic affiliation, whether to modern humans or other hominin' evolutionary paths is still not fully resolved.

Until additional, more indicative human remains are unearthed, emphasis has been given to various behavioral indicators of the different cultural phases that may constitute useful tools when aiming to characterize the holders of the various cultures. These are mostly derived from stone-tool typology and technology, spatial site arrangement and animal remains (e.g., Marks and Friedel 1977; Bar-Yosef 1998; Hovers 2001, 2006, 2009; Henry 2003; Shea 2003; Alperson-Afil and Hovers 2005; Meignen et al. 2006; Yeshurun et al. 2007).

In this discourse, sites that contain both Late Lower Paleolithic and EMP layers may prove most promising for delineating such cultural developments. Misliya Cave, Mount Carmel, is one such rare occasion. In this paper we present the cultural characteristics of the Late Lower Paleolithic (Acheulo-Yabrudian) and EMP cultural assemblages found at the site. The special attributes of the latter will be further highlighted against the picture emerging from the study of Late MP sites in the region.

The Site

Misliya Cave is located on the western slopes of Mount Carmel, slightly to the south of Nahal (Wadi) Sefunim, at an elevation of ca. 90 m, some 12 km south of Haifa

© Springer International Publishing AG 2017
Assaf Marom and Erella Hovers (eds.), *Human Paleontology and Prehistory*,
Vertebrate Paleobiology and Paleoanthropology, DOI: 10.1007/978-3-319-46646-0_14

(Fig. 14.1a). Situated ca. 7 km north of Nahal Me'arot (Wadi el-Mughara) and the caves of Tabun, Jamal, el-Wad and Shkul (Garrod and Bate 1937; McCown 1937; Jelinek et al. 1973; Weinstein-Evron and Tsatskin 1994; Zaidner et al. 2005) it was found to contain rich Middle Paleolithic (Mousterian) and Lower Paleolithic (Acheulo-Yabrudian) layers (Weinstein-Evron et al. 2003a; Zaidner et al. 2006).

Today the site appears as a rock shelter or an over-hang (Fig. 14.2) carved into the limestone cliff of the western escarpment of Mount Carmel. Several small caves (or niches)

extend eastward from the rock shelter and from the continuation of the cliff northward and southward (Weinstein-Evron et al. 2012). The morphologic features of the caves, the remnants of ancient, inactive, flowstones and the form of the central part of the rock shelter indicate that the overhang is a remnant of a large collapsed cave or cave system. Detached blocks of flowstone appear some 20 m west of the cliff within collapsed debris and cemented archaeological sediments. Th/U dating of one of these collapsed flowstone blocks shows that it is older than 650 ky (H. Schwarcz, personal

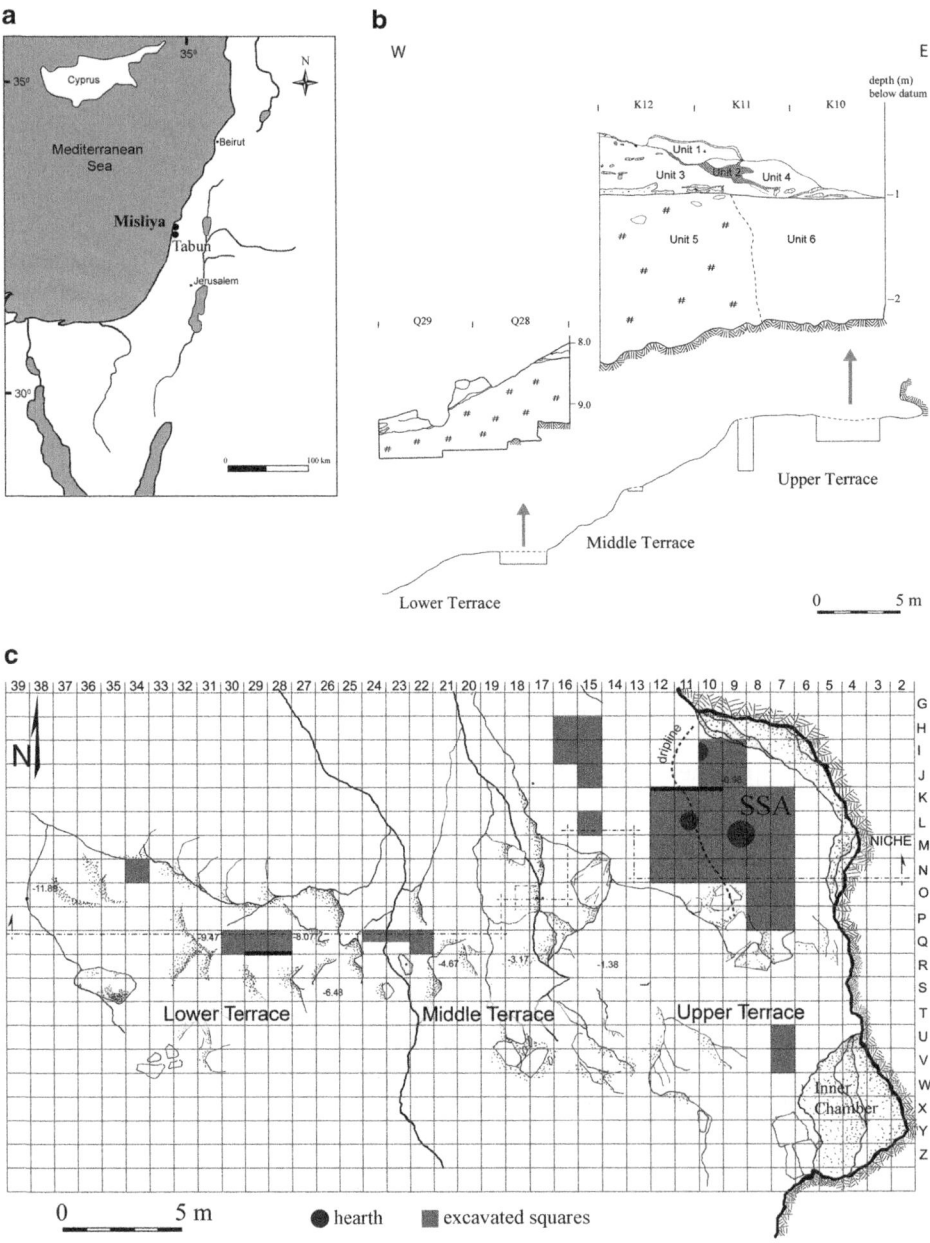

Fig. 14.1 **a**. Location map. **b**. Site section. **c** Site plan

communication 2003). While the date cannot be associated with the archaeological layers, it indicates that cave formation was already in process prior to that time.

Strongly cemented archaeological sediments (breccia) are found on three terrace-like surfaces at the base of the cliff, all sloping gently to the west (henceforth Upper, Middle, and Lower Terraces; see Fig. 14.1b, c; Fig. 14.2). Sub-vertical exposures between the terraces were formed in the course of natural collapse of the cave, and cementation and erosion of its deposits.

Cave collapse was gradual. Its very latest stages occurred during EMP times, as concluded from a detailed geo-archaeological study of a deep sequence in Square L15. During most of the Middle Paleolithic occupation, the Upper Terrace was still enclosed by the cave walls and covered by a roof; the last collapse occurred at the close of hominin habitation of the cave (Weinstein-Evron et al. 2012).

The archaeological excavations were conducted on all three terraces of the cave (Fig. 14.1c). On the Lower Terrace only Acheulo-Yabrudian artifacts were found *in situ*. A small area was excavated on the Middle Terrace of the site but the few unearthed artifacts are not diagnostic. On the Upper Terrace both Acheulo-Yabrudian and Mousterian finds were discovered. The Mousterian layers cover an area

of ca. 70 m^2 on the Upper Terrace. They occur mostly in its northern part, while bedrock is exposed on its southern part, apart from isolated breccia patches. According to a geophysical survey that was conducted at the site prior to excavation the thickness of the sediments on the northern part of the site is about 4 m (Weinstein-Evron et al. 2003b). This observation was validated during excavation with the unearthing of a 3.5 m-deep archaeological sequence in Square L15 on the western part of the Upper Terrace (Fig. 14.1c); the archaeological sediments become shallower towards its eastern part. In the north-eastern area of the Upper Terrace, cemented layers change laterally into softer sediments, forming an area of about 20 m^2, designated as the "Soft Sediments Area" (SSA; Fig. 14.1c). The limit between the lithified and softer sediments lies within the present-day dripline, with the SSA located below the roofed part of the cave. Lying above the natural bedrock, the soft sediments are quite shallow (1.5–2.5 m), apart from the northern part of the excavation near the wall of the cave (squares I9–10) where the layers are deeper (3–3.5 m).

The archaeological sequence of the SSA was divided into six stratigraphic units (Fig. 14.1b). Units 1 and 2 represent eroded surface breccia and later *terra-rossa* intrusion,

Fig. 14.2 Photo of the site, view from the north

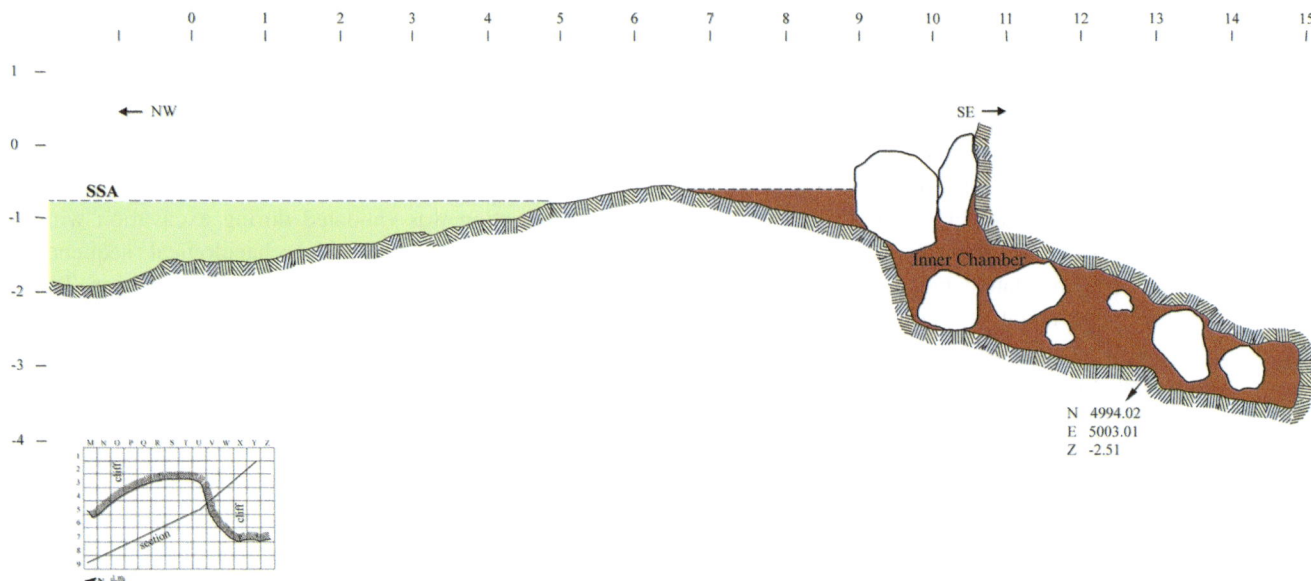

Fig. 14.3 NW-SE section along the Upper Terrace. Note the deep underground Inner Chamber on the SE part of the site

respectively. Units 3-4 represent well-preserved but rather residual MP habitation layers rich in lithics and faunal remains and containing combustion features. Units 3 and 5 that lie slightly to the west of the drip-line are somewhat more lithified than Units 4 and 6. Unit 6 is the richest and best preserved EMP unit of the site. At the bottom of Unit 6, a mixed Acheulo-Yabrudian/Mousterian unit was found in the easternmost squares (K-N/7−9). The unit occurs in the lower part of the SSA within sediments that accumulated in-between large rocks below the MP layers. The unit is ca. 10 cm thick and lays on a rock surface which either constitutes the bedrock or a huge collapsed rock shelf.

On the south-eastern corner of the Upper Terrace, a small underground cavity (henceforth the Inner Chamber) was discovered (Figs. 14.1c, 14.3). It measures ca. 5 × 5 m, with heights varying between 0.7 and 1.5 m, and is filled with mixed archaeological sediments. More than 10,000 artifacts and thousands of bone fragments were retrieved from the coarsely sieved deposits of the Inner Chamber. In spite of the great depth (at the southern end we have reached a depth of 3.40 m below datum) no *in situ* archaeological deposits were found to date. The chamber contains hundreds of Acheulo-Yabrudian handaxes and scrapers mixed with artifacts of Mousterian origin.

The Inner Chamber is the only place on the Upper Terrace where Lower Paleolithic finds occur in significant numbers. Since no *in situ* Lower Paleolithic material was detected on the Upper Terrace, the origin of handaxes and scrapers in the Inner Chamber remains an enigma. It may be postulated that the Upper Terrace originally contained rich Acheulo-Yabrudian layers that were eroded and washed into

the Inner Chamber by post-depositional processes. Since the highest topographical point of the Upper Terrace is located between the SSA and the Inner Chamber, creating a natural barrier that prevented mixture of sediments and finds from both areas (Fig. 14.3), it is clear that artifacts and bones found in the Inner Chamber could not have originated from the SSA. Therefore, it seems that during the Lower Paleolithic (and probably the beginning of the MP) the main living area of the site was located not in what we call now the SSA but at different location/s in the cave. A possible location could be at the south-western part of the Upper Terrace near the entrance to the Inner Chamber, where massive rockfalls and brecciated layers are still present today.

The Lower Paleolithic – Acheulo-Yabrudian

In Misliya Cave, Acheulo-Yabrudian artifacts were found in three contexts: *in situ* layers on the Lower Terrace; mixed Acheulo-Yabrudian/Mousterian material under Unit 6 of the SSA; and mixed Acheulo-Yabrudian/Mousterian material in the Inner Chamber on the southern corner of the Upper Terrace.

In situ Acheulo-Yabrudian layers were excavated on the Lower Terrace of the site. Here the lithified archaeological layers extend over an area of ca 40 m^2 and contain only Acheulo-Yabrudian artifacts. Four square meters were excavated in this part of the site with a total volume of ca. 4.7 m^3 (Fig. 14.1c). A deep section was exposed within the

Table 14.1 Density of artifacts in the Acheulo-Yabrudian of Misliya Cave

Square	Volume	Quantity	Density (m^3)
Q28	1.64	768	468
Q29	1.53	397	259
Q30	0.91	180	197
Total	4.08	1345	329

Table 14.2 General breakdown of the Misliya Acheulo-Yabrudian assemblage

Category	N	%
Flake	859	65.3
Blade	37	2.8
CTE	9	0.7
Biface thinning flakes	35	2.7
Retouch/resharpening flakes	35	2.7
Handaxes	14	1.1
Core	51	3.9
Chunk	195	14.8
Retouched tool	82	6.2
Sub-total	**1317**	**100.0**
Microdebris	1728	
Microdebris (burnt)	740	
Total	**3785**	

strongly lithified layers in squares Q28-Q30 (Fig. 14.1b) and an additional square (N34) was excavated close to the northern limit of the archaeological sediments. The cemented sediments were excavated in 0.5 m^2 squares and 5 cm spits with an electrical hammer, coupled with hand-chiseling. In addition to items collected in the field, this procedure produced lumps of cemented sediments which were further excavated in the laboratory for extraction of archaeological material. Only isolated bone fragments were spotted in the Acheulo-Yabrudian sediments but due to the hardness of the layers we were not able to extract them. Therefore, there is no available faunal data from the Acheulo-Yabrudian layers of the cave.

The Acheulo-Yabrudian layers from Squares Q28-30 were dated using the TL method. In total nine dates were obtained showing a relatively short occupation range (between 273 ± 21 and 238 ± 21 ka). The mean ages are 247 ± 24 ka for Square Q28 and 257 ± 28 ka for Square Q29 (Valladas et al. 2013). These dates place the Acheulo-Yabrudian finds of Misliya Cave at the very end of the Levantine Lower Paleolithic.

The Acheulo-Yabrudian lithic assemblage from the Lower Terrace contains 3785 artifacts. Among them 1317 artifacts are larger than 2.5 cm and 2468 are microdebris. The average density of finds is 329 artifacts larger than 2.5 cm per m^3 (Table 14.1). The assemblage (Table 14.2) is dominated by flakes (859). Blades are rare and Levallois products are absent altogether. Side-scrapers constitute the dominant tool type, with simple, *déjeté*, transverse and

bifacial scrapers being the dominant types. The Misliya handaxes are small (Fig. 14.4: 1), closely resembling handaxes from Layer E of Tabun Cave and probably those of Yabrud I, but differing from Upper Acheulian sites (Chazan and Horowitz 2006; Zaidner et al. 2006). Whether the small size is a common feature of Acheulo-Yabrudian bifaces as a whole, or represents a special trend in handaxe production at the end of the Lower Paleolithic on the Carmel ridge, is still an open question. Many of the Misliya handaxes are made on flat flint pebbles and retain parts of at least one of the cortex surfaces. As a rule, the Misliya knappers focused on shaping the handaxe tip rather than on its entire circumference. In this, the Misliya handaxes differ from some Late Acheulian bifaces that were bifacially flaked all around their circumferences (Zaidner et al. 2006). The presence of biface thinning flakes indicates that handaxes were shaped on site.

One of the most striking features of the Misliya Cave biface assemblage is a continuous range of variation from "true" bifaces (Fig. 14.4: 1), through artifacts fully worked on one face and only partially on another (Fig. 14.4: 3, 5, 8), to real "unifaces" and scrapers (Fig. 14.4: 7). This phenomenon was also observed in Bezez Cave, where "...difficulty was experienced with 36 pieces, which seemed to be intermediate between bifaces and bifacial racloirs" (Copeland 1983: 109). In Misliya Cave, most of the partial bifaces were made on flakes. Usually the dorsal face is almost completely covered by removals, most of which were made after the flake was detached from the core. The ventral face, on the other hand, was poorly retouched, and usually only a few removals were made close to the tip of the handaxe (Fig. 14.4: 3, 8).

While no combustion features were found in the excavated sediments, the evidence for use of fire is inferred from the presence of burnt lithics (16.8% burnt artifacts according to the visual inspection). TL analysis reinforces visual observations, clearly indicating that some of the artifacts were burnt. Among the microdebris, ca. 30% are burnt pot-lids, fragments and chips (Table 14.2), also indicating that fire was used quite intensively.

The Middle Paleolithic – Early Levantine Mousterian

The SSA was the major focus of the excavations. Here approximately 20 square meters of Middle Paleolithic layers were excavated (Fig. 14.1c). The soft sediments were excavated using a three-dimensional recording system for all artifacts and bones larger than 2.5 cm. The squares with hard sediments to the west of the drip-line were excavated using hammers and chisels as well as electric hammers. The artifacts in the brecciated sediments were recorded in sub-squares excavated by 5 cm deep spits. A small area was opened in the brecciated sediments on the north-western part of the Upper Terrace (Squares H-J15; H-I16) and two squares were dug to bedrock immediately to the west of the Inner Chamber (squares U-V/7). In addition, one square meter deep-sounding was excavated in the brecciated sediments of the western part of Upper Terrace (Square L15) (Fig. 14.1c).

The Middle Paleolithic layers of the Upper Terrace were dated using the TL method. In total 23 dates were obtained from squares L15, J15, N12 and L10 (Valladas et al. 2013). The mean ages of the Middle Paleolithic layers range from 212 ± 27 to 166 ± 23 ka, broadly assigning the site to marine isotope stage (MIS) 7 and the early part of MIS6 (Valladas et al. 2013). The lithic analysis of the Middle Paleolithic assemblages of Misliya cave shows that they all belong to the Early Levantine Middle Paleolithic (e.g., Garrod and Bate 1937; Jelinek 1982; Bar-Yosef 1998). The same cultural phase is dated to ca. 250−170 ka in both caves and open-air Levantine Middle Paleolithic sites (Grün and Stringer 2000; Mercier and Valladas 2003; Rink et al. 2003; Mercier et al. 2007). Thus the lithic evidence and TL dates are in general agreement with the known record of the Early Levantine Middle Paleolithic.

Table 14.3 presents data from the initial sorting of ca. 70% of the lithic assemblage of the SSA and the entire assemblage of the deep-sounding (L15). The density of Middle Paleolithic artifacts in the SSA varies between 2000 and 4000 pieces per cubic meter in different squares; the average density is 3017 pieces per cubic meter. Although the assemblages are dominated by flakes (Table 14.3), the industry is blade-oriented. The lithic evidence reveals the use of two major technological systems in Misliya Cave: Levallois and Laminar. Products of both systems occur throughout the site's stratigraphy. The laminar system consists of bi-directional twisted cores, half-pyramidalic cores, crested blades and blades with thick triangular or trapezoid sections. The Levallois system consists of elongated products, with blades comprising ca. 25% of the Levallois assemblage (Zaidner and Weinstein-Evron 2014). Levallois flakes are commonly triangular, similar in shape of the butt and in the use of a unipolar convergent method of core reduction to the Levallois points. A number of true Levallois blades are also present in the assemblage. They are wider and thinner than blades produced by the laminar system and are often produced by a bidirectional method.

The high proportion of elongated retouched points and preference of blades as blanks for tool production are the most characteristic features of the Misliya tool-kit distinguishing it from those of the later MP (Zaidner and Weinstein-Evron 2014; Table 14.4). Points include a variety of well-standardized types (i.e., Levallois, elongated Mousterian, Abu-Sif and Hummal points). In addition, a new type of point was identified at Misliya Cave, the Misliya Point. This is a small point that is characterized by oblique truncation on the distal end (Fig. 14.5: 5−7). The notable metrical and morphological differences between different point types suggest that they likely have been used differently. Some of the pieces of all types exhibited diagnostic impact fractures indicating that they were used as tips of weapons (Yaroshevich et al. 2016). The differences in size may indicate differences in hunting technologies employed by Misliya hominins (Yaroshevich et al. 2016).

The faunal assemblage of Misliya Cave shows similar characteristics in both the SSA and Square L15 and is overwhelmingly dominated by ungulate taxa, especially Mesopotamian fallow deer (*Dama mesopotamica*), mostly prime-aged individuals, and mountain gazelle (*Gazella gazella*) while carnivore remains are absent (Yeshurun et al. 2007; Weinstein-Evron et al. 2012). Aurochs (*Bos primigenius*), wild boar (*Sus scrofa*), red deer (*Cervus elaphus*), roe deer (*Capreolus capreolus*), wild goat (*Capra* sp.) and ostrich (*Struthio camelus*; egg-shell fragments) are present in small numbers. Multivariate taphonomic analysis of the SSA assemblage demonstrated that the assemblage was created solely by humans occupying the cave and was primarily modified by their food-processing activities (Yeshurun et al. 2007). Gazelle carcasses were transported complete to the site, while fallow deer carcasses underwent some field

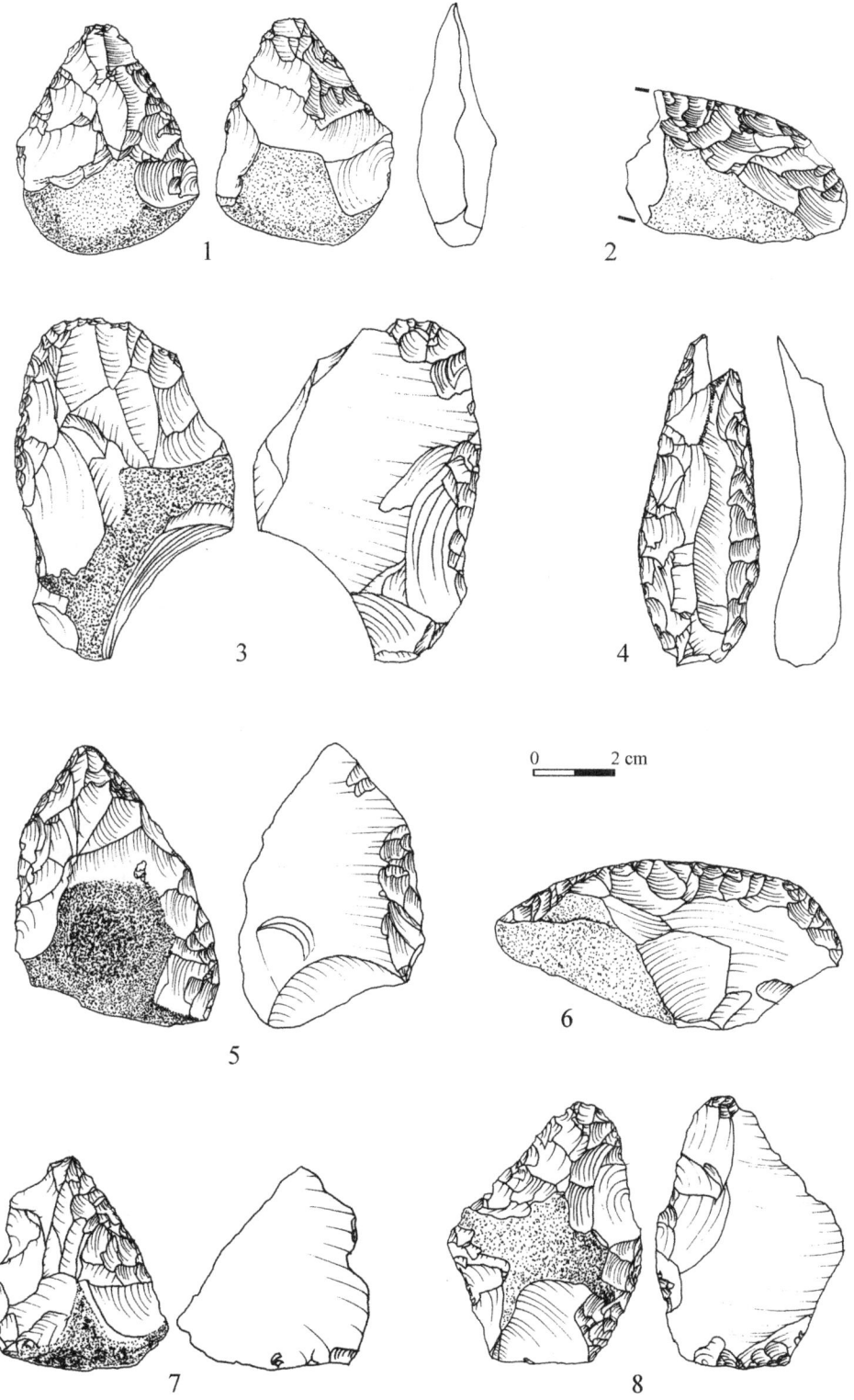

Fig. 14.4 Acheulo-Yabrudian artifacts. 1 – handaxe; 2, 6 – transverse sidescrapers; 3, 5, 8 – sidescrapers with bifacial retouch; 4 – limace; 7 – convergent sidescraper

Table 14.3 General breakdown of the Misliya EMP assemblage

	SSA		L15 (deep sounding)	
Flake	36516	59.3%	1818	71.8%
Blade	10582	17.2%	175	6.9%
Levallois flake	4350	7.1%	109	4.3%
Levallois blade	2038	3.3%	9	0.4%
Levallois point	1331	2.2%	74	2.9%
CTE	665	1.1%	18	0.7%
Burin Spall	121	0.2%	8	0.3%
Core	423	0.7%	14	0.6%
Core (Levallois)	176	0.3%	11	0.4%
Core-on-flake	81	0.1%	10	0.4%
Chunk	2473	4.0%	145	5.7%
Retouched tool	2810	4.6%	142	5.6%
Total	61566	100.0%	2533	100.0%

butchery. Abundance of meat-bearing limb bones that display filleting cut-marks and the acquisition of prime-age prey suggest that the Early Middle Paleolithic people acquired their prey by active and systematic hunting.

During the excavation of the SSA three distinct hearths were discovered. Two of them were found in the soft sediments of Unit 6 (Squares K-L9 and L11), while a small hearth was unearthed in the brecciated Unit 5 (Square I11). An exceptionally well preserved hearth was found in square L11. It is ca. 35 cm in diameter and is clearly differentiated from the surrounding sediment in color (Fig. 14.6 a, b). The hearth lies on a large limestone boulder and consists of three distinct levels, from top to bottom (Fig. 14.6c):

1. Chunks of indurated gray ashes.
2. Black layer 1−2 cm thick rich in burnt bones and flints.
3. Orange (3a) to brown (3b) layer up to 10 cm thick with a lens-like section.

The large hearth found in square L9 is still under micromorphological, mineralogical and archeo-botanical study. In addition to visible hearths, micromorphological and mineralogical evidence points to intensive use of fire in both the SSA and lithified layers, as evident in the geo-archaeological study of the deep sequence in Square L15 (Weinstein-Evron et al. 2012). In the latter, the evidence includes blackened and calcined burnt bones, bedded humified/charred plant material arranged in micro-laminae, reddish lenses probably derived from burnt clayey *terra rossa* and cemented calcite lenses probably originating from partial dissolution and re-precipitation of calcitic wood ash.

The exceptional preservation of vegetal tissues at Misliya is noteworthy. The charred remains were micromorphologically identified in a central part of the collapsed cave, associated with wood ash, burnt bones, and phytoliths (Weinstein-Evron et al. 2012). Similar attributes were recently reported from later MP and Middle Stone Age sites

Table 14.4 Composition of the tool-kit in Early, Middle and Late Levantine Mousterian sites*

	Site	N of retouched tools	Retouched points (%)	Side-scrapers (%)	UP types (%)	Notches/Denticulates (%)	Retouched blades (%)
Early Levantine Mousterian	Misliya Cave*	498	40.4	11.3	8.9	4.2	21.9
	Hummal (Hummalian)[h]	416	35.3	0.2	10.3	7.7	37.0
	Tabun Cave[g]	70	18.6	15.7	12.9	15.7	–
	Rosh Ein Mor[b]	2554	5.3	8.6	27.0	43.7	–
	Qafzeh XV[f]	323	1.2	14	17	24.2	–
Middle and Late Levantine Mousterian	Qafzeh XIII[f]	222	0.5	38.3	6.8	29.3	–
	Hummal (Mousterian)[e]	362	10.5	31.2	8.3	9.7	–
	Quneitra[d]	3011	1.1	31.7	12.5	31.0	–
	Amud Cave[a,c]	251	4.0	13.9	16.7	29.5	–
	Tor Faraj	348	6.6	3.7	17.0	11.8	–

*(data from: Alperson 2001[a]; Crew 1976[b]; Goder 1997[c]; Goren-Inbar 1990[d]; Hauck 2010[e]; Hovers 2009[f]; Jelinek 1975[g]; Wojtczak 2011[h]). *Only complete artifacts

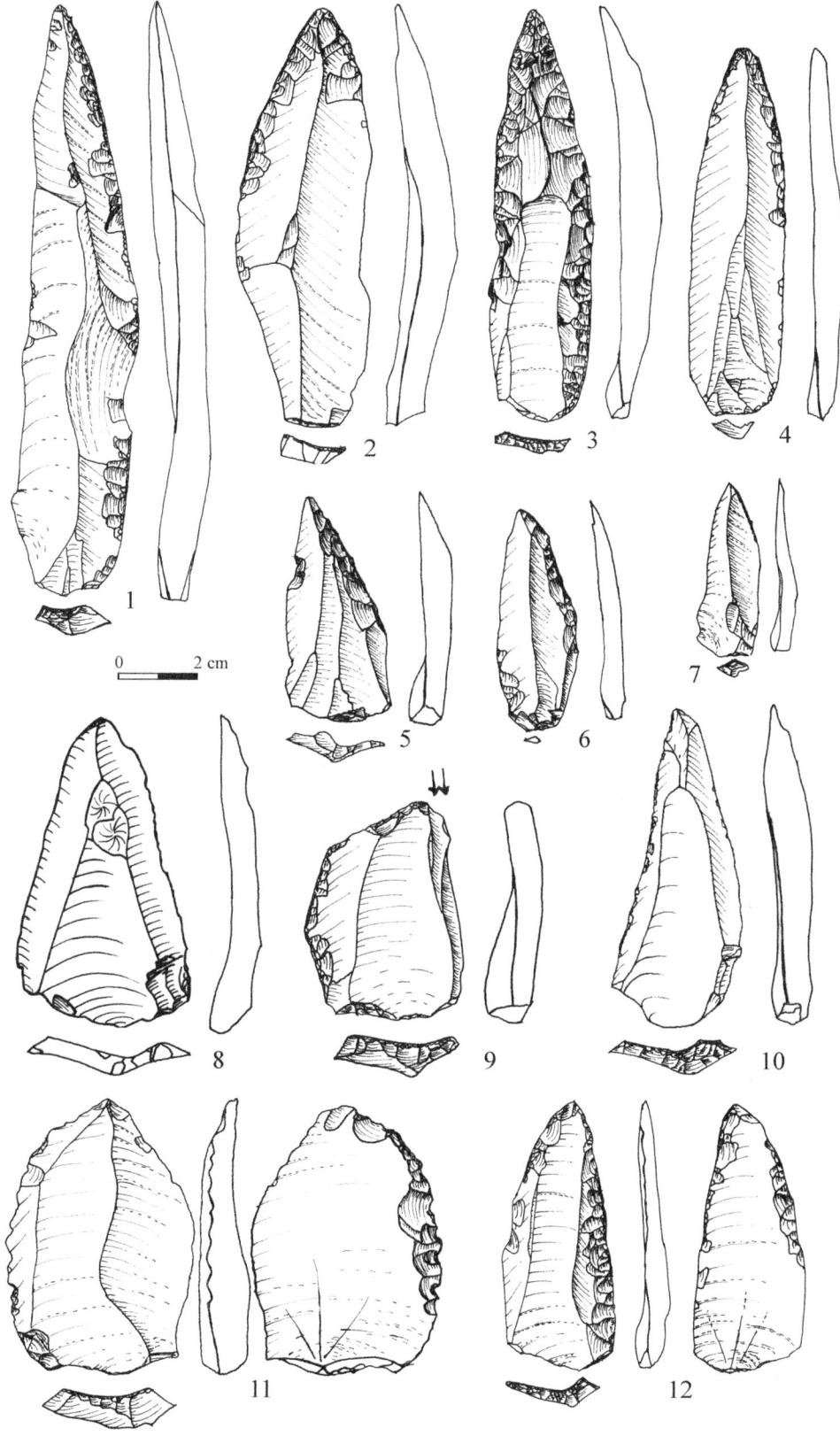

Fig. 14.5 EMP artifacts. 1–2, 4 – Hummal points; 3 – Abu-Sif point; 5–7 – Misliya points; 8, 10 – elongated Levallois points; 9 – burin; 11 – sidescraper on ventral face; 12 – point with bifacial retouch

Fig. 14.6 Small hearth in Square L11

in Spain (Esquilleu Cave; Cabanas et al. 2010) and South Africa (Sibudu rockshelter; Goldberg et al. 2009; Wadley et al. 2011). At Esquilleu Cave, bedded phytoliths have been identified in the central part of the site, associated with remains of wood ash, burnt clay and charred vegetal matter (Cabanes et al. 2010; Mallol et al. 2010). At Sibudu rock-shelter, the bedded phytoliths have been identified along the shelter's wall, associated with wood ash, charred vegetal fibers and burnt bones (Goldberg et al. 2009). Given their early age and cultural affiliation, the Misliya remains of bedding represent the earliest such example to date.

Misliya Cave and the Lower-Middle Paleolithic Levantine Record

The most abundant evidence retrieved from Misliya Cave concerns the EMP. However, it contains significant data regarding the Lower Paleolithic as well. The study of the Lower Paleolithic lithic assemblages, both from the different contexts on the Upper Terrace and the excavated squares on the western edge of the site, clearly indicates that the Acheulo-Yabrudian and the EMP represent different cultural traditions. The Lower Paleolithic Acheulo-Yabrudian industry of the cave is characterized by the production of thick short flakes from cores with unprepared platforms. The flakes were shaped by intensive Quina retouch into typical Acheulo-Yabrudian side-scrapers and bifaces (Weinstein-Evron et al. 2003a; Zaidner et al. 2006). There is no evidence for Levallois or laminar production in the Misliya Acheulo-Yabrudian assemblage. In contrast, the EMP assemblage is dominated by Levallois and laminar reduction sequences (Weinstein-Evron et al. 2003a, 2012; Zaidner and Weinstein-Evron 2014), with elongated points of different types and variably retouched blades dominating the toolkit (Zaidner and Weinstein-Evron 2014).

The dating of the sequence at Misliya, based on a series of dates obtained from 32 burnt flints retrieved from both the Acheulo-Yabrudian and EMP industries and conducted by a single method (TL) places the boundary between these two distinct cultural units at around 250 ka, i.e., at the end of MIS 8, beginning of MIS 7. The marked technological break between these two cultural complexes could have been associated with the arrival of a new population: the bearer of a new laminar blade technology. This new EMP cultural complex developed during MIS 7 and persisted for some 100,000 years.

Major collapses of the cave occurred between these two main episodes of human occupation, thus masking the actual boundary between them. The collapses are attested by the terrace-like configuration of the site (that may be related to a series of now-collapse chambers), the large rock-falls that occasionally include ancient flow-stones, and the maximum extent of the brecciated layers that indicate the extent of the ancient cave. Other rock-falls may be evident at the bottom of the deep L15 sequence (Weinstein-Evron et al. 2012) and below the EMP layers of the SSA, where Acheulo-Yabrudian finds, mixed with EMP material, are typically found. Together with the mixed material washed into the Inner Chamber, these mixed Acheulo-Yabrudian/EMP artifacts most probably indicate the occurrence of an Acheulo-Yabrudian layer that was heavily eroded before the EMP phase of human habitation took place, this time only on the Upper Terrace of the cave.

This major erosional phase may be related to the one postulated for the Mount Carmel at large (Weinstein-Evron 2015). The LP is poorly known on Mount Carmel because of the paucity of sites. The rare occurrence of Lower Paleolithic sites, essentially in caves (Tabun, Jamal, Misliya), may indicate that the ancient Lower Paleolithic landscape had been eroded from the top of the mountain and its upper slopes. Occurrences of Lower Paleolithic finds in taluses underlying those with Middle Paleolithic remains (Weinstein et al. 1975) may indicate repeated processes of erosion and down-sloping. Moreover, the many patches of Middle Paleolithic breccias, habitually found at some distance below extant cliffs, mainly across the western slope of the Mountain, but also within some wadi channels (Olami 1984) attest to a previously much extended cave-system heavily utilized by the Middle Paleolithic inhabitants of the mountain. Cave deterioration continued during the long EMP habitation of the cave, until the final collapse towards its end. Significantly, EMP layers occur immediately on the surface of the Upper Terrace, indicating that the last collapse rendered the cave unattractive for further habitation.

Data about the EMP of the cave are much richer, both concerning the site's layout (mainly related to hearths) and the rich lithic and faunal assemblages. The substantial evidence of the use of fire is one of the outstanding features of Misliya Cave. The site is one of the earliest cases providing solid evidence for the use of fire during the Middle Paleolithic of the Levant. In the Lower Paleolithic the evidence for the use of fire is usually limited to burnt flint artifacts or concentrations of burnt flint micro-flakes (Goren-Inbar et al. 2004; Alperson-Afil and Goren-Inbar 2010). At the end of the Lower Paleolithic the use of fire became more intensive (Karkanas et al. 2007; Shahack et al. 2014; Shimelmitz et al. 2014). In the EMP, remains of hearths were found at Hayonim and Tabun Caves (Garrod and Bate 1937; Goldberg and Bar-Yosef 1998; Stiner 2005). At Misliya, the use of fire is attested in both the Acheulo-Yabrudian, with large numbers of burnt flint flakes and micro-debris and the Early Mousterian, with its abundant, well-defined hearths and ample evidence of burning. The early evidence of bedding or matting, derived from the EMP hearths is also noteworthy.

Besides the apparent break from the Acheulo-Yabrudain, the composition of the Misliya Cave toolkit is also significantly different from Late Mousterian assemblages. It was previously suggested that the major behavioral and cultural change in the course of 200,000 years of the Levantine Middle Paleolithic occurred ca. 160−140 ka during the transition between the early and later Levantine Mousterian (Hovers 2001, 2006, 2009; Shea 2003; Meignen et al. 2006; Hovers and Belfer-Cohen 2013). One facet of this change is in lithic technology that shifted from a system based on a combination of laminar and Levallois reduction strategies (EMP) toward an emphasis on flake and point production, predominantly by the Levallois method in the later MP (Bar-Yosef 1998; Meignen 1998; Kaufman 1999; Hovers 2009; Hauck 2011; Wojtczak 2011). The laminar technological system of the EMP reported from a few sites in the region (Akasawa 1979; Meignen 1994, 1998; Marks and Monigal 1995; Wojtczak 2011) is not yet fully described. At Misliya and Hayonim caves and Hummal spring, the laminar products were obtained from unidirectional prismatic cores or cores with two opposed twisted platforms. Levallois points and triangular flakes produced by unidirectional convergent method are the major products of the Levallois reduction strategy in EMP sites.

The second facet is the possible demographic increase and a change in settlement patterns. On the basis of comparative evidence from Early and Late Middle Paleolithic sites in the Mediterranean zone of the southern Levant and the Judean desert, it was hypothesized that mobility and settlement patterns changed considerably. Drawing on data from Hayonim Cave, Abu Sif, Sahba, Hummal 1 and Tabun Cave, it was suggested that during the EMP the region was occupied by groups with larger home ranges which visited specific localities infrequently and only briefly (Hovers 2001, 2009; Meignen et al. 2006). The data from Hayonim Layer F and lower Layer E suggest that occupations were ephemeral and opportunistic. The occupations at Hayonim Cave are characterized by low densities of artifacts and bones, a high frequency of micromammals, lack of evidence for systematic collection of wood, lack of observed, long-term spatial differentiation in the use of the cave, thin short-lived fireplaces and low intensity exploitation of ungulates and tortoises (Weiner et al. 1995, 2002; Bar-Yosef 1998; Goldberg and Bar-Yosef 1998; Stiner et al. 2000; Albert et al. 2003; Meignen et al. 2006). The pattern of use of raw material reflects a large exploitation territory (Delage et al. 2000). It was suggested that EMP occupations in Hayonim Cave reflect "residential camps of short duration within a strategy of high mobility" (Meignen et al. 2006: 155). The less-detailed available data from other sites (Abu Sif, Sahba, Hummal 1 and Tabun Cave) indicate low artifact densities and high blank to core and waste ratios, which seem to fit the proposed model of

high mobility with short-term occupations (Hovers 2001; Meignen et al. 2006).

By contrast, the LMP record is considered to represent systems of low residential mobility with sites either resembling longer-term repetitive occupations or task-specific localities. The former exhibit thicker deposits, denser clusters of lithics and faunal remains, recurrent use of space over time for similar purposes, and intensification of animal exploitation (Bar-Yosef 1998; Hovers 2001, 2006, 2009; Speth 2004, 2006; Meignen et al. 2006; Speth and Clark 2006; Bar-Yosef and Meignen 2007). Kebara Cave, for example, was reconstructed as a seasonal base camp inhabited during the autumn-spring with a variety of activities performed *in situ* and with a clear partitioning of the domestic space. Amud Cave shows evidence for repetitive use of specific areas as a depository of human remains, for knapping and for activities connected to the hearths (Hovers et al. 1995; Alperson-Afil and Hovers 2005; Shahack-Gross et al. 2008), while in Tor Faraj knapping activities and processing of organic material show consistent spatial patterns (Henry 1998; Henry et al. 2004).

On the basis of the modeled change in settlement pattern, demographic increase from the EMP to the LMP was suggested. It was hypothesized that during the LMP, the Levant was inhabited by larger numbers of people that visited the sites more frequently and stayed for longer periods of time (Hovers 2001; Meignen et al. 2006). This settlement model seemed to hold true for most of the Levant with the exception of the central Negev (Munday 1976; Marks and Freidel 1977; Marks 1988).

Misliya Cave, however, is exceptional in the EMP mainly in the high density of lithic and faunal remains and the presence of a large hearth. The site was occupied during ca. 50,000 TL-years (ca. 212−166 ka), during which between 1.5 and 3.5 m of deposits accumulated, with an average density of ca. 3,000 artifacts per square meter. This density is much higher than those reported from Hayonim Cave, where 300 artifacts were excavated in each cubic meter (Bar-Yosef 1998) and Tabun Cave, where densities of artifacts are generally low (from total 90 m^3 excavated, 44.000 artifacts were unearthed, giving an estimation of 448 artifacts per m^3; Jelinek et al. 1973, 1977). The large hearth in squares K-L9 is ca. 30 cm thick attesting to long-term, repeated use compared to the shallower and more ephemeral hearths of Hayonim and Tabun caves. The wide array of technological systems identified at Misliya Cave, compared, for example, with the nearby Tabun Cave, with its small EMP assemblage, which is dominated by the Levallois reduction system and characterized by a high frequency of retouched tools (Jelinek et al. 1973, 1977; Meignen 2011; Shimelmitz and Kuhn 2013), indicates that Misliya Cave was used intensively and for various tasks. Given the ephemeral nature of occupation in the majority of other known sites, the evidence for intensive and repeated occupations at Misliya Cave is unique in the framework of the proposed settlement and mobility model suggesting that the cave may have been used as an aggregation site for EMP hominin groups of the region, thus indicating that the EMP settlement pattern of the Levant was more varied than previously thought.

The third major facet of the change is in toolkit composition that in EMP sites differs considerably from those of Late Mousterian sites. The composition of the Misliya Cave toolkit is very similar to that of the EMP layers of Hummal (Table 14.4) and possibly also Abu Sif (Neuville 1951: 55). In both sites a variety of point types was found with the most common being Abu Sif and Hummal points (Neville 1951; Copeland 1985). The high frequency of points in the Early Levantine Mousterian sites is especially noticeable in comparison with the later Mousterian sites in which only retouched Levallois points occur in some numbers while Mousterian points are very rare or absent (Table 14.4; e.g., Henry 2003; Hovers 2009), or with preceding Lower Paleolithic techno-complexes in which points were not systematically produced (Garrod and Bate 1937; Rust 1950; Copeland 1983; Hovers 2009; Shimelmitz et al. 2011). This marked change in the toolkit suggests that the range of activities or the way similar activities were carried out changed between the EMP and later MP. Detailed use-wear analysis of the Misliya toolkit is underway and will shed important light on this issue.

The use of laminar and Levallois technology, the high intensity of occupation, the composition of the lithic assemblage that indicates high variety of activities, the complex long-term hearths and large-game hunting, carcass transport and meat processing behaviors altogether highlight the high sophistication of the EMP inhabitants of the cave, more than 160 ka ago. Thus, in many behavioral characteristics, the EMP hominins of Misliya Cave are similar to their late Mousterian counterparts. While the discrete affiliation of the Misliya Cave's inhabitants still eludes us, much is known about their behavior and modes of exploitation of their environments. The stage is all set for their appearance.

Acknowledgments Misliya Cave is located in the Mount Carmel Nature Reserve, managed by the Israel Nature and Parks Authority. The Misliya project is supported by the Dan David Foundation, the Leakey Foundation, the Irene Levi-Sala Care Archaeological Foundation and the Faculty of Humanities, the University of Haifa (by whom it is coined the "Faculty Cave") and the Israel Science Foundation (grant no. 1104/12). LSCE reference: 5003. Special thanks are due to the late Dan David for his enthusiastic support throughout. We thank Daniel Kaufman for his thoughtful remarks. Israel Antiquities Authority permit numbers for the Misliya Cave excavations: G-16/2001, G-39/2002, G-14/2003, G-29/2004, G-12/2005, G-12/2006, G-4/2007, G-54/2008, G-52/2009, G-50/2010.

References

Akazawa, T. (1979). Middle Paleolithic assemblages from Douara Cave. In K. Hanihara & T. Akazawa (Eds.), *Paleolithic site of Douara Cave and paleography of Palmyra basin in Syria* (pp. 1–30). Tokyo: University of Tokyo, The University Museum.

Albert, R. M., Bar-Yosef, O., Meignen, L., & Weiner, S. (2003). Quantitative phytolith study of hearths from the Natufian and Middle Palaeolithic levels of Hayonim Cave (Galilee, Israel). *Journal of Archaeological Science, 30*, 461–480.

Alperson, N. (2001). Differential use of space by Mousterian hominids: Evidence for spatial patterning at Amud Cave, Israel. MA Thesis, The Hebrew University of Jerusalem (in Hebrew).

Alperson-Afil, N., & Goren-Inbar, N. (2010). *The Acheulian site of Gesher Benot Ya'aqov. Volume II: Ancient flames and controlled use of fire.* Dordrecht: Springer.

Alperson-Afil, N., & Hovers, E. (2005). Differential use of space at the Neandertal site of Amud Cave, Israel. *Eurasian Prehistory, 3*, 3–22.

Bar-Yosef, O. (1998). The chronology of the Middle Palaeolithic of the Levant. In T. Akazawa, K. Aoki, & O. Bar-Yosef (Eds.), *Neandertals and modern humans in Western Asia* (pp. 39–56). New York: Plenum.

Bar-Yosef, O., & Meignen, L. (Eds.). (2007). *Kebara Cave Mt. Carmel, Israel. The Middle and Upper Paleolithic archaeology. Part I. American school of prehistoric research bulletin 49.* Cambridge MA: Peabody Museum of Archaeology and Ethnology.

Cabanes, D., Mallol, C., Exposito, I., & Baena, J. (2010). Phytolith evidence for hearths and beds in the late Mousterian occupations of Esquilleu Cave (Cantabria, Spain). *Journal of Archaeological Science, 37*, 2947–2957.

Chazan, M., & Horwitz, L. K. (2006). Finding the message in intricacy: The association of lithics and fauna on Lower Paleolithic multiple carcass sites. *Journal of Anthropological Archaeology, 25*, 436–447.

Copeland, L. (1983). The Paleolithic stone industries. In D. Roe (Ed.), *Adlun in the Stone Age. The excavations of D. A. E. Garrod in the Lebanon 1958–1963* (pp. 89−365). Oxford: BAR.

Copeland, L. (1985). The pointed tools of Hummal Ia (El-Kowm, Syria). *Cahiers de l'Euphrate, 4*, 177–189.

Crew, H. L. (1976). The Mousterian site of Rosh Ein Mor. In A. E. Marks (Ed.), *Prehistory and palaeoenvironments in the Central Negev, Israel, Vol. 1: The Avdat/Aqev area, Part 1* (pp. 75–112) Dallas: SMU Press.

Delage, C., Meignen, L., & Bar-Yosef, O. (2000). Chert procurement and the organization of lithic production in the Mousterian of Hayonim cave (Israel). *Journal of Human Evolution, 38*, A10–A11.

Garrod, D. A. E., & Bate, D. M. A. (1937). *The Stone Age of Mount Carmel* (Vol. I). Excavations at the Wadi Mughara, Oxford: Clarendon Press.

Goder, M. (1997). Technological aspects of the lithic assemblage of Layer B1 at Late Mousterian site of Amud Cave. MA Thesis, The Hebrew University of Jerusalem (in Hebrew).

Goldberg, P., & Bar-Yosef, O. (1998). Site formation processes in Kebara and Hayonim Caves and their significance in Levantine prehistoric caves. In T. Akazawa, K. Aoki & O. Bar-Yosef (Eds.), *Neandertals and modern humans in Western Asia* (pp. 107–125). New-York: Plenum.

Goldberg, P., Miller, C. E., Schiegl, S., Ligouis, B., Berna, F., Conard, N. J., et al. (2009). Bedding, hearths, and site maintenance in the Middle Stone Age of Sibudu Cave, KwaZulu-Natal, South Africa. *Archaeological and Anthropological Sciences, 1*, 95–122.

Goren-Inbar, N. (1990). The Lithic assemblages. In N. Goren-Inbar (Ed.), *Quneitra: A Mousterian site on the Golan Heights. Qedem 31* (pp. 61–167). Jerusalem: Institute of archaeology of the Hebrew University of Jerusalem.

Goren-Inbar, N., Alperson, N., Kislev, M. E., Simchoni, O., Melamed, Y., Ben-Nun, A., et al. (2004). Evidence of hominin control of fire at Gesher Benot Ya'aqov, Israel. *Science, 304*, 725–727.

Grün, R., & Stringer, C. (2000). Tabun revisited: Revised ESR chronology and new ESR and U-series analyses of dental material from Tabun CI. *Journal of Human Evolution, 39*, 601–612.

Hauck, T. (2010). The Mousterian sequence of Hummal (Syria). Ph.D. Dissertation, Universität Basel.

Hauck, T. (2011). The Mousterian sequence of Hummal and its tentative placement in the Levantine Middle Paleolithic. In J.-M. Le Tensorer, R. Jagher & M. Otte (Eds.), *The Lower and Middle Paleolithic of the Middle East and the neighboring regions* (pp. 309–323). *ERAUL 126*. Liège: University of Liège.

Henry, D. O. (1998). Intrasite spatial patterns and behavioral modernity: Indications from the Late Levantine Mousterian rockshelter of Tor Faraj, Southern Jordan. In T. Akazawa, K. Aoki, & O. Bar-Yosef (Eds.), *Neandertals and modern humans in Western Asia 9* (pp. 127–1420). New York: Plenum Press.

Henry, D. O. (2003). *Neanderthals in the Levant: Behavioral organization and the beginnings of human modernity*. London: Continuum.

Henry, D. O., Hietala, H. J., Rosen, A. M., Demidenko, Y. E., Usik, V. I., & Armagan, T. L. (2004). Human behavioral organization in the Middle Paleolithic: Were Neanderthals different? *American Anthropologist, 106*, 17–31.

Hershkovitz, I., Smith, P., Sarig, R., Quam, R., Rodríguez, L., García, R., et al. (2011). Middle Pleistocene dental remains from Qesem Cave (Israel). *American Journal of Physical Anthropology, 144*, 575–592.

Hovers, E. (2001). Territorial behavior in the Middle Paleolithic of the Southern Levant. In N. Conard (Ed.), *Settlement dynamics of the Middle Paleolithic and Middle Stone Age* (pp. 123–152). Tubingen: Kerns Verlag.

Hovers, E. (2006). Neandertals and modern humans in the Middle Paleolithic of the Levant: What kind of interaction? In N. J. Conard (Ed.), *When Neandertals and moderns met* (pp. 650–876). Tubingen: Kerns Verlag.

Hovers, E. (2009). *The lithic assemblages of Qafzeh Cave*. Oxford: Oxford University Press.

Hovers, E., & Belfer-Cohen, A. (2013). On variability and complexity: Lessons from the Levantine Middle Paleolithic record. *Current Anthropology, 54*(S8), S337–S357.

Hovers, E., Rak, Y., Lavi, R., & Kimbel, W. H. (1995). Hominid remains from Amud Cave in the context of the Levantine Middle Paleolithic. *Paléorient, 21*, 47–61.

Jelinek, A. J. (1975). A preliminary report on some Lower and Middle Paleolithic industries from the Tabun Cave, Mount Carmel (Israel). In F. Wendorf & A. E. Marks (Eds.), *Problems in prehistory: North Africa and the Levant* (pp. 279–316). Dallas: SMU Press.

Jelinek, A. (1977). A preliminary study of flakes from the Tabun Cave, Mt. Carmel. In B. Arensburg & O. Bar-Yosef (Eds.), *Eretz-Israel 13: Moshe Stekelis volume* (pp. 87–96). Jerusalem: Israel Exploration Society.

Jelinek, A. J. (1982). The Middle Palaeolithic in the southern Levant, with comments on the appearance of modern *Homo sapiens*. In A. Ronen (Ed.), *The transition from Lower to Middle Palaeolithic and the origin of modern man* (pp. 57–104). BAR International Series. Oxford: BAR.

Jelinek, A. J., Farrand, W. R., Hass, G., Horowitz, A., & Goldberg, P. (1973). New excavations at the Tabun Cave, Mount Carmel, Israel: A preliminary report. *Paléorient, 1*,151–183.

Karkanas, P., Shahack-Gross, R., Ayalon, A., Bar-Matthews, M., Barkai, R., Frumkin, A., et al. (2007). Evidence for habitual use of fire at the end of the Lower Paleolithic: Site formation processes at Qesem Cave, Israel. *Journal of Human Evolution, 53*, 197–212.

Kaufman, D. (1999). *Archaeological perspectives on the origins of modern humans: A view from the Levant*. Westport CT: Bergin & Garvey.

Mallol, C., Cabanes, D., & Baena, J. (2010). Microstratigraphy and diagenesis at the Upper Pleistocene site of Esquilleu Cave (Cantabria, Spain). *Quaternary International, 214*, 70–81.

Marks, A.E. (1988). Early Mousterian settlement patterns in the central Negev, Israel: Their social and economic implications. In M. Otte (Ed.), *L'homme de Neandertal, Vol. 6: La Subsistance* (pp. 115–126). *ERAUL*. Liège: University of Liège.

Marks, A. E., & Freidel, D. (1977). Prehistoric settlement patterns in the Avdat/Aqev area. In A. E. Marks (Ed.), *Prehistory and paleoenvironments in the Central Negev, Israel, Vol. 2: The Avdat/Aqev Area, Part 2* (pp. 131–159). Dallas: SMU Press.

Marks, A. E., & Monigal, K. (1995). Modeling the production of elongated blanks from the early Levantine Mousterian at Rosh Ein Mor. In H. L. Dibble, & O. Bar-Yosef (Eds.), *The definition and interpretation of Levallois technology* (pp. 267–278). Monographs in World archaeology 23. Ann Arbor: Prehistory Press.

McCown, T. D. (1937). Mugharet es-Skhul: Description and excavations. In D. A. E. Garrod & D. M. A. Bate (Eds.), *The Stone Age of Mount Carmel: Excavations at the Wady el-Mughara* (Vol. I, pp. 91–112). Oxford: Clarendon Press.

Meignen, L. (1994). Paleolithique moyen au Proche-Orient: le phenomene laminaire. In S. Révillion, & A. Tuffreau (Eds.), *Les Industries laminaires au paléolithique moyen* (pp. 125–159). *Dossiers de Documentation Archeologique 18*. Paris: Editions CNRS.

Meignen, L. (1998). A preliminary report on Hayonim Cave lithic assemblages in the context of the near Eastern Middle Palaeolithic. In T. Akazawa, K. Aoki, & O. Bar-Yosef (Eds.), *Neandertals and modern humans in Western Asia* (pp. 165–180). New York: Plenum.

Meignen, L. (2011). The contribution of Hayonim Cave assemblages to the understanding of the so-called Early Levantine Mousterian. In J.-M. Le Tensorer, R. Jagher & M. Otte (Eds.), *The Lower and Middle Paleolithic in the Middle East and neighboring regions* (pp. 85–100). *ERAUL 126*. Liège: University of Liège.

Meignen, L., Bar-Yosef, O., Speth, J. D., & Stiner, M. C. (2006). Middle Paleolithic settlement patterns in the Levant. In E. Hovers & S. L. Kuhn (Eds.), *Transitions before the transition: Evolution and stability in the Middle Paleolithic and Middle Stone Age* (pp. 149–169). New York: Springer.

Mercier, N., & Valladas, H. (2003). Reassessment of TL age estimates of burnt flints from the Palaeolithic site of Tabun Cave, Israel. *Journal of Human Evolution, 45*, 401–409.

Mercier, N., Valladas, H., Frojet, L., Joron, J. L., Ryess, J. L., Weiner, S., et al. (2007). Hayonim Cave: A TL-based chronology for this Levantine Mousterian sequence. *Journal of Archaeological Science, 34*, 1064–1077.

Munday, F. (1976). Intersite variability in the Mousterian of the Avdat/Aqev area. In A. E. Marks (Ed.), *Prehistory and paleoenvironments in the central Negev, Israel, Vol. I: The Avdat/Aqev Area, Part 2* (pp. 113–140). Dallas: SMU Press.

Neuville, R. (1951). *Le paléolithique et le mesolithique du désert du Judeé. Archives de Musee de Paleontologie Humaine memoire 24*. Paris: Masson et Cie.

Olami, Y. (1984). *Prehistoric Carmel*. Israel Exploration Society, Jerusalem and M. Stekelis Museum of Prehistory, Haifa

Rink, W. J., Richter, D., Schwarcz, H. P., Marks, A. E., Monigal, K., & Kaufman, D. (2003). Age of the Middle Palaeolithic Site of Rosh Ein Mor, Central Negev, Israel: Implications for the age range of the early Levantine Mousterian of the Levantine corridor. *Journal of Archaeological Science, 30*, 195–204.

Rust, A. (1950). *Die Hohlenfunde von Jabrud (Syrien)*. Neumünster: Karl Wachholtz.

Shahack-Gross, R., Ayalon, A., Goldberg, P., Goren, Y., Ofek, B., Rabinovich, R., et al. (2008). Formation processes of cemented features in karstic cave sites revealed using stable oxygen and carbon isotopic analysis: A case study at Middle Paleolithic Amud Cave, Israel. *Geoarchaeology, 23*, 43–62.

Shahack-Gross, R., Berna, F., Karkanas, P., Lemorini, C., Gopher, A., & Barkai, R. (2014). Evidence for the repeated use of a central hearth at Middle Pleistocene (300 ky ago) Qesem Cave, Israel. *Journal of Archaeological Science, 44*, 12–21.

Shea, J. J. (2003). The Middle Paleolithic of the East Mediterranean Levant. *Journal of World Prehistory, 17*, 313–394.

Shimelmitz, R., & Kuhn, S. L. (2013). Early Mousterian Levallois technology in unit IX of Tabun Cave. *PaleoAnthropology, 2013*, 1–27.

Shimelmitz, R., Barkai, R., & Gopher, A. (2011). Systematic blade production at late Lower Paleolithic (400−200 kyr) Qesem Cave, Israel. *Journal of Human Evolution, 61*, 458–479.

Shimelmitz, R., Kuhn, S. L., Jelinek, A. J., Ronen, A., Clark, A. E., & Weinstein-Evron, M. (2014). 'Fire at will': The emergence of habitual fire use 350,000 years ago. *Journal of Human Evolution, 77*, 196–203.

Speth, J. D. (2004). Hunting pressure, subsistence intensification and demographic change in the Levantine Late Middle Palaeolithic. In N. Goren-Inbar & J. D. Speth (Eds.), *Human palaeoecology in the Levantine corridor* (pp. 149–166). Oxford: Oxbow Press.

Speth, J. D. (2006). Housekeeping, Neandertal-style: Hearth placement and midden formation in Kebara Cave (Israel). In E. Hovers & S. L. Kuhn (Eds.), *Transitions before the transition: Evolution and stability in the Middle Palaeolithic and Middle Stone Age* (pp. 171–188). New York: Springer.

Speth, J. D., & Clark, J. (2006). Hunting and overhunting in the Levantine Late Middle Palaeolithic. *Before Farming, 3*, 1–42.

Stiner, M. C. (2005). *The faunas of Hayonim Cave, Israel: A 200,000 record of Paleolithic diet, demography and society*. American School of Prehistoric Research Bulletin 48. Cambridge: Peabody Museum of Archaeology and Ethnology, Harvard University.

Stiner, M. C., Munro, N. D., & Surovell, T. A. (2000). The tortoise and the hare: Small-game use, the broad-spectrum revolution, and Paleolithic demography. *Current Anthropology, 41*, 39–73.

Valladas, H., Mercier, N., Hershkovitz, I., Zaidner, Y., Tsatskin, A., Yeshurun, R., et al. (2013). Dating the Lower-Middle Paleolithic transition in the Levant: A view from Misliya Cave, Mount Carmel, Israel. *Journal of Human Evolution, 65*, 585–593.

Wadley, L., Sievers, C., Bamford, M., Goldberg, P., Berna, F., & Miller, C. (2011). Middle Stone Age bedding construction and settlement patterns at Sibudu, South Africa. *Science, 334*, 1388–1391.

Weiner, S., Schiegl, S., & Bar-Yosef, O. (1995). Recognizing ash deposits in the archaeological record: A mineralogical study at Kebara and Hayonim caves, Israel. *Acta Anthropologica Sinica, 14*, 340–351.

Weiner, S., Goldberg, P., & Bar-Yosef, O. (2002). Three-dimensional distribution of minerals in the sediments of Hayonim cave, Israel: Diagenetic processes and archaeological implications. *Journal of Archaeological Science, 29*, 1289–1308.

Weinstein, M., Lamdan, M., & Goren-Hitin, I. (1975). An Upper Acheulean industry at Tirat-Carmel. *Journal of the Israel Prehistoric Society, 13*, 36–49.

Weinstein-Evron, M. (2015). The case of Mount Carmel: The Levant and human evolution, future research in the framework of world heritage. In N. Sanz, (Ed.), *Human Origin Sites and the World Heritage Convention in Eurasia,*. World Heritage Papers, 41, HEADS (Human Evolution: Adaptations, Dispersals and Social Developments), (4, Vol. 1, pp. 72–92). Paris: UNESCO.

Weinstein-Evron, M., & Tsatskin, A. (1994). The Jamal Cave is not empty: Recent excavations in the Mount Carmel Caves, Israel. *Paléorient, 20*, 119–128.

Weinstein-Evron, M., Bar-Oz, G., Zaidner, Y., Tsatskin, A., Druck, D., Porat, N., et al. (2003a). Introducing Misliya Cave, Mount Carmel, Israel: A new continuous Lower/Middle Paleolithic sequence in the Levant. *Eurasian Prehistory, 1*, 31–55.

Weinstein-Evron, M., Beck, A., & Ezersky, M. (2003b). Geophysical investigations in the service of Mount Carmel (Israel) prehistoric research. *Journal of Archaeological Science, 30*, 1331–1341.

Weinstein-Evron, M., Tsatskin, A., Bar-Oz, G., Yeshurun, R., Shahack-Gross, R., Weiner, S., et al. (2012). A window onto Early Middle Paleolithic human occupational layers: The case of Misliya Cave, Mount Carmel, Israel. *PaleoAnthropology, 2012*, 202–228.

Wojtczak, D. (2011). Hummal (Central Syria) and its eponymous industry. In J.-M. Le Tensorer, R. Jagher, & M. Otte (Eds.), *The Lower and Middle Paleolithic in the Middle East and neighbouring regions*. (pp. 289–308). *ERAUL 126*. Liège: University of Liège.

Yaroshevich, A., Zaidner, Y., & Weinstein-Evron, M. (2016). Evidence of hunting weapon variability in the Early Middle Paleolithic of the Levant. A view from Misliya Cave, Mount Carmel. In R. Iovita, & K. Sano (Eds.), *Multidisciplinary approaches to the study of Stone Age weaponry* (pp. 119–134). Dodrecht: Springer.

Yeshurun, R., Bar-Oz, G., & Weinstein-Evron, M. (2007). Modern hunting behavior in the early Middle Paleolithic: Faunal remains from Misliya Cave, 316 Mount Carmel, Israel. *Journal of Human Evolution, 53*, 656–677.

Zaidner, Y., & Weinstein-Evron, M. (2014). Making a point: The early Mousterian toolkit at Misliya Cave, Israel. *Before Farming, 2014*, 1–23.

Zaidner, Y., Druck, D., Nadler, M., & Weinstein-Evron, M. (2005). The Acheulo-Yabrudian of Jamal Cave, Mount Carmel, Israel. *Journal of the Israel Prehistoric Society, 35*, 93–116.

Zaidner, Y., Druck, D., & Weinstein-Evron, M. (2006). Acheulo-Yabrudian handaxes from Misliya Cave, Mount Carmel, Israel. In N. Goren-Inbar & G. Sharon (Eds.), *Axe age: Acheulian toolmaking—From quarry to discard* (pp. 243–266). Oxford: Equinox Publishers.

Chapter 15
A 3-D Look at the Tabun C2 Jaw

Katerina Harvati and Elisabeth Nicholson Lopez

Abstract The Tabun cave is among the most important paleoanthropological sites in the Near East. It has yielded a long sequence of archeological record, as well as important fossil human remains, notably the Tabun C1 partial skeleton and the Tabun C2 mandible. The chronology of these specimens, as well as their respective provenience, has been intensely debated. Most recent estimates place the C1 skeleton at oxygen isotope stage 5 or 6, while the C2 mandible is thought to be significantly older. The affinities of the C2 remains are unresolved. While general consensus sees the Tabun C1 skeleton as a lightly built Neanderthal, the Tabun C2 mandible has variably been attributed to early modern humans and to Neanderthals based on both metric and non-metric traits. We conducted a comparative analysis of the three-dimensional shape of the C2 mandible using the methods of geometric morphometrics, with the goal of helping to resolve its taxonomic affinities. Results show that Tabun C2 cannot be easily accommodated either within the early modern human or the Neanderthal sample. This finding is consistent with the proposed great geological age of the specimen.

Keywords Geometric morphometrics • Mandible • Modern human origins • Neanderthals

Introduction

Excavated by Garrod between 1929 and 1934 (Garrod and Bate 1937), the Tabun cave is one of the most important paleoanthropological sites in the Near East. It has yielded both a long sequence of archeological record, now a reference sequence for Levantine Paleolithic archeology, and important fossil human specimens. The most complete of these include the Tabun C1 partial skeleton and the Tabun C2 mandible. Both specimens were recovered from stratigraphic layer C, but their exact provenance and association are uncertain (Garrod and Bate 1937; Bar-Yosef and Pilbeam 1993; see below). The striking differences in the morphology of the Tabun C1 and C2 mandibles were noted early on (see Bar-Yosef and Callander 1999). While Tabun C1 is generally considered a lightly built Neanderthal, most likely a female, opinions differ on the taxonomic affinities of the Tabun C2 specimen (e.g., Quam 1995; Quam and Smith 1998; Stefan and Trinkaus 1998; Rak 1998; see below).

The exact provenience of Tabun C1 and C2, as well as their respective chronology, is not fully resolved. Although both individuals were found within layer C of the stratigraphic sequence, Tabun C1 may have been an intrusive burial from the overlying layer B and was found in the West sector of the excavation (Garrod and Bate 1937; Bar-Yosef and Pilbeam 1993; Bar-Yosef and Callander 1999; Grün and Stringer 2000). Tabun C2 came from the deeper part of layer C and from the East sector of the cave (Garrod and Bate 1937; Bar-Yosef and Callander 1999). Early attempts at absolute dating of layer C of the Tabun cave by radiocarbon dating indicated an age of approximately 50 ka (Jelinek 1982). However, as this date was at the ^{14}C method's limit, it likely underestimated the true age by many millennia. Later dating attempts using more recent dating methods (Electron Spin Resonance [ESR], Thermoluminescence [TL], coupled ESR-Uranium series) have since obtained, for the most part, much older ages for this layer. The Tabun C1 skeleton was initially dated directly through ESR by Schwarcz et al.

K. Harvati (✉)
Paleoanthropology, Senckenberg Center for Human Evolution and Paleoenvironments, Department of Geosciences, Eberhard Karls Universität Tübingen, Tübingen, Germany
e-mail: katerina.harvati@ifu.uni-tuebingen.de

E.N. Lopez
Department of Basic Science and Craniofacial Biology, New York University College of Dentistry, New York, USA
e-mail: el95@nyu.edu

© Springer International Publishing AG 2017
Assaf Marom and Erella Hovers (eds.), *Human Paleontology and Prehistory*,
Vertebrate Paleobiology and Paleoanthropology, DOI: 10.1007/978-3-319-46646-0_15

(1998), who found it to be very young (between 24 ± 5 and 19 ± 2 ka). Schwarcz et al. (1998) concluded that Tabun C1 represented a very late intrusion into layer C, indicating a very late Neanderthal survival in the Levant. However, these late dates are considered problematic based on both methodological and stratigraphic issues (Millard and Pike 1999; Alperson et al. 2000). They are not supported by more recent direct dating of the Tabun C1 specimen to between 112 ± 29 and 143 ± 37 ka (also by ESR; Grün and Stringer 2000). Grün and Stringer (2000) agreed with Schwarcz et al. (1998) that Tabun C1 was likely intrusive from layer B, as initially suggested by Garrod and Bate (1937), albeit an intrusion from a much earlier layer than previously thought. Tabun C1 was therefore probably broadly contemporaneous with the Skhul and Qafzeh early modern human populations roughly between 100 and 130 ka (Grün et al. 2005). If these latest assessments of the chronology of the Tabun sequence are correct, then the Tabun C2 mandible (coming from the lower part of layer C) would likely date to as early as 135–170 ka (Grün and Stringer 2000; Mercier and Valladas 2003).

Because of this specimen's possible association with a commonly recognized Neanderthal and its probable broad contemporaneity with the earliest modern human populations outside of Africa, the interpretation of its taxonomic affiliation and its phylogenetic position play a crucial role in the understanding of modern human origins in the region. We aim to contribute to this discussion by conducting a three-dimensional geometric morphometrics analysis of the shape of the Tabun C2 mandible using a comparative sample of early and Upper Paleolithic modern human, Neanderthal

(including Tabun C1) and *Homo heidelbergensis* mandibular specimens. The use of these methods can potentially be informative, as 3-D geometric morphometrics have several advantages over traditional morphometrics. In addition to providing a means for visualization of shape differences, these techniques enable a better representation of shape than traditional linear and angle measurements and permit the quantitative assessment of traits previously described qualitatively (e.g., Rohlf and Marcus 1993; Harvati 2003). Although the mandible is considered less taxonomically informative than parts of the cranium, an analysis of modern human and Neanderthal mandibular shape was able to discriminate between the two groups and to quantitatively evaluate their described morphological differences (Nicholson and Harvati 2006).

The Tabun C2 Mandible – Previous Interpretations

Tabun C2 is a large, rather robust mandible. On the basis of its size and robusticity it is likely a male. Even though it was recovered in several pieces, it has been reconstructed and it is virtually complete (Garrod and Bate 1937; Fig. 15.1). No agreement exists on its taxonomic placement. In her unpublished field notes, Garrod noted its marked departure from the Tabun C1 mandibular morphology soon after its discovery. This morphological dissimilarity led her to believe that two human taxa were present at the site (see Bar-Yosef and Callander 1999).

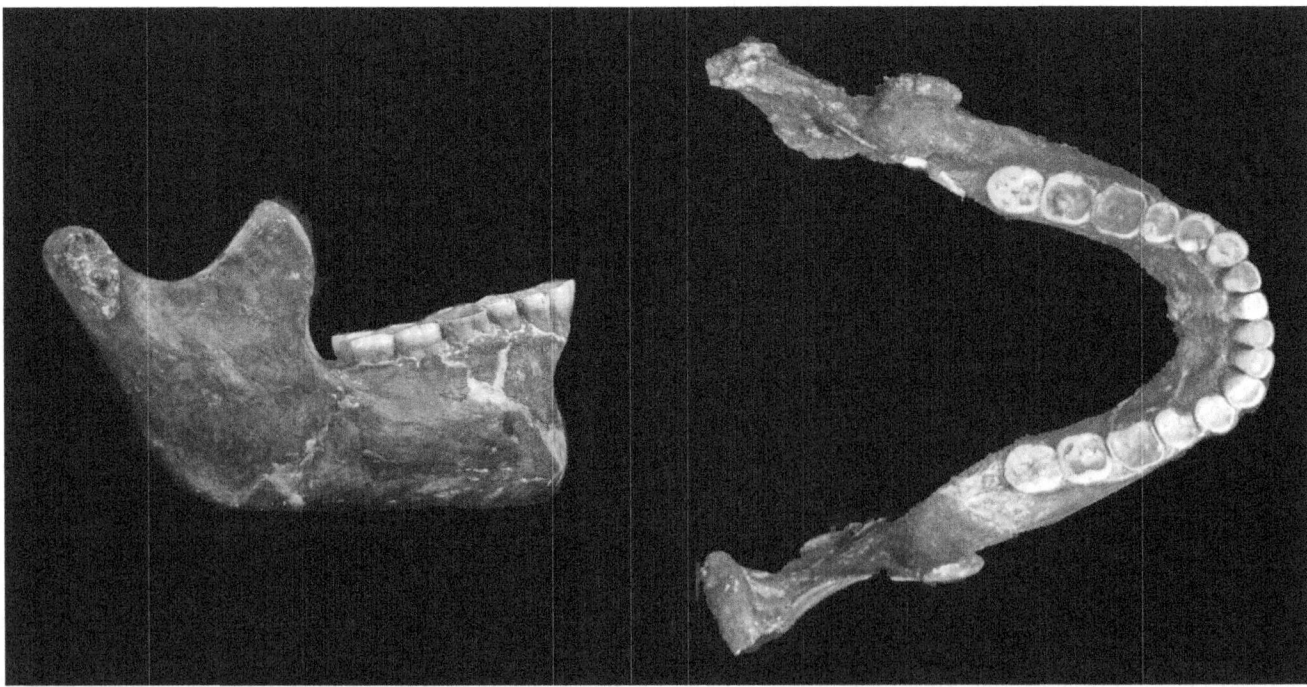

Fig. 15.1 Tabun C2 mandible. Lateral and occlusal views. Image courtesy and copyright © Jeffrey H. Schwartz

More recent analyses have also reached no consensus. Tabun C2 has been attributed to early modern humans by some authors (e.g., Vandermeersch 1981; Bar-Yosef and Pilbeam 1993; Rak 1998; Rak et al. 2002), and has often been described as possessing a distinct chin (McCown and Keith 1939; Vandermeersch 1981; Quam and Smith 1998; Rak 1998), although the lower part of the symphysis is not preserved. Rak (1998) and Rak et al. (2002) pointed out that the ramus and mandibular notch morphology of this specimen lack typical Neanderthal features. He argued that the lateral placement of the mandibular notch crest relative to the condyle, the symmetric shape of the notch, and the presence of a chin align this specimen with early modern humans rather than with Tabun C1 and Neanderthals (Rak 1998). In their analysis of the mandibular notch outline, Rak et al. (2002) found that its morphology clearly distinguishes Neanderthals from modern humans and earlier *Homo erectus* specimens. Tabun C2 fell within the 'generalized' fossil group, including modern humans, early modern specimens from Skhul and Qafzeh, and *H. erectus* specimens, and away from Tabun C1 and the Neanderthal sample.

Other investigations, however, have assigned Tabun C2 to Neanderthals. Stefan and Trinkaus (1998) examined a series of discrete traits and analyzed dental metrics in an effort to elucidate the specimen's affinities. They found that Tabun C2 exhibited an unusual combination of discrete traits, with two features (mental foramen position, mandibular foramen form) aligning it with Neanderthals. Two further features (retromolar space, mandibular notch shape) were found to be ambiguous. *Contra* Rak (1998), Stefan and Trinkaus (1998) considered the notch crest position not to be taxonomically informative. Furthermore, they affirmed that the Tabun C2 symphyseal region is not sufficiently preserved to properly evaluate the presence of a chin. Their analysis of dental crown dimensions, driven by the size of the anterior teeth, classified Tabun C2 as Neanderthal. The authors concluded that their overall results indicate that Tabun C2 should be considered a Neanderthal (Stefan and Trinkaus 1998).

A detailed investigation of the Tabun C2 chin by Schwartz and Tattersall (2000; see also Schwartz and Tattersall 2010) found the preserved portions of the symphysis in this specimen to be neither modern human-like, nor similar to the morphology shown by some of the early modern humans from Skhul. The authors concluded that the specimen is not a *Homo sapiens*, but also hesitated to classify it as a Neanderthal (Schwartz and Tattersall 2000), instead suggesting the possibility of a third taxon. The ambiguous nature of Tabun C2 was also noted by Quam and Smith (1998; see also Quam 1995), who suggested that the ambiguous combination of features might be interpreted as the result of hybridization between Neanderthals and modern humans (Quam and Smith 1998). A more recent analysis of the morphology and size of the anterior dental roots (Le Cabec et al. 2013) was also unable to resolve the controversy. Le Cabec et al. (2013) found that Tabun C2 aligned with Neanderthals in the large size and shape of its anterior tooth roots, but with modern humans in its cynodont molar roots. Since the authors found that Middle Pleistocene specimens show anterior roots similar to those of Neanderthals the morphology exhibited by Tabun C2 could be a primitive retention. However, Le Cabec et al. (2013) could not reject the hypothesis that the specimen might represent a hybrid individual.

Materials and Methods

Samples Our comparative sample comprised 26 fossil mandibles (Table 15.1; see also Nicholson and Harvati 2006). Four European Middle Pleistocene specimens commonly assigned to *Homo heidelbergensis* (HH), eight Neanderthal (NEA), thirteen Upper Paleolithic/Later Stone

Table 15.1 Samples

Comparative fossil samples	Total: 26
Neanderthals (NEA) Amud 1*, Krapina J*, La Ferrassie 1, Shanidar 1*, Tabun C1*, Zafarraya*, Regourdou 1	7
Middle Pleistocene Europeans (MPE) Arago 13*, Mauer 1, Montmaurin, Sima de los Huesos 5*	4
Early Anatomically Modern Humans (EAM) Skhul 5, Qafzeh 9*	2
Upper Paleolithic/Later Stone Age (UP) Grimaldi-Grotte-Des-Enfants 6*†, Isturitz 1950-4-1, Dolní Věstonice 3, 13, 14, 15, 16, Oase, Abri Pataud, Ohalo II, Upper Cave 101* and 103*, Wadi Kubbaniya	13

*Asterisks indicate specimens for which high-quality casts from the AMNH, NYU and MPI-EVA collections were used
†Grimaldi-Grotte-Des-Enfants 6 is a subadult

Age (UP), and two Late Pleistocene early anatomically modern human (EAM) specimens were included. In cases where we were not able to measure the original fossils, high quality casts were measured from the collections of the Division of Anthropology at the American Museum of Natural History (AMNH), the Department of Anthropology, New York University (NYU), and of the Department of Human Evolution, Max Plank Institute for Evolutionary Anthropology (MPI-EVA). All individuals included were adult (with the exception of Grimaldi 6, an adolescent), as determined by a fully erupted permanent dentition. Due to the lack of secure sex assignments for fossil specimens, sexes were pooled in the analysis and shape differences attributed to sexual dimorphism were not explored.

Data The data were collected in the form of three-dimensional coordinates of 26 landmarks using a Microscribe 3DX digitizer, by ENL and KH (Fig. 15.2; for inter- and intra-observer error assessments and landmark definitions see Nicholson and Harvati 2006). Because morphometric analysis does not accommodate missing data, and since many of the fossil specimens were incomplete, some data reconstruction was found to be necessary (see Nicholson and Harvati 2006).

Since Tabun C2 is virtually complete, landmarks were selected to represent the overall shape of the mandible as preserved in this specimen. Although the lower part of the symphysis is missing in Tabun C2, we felt that the

reconstruction of this area was reasonable enough for us to measure gnathion. This point (in conjunction with infradentale) provides an assessment of the corpus supero-inferior height at the symphysis and of symphyseal slope, but does not bear on the evaluation of the chin, one of the proposed modern human-like features of Tabun C2. Since it has been claimed that the impression of a modern human-like chin is partially due to the way that the anterior aspect of the symphysis was reconstructed (Stefan and Trinkaus 1998), we avoided using landmarks or semiland-marks describing the shape of the symphysis in the mid-sagittal plane. Effectively, therefore, we did not include the chin among the features examined, thus removing one of the possible traits indicating modern human affinities for Tabun C2. We also repeated the analysis excluding any landmarks in the symphysis region (for a total of 22 landmarks), so as to assess the impact of this partially reconstructed morphology on our results.

Analysis Landmark coordinates were superimposed using generalized Procrustes analysis (GPA) in Morphologika (O'Higgins and Jones 2004). GPA superimposes the specimen landmark configurations by translating them to a common origin, scaling them to unit centroid size (the square root of the sum of squared distances of all landmarks to the centroid of the object; the measure of size used here), and rotating them according to a best-fit criterion. This procedure allows for the separate analysis of 'shape' and

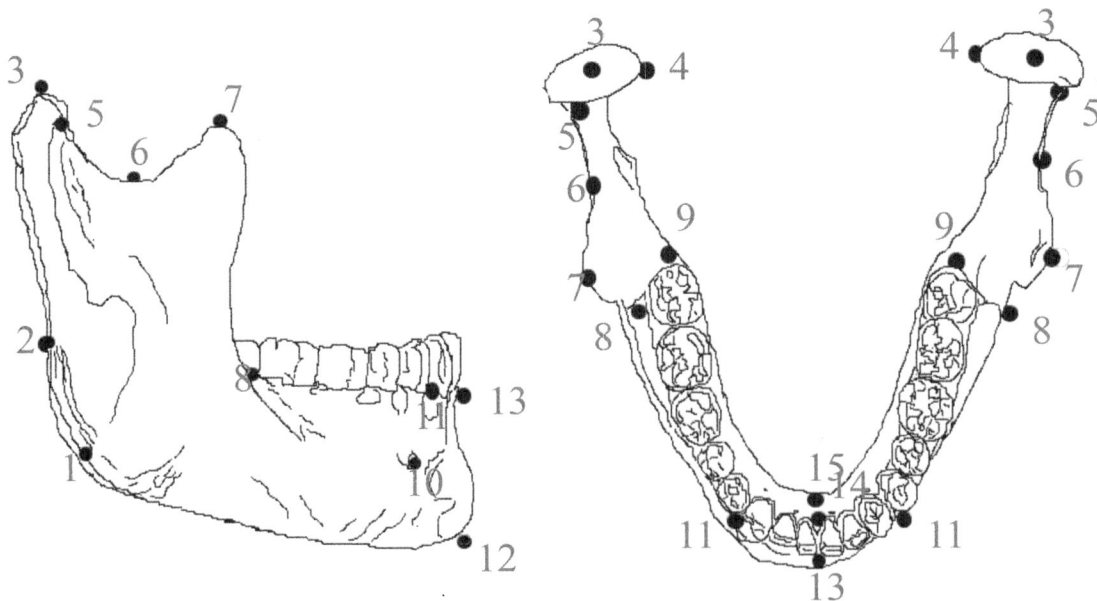

Fig. 15.2 Mandibular landmarks. 1. gonion (right & left), 2. posterior ramus (right & left), 3. condyle tip (right & left), 4. condylion mediale (right & left), 5. root of sigmoid process (right & left), 6. sigmoid notch (right & left), 7. coronion (right & left), 8. anterior ramus (right & left), 9. M3 (right & left), 10. mental foramen (right & left), 11. canine (right & left), 12. *gnathion*, 13. *infradentale*, 14. *mandibular orale*, 15. *superior transverse torus* (Nicholson and Harvati 2006). Landmarks excluded in the 22 landmarks analysis are shown in italics

'size' (although size-related shape differences may remain; Rohlf 1990; Rohlf and Marcus 1993; Slice 1996; O'Higgins and Jones 1998). Procrustes methods have been shown to have higher statistical power than alternative geometric morphometric approaches (Rohlf 2000).

A principal components analysis (PCA) was conducted on the fitted coordinates so as to reduce the variables and explore the patterns of variation present in the data. An ANOVA was performed on centroid size and on the PCA scores to determine the significance of taxonomic effects. For this analysis the two EAM specimens were grouped together with the UP sample as *H. sapiens* (HS). Shape changes along the PC axes were visualized using Morphologika. A discriminant and classification analysis was undertaken using the first 3 principal components (61.07% [26 landmarks] and 63.11% [22 landmarks] of the total variance, chosen on the basis of a scree plot) treating Tabun C2 as unknown, and using UP, EAM, NEA and HH as the a priori groups. Cross-validation classification was performed to evaluate the robustness of the results. All statistical analyses and plots were performed in the Morphologika, SAS and PAST software packages.

Results

Centroid Size

UP and EAM were generally smaller than the two archaic taxa, although the ranges overlapped. Tabun C2's centroid size falls at or close to the upper limit of the UP and EAM range, and within the centroid size range of NEA and HH for both the 26 and the 22 landmarks analyses (Fig. 15.3).

Principal Components Analysis

In the PCA, PC1 (32.83% of the total variance) partially separated the EAM from all other samples, while PC2 (18.64% of total variance) separated the UP from the NEA and HH samples. EAM plotted in an intermediate position but closer to the archaic specimens along these axes (Fig. 15.4, top). Tabun C2 plotted in between Skhul 5 and Qafzeh 9 on the one hand and NEA and HH on the other, though it fell outside the convex hulls of the latter two samples. PC1 was not significant for taxonomic effects. Qafzeh 9 showed a very positive PC1 score and was removed from all other specimens on this axis. The shape changes along PC1 reflected, on the positive end, a narrow, antero-posteriorly (hereafter a-p) elongated mandible with an

anteriorly projecting gnathion and symmetrical mandibular notch; and at the negative end, a wide, a-p shortened mandible, with posteriorly placed gnathion and asymmetric mandibular notch (Fig. 15.4 top).

PC2 was the only axis significant for taxonomic effects (p < 0.0001), separating HS from NEA and HH (although EAM fell with the latter samples along this axis). It was also correlated with centroid size (r = 0.78, p < 0.0001). The correlation between PC2 and centroid size was no longer significant when the taxa were examined separately for either HH (r = −0.07363, p = 0.9264) or NEA (r = 0.58679, p = 0.1661), but remained close to significant for the combined UP and EAM sample (r = 0.50852, p = 0.0529), suggesting that some of the mandibular differences separating the taxa might be allometric. Shape changes along PC2 include many of the described differences between modern humans and Neanderthals, including, on the positive (Neanderthal) end, an asymmetric mandibular notch, a retromolar gap, a more posterior placement of the mental foramen and a posteriorly inclined symphysis (Fig. 15.4 bottom).

Neanderthals were further partially separated from HH along the third principal component (Fig. 15.4 bottom; 9.6% of the total variance). This component approached significance for taxonomic effects (p = 0.07). On these two axes Tabun C2 fell outside the convex hulls of any of the samples, but plotted closest to the HH range, and away from the NEA, EAM and UP (Fig. 15.4, bottom). PC3 was not correlated with centroid size. The shape differences along this axis included, on the negative (HH) end, a more posteriorly inclined symphysis, a (antero-posteriorly) broader ramus, a more posterior placement of gonion and a shallow mandibular notch (Fig. 15.4 bottom).

When the PCA was repeated with the reduced dataset of 22 landmarks, results remained essentially the same (Fig. 15.5).

Discriminant Analysis

When asked to classify to either HH, UP, EAM or NEA, Tabun C2 was classified as HH. However, the cross-validation classification revealed several misclassifications, especially between the HH group and Neanderthals, with two out of seven Neanderthal specimens (La Ferrassie 1 and Tabun C1) being misclassified as HH and two HH (Sima 5 and Arago) misclassified as Neanderthal. One of the two EAM specimens, Skhul 5, was also misclassified as Neanderthal. Results were virtually identical in the 22 landmarks analysis. Summary cross-validation classification results are shown in Table 15.2.

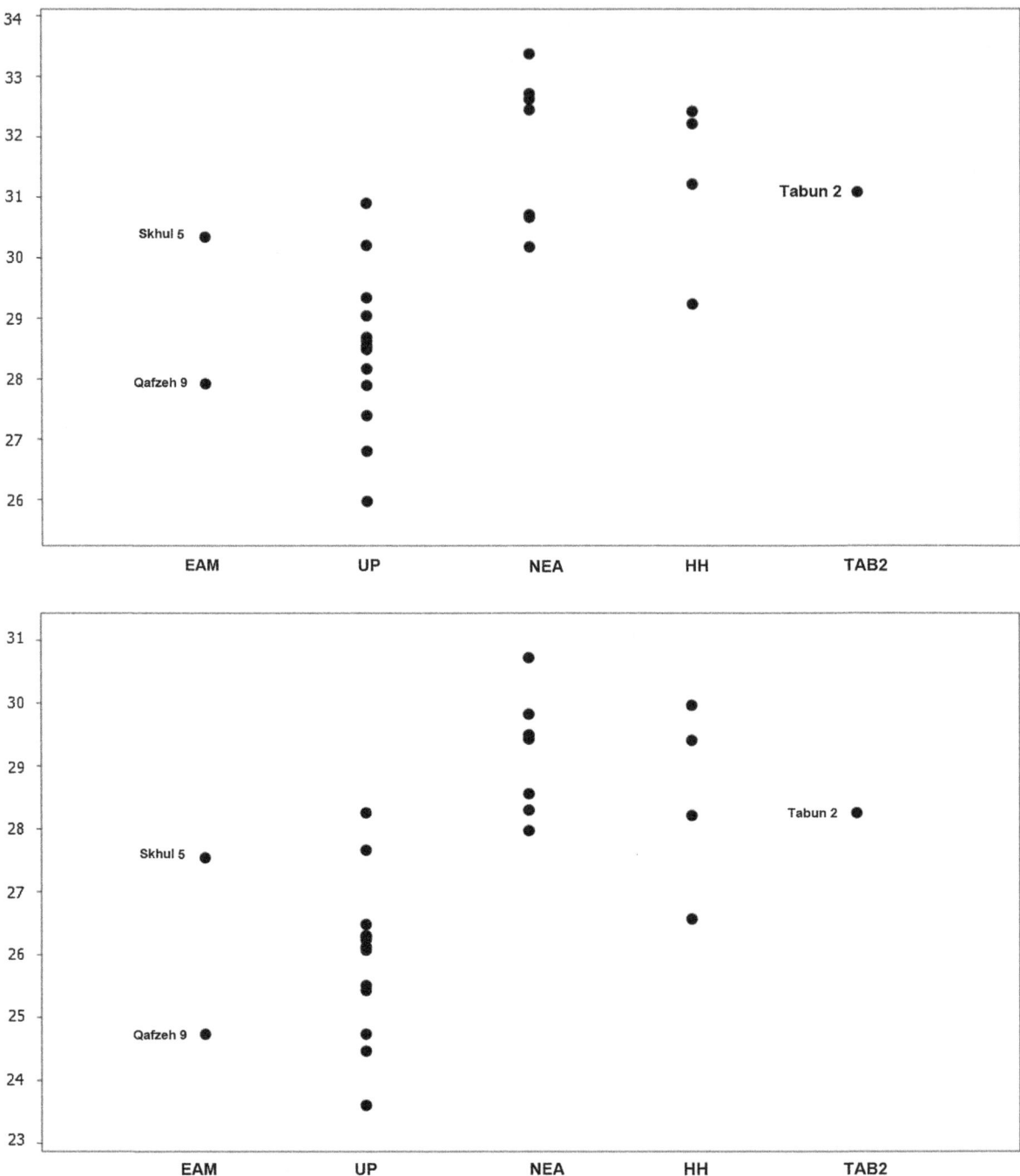

Fig. 15.3 Distribution of centroid size among groups. Labels as in Table 15.1. Top: 26 landmarks analysis; Bottom: 22 landmarks analysis

Procrustes Distances

In terms of Procrustes distances, Tabun C2 was closest to the Sima 5 and the Montmaurin mandibles (0.0880 and 0.1017 respectively), and next closest to Skhul 5 (0.1024). The same specimens were the three closest specimens to Tabun C2 in the 22 landmarks analysis. The Procrustes distances between Tabun C2 and each specimen included in our comparative sample for both the 26 and the 22 landmarks analyses are reported in Table 15.3.

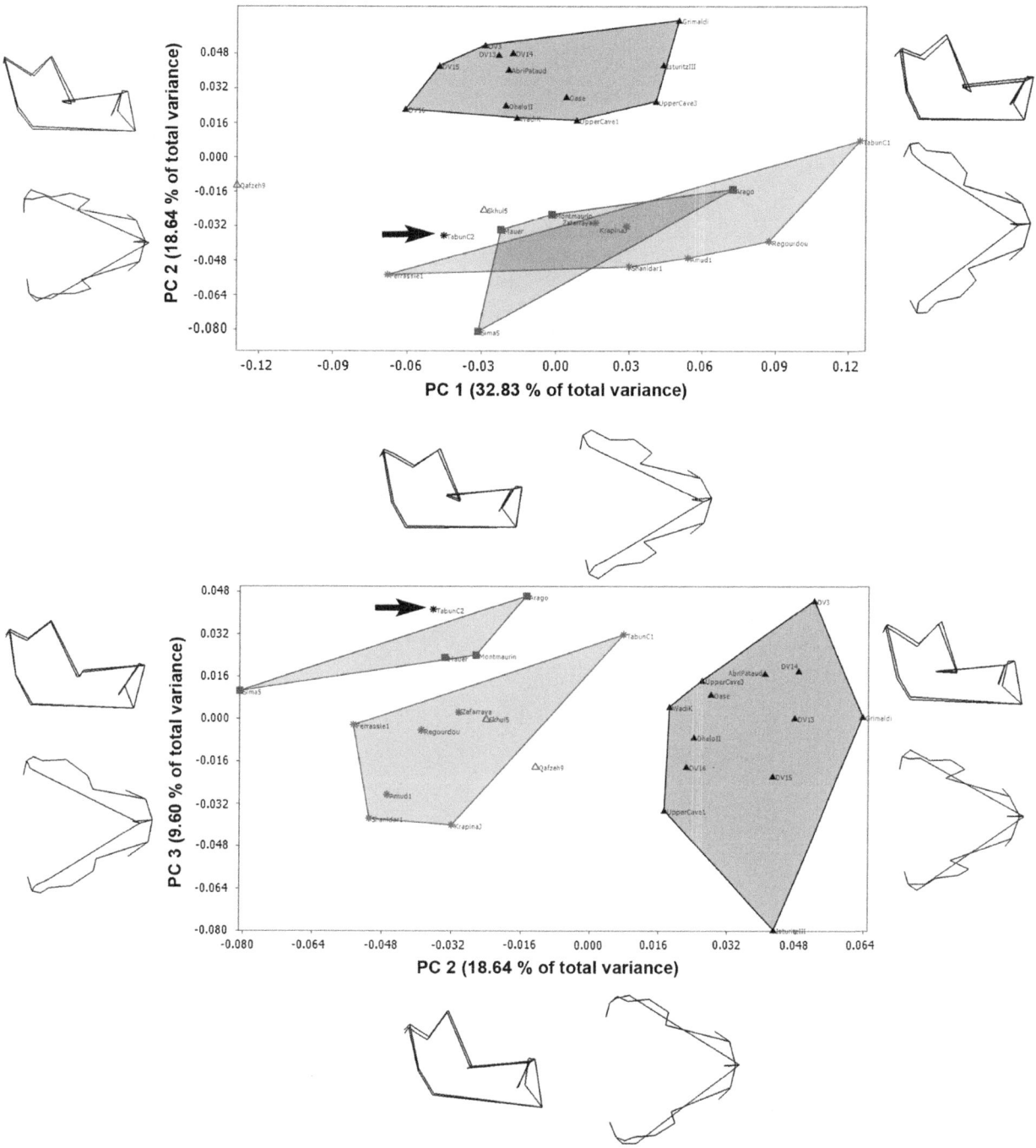

Fig. 15.4 Principal components analysis, 26 landmarks analysis. Top: PC 1 plotted against PC 2. Bottom: PC 2 plotted against PC 3. Shape changes along the principal components are also shown. Black triangles: UP; Grey stars: NEA; Open triangles: EAM; Grey squares: HH; Black start: Tabun C2

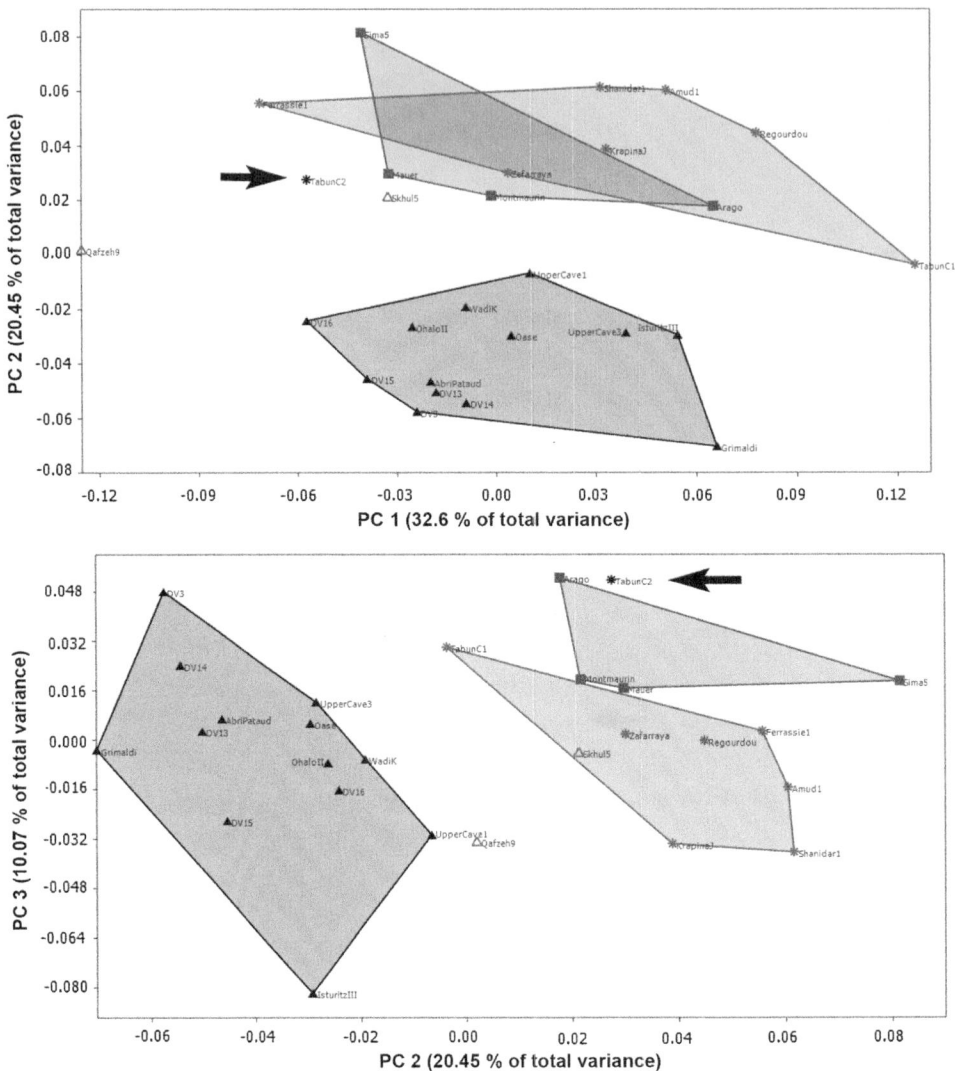

Fig. 15.5 Principal components analysis, 22 landmarks analysis. Top: PC 1 plotted against PC 2. Bottom: PC 2 plotted against PC 3. Symbols as in Fig. 15.4

Table 15.2 Cross validation classification summary. The values are the same for both the 26 and the 22 landmarks analyses

Number of observations and percent classification into group					
From	EAM	HH	NEA	UP	Total
EAM	0	0	1	1	2
	0.00	0.00	50.00	50.00	100.00
HH	0	2	2	0	4
	0.00	50.00	50.00	0.00	100.00
NEA	0	2	5	0	7
	0.00	28.57	71.43	0.00	100.00
UP	0	0	0	13	13
	0.00	0.00	0.00	100.00	100.00
Total	0	4	8	14	

Table 15.3 Procrustes distances between Tabun C2 and each of the specimens included in the comparative sample

26 Landmarks			22 Landmarks		
Specimen	Distance from Tabun 2	Group	Specimen	Distance from Tabun 2	Group
Sima 5	0.0880	HH	Sima 5	0.0903	HH
Montmaurin	0.1017	HH	Montmaurin	0.0945	HH
Skhul 5	0.1024	EAM	Skhul 5	0.1024	EAM
Ohalo II	0.1084	UP	Zafarraya	0.1070	NEA
DV 14	0.1101	UP	Ferrassie 1	0.1084	NEA
DV 16	0.1108	UP	Ohalo II	0.1115	UP
Zafarraya	0.1119	NEA	DV 3	0.1115	UP
Ferrassie 1	0.1122	NEA	DV 16	0.1123	UP
Mauer	0.1124	HH	DV 14	0.1137	UP
DV 3	0.1132	UP	Mauer	0.1142	HH
DV 13	0.1144	UP	DV 13	0.1186	UP
DV 15	0.1220	UP	DV 15	0.1249	UP
Abri Pataud	0.1294	UP	Wadi K.	0.1308	UP
Wadi K.	0.1304	UP	Abri Pataud	0.1320	UP
Oase	0.1306	UP	Upper Cave 101	0.1354	UP
Upper Cave 101	0.1341	UP	Oase	0.1368	UP
Krapina J	0.1343	NEA	Upper Cave 103	0.1396	UP
Upper Cave 103	0.1352	UP	Qafzeh 9	0.1425	EAM
Amud 1	0.1384	NEA	Amud 1	0.1441	NEA
Shanidar 1	0.1441	NEA	Krapina J	0.1458	NEA
Qafzeh 9	0.1477	EAM	Shanidar 1	0.1497	NEA
Arago 13	0.1504	HH	Arago 13	0.1509	HH
Regourdou	0.1557	NEA	Regourdou	0.1605	NEA
Grimaldi	0.1694	UP	Grimaldi	0.1827	UP
Isturitz	0.1825	UP	Isturitz	0.1917	UP
Tabun C1	0.2022	NEA	Tabun C1	0.2055	NEA

Discussion

The results presented should be interpreted with caution. Our approach required relatively complete specimens, and thus limited the sample available for comparison, especially for the early modern humans. It was also based on general mandibular shape, and used relatively few landmarks, thus representing overall, rather than detailed, mandibular shape. Furthermore, the region of the anterior symphysis was not represented by our landmarks, so as to avoid the partially reconstructed chin of Tabun C2. This region is, however, highly informative taxonomically. Finally, the relatively high levels of misclassification between HH and NEA, but also between EAM and NEA, further advise against over-interpretation of our findings.

Nevertheless, our results show that the Tabun C2 overall mandibular shape cannot be easily accommodated either within the Neanderthal or the early modern human range of variation. Although not clearly aligning with either Neanderthals or EAM, Tabun C2 obviously differed from the later UP modern human sample, and generally grouped with the older (including early anatomically modern human) specimens. PC 2, the only axis significant for taxonomic effects,

separated the later UP modern human sample from early anatomically modern humans (EAM), Neanderthals (NEA), European Middle Pleistocene specimens (HH) and Tabun C2, and was correlated with centroid size in both analyses. This indicates that, as also found previously (Nicholson and Harvati 2006), some of mandibular shape differences between modern and archaic humans are influenced by allometry. It also suggests that Tabun C2's large size could be a contributing factor to its archaic-like morphology.

Tabun C2 and Neanderthals Stefan and Trinkaus (1998) concluded that Tabun C2 is best regarded as Neanderthal, although they found its morphology to be ambiguous. Our analyses, however, found no obvious affinity between this specimen and the Neanderthal sample used here. Our PCA could separate Neanderthals from modern humans along PC2, which reflected features commonly described as Neanderthal (e.g., retromolar space, a low condyle relative to the coronoid process, a relatively posterior position of the deepest point of the mandibular notch). Although Tabun C2 generally grouped with the older samples, including HH, EAM and NEA in this analysis, it neither plotted clearly with the NEA sample in the PCA, nor was it classified as Neanderthal in the discriminant analysis.

These findings remained the same when the dataset was reduced to 22 landmarks, indicating that the symphyseal region, which was partly reconstructed in this specimen, plays a minor role in determining our findings. Furthermore, none of the specimens closest to Tabun C2 in total shape, as reflected in Procrustes distance, were Neanderthal (Table 15.2), although some Neanderthals are closer to Tabun C2 in the 22 landmarks analysis. Tabun C2 also showed no particular similarity with Tabun C1, which derives from the same site and, possibly, the same layer (C). It generally plotted away from this specimen in the PCA. Indeed the greatest observed Procrustes distance between Tabun C2 and any of the specimens in both analyses was the Tabun C2 – Tabun C1 distance (0.2022; Table 15.2). While this large distance could be at least in part due to sexual dimorphism (Tabun C1 generally is considered female), it further illustrates the lack of affinities of Tabun C2 for the Neanderthal sample.

Tabun C2 and the Early Modern Human Sample Tabun C2 has also been interpreted as an early modern human (e.g., Rak 1998). Perhaps the clearest result reported here is that Tabun C2 differs from the Upper Paleolithic/ Later Stone Age sample included in our analysis. However, the relationship of Tabun C2 with the early modern human specimens from Skhul and Qafzeh is more difficult to evaluate. This is due in part to the very small number of specimens that could be included in our analysis: only two, Skhul 5 and Qafzeh 9. The interpretation of our results with respect to this sample is further complicated by the extreme position of Qafzeh 9 on PC1, which suggests that distortion affects this mandible's shape (Vandermeersch 1981; see also Nicholson and Harvati 2006). Beyond taphonomic considerations, Qafzeh 9 has recently been described as exhibiting severe malocclusion (Sarig et al. 2013), which may have also affected its shape. Tabun C2 did not plot consistently with the EAM specimens in the PCA. It was also not classified as EAM in the discriminant analysis. Nevertheless, the third closest specimen to Tabun C2 in Procrustes distance was Skhul 5 in both analyses.

Our results therefore do not clearly support an affinity with early modern humans either. However, given the extremely small number of EAM specimens that could be included here, as well as the likely distorted nature of one of them, we consider this outcome inconclusive.

Tabun C2 and the European Middle Pleistocene Sample Our most surprising result was the alignment of Tabun C2 with the European Middle Pleistocene Homo heidelbergensis sample included in our study. It fell closest to HH along PC3 and was classified as HH in both analyses. It also showed the two smallest Procrustes distances, and therefore closest similarity in total shape, with two of the four HH specimens included here (Sima 5 and Montmaurin;

Table 15.2). This result is perplexing, as the HH specimens are much older than the purported possible age of Tabun C2. However, it suggests that Tabun C2's morphology might best be regarded as preserving primitive features. Indeed the features that are reflected by the extreme PC3 scores characteristic of Tabun C2 and HH mandibles include a relatively long ramus antero-posteriorly, a subequal height of the coronoid and condyle, and a relatively receding symphyseal orientation (Fig. 15.5), traits described as characteristic for middle and early Pleistocene specimens (e.g., Mounier et al. 2009). Such a finding is consistent with that of Le Cabec et al. 2013, who also described a mix of archaic, likely primitive, and modern traits in its dental root morphology. A similar pattern of mosaic morphology has been reported for the Middle Pleistocene human dental remains from Qesem cave (Hershkovitz et al. 2011), while generalized primitive morphology has also been proposed for the partial cranial remains from Zuttiyeh (Freidline et al. 2012).

Tabun C2 as a Neanderthal-Early Modern Human Hybrid? Tabun C2 has also been proposed to reflect admixture between Neanderthals and early modern humans. This hypothesis is difficult to evaluate, as there are no clear expectations of how hybridization may be reflected in skeletal morphology (see Harvati et al. 2007; Ackermann 2010; Kelaita and Cortes-Ortiz 2013). Although large, Tabun C2 is not greatly different in size than either proposed parent populations, as might be expected from a hybrid (Ackermann et al. 2006; Harvati et al. 2007) nor does it show any peculiar dental anomalies (Ackermann et al. 2006; Ackermann 2010). Although it has been described as showing a mixture of Neanderthal-like and modern human-like features (e.g., Quam and Smith 1998; Le Cabec et al. 2013), it was not consistently intermediate in overall mandibular shape between the two taxa in our study. We feel, however, that our analysis cannot adequately address this hypothesis.

Conclusions

Our results do not indicate a clear affinity of Tabun C2 with either Neanderthals or early modern humans, and therefore do not support assignment to either taxon. Rather, our findings point to similarity of Tabun C2 with geologically older specimens, and suggest that the large size of the specimen may be a contributing factor to its archaic morphology. We tentatively conclude that Tabun C2 may retain a primitive overall mandibular shape, as might be consistent with its proposed great geological age. Our findings also suggest a possible presence of a third taxon in this region during the later part of the middle Pleistocene.

Acknowledgments We thank Jeffrey Schwartz for kindly providing the photographs of Tabun C2 used in Fig. 15.1, Kieran McNulty and Karen Baab for sharing their SAS routines, and Hugo Reyes-Centeno for help with PAST. We also thank Erella Hovers and Assaf Marom for their invitation to participate in this volume. We are honored to join in the celebration of Yoel Rak's lifelong contributions to paleoanthropology, and take this opportunity to thank him for his friendship and support.

References

Ackermann, R. R. (2010). Phenotypic traits of primate hybrids: Recognizing admixture in the fossil record. *Evolutionary Anthropology, 19*, 258–270.

Ackermann, R. R., Rogers, J., & Cheverud, J. M. (2006). Identifying the morphological signatures of hybridization in primate and human evolution. *Journal of Human Evolution, 51*, 632–645.

Alperson, N., Barzilai, O., Dag, D., Hartman, G., & Matskevich, Z. (2000). The age and context of the Tabun I skeleton: A reply to Schwarcz et al. *Journal of Human Evolution, 38*, 849–853

Bar-Yosef, O., & Callander, J. (1999). The woman from Tabun: Garrod's doubts in historical perspective. *Journal of Human Evolution, 37*, 879–885.

Bar-Yosef, O., & Pilbeam, D. (1993). Dating hominid remains. *Nature, 366*, 415.

Freidline, S. E., Gunz, P., Jankovic, I., Harvati, K., & Hublin, J.-J. (2012). A comprehensive morphometric analysis of the frontal and zygomatic bone of the Zuttiyeh fossil from Israel. *Journal of Human Evolution, 62*, 225–241.

Garrod, D. A. E., & Bate, D. M. A. (1937). *The Stone Age of Mount Carmel*. Oxford: Clarendon Press.

Grün, R., & Stringer, C. (2000). Tabun revisited: Revised ESR chronology and new ESR and U-series analyses of dental material from Tabun C1. *Journal of Human Evolution, 39*, 601–612.

Grün, R., Stringer, C., McDermott, F., Nathan, R., Porat, N., Robertson, S., et al. (2005). U-series and ESR analyses of bones and teeth relating to the human burials from Skhul. *Journal of Human Evolution, 49*, 316–334.

Harvati, K. (2003). Quantitative analysis of Neanderthal temporal bone morphology using 3-D geometric morphometrics. *American Journal of Physical Anthropology, 120*, 323–338.

Harvati, K., Gunz, P., & Grigorescu, D. (2007). Cioclovina (Romania): Affinities of an early modern European. *Journal of Human Evolution, 53*, 732–746.

Hershkovitz, I., Smith, P., Sarig, R., Quam, R., Rodríguez, L., García, R., et al. (2011). Middle Pleistocene dental remains from Qesem Cave (Israel). *American Journal of Physical Anthropology, 144*, 575–592.

Jelinek, A. J. (1982). The Tabun cave and Paleolithic man in the Levant. *Science, 216*, 1369–1375.

Kelaita, M. A., & Cortes-Ortiz, L. (2013). Morphological variation of genetically confirmed *Alouatta pigra* × *A. palliata* hybrids from a natural hybrid zone in Tabasco, Mexico. *American Journal of Physical Anthropology, 150*, 223–234.

Le Cabec, A., Gunz, P., Kupczik, K., Braga, J., & Hublin, J.-J. (2013). Anterior tooth root morphology and size in Neanderthals: Taxonomic and functional implications. *Journal of Human Evolution, 64*, 69–193.

McCown, T. D., & Keith, A. (1939). *The Stone Age of Mount Carmel II: The fossil human remains from the Levalloiso-Mousterian*. Oxford: Clarendon Press.

Mercier, N., & Valladas, H. (2003). Reassessment of TL age estimates of burnt flints from the Paleolithic site of Tabun Cave, Israel. *Journal of Human Evolution, 45*, 401–409.

Millard, A. R., & Pike, A. W. (1999). Uranium-series dating of the Tabun Neanderthal: A cautionary note. *Journal of Human Evolution, 36*, 581–585.

Mounier, A., Marchal, F., & Condemi, S. (2009). Is *Homo heidelbergensis* a distinct species? New insights on the Mauer mandible. *Journal of Human Evolution, 56*, 219–246.

Nicholson, E., & Harvati, K. (2006). Quantitative analysis of human mandibular shape using three-dimensional geometric morphometrics. *American Journal of Physical Anthropology, 131*, 368–383.

O'Higgins, P., & Jones, N. (1998). Facial growth in *Cercocebus torquatus*: An application of three-dimensional geometric morphometric techniques to the study of morphological variation. *Journal of Anatomy, 193*, 251–272.

O'Higgins, P., & Jones, N. (2004). *Morphologika*. York: University of York.

Quam, R. M. (1995). Tabun Too? A morphometric comparison of Upper Pleistocene mandibles. Ph.D. Dissertation, Northern Illinois University.

Quam, R. M., & Smith, F. H. (1998). Reassessment of the Tabun C2 mandible. In T. Akazawa, K. Aoki, & O. Bar-Yosef (Eds.), *Neandertals and modern humans in western Asia* (pp. 405–421). New York: Plenum Press.

Rak, Y. (1998). Does any Mousterian cave present evidence of two hominid species? In T. Akazawa, K. Aoki, & O. Bar-Yosef (Eds.), *Neandertals and modern humans in western Asia* (pp. 353–366). New York: Plenum Press.

Rak, Y., Ginzburg, A., & Geffen, E. (2002). Does *Homo neanderthalensis* play a role in modern human ancestry? The mandibular evidence. *American Journal of Physical Anthropology, 119*, 199–204.

Rohlf, F. J. (1990). Rotational fit (Procrustes) methods. In F. J. Rohlf & F. L. Bookstein (Eds.), *Proceedings of the Michigan morphometrics workshop* (pp. 227–236). Ann Arbor: University of Michigan Museum of Zoology.

Rohlf, F. J. (2000). Statistical power comparisons among alternative morphometric methods. *American Journal of Physical Anthropology, 111*, 463–478.

Rohlf, F. J., & Marcus, L. F. (1993). A revolution in morphometrics. *Trends in Ecology and Evolution, 8*, 129–132.

Sarig, R., Slon, V., Abbas, J., May, H., Shpack, N., Vardimon, A. D., et al. (2013). Malocclusion in early anatomically modern human: A reflection on the etiology of modern dental misalignment. *PLoS ONE, 8*, e80771.

Schwarcz, H. P., Simpson, J. J., & Stringer, C. B. (1998). Neanderthal skeleton from Tabun: U-series data by gamma-ray spectrometry. *Journal of Human Evolution, 35*, 635–645.

Schwartz, J. H., & Tattersall, I. (2000). The human chin revisited: What is it and who has it? *Journal of Human Evolution, 38*, 367–409.

Schwartz, J. H., & Tattersall, I. (2010). Fossil evidence for the origin of *Homo sapiens*. *American Journal of Physical Anthropology, 143*, 94–121.

Slice, D. E. (1996). Three-dimensional generalised resistant fitting and the comparison of least-squares and resistant fit residuals. In L. F. Marcus, M. Corti, A. Loy, Naylor, S. J. P., & D. Slice (Eds.). *Advances in morphometrics* (pp. 179–199). New York: Plenum Press.

Stefan, V., & Trinkaus, E. (1998). La Quina 9 and Neandertal mandibular variability. *Mémoires de la Société d'Anthropoloige de Paris, 10*, 293–324.

Vandermeersch, B. (1981). *Les hommes fossiles de Qafzeh (Israël)*. Paris: Editions CNRS.

Chapter 16
The Dentition of the Earliest Modern Humans: How 'Modern' Are They?

Shara E. Bailey, Timothy D. Weaver, and Jean-Jacques Hublin

Abstract African and Western Asian contemporaries of Neanderthals, generally considered to be the earliest *Homo sapiens*, are not particularly 'modern' looking in their cranial anatomy. Here we test whether the dental morphological signal agrees with this assessment. We used a Bayesian statistical approach to classifying individuals into 'modern' and 'non-modern' groups based on dental non-metric traits. The classification was based on dental trait frequencies for two 'known' samples of 109 Upper Paleolithic *H. sapiens* and 129 Neanderthal individuals. A cross-validation test of these individuals correctly classified them 95% of the time. Our early *H. sapiens* sample included 41 individuals from Southern Africa, Northern Africa and Western Asia. We treated our early *H. sapiens* individuals as 'unknown' and calculated the probability that each belonged to either the Upper Paleolithic or Neanderthal sample. We hypothesized that if the earliest *H. sapiens* were already dentally modern, then they would be assigned to the Upper Paleolithic *H. sapiens* group. We also hypothesized that if there had been significant admixture in Western Asia during the initial dispersal out of Africa, these samples would have the largest proportion of individuals classified as Neanderthal. Our results indicated that the latter was not the case. The smallest proportion of misclassified individuals came from Western Asia (7%) and the highest proportion of misclassified individuals came from Northern Africa (38%). In most cases it appears to be the predominance of primitive features, rather than derived Neanderthal traits that drove the classification. We conclude (1) by the time the earliest *H. sapiens* dispersed from Africa they had already attained a more-or-less modern dental pattern; (2) in the past, as is the case today, Late Pleistocene Africans were not a homogeneous group, some retained primitive dental traits in higher proportions than others. Furthermore, we acknowledge that while our method is an excellent tool for discriminating between Upper Paleolithic *H. sapiens* and Neanderthals, it may not be appropriate for testing Neanderthal – *H. sapiens* admixture because all traits (primitive and derived) are weighed equally. Moreover, to best assess admixture it is likely necessary to incorporate a model for how the traits track population history and/or gene flow.

Keywords Bayesian approach • Dental modernity • *Homo sapiens* • Neanderthals • Qafzeh • Skhul

Introduction

While most anthropologists believe they know what it means to be 'modern', morphological (Pearson 2008) and behavioral (Henshilwood and Marean 2003) 'modernity' have been difficult concepts to pin down. Not all recent *Homo sapiens* have all the hallmarks of morphological modernity and some 'non-modern' humans have traits that are generally considered to be modern (Frayer et al. 1993; Arensburg and Belfer-Cohen 1998). Part of the problem lies in the fact that *Homo sapiens* (past and present) exhibit a wide range of morphological variability (Gunz et al. 2009), which makes it difficult to develop a universal definition of what is (skeletally) modern.

Most studies of morphological modernity have focused on the skull. However, skull morphology, especially facial shape, is known to be significantly influenced by factors

S.E. Bailey (✉)
Department of Anthropology, New York University, New York, NY 10003, USA
e-mail: sbailey@nyu.edu

T.D. Weaver
Department of Anthropology, University of California, Davis, CA 95616, USA
e-mail: tdweaver@ucdavis.edu

J.-J. Hublin
Department of Human Evolution, Max Planck Institute for Evolutionary Anthropology, 04103 Leipzig, Germany
e-mail: hublin@eva.mpg.de

© Springer International Publishing AG 2017
Assaf Marom and Erella Hovers (eds.), *Human Paleontology and Prehistory*,
Vertebrate Paleobiology and Paleoanthropology, DOI: 10.1007/978-3-319-46646-0_16

other than population history, such as climate (Harvati and Weaver 2006). Dental morphological traits have a strong genetic component and they appear to track population history/phylogeny well (Baume and Crawford 1978; Lukacs 1983; Turner 1987, 1992; Hanihara 1990; Scott and Turner 1997; Vargiu et al. 1997; Hawkey 1998; Bailey 2000; Irish 2005). Teeth also preserve exceptionally well in the fossil record.

We recently examined whether or not a universal criterion for dental modernity could be defined (Bailey and Hublin 2013). Like cranial morphology, dental morphology shows a marked range of variation; so much that multiple geographic dental patterns (e.g., Mongoloid, Proto-Sundadont, Indodont, Sub-Saharan African, Afridont, Caucasoid, Eurodont, Sundadont, Sinodont) have been identified in recent humans (Hanihara 1969, 1992; Mayhall et al. 1982; Turner 1990; Hawkey 1998; Irish 1998, 2013; Scott et al. 2013). Our analysis confirmed that, while some populations retain higher frequencies of ancestral (i.e., primitive) dental traits [e.g., *Dryopithecus* molar, moderate incisor shoveling (Irish 1997)] and others show higher frequencies of recently evolved (i.e., derived) dental traits [e.g., double shoveling, four-cusped lower molars (Turner 1983; Irish and Guatelli-Steinberg 2003)], all recent humans show some combination of both primitive and derived traits (Bailey and Hublin 2013). When we expanded our analysis to include fossil hominins we found that, in general, the *H. sapiens* dentition is predominantly characterized by retention of primitive features (Bailey and Hublin 2013): nearly all recent human dental traits are present in varying frequencies in fossil hominins. We did find that some traits appear in *H. sapiens* to the exclusion of other *Homo* species, and this information is useful in distinguishing modern from non-modern. However, few of these appear at the origin of our lineage and most evolve around the time of the Upper Paleolithic (e.g., loss of hypocone, double shoveling). Postcanine dental reduction is a trend that starts in the Pleistocene (Brace and Mahler 1971; Brace et al. 1987) and diminutive postcanine teeth have sometimes been considered a derived feature of *H. sapiens;* however, evidence from the Sima de los Huesos (Bermúdez de Castro and Nicolas 1995; Martinón-Torres et al. 2012), Qesem Cave (Hershkovitz et al. 2010) and perhaps Liang Bua[1] (Brown et al. 2004) suggest that postcanine dental reduction has evolved independently multiple times during the course of human evolution.

Although our attempt to find a unique pattern of dental trait frequencies that characterizes *all H. sapiens* was unsuccessful (Bailey and Hublin 2013), it remains possible that a pattern of trait frequencies can be used to statistically distinguish *H. sapiens* from non-*H. sapiens*. In an earlier study we tested

whether dental morphology alone could be used to distinguish Upper Paleolithic *H. sapiens* from Neanderthals (Bailey et al. 2009). Results of this study were very promising: we demonstrated that dental morphology was 89% successful in correctly identifying Neanderthals and Upper Paleolithic *H. sapiens*. At the time, our goal was to test the claim that it was impossible to identify the makers of early Upper Paleolithic industries because many sites preserved only 'undiagnostic' material (e.g., isolated teeth) (Conard et al. 2004; Henry-Gambier et al. 2004). For that reason, we limited our first study to *H. sapiens* spanning 40–20 ka.

However, the legitimate question has been raised regarding whether or not this same method could be used to distinguish the earliest *H. sapiens* from Neanderthals. Here, the assumption is that the dentitions of earlier *H. sapiens* retain a higher frequency of primitive traits than do those of later *H. sapiens* and may, therefore, be more difficult to distinguish from Neanderthals. Indeed, studies of cranial morphology (Schwartz and Tattersall 2000; Trinkaus 2005) have suggested that the earliest *Homo sapiens* do not have especially modern looking cranial anatomy. If these early modern humans lack many of the derived cranial features found in later *H. sapiens,* perhaps this is true also of their dentition. Higher frequencies of primitive traits in early *H. sapiens* might lead to greater overlap and therefore poorer discrimination between *H. sapiens* and Neanderthals, especially if the Neanderthal dentition was primarily comprised of primitive rather than derived features.

The Western Asian sites of Qafzeh and Skhūl have provided some of the earliest evidence of *H. sapiens* and have historically played an important role in discussion of modern human origins. For several reasons Western Asia holds an important, if not unique, place in modern human evolution. First, it served as one of the most important corridors for the dispersals of modern humans into Eurasia. Second, it was occupied by both Neanderthals and *H. sapiens*. Third, recent interpretation of the genetic evidence for Neanderthal admixture with *H. sapiens* has suggested that an admixture event may have occurred in Western Asia with the earliest *H. sapiens* dispersals (Green et al. 2010).

While few would suggest that all recent human variation is the result of a single out-of-Africa event, there are differences of opinion regarding how many dispersals there were and when these dispersals took place (Lahr and Foley 1998; Stringer 2003). While some have argued for local origin for the earliest *H. sapiens* in Western Asia (Hershkovitz et al. 2010), it has been more commonly agreed that the early *H. sapiens* in Western Asia are the result of the first successful African exodus of *H. sapiens* occurring between 100–90 ka (Schwarcz et al. 1988).

In the earliest publication of the Mt. Carmel human remains, McCown and Keith (1939) recognized the variability in the Western Asian sample. They favored viewing all remains as a local population with mixed characteristics.

---

[1]If *H. floresiensis* is, in fact, derived from large toothed early *Homo*: see Brown and Maeda 2009.

Later, two morphs were recognized ("Neanderthal" and "proto-Cromagnons"), with the belief that the former was continuous with the latter, conforming to the popular multiregional view for modern humans (Hrdlička 1927). The most recent chronological evidence points to intermittent occupation of both Neanderthals and *H. sapiens* with perhaps the earliest occupation by Neanderthals at Tabun, and subsequent occupations by *H. sapiens* (Qafzeh and Skhūl) and later Neanderthals (e.g., Kebara, Amud). These chronological revisions have led most researchers to agree that Neanderthals did not evolve into *H. sapiens* in Western Asia and to treat them as genetically distinct groups (Zilberman et al. 1992; Minugh-Purvis 1998; Rak 1998; Schwartz and Tattersall 2000).

Even if *H. neanderthalensis* is a good evolutionary species (Simpson 1943), molecular evidence suggests they shared a common ancestor with *H. sapiens* fairly recently - between 800 and 300 ka (Krings et al. 1997; Green et al. 2010; Langergraber et al. 2012). Given this relatively recent divergence, there is reason to believe that Neanderthals and *H. sapiens* had the potential to form a syngameon [hybridizing species: (Holliday 2003). Thus, in areas where Neanderthals and *H. sapiens* co-existed, we should expect that they interbred on occasion. Furthermore, if we accept the claims for Neanderthal-*H. sapiens* hybrids (Duarte et al. 1999; Trinkaus et al. 2003; Andrei et al. 2007), it would suggest that at least some of these matings were fertile.

Interestingly, recent interpretation of Neanderthal nuclear DNA has suggested that 1–4% of *H. sapiens* DNA comes from Neanderthals (Green et al. 2010; Sanchéz-Quinto et al. 2012). One explanation posits a Neanderthal-*H. sapiens* admixture event in Western Asia upon the initial *H. sapiens* expansion (Green et al. 2010), although more recent studies suggest the admixture could have occurred later (Currat and Excoffier 2011; Sankararaman et al. 2012; Fu et al. 2014).

While the chronological and archaeological record is of insufficient detail to indicate whether or not these two species actually coexisted and encountered one another, if Neanderthals and *H. sapiens* did interbreed extensively in Western Asia, we might expect to find morphological evidence of such admixture in the dentition. It is difficult to predict exactly what kind of dental pattern to expect in hybridized species of *Homo*; however, previous studies of admixture between dentally divergent recent human groups suggest that an admixture 'signal' shows up in intermediate trait frequencies and expression in 'hybrids' (Hanihara 1963; Baume and Crawford 1978). These intermediate frequencies are predicted to result in poor classification of the hybrids to either parental population or higher misclassification of 'mixed' individuals (Baume and Crawford 1978).

This study had two goals. The first was to apply our previous Bayesian method, used to distinguish Upper Paleolithic *H. sapiens* and *H. neanderthalensis*, to the earliest *H. sapiens*. Although we understand that early *H. sapiens* and Upper Paleolithic *H. sapiens* represent different populations in time and space, we hypothesized that if an identifiable modern human dental pattern emerged early in our lineage, then the earliest *H. sapiens* should classify predominantly as *H. sapiens*, even if Upper Paleolithic modern humans are used to define what is modern. If, on the other hand, the earliest *H. sapiens* are characterized by a primitive dental pattern, lacking the modern trait frequencies of *H. sapiens,* then their classification should be ambiguous.

Second, because we were able to include samples of early *H. sapiens* from different regions of Africa as well as from Western Asia, we predicted that if the above hypothesis was not true (not all early *H. sapiens* classified as such) then the highest frequency of misclassified individuals would be in places where there was the highest likelihood of admixture with Neanderthals. Specifically, if there had been a major admixture event in Western Asia (as proposed by some interpretations of the recent genetic evidence) then a higher percentage of *H. sapiens* in Western Asia would be misclassified as *H. neanderthalensis*.

Materials

Our initial steps were to (1) establish a baseline for modernity, and (2) establish that modern could be differentiated from non-modern (represented by Neanderthals). Recent *H. sapiens* has undergone substantial micro-evolutionary dental changes since dispersing from Africa. These include trait losses (e.g., lower molar hypoconulid), trait additions (e.g., incisor double shoveling) and redistribution of trait frequencies that have resulted in the geographic patterns noted above (Bailey and Hublin 2013). For this reason, we use fossil *H. sapiens* as a baseline for modernity. These fossil *H. sapiens* include samples from approximately 40–11 ka. For simplicity's sake we refer to them as Upper Paleolithic *H. sapiens* even though some – like Oase – are not associated with artifacts (Trinkaus et al. 2004). Our cross-validation test, used to establish that modern and non-modern could be differentiated, was based on sample of 129 Neanderthal and 109 Upper Paleolithic modern human individuals (see Appendix A). Individuals were represented by one to several teeth, depending on preservation. Some individuals were composites of teeth that likely belonged together [like the Krapina 'dental people' (Radovčić et al. 1988)] based on wear, morphology or stratigraphic association. These known samples were used to establish the criteria for *H. sapiens* (a.k.a. modern) and *H. neanderthalensis* (a.k.a. non-modern).

Our early *H. sapiens* sample included 41 individuals from three geographically distinct regions: Qafzeh and Skhūl representing Western Asia; Die Kelders, Equus Cave and Klasies River Mouth representing Southern Africa; and Temara, Dar es Soltan and El Harhoura representing Northern Africa. The age of these Northern African Aterian fossils has been the subject of debate, but the most recent evidence places them in a comparable time period as other early *H. sapiens* (90–35 Kya: Bouzouggar and Barton 2012; Raynal and Occhietti 2012; Richter et al. 2012). The early modern *H. sapiens* sample was our test sample and was treated as 'unknown'.

Methods

The data used in this analysis included 81 dental non-metric traits (Table 16.1). However, given the fragmentary nature of the fossil record, the highest number of traits that could be scored in any one individual was 66, and most individuals preserved fewer than half these traits. The high number of dental traits reflects the fact that our method included scores for the same trait on multiple teeth within a dental field (e.g., first, second and third molars). Normally, this practice would be avoided because it is known that in recent humans the expressions of non-metric dental traits may be moderately correlated within a tooth field (Scott and Turner 1997: 111). Such 'inter-class' correlations have not been investigated in fossil groups, although we have reason to believe that they would differ from those observed in *H. sapiens* (Bailey et al. 2009). Unfortunately, because few fossil individuals preserve multiple teeth within a field (e.g., only 8% of Neanderthals possessed both central and lateral incisors), it was not possible to calculate inter-class correlations in our fossil sample.

One might argue that we should assume non-independence and use a method based only on one (the most stable) tooth in the dental morphogenetic field [the 'key' tooth of Dahlberg (1945), see also Butler (1939)]. It is thought that these teeth (e.g., the central incisor, canine, first premolar and first molar) most accurately reflect the underlying genotype for a particular trait. In an earlier study the 'key tooth' method resulted in a loss of between 9 and 17% of the sample because the tooth of interest was not preserved (Bailey et al. 2009). This same study found that classification using a 'modified key tooth' method (when the key tooth is missing we substituted another tooth within the field) provided highly accurate classification (~91% accuracy), but it still resulted in loss of data points. The 'complete trait' method, which relies on all available information,

was nearly as accurate (~89%) and had the advantage of being able to include a greater number of individuals in the analysis. For this reason, we employed the complete-trait method in this study.

Most traits were scored according to the Arizona State University Dental Anthropology System [ASUDAS: (Turner et al. 1991; Burnett et al. 2010)]. Expression of ASUDAS traits are scored with reference to standard plaques and written descriptions (Turner et al. 1991). For those traits that are not part of the ASUDAS standards outlined by Bailey (2002b) were used. These traits do not have associated reference plaques for scoring expression, but are predominantly scored as present or absent (e.g., LP4 asymmetry). Although trait expressions were scored on both antimeres (if present), only the side with the strongest expression was used in the analysis (the 'individual count' method). It is assumed that this side represents the underlying genetic potential for an individual (Turner and Scott 1977; Scott and Turner 1997). In the analysis, trait expression was dichotomized into 'presence' and 'absence' according to standard breakpoints (Irish 1998; Bailey 2002b). However, the relative occlusal polygon area – a metric trait – was also included because it discriminates well between Neanderthals and *H. sapiens*. In this case, trait presence (reduced relative occlusal polygon area) was defined as being less than 30% of the total crown area.

Classification

Our classification used a Bayesian statistical approach. For each unknown tooth, we calculated probabilities of group membership – either Neanderthal or Upper Paleolithic modern human. These probabilities are based on the frequencies and sample sizes of particular dental traits in our "known" sample of Neanderthals and Upper Paleolithic modern humans. We classified each unknown tooth in whichever group had the higher probability. In addition, we want to point out that:

1. Our approach is weighted by sample sizes for each particular dental trait in the "known" sample. For example, a frequency of 100% in the "known" Neanderthal sample based on one individual is less informative than if it were based on 10 individuals.
2. The calculated probabilities of group membership are posterior probabilities. This means that each tooth is classified as either a Neanderthal or an Upper Paleolithic modern human, even though we know – in the case of the earliest *H. sapiens* – they belong to a different group. We assume that the earliest *H. sapiens* should classify as Upper Paleolithic *H. sapiens* rather than Neanderthals if they have evolved a basically modern human dental pattern.

Table 16.1 List of non-metric dental traits used in this study. 'Presence' score indicated

Tooth/presence	Maxilla	Tooth	Mandible
I^1			
ASU 2-4	Labial convexity		
ASU 2-7	Shoveling		
ASU 1-6	Double Shoveling		
ASU 2-6	*Tuberculum dentale*		
I^2			
ASU 2-4	Shoveling		
ASU 2-6	*Tuberculum dentale*		
$\underline{C}$		C	
ASU 2-4	Shoveling	ASU 2-5	Distal accessory ridge
ASU 2-6	*Tuberculum dentale*		
ASU 1-3	Bushman canine		
ASU 2-5	Distal accessory ridge		
P^3		P_3	
SEB +	Buccal medial ridge	ASU 2-9	Lingual cusp number
SEB +	Lingual medial ridge	SEB 1-2	Transverse crest
SEB bifurcated	Buccal medial ridge form	SEB 2-3	Distal accessory ridge
SEB bifurcated	Lingual medial ridge form	SEB 2-3	Mesial accessory ridge
Burnett +	MxPAR (B)	SEB +	Mesial lingual groove
Burnett +	MxPAR (L)	SEB +	Distal lingual groove
ASU +	Distal accessory cusp	SEB 1-2	Asymmetry
ASU +	Mesial accessory cusp		
	Transverse crest		
P^4		P_4	
	Buccal medial ridge	ASU 2-9	Lingual cusp number
	Lingual medial ridge	SEB 1-2	Transverse crest
SEB bifurcated	Buccal medial ridge form	SEB 2-3	Distal accessory ridge
SEB bifurcated	Lingual medial ridge form	SEB 2-3	Mesial accessory ridge
Burnett +	MxPAR (B)	SEB +	Mesial lingual groove
Burnett +	MxPAR (L)	SEB 1-2	Asymmetry
ASU +	Distal accessory cusp		
ASU +	Mesial accessory cusp		
	Transverse crest		
M^1		M_1	
ASU 1-5	Cusp 5	ASU Y	Y-pattern
ASU 3-7	Carabelli's cusp	ASU 4	Cusp number
SEB < 30%	Occusal polygon area	ASU 2-3	Deflecting wrinkle
SEB +	Mesial accessory cusps	ASU +	Distal trigonid crest
		SEB 1-3	Mid-trigonid crest
		ASU 1-6	Cusp 6
		ASU 2-4	Cusp 7
M^2		M_2	
ASU 0-2	Hypocone reduction	ASU Y	Y-pattern
ASU 1-5	Cusp 5	ASU 4	Four cusped
ASU 3-7	Carabelli's cusp	ASU 2-3	Deflecting wrinkle
SEB +	Mesial accessory cusps	ASU +	Distal trigonid crest
		SEB 1-3	Mid-trigonid crest
		ASU 1-6	Cusp 6
		ASU 2-4	Cusp 7
M^3		M_3	
ASU 0-3	Metacone reduction	ASU Y	Y-pattern
ASU 0-2	Hypocone reduction	ASU 4	Four cusped
ASU 1-5	Cusp 5	ASU 2-3	Deflecting wrinkle
ASU 3-7	Carabelli's cusp	ASU +	Distal trigonid crest
SEB +	Mesial accessory cusp	SEB 1-3	Mid-trigonid crest
ASU P or R	Reduced/peg	ASU 1-6	Cusp 6
		ASU 2-4	Cusp 7

3. In calculating the probabilities, we assumed the dental traits were independent. Intertrait correlations between different traits on different teeth are generally low and differ among populations. Intertrait correlations for the same trait on different teeth are higher (Scott and Turner 1997: 113). However, since we are dealing with fossils (and often isolated teeth) there were few individuals for whom the same trait was scored on more than one tooth (e.g., four of the early *H. sapiens* individuals preserved both upper central and lateral incisors or both lower M1 and M2).

We used cross-validation to test the accuracy of our method in classifying unknown individuals. Performance is a practical, rather than theoretical, problem. Therefore, on some level, it does not really matter what theoretical assumptions the method is based on – i.e., trait independence – as long as we can show empirically that it works. The way cross-validation works is:

1. Select one individual from the "known" sample
2. Classify this individual based on all the OTHER individuals in the "known" sample. This approach is important, because the selected individual is not included in the sample used to calculate the classification; thus it mimics an "unknown" individual.
3. Repeat for all individuals in the "known" sample.
4. Calculate the number of individuals who were correctly classified.

Results

The results of our cross validation test, with the expanded samples, were higher than those of the previous study (Bailey et al. 2009). Here, 93% of the 'known' Neanderthals and 97% of the known Upper Paleolithic *H. sapiens* were correctly assigned to their respective groups with a total of 95% correct discrimination (Table 16.2). Ninety percent of the correctly classified individuals were assigned posterior probabilities above 0.65, and 86% were classified with posterior probabilities above 0.80. Of those correctly classified with posterior probabilities below 0.65, nearly all (91% or 22/24) possessed fewer than 10 traits.

Table 16.3 presents the misclassified individuals. Five of the nine Neanderthals misclassified as Upper Paleolithic *H. sapiens* had posterior probabilities below 0.65. All three of the Upper Paleolithic *H. sapiens* misclassified as Neanderthals were classified with high posterior probabilities (0.88–0.96).

Table 16.4 presents the classification of the early *H. sapiens* individuals. As predicted, most (31/41) were classified as *H. sapiens* based on Upper Paleolithic *H. sapiens* 'standards'. However, compared to results of the cross validation study, a higher than expected number (10/41: 24%) were classified as Neanderthals. These classifications are associated with moderately high posterior probabilities (>0.65). Of the three samples, the Northern African sample had the highest number of misclassified individuals (5/13: 38%), while the Western Asia sample had the fewest (1/14: 7%).

Table 16.2 Cross validation test of Neandertals and Upper Paleolithic *H. sapiens*

	Correct	Incorrect
Neandertal	93% (n = 120)	7% (n = 9)
Upper Paleolithic *H. sapiens*	97% (n = 106)	3% (n = 3)
Total	95% (n = 226)	5% (n = 12)

Table 16.3 Misclassified individuals from the cross-validation test. Posterior probabilities are given with number of traits used in parentheses

Individual	Neandertal	Upper Paleolithic *H. sapiens*
Neandertals		
Krapina 40		0.81 (6)
Krapina DP8		0.75 (5)
Krapina DP25		0.55 (2)
Arcy #9 (Grotte Hyene)		0.62 (6)
Arcy #17 (Grotte Renne)		0.51 (2)
Vindija Vi259		0.98 (4)
Vindija Vi287 (G1)		0.56 (4)
Marillac		0.59 (9)
Tabun C2		0.95 (5)
Upper Paleolithic *H. sapiens*		
Vindija 289 (level Fd)	0.88 (2)	
Mladeč isolated teeth	0.95 (16)	
Oase 2	0.97 (6)	

Table 16.4 Classification of early *H. sapiens* individuals. Posterior probabilities are given with number of traits used in parentheses

	Neandertal	H. sapiens
Southern Africa		
Die Kelders 6242		0.90 (6)
Die Kelders 6258		0.96 (6)
Die Kelders 6277		0.96 (7)
Die Kelders 6279		0.97 (6)
Die Kelders 6280, 6281, 6282, 6275, 6264[a]		0.98 (19)
Equus Cave EQ-H1		0.71 (1)
Equus Cave EQ-H4, EQ-H3[a]	0.74 (11)	
Equus Cave EQ-H5, EQ-H6[a]	0.99 (9)	
Equus Cave EQ-H7, EQ-H12[a]	0.66 (2)	
Equus Cave EQ-H10, EQ-H11[a]	0.77 (6)	
Klasies River Mouth AP-6225		0.99 (13)
Klasies River Mouth AP-6226		1.0 (12)
Klasies River Mouth AP-6227		0.93 (6)
Klasies River Mouth AP-6228, 6229, 2230[a]		1.0 (14)
South Africa classification	*4 (28%)*	*10 (72%)*
Western Asia		
Qafzeh 3		0.59 (17)
Qafzeh 4		0.99 (11)
Qafzeh 5		0.94 (10)
Qafzeh 6		0.51 (10)
Qafzeh 7		0.99 (33)
Qafzeh 8		0.94 (12)
Qafzeh 9		1.0 (73)
Qafzeh 10		0.62 (11)
Qafzeh 11		1.0 (61)
Qafzeh 15		0.88 (7)
Skhūl 1	0.89 (7)	
Skhūl 4		0.95 (9)
Skhūl 5		0.98 (16)
Skhūl 6		0.88 (12)
Western Asia classification	*1 (7%)*	*13 (93%)*
Northern Africa		
El Harhoura		1.0 (19)
Temara mandible		1.0 (29)
Temara 1b-19		0.65 (3)
Temara H7	0.72 (2)	
Temera T4	0.67 (2)	
Temara T3b		0.58 (1)
Temara 3a		0.95 (7)
Temara T1		0.92 (5)
Dar es Soltan II-5	0.75 (5)	
Dar es Soltan II-H4		0.99 (25)
Dar es Soltan H6	0.70 (3)	
Dar es Soltan H9	0.72 (2)	
Dar est Soltan H10		0.53 (2)
Northern Africa classification	*5 (38%)*	*8 (62%)*
Total	**10 (24%)**	**31 (66%)**

[a]*Note* composite individuals based on review of primary literature. Specimens were grouped based on special proximity, state of preservation and inferred age

Discussion

In our earlier study (Bailey et al. 2009), the goal was to develop a method that could correctly assign Neanderthals and Upper Paleolithic *H. sapiens* based on dental non-metric traits alone. We found the method to be very successful and we used it to classify fragmentary remains associated with early Upper Paleolithic industries. In the present study, we applied the same method to early *H. sapiens*. First, we wanted to determine if the method worked on earlier members of our species, hypothesizing that it may be more difficult to classify early *H. sapiens* if they retain a higher frequency of primitive dental traits. Second, we wanted to determine if there was a higher misclassification rate in those early *H. sapiens* samples thought to have co-existed with Neanderthals. If this were the case, significant gene flow between the groups would be supported.

In the earlier study, we found that Upper Paleolithic *H. sapiens* individuals may misclassify as Neanderthals for two reasons: (1) if they possess any derived Neanderthal dental traits (Bailey 2002a, b, 2004; Martinón-Torres et al. 2006; Gómez-Robles et al. 2007) and (2) if they possess a suite of primitive traits found more frequently in Neanderthals than in *H. sapiens* and the most diagnostic teeth (upper M1, lower P4, etc.) are missing or are unscorable. For example, the early European *H. sapiens* Oase 2 maxilla (but not Oase 1 mandible) classified as Neanderthal in the absence of any derived Neanderthal traits (the most diagnostic teeth were absent or could not be scored).

We also found that there was a relationship between posterior probabilities and correct classification (Bailey et al. 2009). In the previous study, about 95% of the individuals were classified correctly when posterior probabilities were 0.65 or greater; the misclassification rate was much higher when posterior probabilities were below 0.65. We also found that low posterior probabilities were often associated with few observable traits. In the present study, nearly all misclassified early *H. sapiens* possessed fewer than 10 traits.

North Africa

The Northern African Aterian sample had the highest frequency (38%) of misclassified individuals, all of which possessed fewer than 10 traits. However, all misclassified individuals had posterior probabilities above 0.65, which suggests that trait number alone may not explain the misclassification. Upon examination, the Northern African material shows a generally primitive morphology with mass additive traits (e.g., large hypocone, upper molar Cusp 5) and do not resemble Neanderthals per se (Fig. 16.1). In fact, the split hypocone and multiple distal cusps observed on several of the North African upper molars are not observed on Neanderthal upper molars (SEB personal observation).

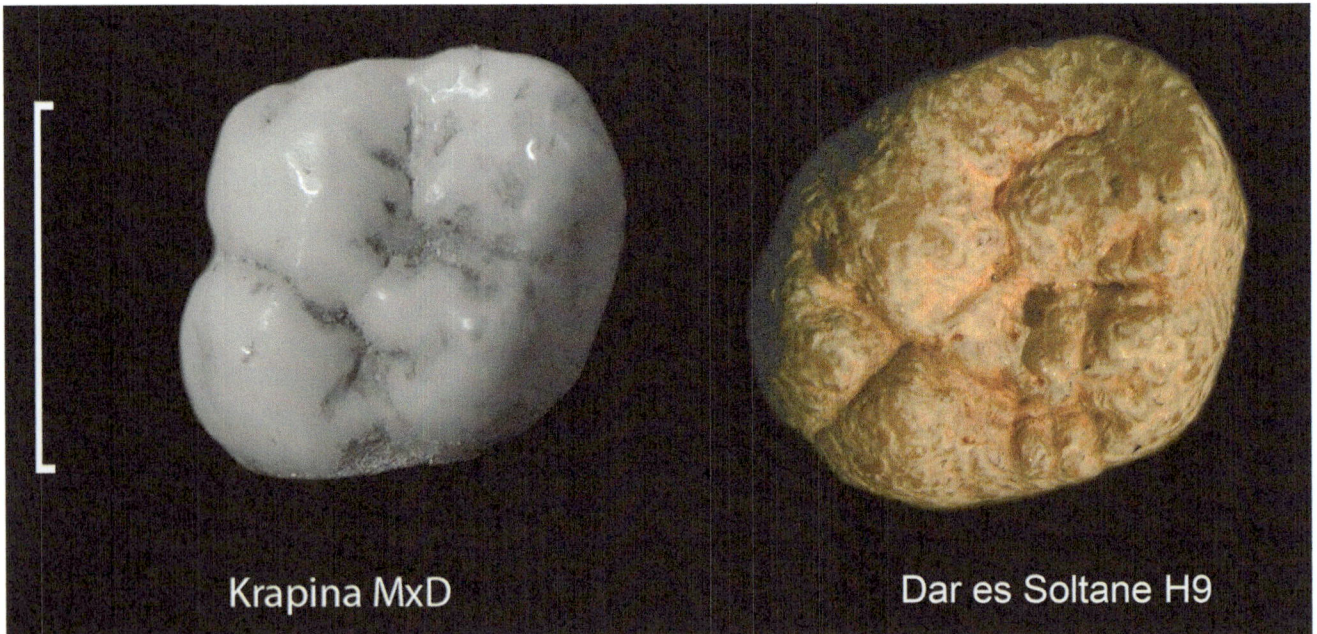

Fig. 16.1 Comparison of a Neanderthal upper first molar (left) to that of a North African Aterian (right). Both are morphologically complex but the split hypocone and double distal cusps are typical in Aterians and not found in Neanderthals. Scale bar is 1 cm

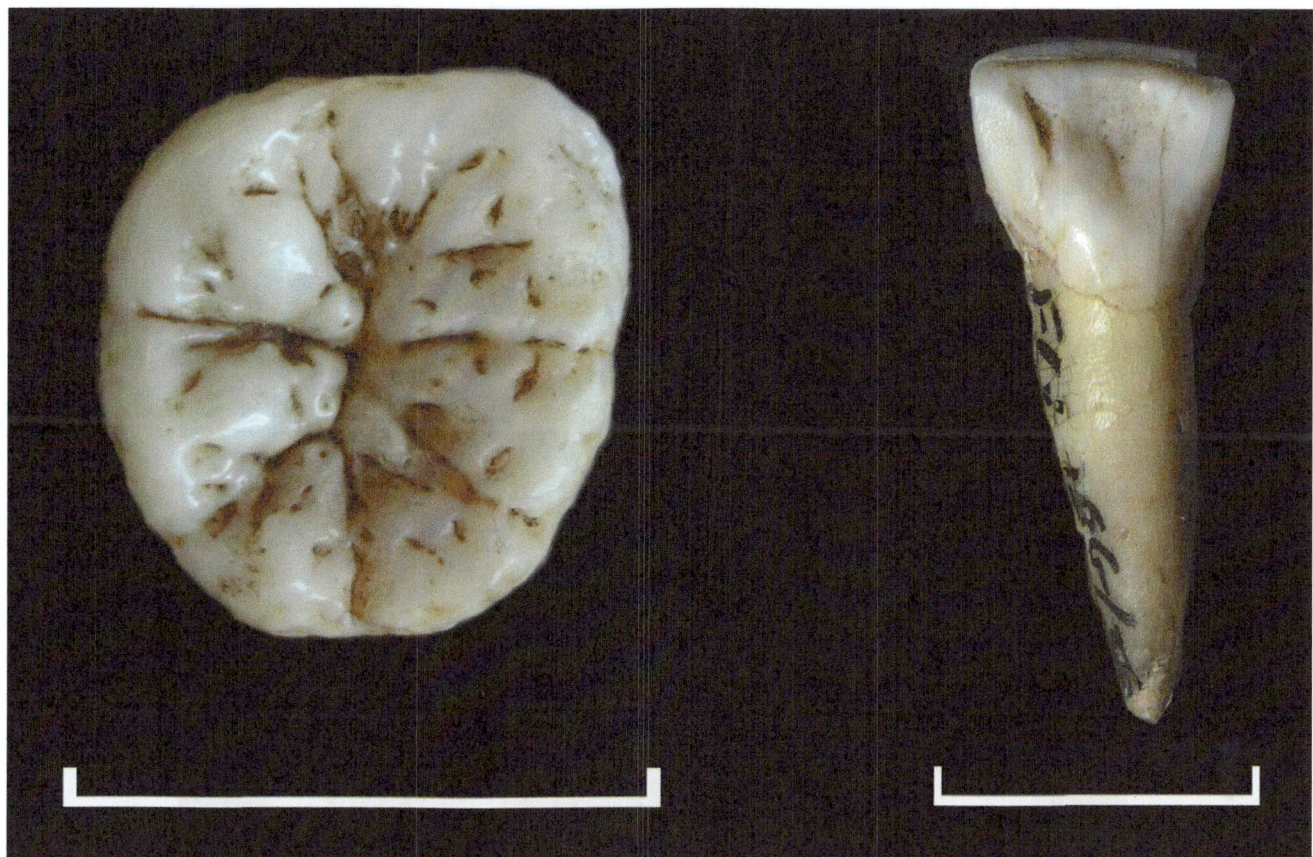

Fig. 16.2 The composite individual from Equus Cave, South Africa (EQ H4 and EQ-H3) that classified as a Neanderthal with a posterior probability of 0.75. The morphologically complex M3 (left) lacks a middle trigonid crest, which is typical of Neanderthals; however, the incisor (right), possesses well developed shoveling, lingual tubercles and labial convexity, which are observed in Neanderthals in high frequencies. Each scale bars represents 1 cm

Southern Africa

The relatively high frequency (29%) of misclassified individuals in the Southern African sample is more difficult to interpret. Closer inspection (Table 16.4) shows that the sample comes from a single site: Equus Cave. All but one of these individuals were classified with fewer than 10 traits. But as was the case with the Northern African material, all had posterior probabilities of 0.65 or higher. A composite individual (EQ H4 and EQ-H3) was classified as a Neanderthal with 11 traits and a posterior probability of 0.75. This individual consists of an isolated incisor and lower third molar (Fig. 16.2). The incisor possesses a combination of characters most often seen in Neanderthals: well-developed shoveling and lingual tubercles, moderate convexity. The lower third molar is morphologically complex with a large Cusp 6. If classified on, its own the third molar would likely have classified as modern given its lack of a middle trigonid crest,

which is nearly ubiquitous in Neanderthals (Bailey 2002a; Bailey et al. 2011). The incisor, however, would have classified as a Neanderthal with high posterior probability. The incisor, and to some degree the molar, contrasts sharply with teeth from Die Kelders and Klasies River Mouth (Fig. 16.3), which exhibit more simplified, and thus modern, morphology.

Western Asia

We hypothesized that if the 1–4% Neanderthal contribution to the recent human gene pool was due to admixture that occurred when the first *H. sapiens* left Africa about 100,000 years ago (Green et al. 2010), then the early *H. sapiens* representing that dispersal would have the highest misclassification rate. In fact, these fossils (represented by Qafzeh and Skhūl) had the lowest misclassification rate of all early *H. sapiens*. All but three individuals classified as

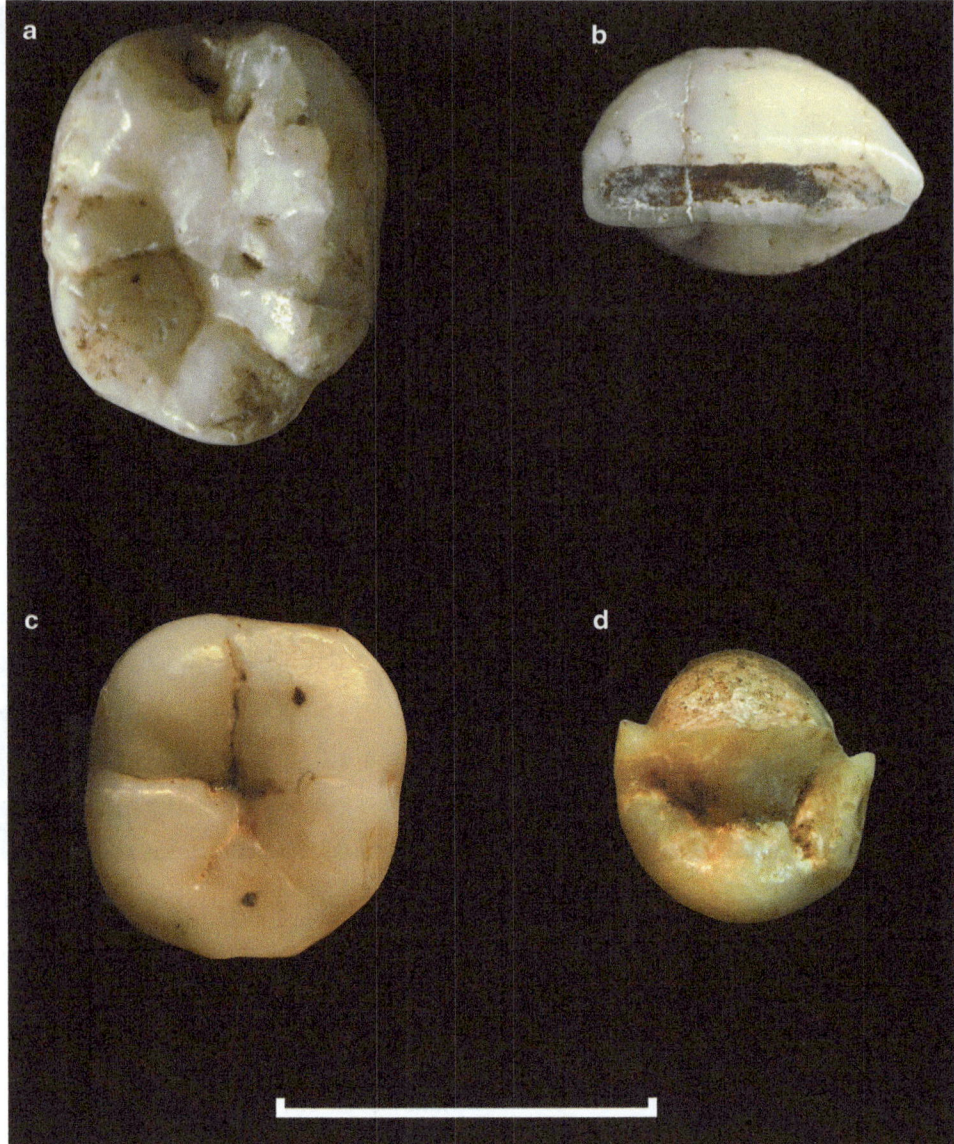

Fig. 16.3 Morphologically simplified teeth from Die Kelders (**a**, **b**) and Klasies River Mouth (**c**, **d**) contrast with the more complex teeth from Equus Cave (see Fig. 16.2). Scale bar is 1 cm

modern with high posterior probabilities (>0.88). The single misclassification was the Skhūl 1 child, which classified as a Neanderthal with a high posterior probability (0.89).

It is often assumed that Neanderthals and *H. sapiens* coexisted in Western Asia. However, given the contentiousness of identification and stratigraphic association of the material from Tabun layer C (Vandermeersch 1989; Quam and Smith 1998; Rak 1998; Stefan and Trinkaus 1998), it is unknown precisely when humans (Neanderthal or *H. sapiens)*

first appeared in the area. It is also unclear whether Neanderthals and *H. sapiens* occupied the region intermittently or continuously (Valladas et al. 1998). The overwhelming modern dental signal in the Qafzeh/Skhūl material might be expected if the two species (a) occupied the region intermittently and never encountered one another or (b) coexisted but did not choose each other as mates. Either way, the dental material from these two sites does not provide strong evidence of interbreeding between groups. That said, the Skhūl 1 child is an interesting case considering its classification can be attributed, in part, to the middle-trigonid crest on the lower M1 and its skewed upper M1. The mandible of this specimen exhibits primitive morphology most often attributed to Neanderthals (lack of mental eminence and strongly parabolic inferior aspect of the mandible: personal observation; see also Schwartz and Tattersall 2000).

Unexpectedly, the least modern signal comes from the Northern and Southern African samples, where there is no evidence of Neanderthals and *H. sapiens* coexisting. In Northern Africa the non-modern signal can be primarily attributed to high frequencies of primitive dental traits. But in Southern Africa several of the individuals possess traits that are thought to have uniquely high frequencies in Neanderthals. Rather than invoke excursions into Africa by Neanderthals, we believe the most parsimonious explanation may be that some of the traits previously thought to be unique to Neanderthals simply are not. While the diagnostic Neanderthal incisor morphology (combination of strong labial convexity, shoveling, and tubercle development) is not found in Early and Middle Pleistocene Asians, nor in Early Pleistocene Africans (Bailey 2002b), it is present in moderately high frequencies in Middle Pleistocene Europeans. The possibility remains that these traits were also present in moderate-to-high frequencies in some Middle Pleistocene Africans.

Unfortunately, the African Middle Pleistocene dental sample is not well documented (Manzi 2004; Rightmire 2008) and cross validation tests require relatively large samples to accurately reflect misclassification rates. Therefore it was not possible to include a large non-Neanderthal archaic *Homo* sample in our analysis that would be necessary to test this hypothesis. But even without this sample, the results of this study suggest that some dental features previously identified as Neanderthal autapomorphies (e.g., lower molar middle trigonid crest, strong incisor convexity/tubercle/marginal ridge configuration) may not be unique to Neanderthals and their direct ancestors. Although *H. antecessor* was not considered in this study, the skewed outline shape of the Neanderthal upper first molar has been found in >780,000 year old specimens from Gran Dolina (Gómez-Robles et al. 2007). Larger samples and/or improved access to Middle Pleistocene hominin fossils will allow us better ascertain trait polarity.

The presence of a strong modern signal at two of the earliest *H. sapiens* sites (Qafzeh and Skhūl) suggests that dental modernity does appear early in our lineage. However, the marked heterogeneity in the African sample – even between sites in the same geographic region – suggests that Late Pleistocene Africans are not a dentally homogeneous group. Some populations appear to have retained higher frequencies of primitive characteristics than others.

Conclusions

We found that 66% of the earliest *H. sapiens* classified as modern. This rate suggests that a basic modern human dental pattern (combination of primitive and derived traits) does exist and that it emerged early during the evolution of our lineage. However geographic diversity also exists: the Northern African sample showed the weakest modern signal, and the Western Asian sample showed the strongest. The strong modern signal in Western Asia argues against significant admixture between Neanderthals and *H. sapiens* in this region. In Southern Africa, two of the three sites had an overwhelming modern signal (Klasies River Mouth and Die Kelders), while one (Equus Cave) had a more mixed signal with a majority of the individuals classifying as Neanderthal. We conclude that in the past, as it is the case today, Late Pleistocene Africans were not a homogeneous group and some retained primitive dental traits in higher proportions than others.

In the face a moderate frequency of African material classifying as Neanderthal, we conclude that our method may not be the best to address the hybridization question since fragmentary remains may not preserve the most diagnostic traits and because we may have to consider that some Neanderthal dental traits may be more primitive than we once believed.

Finally, although classifications using more than 10 traits was much more accurate than those using fewer, if we had limited ourselves to individuals possessing more than 10 traits we would have gained accuracy at the loss of applicability. In this study nearly half of the individuals possessed fewer than 10 traits and would have to have been eliminated from the analysis. While it is true that many of our correctly classified individuals in the cross validation study possessed fewer than 10 traits, the success of our method depends on which teeth are preserved. Individuals correctly classified with fewer than 10 traits tended to preserve teeth that are diagnostic for Neanderthals (upper M1, lower P4, lower molars). Therefore, we recommend all individuals (regardless of trait number) should be considered, but that individuals with few traits that do not include key teeth should be interpreted cautiously.

Appendix A
List of Neanderthal and Upper Paleolithic Specimens, Number of Traits Preserved, Posterior Probabilities and Classification

Specimen	No. traits used	Neanderthal	UP Modern	Classification
Krapina DP#1	24	1	0	Neanderthal
Krapina DP#2	23	1	0	Neanderthal
Krapina DP#3	22	1	0	Neanderthal
Krapina DP#4	62	1	0	Neanderthal
Krapina DP#5	39	1	0	Neanderthal
Krapina DP#6	43	1	0	Neanderthal
Krapina DP#8	5	0.25	0.75	UP Modern
Krapina DP#10	20	1	0	Neanderthal
Krapina DP#11	11	0.94	0.06	Neanderthal
Krapina DP#12	18	1	0	Neanderthal
Krapina DP#13	16	1	0	Neanderthal
Krapina DP#17	9	0.98	0.02	Neanderthal
Krapina DP#18	33	1	0	Neanderthal
Krapina DP#19	32	1	0	Neanderthal
Krapina DP#20	17	1	0	Neanderthal
Krapina DP#21	5	0.97	0.03	Neanderthal
Krapina DP#22	8	0.99	0.01	Neanderthal
Krapina DP#23	53	1	0	Neanderthal
Krapina DP#24	8	0.97	0.03	Neanderthal
Krapina DP#25	2	0.45	0.55	UP Modern
Krapina DP#27	26	1	0	Neanderthal
Krapina DP#28	7	0.99	0.01	Neanderthal
Krapina DP#29	6	0.99	0.01	Neanderthal
Krapina DP#30	10	1	0	Neanderthal
Krapina DP#31	13	0.99	0.01	Neanderthal
Krapina DP#32	12	0.89	0.11	Neanderthal
Krapina DP#33	9	0.86	0.14	Neanderthal
Krapina DP#34	10	0.99	0.01	Neanderthal
Krapina DP#35	6	0.99	0.01	Neanderthal
Krapina 40	6	0.19	0.81	UP Modern
Krapina Maxilla B	3	0.84	0.16	Neanderthal
Krapina Maxilla and mandible C	24	1	0	Neanderthal
Krapina Composite 1	52	1	0	Neanderthal
Krapina Composite 2	48	1	0	Neanderthal
Krapina Composite 3	14	0.99	0.01	Neanderthal
Malarnaud	6	0.96	0.04	Neanderthal
Monsempron 1953-1	37	1	0	Neanderthal
Monsempron miscellaneous	6	0.92	0.08	Neanderthal
Regourdou	23	1	0	Neanderthal
Arcy-sur-Cure #40 (Renne)	8	0.81	0.19	Neanderthal
Arcy-sur-Cure #39 (Renne)	4	0.78	0.22	Neanderthal
Arcy-sur-Cure #41 (Renne)	6	0.92	0.08	Neanderthal
Arcy-sur-Cure #43 (Renne)	5	0.99	0.01	Neanderthal
Arcy-sur-Cure #45 (Renne)	4	0.90	0.10	Neanderthal
Arcy-sur-Cure #IVb6B11-Z11 (Hyène)	7	0.77	0.23	Neanderthal
Arcy-sur-Cure #9 (Hyène)	6	0.38	0.62	UP Modern
Arcy-sur-Cure #13 (Renne)	7	1	0	Neanderthal
Arcy-sur-Cure #4 (Renne)	6	0.85	0.15	Neanderthal
Arcy-sur-Cure #16 (Renne)	1	0.62	0.38	Neanderthal

(continued)

Specimen	No. traits used	Neanderthal	UP Modern	Classification
Arcy-sur-Cure #19 (Renne)	2	0.85	0.15	Neanderthal
Arcy-sur-Cure #20 (Renne)	8	0.89	0.11	Neanderthal
Arcy-sur-Cure #7 (Renne)	1	0.62	0.38	Neanderthal
Arcy-sur-Cure #5 (Renne)	7	0.99	0.01	Neanderthal
Arcy-sur-Cure #35 (Renne)	7	0.87	0.13	Neanderthal
Arcy-sur-Cure #6 (Renne)	1	0.61	0.39	Neanderthal
Arcy-sur-Cure #17 (Renne)	2	0.49	0.51	UP Modern
Arcy-sur-Cure #23 (Renne)	2	0.85	0.15	Neanderthal
Arcy-sur-Cure #24 (Renne)	7	1	0	Neanderthal
Arcy-sur-Cure #21 (Renne)	7	0.97	0.03	Neanderthal
Arcy-sur-Cure #32 (Renne)	1	0.61	0.39	Neanderthal
Arcy-sur-Cure #30 (Renne)	7	0.98	0.02	Neanderthal
Valgadoba 1	34	1	0	Neanderthal
Valgadoba 2	12	1	0	Neanderthal
Devils Tower (Gibraltar II)	6	0.61	0.39	Neanderthal
Ochoz	9	0.65	0.35	Neanderthal
Kůlna	23	1	0	Neanderthal
Petit-Puymoyen 211	14	0.98	0.02	Neanderthal
Petit-Puymoyen 1975-30-5	26	1	0	Neanderthal
Petit-Puymoyen 3	10	0.99	0.0116	Neanderthal
Petit-Puymoyen 2	11	0.94	0.0648	Neanderthal
Petit-Puymoyen 4B	7	1	0	Neanderthal
Petit-Puymoyen 4A	7	0.99	0.01	Neanderthal
Petit-Puymoyen 1	18	0.98	0.02	Neanderthal
Pech de l'Azé	1	0.65	0.35	Neanderthal
Hortus III	23	1	0	Neanderthal
Hortus II	15	1	0	Neanderthal
Hortus IV	16	1	0	Neanderthal
Hortus V	22	1	0	Neanderthal
Hortus VI	12	0.98	0.02	Neanderthal
Hortus VII	6	0.99	0.01	Neanderthal
Taubach	6	0.96	0.04	Neanderthal
La Fate VI	5	0.97	0.03	Neanderthal
La Fate XII	4	0.93	0.07	Neanderthal
La Fate 2	7	0.98	0.02	Neanderthal
Roc du Marsal	3	0.98	0.02	Neanderthal
Ciota Ciara #2	7	0.85	0.15	Neanderthal
Ciota Ciara #3	9	0.89	0.11	Neanderthal
Grotte Taddeo Rep H	5	0.92	0.08	Neanderthal
Grotte Taddeo Rep L	8	0.98	0.02	Neanderthal
Guattari III	14	1	0	Neanderthal
Saccopastore 2	14	0.96	0.04	Neanderthal
Saccopastore 1	4	0.62	0.38	Neanderthal
Scladina	27	1	0	Neanderthal
Vindija 2-Vi149	2	0.53	0.47	Neanderthal
Vindija - Vi146 (231)	9	0.76	0.24	Neanderthal
Vindija - Vi259	4	0.02	0.98	UP Modern
Vindija - Vi148 (266)	16	1	0	Neanderthal
Vindija 287 (level G1)	4	0.44	0.56	UP Modern
Vindija 290 (level G1)	4	0.95	0.05	Neanderthal
Vindija - Vi 76 (229)	1	0.61	0.39	Neanderthal
Spy 1	9	0.96	0.04	Neanderthal
Spy 2	5	0.60	0.40	Neanderthal
Le Moustier	66	1	0	Neanderthal

(continued)

Specimen	No. traits used	Neanderthal	UP Modern	Classification
La Quina 5	17	1	0	Neanderthal
La Quina 9	16	0.92	0.08	Neanderthal
La Quina 18	8	1	0	Neanderthal
Montgaudier 5	7	0.76	0.24	Neanderthal
Combe Grenal (10&11)	13	1	0	Neanderthal
Combe Grenal 5	3	0.55	0.45	Neanderthal
Combe Grenal 4	11	0.95	0.05	Neanderthal
Combe Grenal 1	21	1	0	Neanderthal
Combe Grenal 29	6	0.85	0.15	Neanderthal
Châteauneuf 2	6	1	0	Neanderthal
Marillac	9	0.41	0.59	UP Modern
La Ferrassie 10	7	0.68	0.32	Neanderthal
Suba-lyuk 1	18	1	0	Neanderthal
Suba-lyuk 2	4	0.90	0.10	Neanderthal
Cova Negra	16	1	0	Neanderthal
St Césaire 1	49	0.96	0.04	Neanderthal
St Césaire 2	3	0.56	0.44	Neanderthal
Obi Rakhmat	27	1	0	Neanderthal
Amud 2	25	1	0	Neanderthal
Kebara 14	7	0.98	0.02	Neanderthal
Kebara 4	7	0.92	0.08	Neanderthal
Kebara 2	22	1	0	Neanderthal
Shanidar 2	25	0.94	0.06	Neanderthal
Tabun C1	36	1	0	Neanderthal
Tabun C2	5	0.05	0.95	UP Modern
Tabun Ser III Harvard	14	1	0	Neanderthal
Les Vachons	11	0.17	0.83	UP Modern
Roc de Combe 4	2	0.18	0.82	UP Modern
Lagar Velho	49	0	1	UP Modern
Dolní Věstonice 13	35	0	1	UP Modern
Dolní Věstonice 14	37	0	1	UP Modern
Dolní Věstonice 15	47	0	1	UP Modern
Dolní Věstonice 31	6	00	1	UP Modern
Dolní Věstonice 37	5	0.02	0.98	UP Modern
Dolní Věstonice 36	9	0.14	0.86	UP Modern
Vindija 289 level Fd	2	0.88	0.12	Neanderthal
Parpalló	33	0	1	UP Modern
Pavlov 3	7	0.11	0.89	UP Modern
Pavlov 2	8	0.48	0.52	UP Modern
Abri Pataud 1	51	0	1	UP Modern
Abri Pataud 2	3	0.14	0.86	UP Modern
Abri Blanchard 1956-46	7	0.01	0.99	UP Modern
Abri Labatut 1956-47	10	0.30	0.70	UP Modern
Mieslingtal	6	0.05	0.95	UP Modern
Grotte des Abeilles	11	0	1	UP Modern
Grotte des Abeilles 3	4	0.01	0.99	UP Modern
Lespugue	13	0.01	0.99	UP Modern
La Gravette	3	0.49	0.51	UP Modern
Balla-barlang 68.145.1	7	0.08	0.92	UP Modern
Bervavolgy 68.142.1	10	0.01	0.99	UP Modern
Gruta do Caldeirão 1	8	0.17	0.83	UP Modern
Cisterna 1	17	0	1	UP Modern
La Madeleine	11	0.01	0.99	UP Modern
Pech de la Boissière 1	7	0.05	0.95	UP Modern

(continued)

Specimen	No. traits used	Neanderthal	UP Modern	Classification
Pech de la Boissière 2	6	0.40	0.60	UP Modern
Farincourt	4	0.30	0.70	UP Modern
Farincourt	27	0	1	UP Modern
Laugerie Basse	48	0	1	UP Modern
St. Germaine-la Rivière 19,20	13	0.22	0.78	UP Modern
St. Germaine-la-Rivière (unnumbered)	4	0.48	0.52	UP Modern
St Germaine-la-Rivière B4	14	0	1	UP Modern
St Germaine-la-Rivière B3	7	0.16	0.84	UP Modern
St Germaine-la-Rivière B5	7	0.02	0.98	UP Modern
St Germaine-la-Rivière B6&B7	4	0.05	0.95	UP Modern
St Germaine-la-Rivière 3 (1970-7)	7	0	1	UP Modern
St Germaine-la-Rivière 6 (1970-7)	2	0.27	0.73	UP Modern
St Germaine-la-Rivière 21 (1970-7)	4	0	1	UP Modern
St Germaine-la-Rivière 9 (1970-7)	1	0.35	0.65	UP Modern
St Germaine-la-Rivière 10 (1970-7)	4	0.16	0.84	UP Modern
St Germaine-la-Rivière 11 (1970-7)	4	0.01	0.99	UP Modern
St Germaine-la-Rivière 12 (1970-7)	3	0.12	0.88	UP Modern
St Germaine-la-Rivière 14 (1970-7)	6	0.45	0.55	UP Modern
St Germaine-la-Rivière 15 (1970-7)	4	0.25	0.75	UP Modern
St Germaine-la-Rivière 16 (1970-7)	3	0.17	0.83	UP Modern
St Germaine-la-Rivière 7 (1970-7)	6	0.18	0.82	UP Modern
St Germaine-la-Rivière 18 (1970-7)	4	0.21	0.79	UP Modern
St Germaine-la-Rivière 2 (1970-7)	2	0.24	0.76	UP Modern
St Germaine-la-Rivière 1 (1970-7)	3	0.08	0.92	UP Modern
Oberkassel D999	33	0	1	UP Modern
Oberkassel unnumbered	6	0.01	0.99	UP Modern
Isturitz 1950-6	6	0.06	0.94	UP Modern
Isturitz 1950-10-3	7	0.10	0.90	UP Modern
Isturitz 1950-9 (IV-105)	6	0.07	0.93	UP Modern
Isturitz 1950-10-2	5	0.05	0.95	UP Modern
Isturitz IV 1942/1950	7	0.05	0.95	UP Modern
Istruitz Ser. 7B 1950-4-1	7	0.16	0.84	UP Modern
Kostenki 14	39	0	1	UP Modern
Kostenki 15	6	0.01	0.99	UP Modern
Kostenki 17	3	0.08	0.92	UP Modern
Kostenki 18	43	0	1	UP Modern
Sunghir 2	60	0	1	UP Modern
Sunghir 3	31	0	1	UP Modern
La Chaud 4	17	0	1	UP Modern
La Chaud 5	26	0	1	UP Modern
La Chaud 3	19	0	1	UP Modern
La Chaud 83	6	0	1	UP Modern
Mladeč (misc teeth)	16	0.95	0.05	Neanderthal
Mladeč 2	10	0.11	0.89	UP Modern
Mladeč 1	6	0.02	0.98	UP Modern
Fontéchevade 1954-54 #1	4	0.09	0.91	UP Modern
Fontéchevade 1954-53 #2	7	0.17	0.83	UP Modern
Font de Gaume 2	7	0.05	0.95	UP Modern
Grotta del Fossellone	6	0.01	0.99	UP Modern
La Ferrasie 7	3	0.30	0.70	UP Modern
La Ferrasie 8&9	9	0.01	0.99	UP Modern
Derava Skala	9	0	1	UP Modern
Istállóskő	7	0.01	0.99	UP Modern
Brassempouy #16, 884, 542, 1046, 2206, 262, 441	19	0	1	UP Modern
Oase 2	6	0.97	0.03	Neanderthal

(continued)

Specimen	No. traits used	Neanderthal	UP Modern	Classification
Oase 1	12	0.02	0.98	UP Modern
Grotte des Rois mandible A	15	0.04	0.96	UP Modern
Grotte des Rois R50 mandible B	14	0.45	0.55	UP Modern
Grotte des Rois unnumbered	3	0.03	0.97	UP Modern
Grotte des Rois 19	3	0.03	0.97	UP Modern
Grotte des Rois 24	6	0.37	0.63	UP Modern
Grotte des Rois 15	6	0.37	0.63	UP Modern
Grotte des Rois 5a	4	0.48	0.52	UP Modern
Grotte des Rois 32	3	0.20	0.80	UP Modern
Grotte des Rois 45	4	0.25	0.75	UP Modern
Grotte des Rois 5	4	0.25	0.75	UP Modern
Grotte des Rois unnumbered 1	4	0	1	UP Modern
Grotte des Rois unnumbered 2	4	0.0	1	UP Modern
Grotte des Rois unnumbered C1	3	0.37	0.63	UP Modern
Grotte des Rois 29	6	0	1	UP Modern
Grotte des Rois 18	5	0.09	0.91	UP Modern
Grotte des Rois R8	7	0.14	0.86	UP Modern
Grotte des Rois R51-30	7	0.02	0.98	UP Modern
Grotte des Rois R16	7	0.05	0.95	UP Modern
Grotte des Rois 1955-148	7	0.05	0.95	UP Modern
Grotte des Rois R31	2	0.06	0.94	UP Modern
Grotte des Rois R51-14	7	0.05	0.95	UP Modern
Grotte des Rois 55	7	0.10	0.90	UP Modern
Grotte des Rois 50	7	0.05	0.95	UP Modern
Grotte des Rois 3	7	0.04	0.96	UP Modern
Grotte des Rois 6	7	0	1	UP Modern

References

Andrei, S., Petrea, C., Doboş, A., & Trinkaus, E. (2007). The Human cranium from the Peştera Cioclovina Uscată, Romania: Context, age, taphonomy, morphology, and paleopathology. *Current Anthropology, 48*, 611–619.

Arensburg, B., & Belfer-Cohen, A. (1998). Sapiens and Neanderthals. In T. Akazawa, K. Aoki, & O. Bar-Yosef (Eds.), *Neanderthals and modern humans in Western Asia* (pp. 311–322). New York: Plenum Press.

Bailey, S. E. (2000). Implications of dental morphology for population affinity among Late Pleistocene and recent humans. *Journal of Humam Evolution, 38*, A5–A6.

Bailey, S. E. (2002a). A closer look at Neanderthal postcanine dental morphology. I. The mandibular dentition. *The Anatomical Record, 269*, 148–156.

Bailey, S. E. (2002b). Neanderthal dental morphology: Implications for modern human origins. Ph.D. Dissertation, Arizona State University.

Bailey, S. E. (2004). Derived morphology in Neanderthal maxillary molars: Insights from above. *American Journal of Physical Anthropology, 123*, 57.

Bailey, S. E., & Hublin, J.-J. (2013). What does it mean to be dentally 'modern'? In G. R. Scott & J. D. Irish (Eds.), *Anthropological perspectives on tooth morphology: Genetics, evolution, variation* (pp. 222–249). Cambridge: Cambridge University Press.

Bailey, S. E., Weaver, T. D., & Hublin, J.-J. (2009). Who made the Aurignacian and other early Upper Paleolithic industries? *Journal of Human Evolution, 57*, 11–26.

Bailey, S. E., Skinner, M. M., & Hublin, J.-J. (2011). What lies beneath? An evaluation of lower molar trigonid crest patterns based on both dentine and enamel expression. *American Journal of Physical Anthropology, 45*, 505–518.

Baume, R. M., & Crawford, M. H. (1978). Discrete dental traits in four Tlaxcaltecan Mexican populations. *American Journal of Physical Anthropology, 49*, 351–360.

Bermúdez de Castro, J. M., & Nicolas, M. E. (1995). Posterior dental size reduction in hominids: The Atapuerca evidence. *American Journal of Physical Anthropology, 96*, 335–356.

Bouzouggar, A., & Barton, R. N. E. (2012). The identity and timing of the Aterian in Morocco. In J.-J. Hublin & S. McPherron (Eds.), *Modern origins: A North African perspective* (pp. 93–106). Dordrecht: Springer.

Brace, C. L., & Mahler, P. E. (1971). Post-Pleistocene changes in the human dentition. *American Journal of Physical Anthropology, 34*, 191–204.

Brace, C. L., Rosenberg, K. R., & Hunt, K. D. (1987). Gradual change in human tooth size in the late Pleistocene and post-Pleistocene. *Evolution, 41*, 705–720.

Brown, P., & Maeda, T. (2009). Liang Bua *Homo floresiensis* mandibles and mandibular teeth: A contribution to the comparative morphology of a new hominin species. *Journal of Human Evolution, 57*, 571–596.

Brown, P., Sutikna, T., Morwood, M., Soejono, R. P., Jatmiko, Saptomo, E. W., et al. (2004). A new small-bodied hominin from the Late Pleistocene of Flores, Indonesia. *Nature, 431*, 1055–1061.

Burnett, S. E., Hawkey, D. E., & Turner, C. G. (2010). Brief communication: Population variation in human maxillary premolar accessory ridges (MxPAR). *American Journal of Physical Anthropology, 141*, 319–324.

Butler, P. M. (1939). Studies of the mammalian dentition. Differentiation of the post-canine dentition. *Proceedings of the Zoological Society of London B, 109*, 1–36.

Conard, N. J., Grootes, P. M., & Smith, F. H. (2004). Unexpectedly recent dates for human remains from Vogelherd. *Nature, 430*, 198–201.

Currat, M., & Excoffier, L. (2011). Strong reproductive isolation between humans and Neanderthals inferred from observed patterns of introgression. *Proceedings of the National Academy of Sciences USA, 108*, 15129–15134.

Dahlberg, A. (1945). The changing dentition of man. *Journal of the American Dental Association, 32*, 676–680.

Duarte, C., Maurício, J., Pettitt, P., Souto, P., Trinkaus, E., van der Plicht, H., et al. (1999). The early Upper Paleolithic human skeleton from the Abrigo do Lagar Velho (Portugal) and modern human emergence in Iberia. *Proceedings of the National Academy of Sciences USA, 96*, 7604–7609.

Frayer, D., Wolpoff, M., Thorne, A., Smith, F., & Pope, G. (1993). Theories of modern human origins: The paleontological test. *American Anthropologist, 95*, 14–50.

Fu, Q., Li, H., Moorjani, P., Jay, F., Slepchenko, S. M., Bondarev, A. A., et al. (2014). Genome sequence of a 45,000-year-old modern human from western Siberia. *Nature, 514*, 445–449.

Gómez-Robles, A., Martinón-Torres, M., Bermúdez De Castro, J. M., Margvelashvili, A., Bastir, M., Arsuaga, J.-L., et al. (2007). A geometric morphometric analysis of hominin upper first molar shape. *Journal of Human Evolution, 55*, 627–638.

Green, R. E., Krause, J., Briggs, A. W., Maricic, T., Stenzel, U., Kircher, M., et al. (2010). A draft sequence of the Neanderthal genome. *Science, 328*, 710–722.

Gunz, P., Bookstein, F. L., Mitteroecker, P., Stadlmayr, A., Seidler, H., & Weber, G. W. (2009). Early modern human diversity suggests subdivided population structure and a complex out-of-Africa scenario. *Proceedings of the National Academy of Sciences USA, 106*, 6094–6098.

Hanihara, K. (1963). Crown characters of the deciduous dentition of the Japanese-American hybrids. In D. R. Brothwell (Ed.), *Dental anthropology* (pp. 105–124). New York: Pergamon Press.

Hanihara, K. (1969). Mongoloid dental complex in the permanent dentition. *Proceedings of the VIIIth International Congress of Anthropological and Ethnological Sciences, Tokyo and Kyoto, 1968* (pp. 298–300). Tokyo: Science Council of Japan.

Hanihara, T. (1990). Dental anthropological evidence of affinities among the Oceania and the Pan-Pacific populations: The basic populations of East Asia, II. *Journal of the Anthropological Society of Nippon, 98*, 233–246.

Hanihara, T. (1992). Negritos, Australian aborigines and the "Proto-Sundadont" dental pattern: The basic populations in East Asia, V. *American Journal of Physical Anthropology, 83*, 182–196.

Harvati, K., & Weaver, T. D. (2006). Human cranial anatomy and the differential preservation of population history and climate signatures. *The Anatomical Record, 288A*, 1225–1233.

Hawkey, D. (1998). Out of Asia: Dental evidence for affinities and microevolution of early populations from India/Sri Lanka. Ph.D. dissertation, Arizona State University.

Henry-Gambier, D., Maureille, B., & White, R. (2004). Vestiges humains des niveaux de l'Aurignacien ancien du site de Brassempouy (Landes). *Bulletin et Mémoirs de Société Anthropologique de Paris, 16*, 49–87.

Henshilwood, C. S., & Marean, C. W. (2003). The origin of modern human behavior. Critique of the models and their test implications. *Current Anthropology, 44*, 627–651.

Hershkovitz, I., Smith, P., Sarig, R., Quam, R. M., Rodriguez, L., Garcia, R., et al. (2010). Middle Pleistocene dental remains from Qesem Cave (Israel). *American Journal of Physical Anthropology, 144*, 575–592.

Holliday, T. W. (2003). Species concepts, reticulation, and human evolution. *Current Anthropology, 44*, 653–673.

Hrdlička, A. (1927). The Neanderthal phase of man. *Journal of the Royal Anthropological Institute, 57*, 249–274.

Irish, J. D. (1997). Characteristic high and low frequency dental traits in sub-Saharan African populations. *American Journal of Physical Anthropology, 102*, 455–467.

Irish, J. D. (1998). Ancestral dental traits in recent sub-Saharan Africans and the origins of modern humans. *Journal of Human Evolution, 34*, 81–98.

Irish, J. D. (2005). Population continuity vs. discontinuity revisited: Dental affinities among late Paleolithic through Christian-era Nubians. *American Journal of Physical Anthropology, 128*, 520–535.

Irish, J. D. (2013). Afridonty: The "Sub-Saharan African Dental Complex" revisited. In G. R. Scott & J. D. Irish (Eds.), *Anthropological perspectives on tooth morphology: Genetics, evolution, variation* (pp. 278–295). Cambridge: Cambridge University Press.

Irish, J. D., & Guatelli-Steinberg, D. (2003). Ancient teeth and modern human origins: An expanded comparison of African Plio-Pleistocene and recent world dental samples. *Journal of Human Evolution, 45*, 113–144.

Krings, M., Stone, A., Schmitz, R. W., Krainitzki, H., Stoneking, M., & Pääbo, S. (1997). Neanderthal DNA sequences and the origin of modern humans. *Cell, 90*, 19–30.

Lahr, M. M., & Foley, R. A. (1998). Towards a theory of modern human origins: Geography, demography, and diversity in recent human evolution. *American Journal of Physical Anthropology, Supplement, 27*, 137–176.

Langergraber, K. E., Prüfer, K., Rowney, C., Boesch, C., Crockford, C., Fawcett, K., et al. (2012). Generation times in wild chimpanzees and gorillas suggest earlier divergence times in great ape and human evolution. *Proceedings of the National Academy of Sciences USA, 109*, 15716–15721.

Lukacs, J. (1983). Dental anthropology and the origins of two Iron Age populations from northern Pakistan. *Homo, 34*, 1–15.

Manzi, G. (2004). Human evolution at the Matuyama-Bruhnes boundary. *Evolutionary Anthropology, 13*, 11–24.

Martinón-Torres, M., Bermúdez de Castro, J. M., Gómez-Robles, A., Prado-Simón, L., & Arsuaga, J.-L. (2012). Morphological description and comparison of the dental remains from Atapuerca-Sima de los Huesos site (Spain). *Journal of Human Evolution, 62*, 7–58.

Martinón-Torres, M., Mastir, M., Bermúdez De Castro, J. M., Gómez, A., Sarmiento, S., Muela, A., et al. (2006). Hominin lower second premolar morphology: Evolutionary inferences through geometric morphometric analysis. *Journal of Human Evolution, 50*, 523–533.

Mayhall, J., Saundersm, S., & Belier, P. (1982). The dental morphology of North American whites: A reappraisal. In B. Kurten (Ed.), *B, Teeth: Form, function, and evolution* (pp. 235–258). New York: Columbia University Press.

McCown, T., & Keith, A. (1939). *The Stone Age of Mount Carmel*. Oxford: Clarendon Press.

Minugh-Purvis, N. (1998). The search for the earliest modern Europeans. In T. Akazawa, K. Aoki, & O. Bar-Yosef (Eds.), *Neanderthals and modern humans in Western Asia* (pp. 339–352). New York: Plenum Press.

Pearson, O. M. (2008). Statistical and biological definitions of "anatomically modern" humans: Suggestions for a unified approach to modern morphology. *Evolutionary Anthropology, 17*, 38–48.

Quam, R., & Smith, F. (1998). A reassessment of the Tabun C2 mandible. In T. Akazawa, K. Aoki, & O. Bar-Yosef (Eds.), *Neanderthals and modern humans in Western Asia* (pp. 405–421). New York: Plenum Press.

Radovčić, J., Smith, F. H., Trinkaus, E., & Wolpoff, M. H. (1988). *The Krapina hominids: An illustrated catalog of skeletal collection.* Zagreb: Mladost Publishing, Croatian Hatural History Museum.

Rak, Y. (1998). Does any Mousterian cave present evidence of two hominid species? In T. Akazawa, K. Aoki, & O. Bar-Yosef (Eds.), *Neanderthals and modern humans in Western Asia* (pp. 353–366). New York: Plenum Press.

Raynal, J.-P., & Occhietti, S. (2012). Amino-chronology and an earlier age for the Aterian. In J.-J. Hublin & S. McPherron (Eds.), *Modern origins: A North African perspective* (pp. 79–92). Dordrecht: Springer.

Richter, D. J., Moser, J., & Nami, M. (2012). New data from the site of Ifri n'Ammar (Morocco) and some remarks on the chronometric status of the Middle Paleolithic in the Maghreb. In J.-J. Hublin & S. McPherron (Eds.), *Modern origins: A North African perspective* (pp. 61–78). Dordrecht: Springer.

Rightmire, G. P. (2008). *Homo* in the Middle Pleistocene: Hypodigms, variation and species recognition. *Evolutionary Anthropology, 17*, 8–21.

Sanchéz-Quinto, F., Botigué, L. R., Civit, S. A. C, Ávila-Arcos, M. C., Bustamante, C. D., Comas, D., et al. (2012). North African populations carry the signature of admixture with Neanderthals. *PLoS ONE, 2012*, e47765.

Sankararaman, S., Patterson, N., Li, H., Pääbo, S., & Reich, D. E. (2012). The date of interbreeding between Neanderthals and modern humans. *PLoS ONE, 2012*, e1002947.

Schwarcz, H. P., Grün, R., Vandermeersch, B., Bar-Yose, O., Valladas, H., & Tchernov, E. (1988). ESR dates for the hominid burial site of Qafzeh in Israel. *Journal of Human Evolution, 17*, 733–737.

Schwartz, J. H., & Tattersall, I. (2000). The human chin revisited: What is it and who has it? *Journal of Human Evolution, 38*, 367–409.

Scott, G. R., & Turner, C. G., II. (1997). *The anthropology of modern human teeth. Dental morphology and its variation in recent human populations.* Cambridge: Cambridge University Press.

Scott, G. R., Anta, A., Schomberg, R., & de la Rua, C. (2013). Basque dental morphology and the "Eurodont" dental pattern. In G. R. Scott & J. D. Irish (Eds.), *Anthropological perspectives on tooth morphology: Genetics, evolution, variation* (pp. 296–318). Cambridge: Cambridge University Press.

Simpson, G. G. (1943). Criteria for genera, species and subspecies in zoology and paleontology. *Annals New York Academy of Science, 44*, 145–178.

Stefan, V., & Trinkaus, E. (1998). Discrete trait and dental morphometric affinities of the Tabun 2 mandible. *Journal of Human Evolution, 34*, 443–468.

Stringer, C. (2003). Human evolution: Out of Ethiopia. *Nature, 423*, 692–695.

Trinkaus, E. (2005). Early modern humans. *Annual Review of Anthropology, 34*, 207–230.

Trinkaus, E., Milota, S., Rodrigo, R., Mircea, G., & Moldovan, O. (2003). Early modern human cranial remains from the Peştera cu Oase, Romania. *Journal of Human Evolution, 45*, 245–253.

Trinkaus, E., Milota, S., Gerhase, M., Sarcina, L., Bilgar, A., Moldovan, O., et al. (2004). Bones, bodies and bears in the Peştera cu Oase, Romania. *Paper Presented at the paleoanthropology society meetings*, Montreal ON.

Turner, C. G., II. (1983). Sinodonty and Sundadonty: A dental anthropological view of Mongoloid microevolution, origin, and dispersal into the Pacific basin, Siberia, and the Americas. In R. Vasilievsky (Ed.), *Late Pleistocene and Early Holocene cultural connections of Asia and America* (pp. 72–76). Novosibirsk: USR Academy of Science, Siber.

Turner, C. G., II. (1987). Late Pleistocene and Holocene population history of East Asia based on dental variation. *American Journal of Physical Anthorpology, 73*, 305–321.

Turner, C. G., II. (1990). Major features of Sundadonty and Sinodonty, including suggestions about East Asian microevolution, population history, and late Pleistocene relationships with Australian aboriginals. *American Journal of Physical Anthropology, 82*, 295–317.

Turner, C. G., II. (1992). Microevolution of East Asian and European populations: A dental perspective. In T. Akazawa, K. Aok & T. Kimura (Eds.), *The evolution and dispersal of modern humans in Asia* (pp. 415–438). Tokyo: Hokusen-Sha.

Turner, C. G., II, & Scott, G. R. (1977). Dentition of Easter Islanders. In A. Dahlberg & T. Graber (Eds.), *Orofacial growth and development* (pp. 229–249). The Hague: Mouton Publishers.

Turner, C. G., II, Nichol, C. R., & Scott, G. R. (1991). Scoring procedures for key morphological traits of the permanent dentition: The Arizona State University dental anthropology system. In M. Kelley & C. Larsen (Eds.), *Advances in dental anthropology* (pp. 13–31). New York: Wiley Liss.

Valladas, H., Mercier, N., Joron, J. L., & Reyss J. L. (1998). GIF Laboratory dates for the Middle Paleolithic Levant. In T. Akazawa, K. Aoki & O. Bar-Yosef (Eds.), *Neanderthals and modern humans in Western Asia* (pp. 69–75). New York: Plenum.

Vandermeersch, B. (1989). The evolution of modern humans: Recent evidence from Southwest Asia. In P. Mellars & C. Stringer (Eds.), *The Human revolution* (pp. 155–164). Edinburgh: Edinburgh University Press.

Vargiu, R., Coppa, A., Lucci, M., Mancinelli, D., Rubini, M., & Calcagno, J. (1997). Population relationships and non-metric dental traits in Copper and Bronze Age Italy. *American Journal of Physical Anthropology, 24*, 232.

Zilberman, U., Skinner, M., & Smith, P. (1992). Tooth components of mandibular deciduous molars of *Homo sapiens sapiens* and *Homo sapiens neanderthalensis*: A radiographic study. *American Journal of Physical Anthropology, 87*, 255–262.

Chapter 17
Talking Hyoids and Talking Neanderthals

David W. Frayer

Abstract Yoel Rak and others published the first known Neanderthal hyoid bone in 1989. Contrary to expectations, the ∼60 ka Kebara hyoid was completely within modern human variation and led them to conclude, "the assumed speech limitations of Neanderthals… would seem to require revision." Subsequently two more fragmentary hyoid bones from Sima de los Huesos (Atapuerca), dating to over 400,000 years ago were determined to be not different from anatomically modern morphology. Most recently, the hyoid of the Dikika child (*Au. afarensis*), dated much earlier at ∼3.3 Ma, was found to clearly resemble that of an ape. The time span represented by these three sites shows that at least part of the anatomy surrounding the vocal tract was of a modern morphology in Neanderthals and their likely ancestors, but not in the much earlier *Australopithecus*. It was the Kebara hyoid which marks the beginning of a modern understanding of Neanderthal speech capability. This paper reviews the controversy surrounding the interpretation of the Kebara hyoid and other evidence from fossil anatomy, archaeology and paleogenetic data accumulated since 1989, which convincingly shows that Neanderthals possessed the ability to speak like us.

Keywords Kebara cave • Language • Sima de los Huesos • Speech

In 1989, Yoel Rak announced with others the discovery of a hyoid bone associated with the Kebara 2 Neanderthal skeleton. This was the first hyoid found in a fossil hominid context and its morphology and metrics indicated that, unlike the associated mandible, the hyoid was completely modern. They concluded that "the assumed speech limitations of the Neanderthals, that have hitherto been based primarily on studies of basicranial morphology, would seem to require revision" (Arensburg et al. 1989: 760) since the hyoid showed no fundamental differences compared to 67 modern hyoids. In a more detailed account, based on the Kebara hyoid's modern-looking metrics and anatomy, Yoel and his colleagues concluded Neanderthals "appear to be as 'anatomically capable' of speech as modern humans." (Arensburg et al. 1990: 145). And, in a *Current Anthropology* paper Yoel and others argued the modern morphology of the hyoid "strongly suggests that Middle Paleolithic hominids were equally capable of speech when hyoid positioning and supralaryngeal space are the criteria considered" (Bar-Yosef et al. 1992: 530). Thus, the discovery at Kebara completely changed the atmosphere about Neanderthal language capacity, and Yoel and his colleagues' publications serve as a milestone in the slow acceptance of Neanderthals having language ability like modern humans.

Despite its significance, the Kebara 2 hyoid met with a barrage of skepticism by a few, especially those who had a history of denying Neanderthals the ability to speak like us. For example, just a decade earlier, in a review article, Laitman et al. (1979: 15) stated, "bony landmarks, such as the hyoid bone or styloid process which give clues to the position and shape of the upper respiratory structures are often missing." One might have anticipated that Kebara 2 would have provided a welcome resolution for some issues about Neanderthal communicative abilities, but on the contrary, at the American Association of Physical Anthropologists meeting in Miami, Laitman et al. (1990: 254) claimed "[a]s we do not know what the hyoids of other fossil hominids looked like, it is possible that hyoid morphology was similar as far back as early members of *Homo*, if not earlier. If so, then the hyoid would be an irrelevant indicator of vocal tract evolution." They further asserted, "hyoids of

D.W. Frayer (✉)
Department of Anthropology, University of Kansas, Lawrence, KS 66044, USA
e-mail: frayer@ku.edu

© Springer International Publishing AG 2017
Assaf Marom and Erella Hovers (eds.), *Human Paleontology and Prehistory*, Vertebrate Paleobiology and Paleoanthropology, DOI: 10.1007/978-3-319-46646-0_17

mammals with vocal tracts clearly unlike those of modern humans also show metric features which would, by themselves, identify them as 'human.' For example, suid hyoids are metrically more similar to those of modern humans than Kebara." The former assertion assumes its conclusion and the latter is demonstrably incorrect.

The implications of the Kebara hyoid also drew strong criticism from Lieberman (1992) who originally speculated (Lieberman and Crelin 1971; Lieberman et al. 1972) that Neanderthals could not have produced essential vowels "a", "i" and "u." Lieberman (1992) argued in detail that the modern looking Kebara hyoid tells nothing about the supralaryngeal space, since the Kebara 2 base was missing and other Neanderthal crania had flat cranial bases like apes. A similar denouncement of the hyoid's importance was given in a *Nature* response (Lieberman et al. 1989) when the Kebara hyoid was first published (Arensburg et al. 1989; Marshall 1989). Here, Lieberman et al. (1989: 486) maintained in almost the identical sentences as above, that "[a]s we do not know what the hyoids of other fossil hominids looked like, it is possible that hyoid morphology was similar as far back as early members of *Homo*, if not earlier. If so, then the hyoid would be an irrelevant indicator of vocal tract evolution".

In the intervening years more Neanderthal and pre-Neanderthal hyoids have been found, at El Sidrón cave (Rodríguez et al. 2002) and in the Sima de los Huesos at Atapuerca (Martínez et al. 2008), both in Spain. These bones, like the Kebara hyoid, are completely modern in their metrics and morphology and undeniably confirm that fossil *Homo* hyoids from Europe conform to the modern pattern. An apparent exception to this is the 'hyoid' from Castel di Guido, described by Capasso and D'Anastasio (2008), but the bone is now identified as the dorsal rim of a first cervical vertebra (Capasso et al. 2016). Interestingly, the hyoid associated with the 3.3 myr-old *Au. afarensis* child from Dikika differs significantly from the modern and fossil *Homo* condition. Comparing the morphology and metrics of the hyoid to gorillas, chimpanzees and living humans Alemseged et al. (2006) concluded the Dikika hyoid has corpus metrics completely outside the human range and completely within the ape range. The morphology of the Dikika hyoid suggests (Alemseged et al. 2006) that early *Australopithecus* had a vocal tract similar to apes with a functioning air sac. deBoer (2012) has argued that air sacs interfere with vowel-like articulations, and their presence in Dikika indicates it lacked modern human-like supra-laryngeal sound production.

Evidence for an air sac is clearly absent in any of the fossil European hyoids, which resemble neither Dikika nor suids. As for suids, their hyoids in actuality bear no relevant similarity to Kebara (Fig. 17.1). Twenty years ago I reported that suids have a tall/thick corpus and tall greater horns

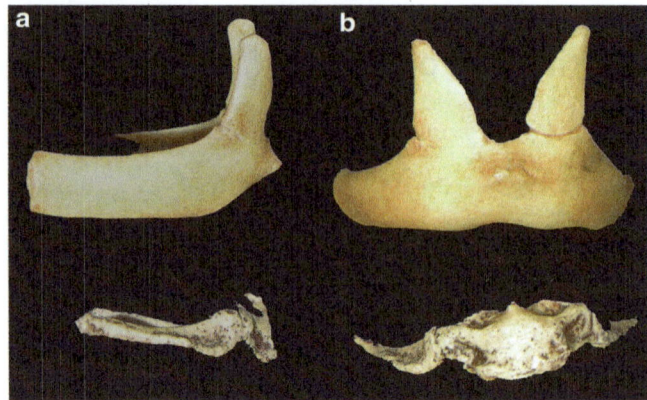

Fig. 17.1 Lateral (**a**) and anterior (**b**) views of a modern domestic pig (above) and Kebara 2 hyoid (below). The two show no anatomically or evolutionary relevant similar features

along with a large central tubercle and massive lesser horns. The anatomy and metrics of suids are not 'more similar' to Kebara (Frayer 1993). Whatever the logic of finding an affinity between a suid and the Kebara hyoid, any important similarity should be forever dismissed. This assertion by Laitman et al. (1992) violates what Le Gros Clark called the principle of morphological equivalence in making statistical comparisons, a mistake Le Gros Clark considered to be "one of the most serious sources of fallacy likely to affect statistical studies by those who are not thoroughly acquainted with the skeletal elements with which they are dealing" (Le Gros Clark 1964: 32).

While the hyoid reveals little about the supra-laryngeal space, if Neanderthal hyoids looked like apes or Dikika, an air sac would be a reasonable interpretation. Had this been the case, Neanderthals inarguably would have had vocal shortcomings. Most recently, d'Anastasio et al. (2013: 6) have confirmed Kebara 2's similarity with moderns and commented on how this relates to a modern vocal tract. They write:

…the presence of modern-human-like histological features and micro-biomechanical behavior in the Kebara 2 hyoid indicates that this bone not only resembled that of a modern human, but that it was used in very similar ways.

So, despite assertions to the contrary, hyoid morphology does reveal something about the linguistic capacity of a hominid and the Kebara hyoid, along with others from Sima de los Huesos and El Sidrón are morphologically and histologically equivalent to moderns.

What about additional evidence for Neanderthal linguistic ability since the discovery of the Kebara hyoid? For this, there has been a sea change of new evidence from anatomy to archaeology to paleogenetics. We now know that reconstructions of the cranial base are not flat, but arched like in us (Heim 1989; Lieberman 1998; Boë et al. 1999, 2002;

Frayer and Nicolay 2000) and that the Neanderthal vocal tract is capable of producing vowels very similar or identical to modern Europeans (Barney et al. 2012; Dediu and Levinson 2013). We also know that Neanderthal ear ossicles are similar to modern humans, thanks again to Yoel's work (Quam and Rak 2008) and that modern auditory anatomy stretches back to more than 0.5 million years ago (Martínez et al. 2004).

Holloway (1985) argued that Neanderthal brains were lateralized like modern humans, a likely signature of language ability. Subsequent work by Holloway et al. (2005) stressed again the importance of paleoneurological data, which clearly showed that Neanderthals had brain lateralization and regional specialization like living people. Brain lateralization is a key component of language capacity and work by Gotts et al. (2013: 1) has confirmed with fMRI the importance of the left hemisphere in its "cortical regions involved in language and fine motor control."

Some of my joint work with Italian, French and Spanish colleagues has shown that Neanderthals and their likely European ancestors were predominately right-handed like modern humans based on obliquity of scratches found on the labial face of incisors and canines (Fig. 17.2; Frayer et al. 2012; Volpato et al. 2012). Since handedness is a reflection of laterality, our data from tooth scratches and Holloway's observations from endocasts are completely concordant. We also know that apes are not lateralized like humans and certainly not handed in the way of humans and Neanderthals (McGrew and Marchant 1997).

For archeological discoveries pointing to linguistic competence we know that Neanderthals had ornaments (Zilhão et al. 2010), decorated themselves with paint (Cârciumaru and Țuțuianu-Cârciumaru 2009), feathers (Soressi and d'Errico 2007; Peresani et al. 2011) and eagle talons (Morin and Laroulandie 2012; Radovčić et al. 2015), practiced seafaring (Ferentinos et al. 2012), had complex site structures (Henry et al. 2004; Vallverdú et al. 2010) with resource scheduling, including marine foods (Daujeard and Moncel 2010; Cortés-Sánchez et al. 2011). Consumption of plant materials has been documented through analysis of plant seeds and debris (Lev et al. 2005; Henry et al. 2010) based on starches preserved in dental calculus and residue on tools (Hardy and Moncel 2011). There is even evidence of Neanderthals consuming plants of no nutritional, but pharmacological, value (Hardy et al. 2012). Neanderthals made bone tools for leather working (Soressi et al. 2013), transported or exchanged raw materials over long distances (Slimak and Giraud 2007; Peresani et al. 2013) and had complex site arrangements as seen in moderns (Henry et al. 2004; Vallverdú et al. 2010). For ritual behavior there is no doubt they buried their dead of all ages (Maureille and Vandermeersch 2007; Pettit 2012) and at least in one site there appears to be other types of ritual treatment of the dead (Frayer et al. 2008).

But, perhaps, the most wondrous new evidence addressing language ability comes with discovery of Neanderthal nuclear DNA from a number of specimens and sites (Green et al. 2010). From these sequences we know that unique

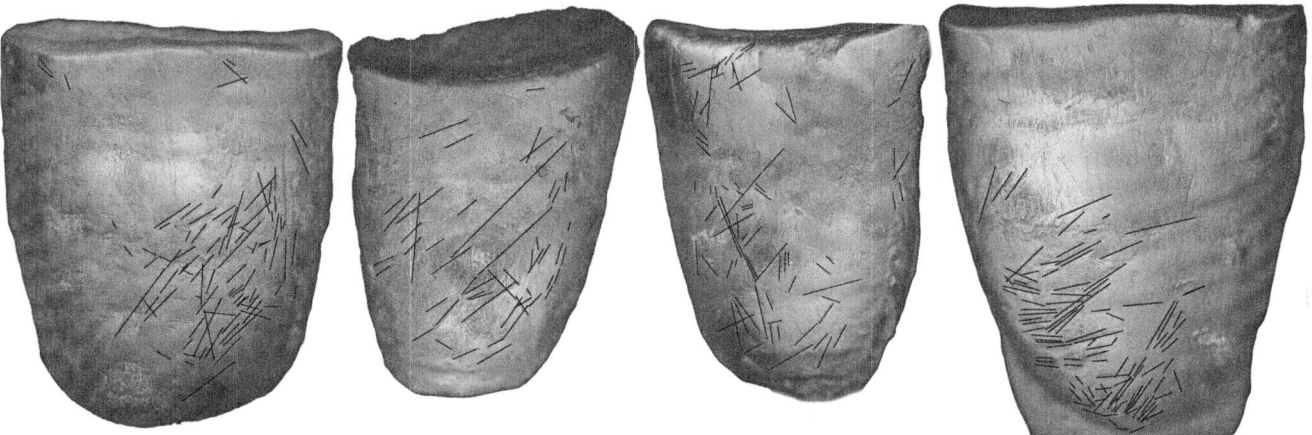

Fig. 17.2 Four incisors from the Neanderthal mandible Regourdou, dated to OIS 4, ca. 70 ka. Obliquity of marks on the two lateral incisors and the right I$_1$ are typical of right-handed scratches found in many other Neanderthal teeth (Volpato et al. 2012). Arm chain remains from the skeleton more than 3 decades ago identified it as right-handed (Vandermeersch and Trinkaus 1995)

Neanderthal genes are found in at least 2–4% in living Europeans (Green et al. 2010), if not double this (Lohse and Frantz 2013). One of the genetic sequences is the FOXP2 gene. This gene is linked to language production, in that those who possess mutations in it have grammar, syntax and vocal deficiencies (Hurst et al. 1990; Lai et al. 2001) and numerous other factors affecting development. The fact that Neanderthals share two key FOXP2 nucleotide sequences with humans, which distinguish us from apes (Krause et al. 2007) completes the circle of evidence for Neanderthals having linguistic ability like us. Following paleogenetic estimates (Green et al. 2010), this marker of language capacity may extend back to more than 0.5 mya. This would make language old, not young as some have argued (Lieberman et al. 1972; Laitman et al. 1979; Diamond 1989).

Yoel's work on the Kebara hyoid triggered the re-thinking and re-analysis of the perception of Neanderthal vocal ability. Yoel and his colleagues concluded in their *American Journal of Physical Anthropology* (Arensburg et al. 1990: 145) article:

> Hopefully the hyoid and related bones of other fossil hominids will be recovered in the future, and we believe this will add to our understanding of the vocal and upper respiratory organs of fossil humans.

One doubts he could have fully anticipated the chain of evidence now leading from this little bone in the throat to a full appreciation of the modern language capacities of Neanderthals.

Acknowledgement and Nota Bene Milford Wolpoff (Michigan) provided some important suggestions. As for Yoel, we first met in Berkeley in 1979. In those days the single species hypothesis as applied to *Australopithecus* was still being debated and Yoel wanted me to know for sure that what my adviser, Milford Wolpoff, wrote could not be correct. Since that time we have met many times at professional meetings or overseas and, while we often disagreed about some things, Yoel could not have been more congenial nor as courteously dismissive of my ideas. All my best to a first class person and scholar.

References

Alemseged, Z., Spoor, F., Kimbel, W. H., Bobe, R., Geraads, D., Reed, D., et al. (2006). A juvenile early hominin skeleton from Dikika, Ethiopia. *Nature, 443*, 296–301.

Arensburg, B., Tillier, A.-M., Vandermeersch, B., Duday, H., Schepartz, L. A., & Rak, Y. (1989). A Middle Paleolithic human hyoid bone. *Nature, 338*, 758–760.

Arensburg, B., Schepartz, L. A., Tillier, A.-M., & Vandermeersch, B. (1990). A reappraisal of the anatomical basis for speech in Middle Paleolithic hominids. *American Journal of Physical Anthropology, 83*, 137–146.

Bar-Yosef, O., Vandermeersch, B., Arensburg, B., Belfer-Cohen, A., Goldberg, P., Laville, H., et al. (1992). The excavations in Kebara. Mt. Carmel. *Current Anthropology, 33*, 497–550.

Barney, A., Martelli, S., Serrurier, A., & Steele, J. (2012). Articulatory capacity of Neanderthals, a very recent and hominin-like fossil. *Philosophical Transactions of the Royal Society B, 367*, 88–102.

Boë, L.-J., Maeda, S., & Heim, J.-L. (1999). Neanderthal man was not morphologically handicapped for speech. *Evolution of Communication, 3*, 49–77.

Boë, L. J., Heim, J.-L., Honda, K., & Madea, S. (2002). The potential Neanderthal vowel space was as large as that of modern humans. *Journal of Phonetica, 30*, 465–484.

Cârciumaru, M., & Ţuţuianu- Cârciumaru, M. (2009). *L'ocre et les récipients pour ocre de la grotte Cioarei* (pp. 7–19). XI: Annals d'Université Val Targ.

Capasso, L., Michett,i E., & D'Anastasio, R. (2008). A *Homo erectus* hyoid bone: Possible implications for the origin of the human capability for speech. *Collegium Anthropologicum, 32*, 107–112.

Capasso, L., d'Anastasio, R., Mancini, L., Tuniz, C., & Frayer, D. W. (2016). New evaluation of the Castel di Guido 'hyoid.' *Journal of Anthropological Sciences, 94*, 231–235.

Cortés-Sánchez, M., Morales-Muiz, A., Simón-Vallejo, M. D., Lozano-Francisco, M. C., Vera-Peláez, J. L., Finlayson, C., et al. (2011). Earliest known use of marine resources by Neanderthals. *PLoS ONE, 6*, e24026.

D'Anastasio, R., Wroe, S., Tuniz, C., Mancin, i. L., Cesana, D. Y., Dreossi, D., et al. (2013). Micro-biomechanics of the Kebara 2 hyoid and its implications for speech in Neanderthals. *PLoS ONE, 8*, e82261.

Daujeard, C., & Moncel, M.-H. (2010). On Neanderthal subsistence strategies and land use: A regional focus on the Rhone Valley area in southeastern France. *Journal of Anthropological Archaeology, 29*, 368–391.

deBoer, B. (2012). Loss of air sacs improved hominin speech abilities. *Journal of Human Evolution, 62*, 1–6.

Dediu, D., & Levinson, S. (2013). On the antiquity of language: The re-interpretation of Neanderthal linguistic capacities and its consequences. *Frontiers in Psychology, 4*, 1–17.

Diamond, J. (1989). The great leap forward. *Discover Magazine, 10*, 50–60.

Ferentinos, G., Gkioni, M., Geraga, M., & Papatheodorou, G. (2012). Early seafaring activity in the southern Ionian Islands, Mediterranean Sea. *Journal of Archaeological Science, 39*, 2167–2176.

Frayer, D. W. (1993). The Kebara 2 hyoid only resembles humans. *American Journal of Physical Anthropology, Supplement, 16*, 88.

Frayer, D. W., & Nicolay, C. (2000) Fossil evidence for the origin of speech sounds. In N. L.Wallin, B. Merker, & S. Brown (Eds.), *The origins of music* (pp. 217–243). Cambridge: MIT Press.

Frayer, D. W., Orschiedt, J., Cook, J., Russell, M. D., & Radovčić, J. (2008). Krapina 3: Cut marks and ritual behavior? In J. Monge, A. Mann, & D. W. Frayer (Eds.), *New insights on the Krapina Neanderthals: 100 years after Gorjanović-Kramberger* (pp. 285–290). Zagreb: Croatian Natural History Museum.

Frayer, D. W., Lozano, M., Bermúdez, de Castro J.-M., Carbonell, E., Arsuaga, J.-L., Radovčić, J., et al. (2012). More than 500,000 years of right-handedness in Europeans. *Laterality, 17*, 51–69.

Gotts, S. J., Jo, H. J., Wallace, G. L., Ziad, S. S., Cox, R. W., & Martin, A. (2013). Two distinct forms of functional lateralization in the human brain. *Proceedings of the National Academy of Sciences USA, 110*, 3435–3444.

Green, R. E., Krause, J., Briggs, A. W., Maricic, T., Stenzel, U., Kircher, M., Patterson, N., et al. (2010). A draft sequence of the Neanderthal genome. *Science, 328*, 710–722.

Hardy, B. L., & Moncel, M.-H. (2011). Neanderthal use of fish, mammals, birds, starchy plants and wood 125-250,000 years ago. *PLoS ONE, 6*, e23768.

Hardy, K., Buckley, S., Collins, M. J., Estalrrich, A., Brothwell, D., Copeland, L., et al. (2012). Neanderthal medics? Evidence for food, cooking, and medicinal plants entrapped in dental calculus. *Naturwissen, 99*, 617–626.

Heim, J.-L. (1989). La nouvelle reconstitution du crâne néandertalien de La Chapelle-aux-Saints: méthode et résultats. *Bulletin ets Mémoires de Société d'Anthropologie Paris, 1*, 95–118.

Henry, A. G., Brooks, A. S., & Piperno, D. R. (2010). Microfossils in calculus demonstrate consumption of plants and cooked foods in Neanderthal diets (Shanidar III, Iraq; Spy I and II, Belgium). *Proceedings of the National Academy of Sciences USA, 108*, 486–491.

Henry, D. O., Hietala, H. J., Rosen, A. M., Demidenko, Y. E., Usik, V. I., & Armagan, T. L. (2004). Human behavioral organization in the Middle Paleolithic: Were Neanderthals different? *American Anthropologist, 106*, 17–31.

Holloway, R. L. (1985). The poor brain of *Homo sapiens* neanderthalensis: See what you please. In E. Delson (Ed.), *Ancestors: The hard evidence* (pp. 319–324). New York: Alan R. Liss.

Holloway, R. L., Broadfield, D. C., & Yuan, M. S. (2005). *The human fossil record: Brain endocasts—The paleoneurological evidence.* New York: Wiley-Liss.

Hurst, J. A., Baraitser, M., Auger, E., Graham, F., & Norell, S. (1990). An extended family with a dominantly inherited speech disorder. *Developmental Medicine and Child Neurology, 32*, 352–355.

Krause, J., Lalueza-Fox, C., Orlando, L., Enard, W., Green, R. E., Burbano, H. A., et al. (2007). The derived FOXP2 variant of modern humans was shared with Neanderthals. *Current Biology, 17*, 1908–1912.

Lai, C. S., Fisher, S. E., Hurst, J. A., Vargha-Khadem, F., & Monco, A. P. (2001). A forkhead-domain gene is mutated in a severe speech and language disorder. *Nature, 413*, 519–523.

Laitman, J. T., Heimbuch, R. C., & Crelin, E. S. (1979). The basicranium of fossil hominids as an indicator of their upper respiratory systems. *American Journal of Physical Anthropology, 51*, 15–34.

Laitman, J. T., Reidenberg, J. S., & Gannon, P. J. (1992). Fossil skulls and hominid vocal track: New applications to charting the evolution of human speech. In J. Wind, B. Chiarelli, C. Bichakjian, A. Nocentini, & A. Jonker (Eds.), *Language origin: A multidisciplinary approach* (pp. 385–419). Amsterdam: Kluwer.

Laitman, J. T., Reidenberg, J. S., Gannon, P. J., Johansson, B., Landahl, K., & Lieberman, P. (1990). The Kebara hyoid: What can it tell us about the evolution of the vocal tract. *American Journal of Physical Anthropology, 81*, 254.

Le Gros Clark, W. E. (1964). *The fossil evidence for human evolution.* Chicago: University of Chicago Press.

Lev, E., Kislev, M. E., & Bar-Yosef, O. (2005). Mousterian vegetal food in Kebara Cave, Mt. Carmel. *Journal of Archaeological Science, 32*, 475–484.

Lieberman, D. E. (1998). Sphenoid shortening and the evolution of modern human cranial shape. *Nature, 393*, 158–162.

Lieberman, P. (1992). On the evolutionary biology of speech and syntax. In J. Wind, B. Chiarelli, C. Bichakjian, A. Nocentini, & A. Jonker (Eds.), *Language origin: A multidisciplinary approach* (pp. 391–397). Amsterdam: Kluwer.

Lieberman, P., & Crelin, E. S. (1971). On the speech of Neanderthal man. *Linguistic Inquiry, 2*, 203–222.

Lieberman, P., Crelin, E. S., & Klatt, D. H. (1972). Phonetic ability and related anatomy of the newborn and adult human, Neanderthal man, and the chimpanzee. *American Anthropologist, 74*, 287–307.

Lieberman, P., Laitman, J. T., Reidenberg, J. S., Landahl, K., & Gannon, P. J. (1989). Folk physiology and talking hyoids. *Nature, 342*, 486.

Lohse, K., & Frantz., L. A. F. (2013). Maximum likelihood evidence for Neanderthal admixture in Eurasian populations from three genomes. *arXiv* 1307.8263.

Marshall, J. C. (1989). The descent of the larynx. *Nature, 338*, 702–703.

Martínez, I., Rosa, M., Arsuaga, J.-L., Jarabo, P., & Quam, R. (2004). Auditory capacities in Middle Pleistocene humans from the Sierra de Atapuerca in Spain. *Proceedings of the National Academy of Sciences, 101*, 9976–9981.

Martínez, I., Arsuaga, J.-L., Quam, R., Carretero, J. M., Gracia, A., & Rodríguez, L. (2008). Human hyoid bones from the Middle Pleistocene site of the Sima de los Huesos (Sierra de Atapuerca, Spain). *Journal of Human Evolution, 54*, 18–24.

Maureille, B., & Vandermeersch, B. (2007). Les sepultures néandertaliennes. In B. Vandermeersch & B. Maureille (Eds.), *Les Néandertaliens: biologie et cultures* (pp. 311–322). Paris: Éditions du CTHS.

McGrew, W. C., & Marchant, L. F. (1997). On the other hand: Current issues and meta-analysis of the behavioral laterality of hand function in nonhuman primates. *Yearbook of Physical Anthropology, 40*, 201–232.

Morin, E., & Laroulandie, V. (2012). Presumed symbolic use of diurnal raptors by Neanderthals. *PLoS ONE, 7*, e32856.

Peresani, M., Fiore, I., Gala, M., Romandini, M., & Tagliacosso, A. (2011). Late Neanderthals and the intentional removal of feathers as evidenced from bird bone taphonomy at Fumane Cave 44ky B.P., Italy. *Proceedings of the National Academy of Sciences USA, 108*, 3888–3893.

Peresani, M., Vanhaeren, M., Quaggiotto, E., Queffelec, A., & d'Errico, F. (2013). An ochered fossil marine shell from the Mousterian of Fumane Cave, Italy. *PLoS ONE, 8*, e68572.

Pettitt, P. (2012). Religion and ritual in the Lower and Middle Paleolithic. In T. Insoll (Ed.), *The Oxford handbook of the archaeology of ritual and religion* (pp. 329–343.) New York: Oxford University Press.

Quam, R., & Rak, Y. (2008). Auditory ossicles from southwest Asian Mousterian sites. *Journal of Human Evolution, 54*, 414–433.

Radovčić, D., Sršen, A. O., Radovčić, J., & Frayer, D. W. (2015). Evidence for Neandertal jewelry: Modified white-tailed eagle claws at Krapina. *PLoS ONE*, e0119802.

Rodríguez, L., Cabo, L. L., & Egocheaga, J. E. (2002). Breve nota sobre el hyoides Neanderthalense de Sidron (Piloña, Asturias). *Antropología y Biodiversidad, 1*, 484–493.

Slimak, L., & Giraud, Y. (2007). Circulations sur plusieurs centaines de kilométres durant le Paléolithique moyen. Contribution à la connaissance des sociétés néandertaliennes. *Comptes Rendus Palevol, 6*, 359–368.

Soressi, M., & d'Errico, F. (2007) Pigments, gravures, parures: les comportements symboliques controversies des Néandertaliens. In B. Vandermeersch & B. Maureille (Eds.), *Les Néandertaliens: biologie et cultures* (pp. 297–309). Paris: Éditions du CTHS.

Soressi, M., McPherron, S. P., Lenoir, M., Dogandžić, T., Goldberg, P., Jacobs, Z., et al. (2013). Neanderthals made the first specialized bone tools in Europe. *Proceedings of the National Academy of Sciences USA, 110*, 14186–14190.

Vallverdú, J., Vaquero, M., Cáceres, I., Allué, E., Rosell, J., Saladié, P., et al. (2010). Sleeping activity area within the site structure of archaic human groups. *Current Anthropology, 51*, 137–145.

Vandermeersch, B., & Trinkaus, E. (1995). The postcranial remains of the Régourdou 1 Neanderthal: The shoulder and arm remains. *Journal of Human Evolution, 28*, 439–476.

Volpato, V., Macchiarelli, R., Guatelli-Steinberg, D., Fiore, I., Bondioli, L., & Frayer, D. W. (2012). Hand to mouth in a Neanderthal: Right-handedness in Regourdou 1. *PLoS ONE, 7*, e43949.

Zilhão, J., Angelucci, D. E., Badal-García, E., d'Errico, F., Daniel, F., Dayet, L., et al. (2010). Symbolic use of marine shells and mineral pigments by Iberian Neanderthals. *Proceedings of the National Academy of Sciences USA, 107*, 1023–1028.

Chapter 18
3D Reconstruction of Spinal Posture of the Kebara 2 Neanderthal

Ella Been, Asier Gómez-Olivencia, Patricia A. Kramer, and Alon Barash

Abstract Spinal posture has vast biomechanical, locomotor and pathological implications in hominins. Assessing the curvatures of the spine of fossil hominins can provide important information towards the understanding of their paleobiology. Unfortunately, complete hominin spines are very rarely preserved in the fossil record. The Neanderthal partial skeleton, Kebara 2 from Israel, constitutes a remarkable exception, representing an almost complete spine and pelvis. The aim of this study is, therefore, to create a new 3D virtual reconstruction of the spine of Kebara 2. To build the model, we used the CT scans of the sacrum, lumbar and thoracic vertebrae of Kebara 2, captured its 3D morphology, and, using visualization software (Amira 5.2©), aligned the 3D reconstruction of the original bones into the spinal curvature. First we aligned the sacrum and then we added one vertebra at a time, until the complete spine (T1-S5) was intact. The amount of spinal curvature (lordosis and kyphosis), the sacral orientation, and the coronal plane deviation was determined based on the current literature or measured and calculated specifically for this study based on published methods. This reconstruction provides, for the first time, a complete 3D virtual reconstruction of the spine of an extinct hominin. The spinal posture and spinopelvic alignment of Kebara 2 show a unique configuration compared with that of modern humans, suggesting locomotor and weight-bearing differences between the two. The spinal posture of Kebara 2 also shows slight asymmetry in the coronal plane. Stature estimation of Kebara 2 based on spinal length confirms that the height of Kebara 2 was around 170 cm. This reconstruction can now serve as the basis for a more complete reconstruction of the Kebara 2 specimen, which will include other parts of this remarkable fossil, such as the pelvis, the rib cage and the cervical spine.

Keywords 3D reconstruction • Locomotor differences • Modern humans • Spinal posture

E. Been (✉)
Faculty of Health Professions, Physical Therapy Department, Ono Academic College, 55000 Kiryat Ono, Israel
e-mail: beenella1@gmail.com

E. Been
Sackler Faculty of Medicine, Department of Anatomy and Anthropology, Tel Aviv University, 69978 Tel Aviv, Israel

A. Gómez-Olivencia
IKERBASQUE. Basque Foundation for Science & Facultad de Ciencia y Tecnología, Department o de Estratigrafía y Paleontología, Euskal Herriko Unibertsitatea, UPV-EHU, P.O. Box 64448080 Bilbao, Spain
e-mail: asier.gomezo@ehu.eus

A. Gómez-Olivencia
Département de Préhistoire, Muséum National d'Histoire Naturelle, Musée de l'Homme, 75016 Paris, France

A. Gómez-Olivencia
Centro UCM-ISCIII de Investigación sobre Evolución y Comportamiento Humanos, 28029 Madrid, Spain

P.A. Kramer
Departments of Anthropology and Orthopaedics and Sports Medicine, University of Washington, Seattle, WA 98195-3100, USA
e-mail: pakramer@u.washington.edu

A. Barash
Faculty of Medicine in the Galilee, Bar-Ilan University, 1311502 Zefat, Israel
e-mail: Alon.Barash@biu.ac.il

Introduction

The morphology and evolution of the vertebral column is of considerable interest in paleoanthropology. The number of vertebrae, their specific shape and the overall morphology of the vertebral column have key functional and postural ramifications. In the case of the spinal curvatures, they have crucial functional and pathological implications for bipedal walking and weight-bearing (Gracovetsky and Iacono 1987;

Farfan 1995). Spinal curvature serves as a shock absorber, helps maintain minimal perturbations of the head, helps to keep the weight of the upper body in line with the pelvis, and helps to create the torque that moves the pelvis and the legs during bipedal walking (Gracovetsky and Iacono 1987; Farfan 1995; Booth et al. 1999; Adams et al. 1999; Harrison et al. 2002; Hosman et al. 2002; Hart et al. 2007; Jang et al. 2009). In the sagittal plane (i.e., viewed from the side), the human spine shows four spinal curvatures: two with a dorsal concavity (lordosis) at the cervical and lumbar levels and two with a ventral concavity at the thoracic and sacral levels (Fig. 18.1). The curvatures of the spine are influenced by the orientation of the sacrum and by vertebral and intervertebral disc morphology (Korovessis et al. 1998; Kimura et al. 2001; Vaz et al. 2002; Vialle et al. 2005; Been et al. 2007, 2010). Finally, there is an ontogenetic component in the development of human spinal curvatures, as the degree of curvature increases to reach maximum values in the adulthood (Cil et al. 2005; Shefi et al. 2013).

In the coronal plane (i.e., viewed from behind), the human spine is straight. Vertebrae are situated one above another. Lateral deviation from this morphology, i.e., when the vertebrae are not aligned vertically (in a straight line), is called scoliosis. In the orthopedic literature, a deviation of <10° (measured by the Cobb method, Cobb 1948) from the vertical line is considered within the normal range. A deviation between 10° and 20° is considered as mild scoliosis, and a higher deviation would be considered pathological scoliosis (Negrini et al. 2012; Scherrer et al. 2013).

Assessing the curvatures of the spine in fossil hominins can, therefore, provide important information to inform the reconstruction of their paleobiology and can also provide important insights on their postural abnormalities. Unfortunately, complete hominin spines are very rarely preserved in the fossil record, and thus we have limited information on this interesting subject. The Neanderthal partial skeleton Kebara 2 from Israel constitutes a remarkable exception, representing an almost complete *in situ* burial (Arensburg et al. 1985). The skeleton of Kebara 2, among other elements, possesses a complete pelvis (Rak and Arensburg 1987) and lumbar and thoracic spine (Arensburg 1991), and therefore, it enables us to fully reconstruct its 3D posture.

The main challenge in reconstructing spinal posture in extinct hominins is the fragmentary nature of the record. The complete spine of Kebara 2 provides rare opportunity to reconstruct spinal posture in a hominin. This paper aims to

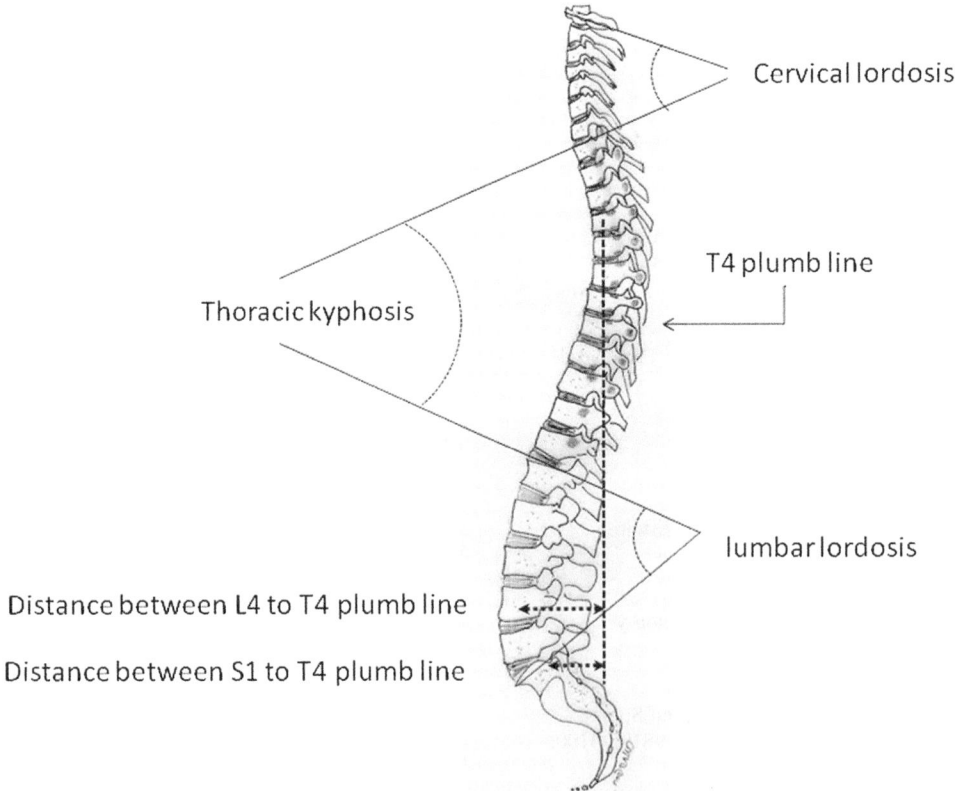

Fig. 18.1 A schematic drawing of the spinal column of an adult modern human in lateral view. Note the spinal curvatures (lordosis and kyphosis) and the position of the T4 plumb line

present for the first time a 3D virtual reconstruction of the spine of an adult Neanderthal, using advanced imaging and virtual reconstruction techniques. The first objective is to evaluate the spinal posture of Kebara 2, including the sacral, lumbar and thoracic spine, in both sagittal and coronal plane.

The study of the Kebara 2 material has revealed the existence of asymmetry in the height of the vertebral bodies of the thoracic vertebrae and the deviation from the mid-plane of some spinous processes, both of which constitute signs of scoliosis. Thus, as a secondary objective, the spinal coronal posture of Kebara 2 will be assessed to investigate whether or not the potential deviation constitutes pathological scoliosis.

Materials and Methods
Fossil Sample

The original specimens, radiographs and CT scans of the sacrum, lumbar and thoracic vertebrae of Kebara 2 were used in the study. The results for Kebara were compared with published values for modern humans.

Sagittal Plane Reconstruction

Until recently, few studies offered a reliable method for measuring and calculating spinal posture based on osteological material, but this situation has been rectified. Peleg et al. (2007) demonstrated how to establish sacral orientation within the pelvic girdle. Been et al. (2007, 2012, 2013, 2014) established a method for calculating the lordotic curvature of the lumbar spine and Goh et al. (1999) offered a way to reconstruct thoracic kyphosis. Based on these methods, we measured the pelvis and the vertebrae of Kebara 2 and reconstructed its sagittal spinal posture.

Overview of the reconstruction process: To build the model, we used the CT scans of the sacrum, lumbar and thoracic vertebrae of Kebara 2. Scanning of the fossils was done on a Phillips Brilliance 64, with the standard settings of 120 kV, 30 mA and 1.5 mm slice thickness. DICOM output was imported into Amira software for segmentation. Following that, we captured its 3D morphology and using visualization software (Amira5.2©), aligned the 3D reconstruction of the original bones into the spinal curvatures. First, we aligned the sacrum and subsequently added each vertebra one at a time. We ensured maximum congruency between the superior and inferior vertebrae, using several reference features. These included the vertebral body endplates, articular facets, spinal and transverse processes, and the spinal canal. The intervertebral disc heights of modern humans (Zhou et al. 2000; Kunkel et al. 2011) were used to

establish the vertical distance between consecutive vertebral bodies and the congruency between the articular processes was checked using the built-in measurement features of Amira. The alignment of the complete lumbar spine and the two caudal-most thoracic vertebrae (T11 – S1) was straightforward as the congruency between the articular facets constrains the positioning. The alignment of T2 – T10 was more difficult because the vertebrae were not as complete as in T11 – S1 and the fit between the articular processes was not as clear.

Reconstruction of Sacral Orientation: We used two approaches to reconstruct the spatial orientation of the sacrum:

(1) Pelvic incidence – which measures the orientation of the sacral endplate in relation to the acetabulum (Fig. 18.2) (Peleg et al. 2007; Been et al. 2013, 2014). The pelvic incidence of Kebara 2 was taken from Been et al. (2013).
(2) Sacral anatomical orientation – (angle γ after Peleg et al. 2007, Fig. 18.2), similar to sacral slope in living modern humans. This angle measures the orientation of the sacral endplate when the pelvis is held in anatomical position. The measurements for sacral anatomical orientation were conducted with the device and methods described by Peleg et al. 2007 (angle γ).

Both the pelvic incidence and sacral anatomical orientation of Kebara 2 indicate a position of the sacrum that is 20–22° less than that of modern humans (Table 18.1). Given that, the sacral endplate of modern humans is aligned at an angle of 39–41° to the horizontal plane (Boulay et al. 2006; Legaye 2007; Peleg et al. 2007; Mac-Thiong 2010), we aligned the sacrum of Kebara 2 at 21° to the horizontal plane (Fig. 18.3).

Reconstruction of lumbar spine: Lumbar lordosis is defined here as the angle between the superior endplate of the sacrum and the superior endplate of the first lumbar vertebra (Fig. 18.1). Two values for the lordosis angle of Kebara 2 were recently published. The first one (25°) is based on the correlation between the degree of lordosis and the orientation of the inferior articular processes of the lumbar vertebrae (Been et al. 2012). The second one (29°) is based on correlation between the degree of lordosis and the pelvic incidence (Been et al. 2013, 2014). Lumbar lordosis results from the wedging of the lumbar vertebral bodies and the morphology of the intervertebral discs (Korovessis et al. 1998; Kimura et al. 2001; Vialle et al. 2005; Been et al. 2010). Both equally influence the lordosis of the lumbar spine (Been et al. 2010). Consequently, the major challenge in reconstructing the lumbar spine of Kebara 2 was determining how to overcome the absence of the intervertebral discs. For the present reconstruction, we used the average of the two published values or 27° as the actual lordosis in our lumbar reconstruction.

In order to reconstruct the intervertebral disc height of Kebara 2, we used the disc heights of modern humans (Goh et al. 1999; Zhou et al. 2000). Beginning from the sacrum, we aligned each vertebra in the sagittal, coronal, and horizontal planes to the vertebra inferior to it. Using Amira software, each vertebra was positioned such that the articular processes of the inferior and superior facets of the adjacent vertebra were parallel to each other and the distance between them was 1–2 mm, which is similar to the value seen in modern humans (Simon et al. 2012). The reconstruction of the lumbar spine is shown in Fig. 18.4.

Reconstruction of thoracic spine: Thoracic kyphosis is defined here as the angle between the inferior endplate of the twelfth thoracic vertebra (T12) and the superior endplate of the first thoracic vertebra (T1) (Fig. 18.1). Similar to lumbar lordosis, thoracic kyphosis is formed by wedging of the thoracic vertebral bodies and of the intervertebral discs. Thoracic kyphosis has been shown to be associated with vertebral body morphology to a greater extent than to intervertebral disc morphology (Goh et al. 1999). To calculate the thoracic kyphosis of Kebara 2, we used the method developed by Goh et al. (1999). In this method, we measured the anterior and the posterior vertebral heights of the twelve thoracic vertebrae (all measurements were made on the fossils) and calculated the

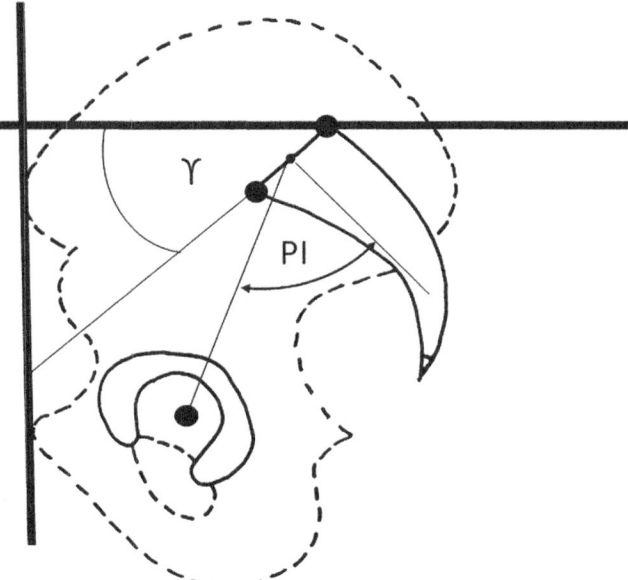

Fig. 18.2 A schematic drawing of the pelvis and sacrum of an adult modern human showing the pelvic measurements used in the study: pelvic incidence (PI) and sacral anatomical orientation (angle γ) similar to sacral slope in living modern humans. The sacral anatomical orientation (angle γ) is the angle created between a line parallel to the superior surface of the sacrum and the horizontal line. The horizontal line is 90° to the line running between the anterior superior iliac spine (ASIS) and the anterior-superior edge of the symphysis pubis (after Peleg et al. 2007)

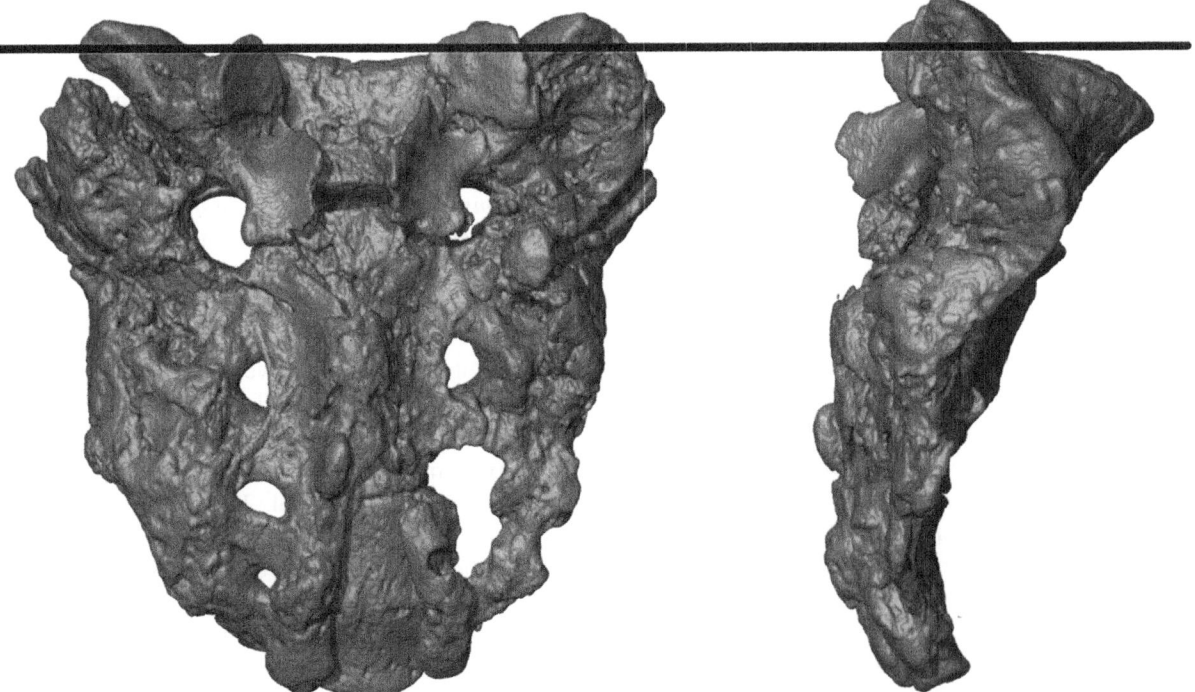

Fig. 18.3 The 3D reconstruction of the sacrum of Kebara 2 in lateral (right) and posterior (left) views. The black line represents the horizontal plane

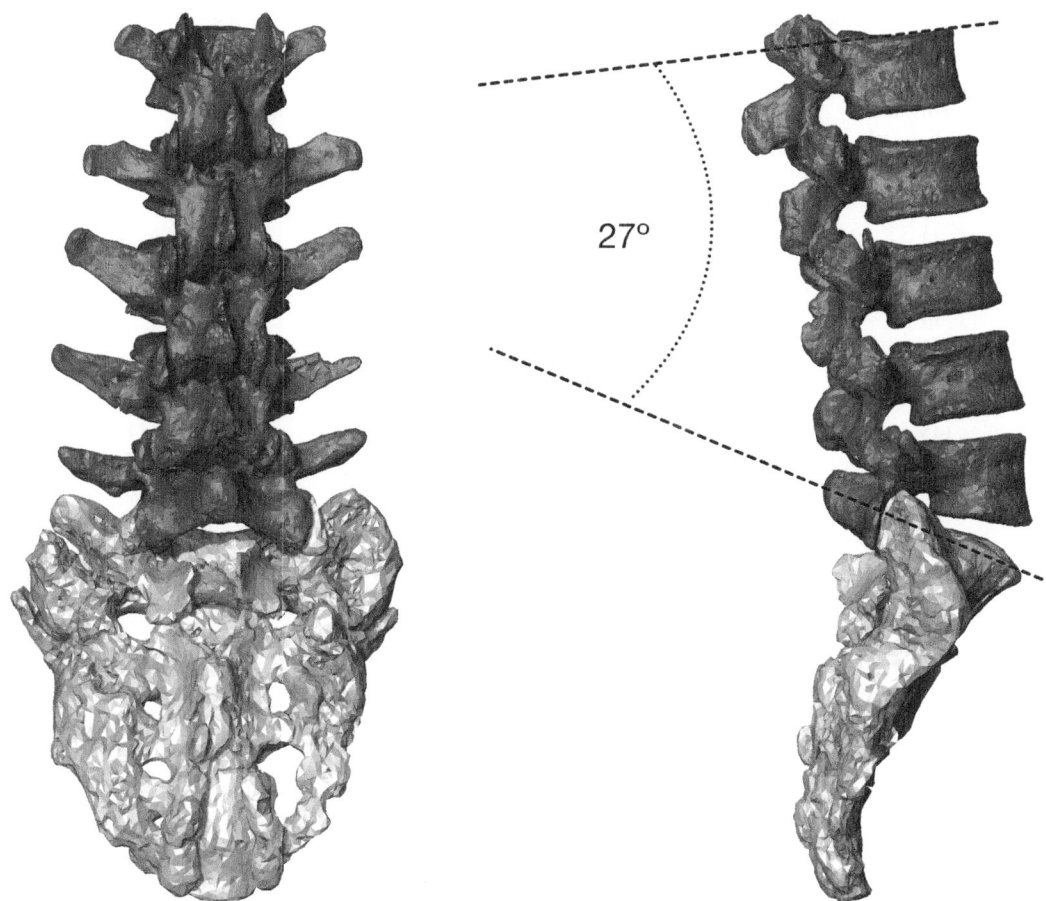

Fig. 18.4 The 3D reconstruction of the lumbar spine (dark gray) and sacrum (medium gray) of Kebara 2 in lateral (right) and posterior (left) views. The lumbar lordosis angle (Cobb angle, 27°) is shown on the picture

ratio between the anterior and posterior heights (Table 18.2 and Fig. 18.5). Based on this ratio, we calculated the expected thoracic kyphosis for Kebara 2.[1]

Utilizing the expected thoracic kyphosis for Kebara 2 as the target, we used the disc heights of modern humans (Goh et al. 1999; Kunkel et al. 2011) for the reconstruction, similar to what we did for the lumbar spine. Beginning from the L1, we aligned each vertebra in the sagittal, coronal, and horizontal planes to the vertebra inferior to it. Using Amira software, each vertebra was positioned such that the articular processes of the inferior and superior facets of the adjacent vertebra were parallel to each other and the distance between them was 1–2 mm, which is similar to the value seen in modern humans (Simon et al.

2012). The alignment of T2 – T10 was more difficult because the vertebrae were not as complete as in T11 – S1 and the fit between the articular processes was not as clear. Consequently, in the thoracic region (T2 – T10), we aligned the vertebra based on our judgment of its position relative to its neighbor, using modern human dimensions (Kunkel et al. 2011) as a guide, and on thoracic kyphosis calculations. The complete reconstruction of the spine of Kebara 2, including the sacrum lumbar and thoracic spine, is shown in Fig. 18.6.

Coronal Plane Orientation

The orientation of the sacrum in the coronal plane is dependent on the spatial orientation of the pelvis. As we have no way to verify the orientation of the pelvis in the coronal plane, we assumed that, as in the majority of modern humans, the right and left sides of the sacral endplate of Kebara 2 were level. We have to bear in mind that Duday and Arensburg (1991) described certain pathologies and anomalies in the Kebara 2 skeleton. Among them these

---

[1]Thoracic kyphosis calculation: Anterior vertebral body height for Kebara 2 is 226.01 mm while posterior vertebral body height for Kebara 2 is 243.38 mm, and the ratio between the two of 0.9286 (Table 18.2). The regression model of Goh et al. (1999) yields a thoracic kyphosis of 44° (=297.114 − 272.31 * 0.9286). The average thoracic kyphosis in modern humans is 46–53° (Table 18.1), so the 44° of kyphosis in Kebara 2 is within the normal range of kyphosis for humans, but slightly below the modern human average.

Table 18.1 Spinal posture (sacral slope, pelvic incidence, lumbar lordosis, and thoracic kyphosis) of modern humans and Kebara 2

Reference	Number of individuals	Age	Method	Sacral slope/sacral anatomical angle (°)	Pelvic incidence	Lumbar lordosis (L1-S1)	Thoracic kyphosis (T1-T12)
Gelb et al. (1995)	100	Adults	Radiographs			62 ± 10	48
Chen (1999)	16	Adults	Radiographs			48 ± 11	
Goh et al. (1999)	93	Adults	Radiographs				50
Jackson and Hales (2000)	75	Adults (20–63 year)	Radiographs			60 ± 12	46 ± 11
Harrison et al. (2002)	15	Young adults	Radiographs	41 ± 10		60 ± 12	48 ± 10
Boulay et al. (2006)	149	Adults (20–50 year)	Radiographs	41 ± 7	53 ± 9		
Legaye (2007)	145	Adults	Radiographs	39 ± 7	50 ± 11	62 ± 8	
Peleg et al. (2007)	424	Adults	Osteological material	41 ± 10	54 ± 12		
Mac-Thiong et al. (2010)	709	Adults	Radiographs	40 ± 8	53 ± 10		
Been et al. (2010, 2013)	106	Adults	Radiographs, osteological material		54 ± 10	51 ± 11	
Cil et al. (2005)	31	Adolescents (13–15 year)	Radiographs			55 ± 10	53 ± 9
Been et al. (2012, 2013, 2014), present study	Kebara 2, Neanderthal	Adult	Osteological material (Sacral slope, Pelvic incidence)	19	34	25/29	44

authors describe an asymmetry of the sacro-iliac articulation. Rak (1991) also described the existence of asymmetry between the superior articular facets of the sacrum, in both size and inclination, being the right side larger and more coronally oriented (Fig. 18.3).

When we level the sacrum in the horizontal plane, the alignment of the lumbar vertebrae in the coronal plane is dictated by the relationship between the superior articular process of the inferior vertebra and that of the inferior articular process of its neighboring superior vertebra, leaving little room for error. The lumbar vertebrae form a straight spine in the coronal plane, similar to non-pathological modern humans.

While working with the thoracic vertebrae, we noticed slight asymmetry in the lower ones. A small height difference between the right and left lateral walls of the vertebral bodies (T11, T12) and slight lateral deviation from the sagittal plane of the tip of the spinous processes in T8-T12 was apparent. Because both of these signs might indicate scoliotic deformity of the spine (Coillard and Rivard 1996; Modi et al. 2008; Stokes and Aronsson 2001), on the

specimen we measured the heights of the left and right walls of the vertebral bodies, to determine whether or not there was a scoliotic anomaly (see side bar for detailed description). Due to the fragmentary preservation of some of the vertebrae, we were only able to measure the lateral walls of T1, T5–T8 and T11–12. We also calculated the lateral wedging of the vertebral bodies of Kebara 2 (T11 and T12) (Table 18.2).

The heights of the lateral walls of the thoracic vertebral bodies of Kebara 2 are within the normal range for modern humans (Masharawi et al. 2008, Table 18.2). The lateral wedging of the vertebral bodies of T11 and T12 is also within the normal range for modern humans (Schiess et al. 2014, Table 18.2). All of the above combines to indicate that the asymmetry shown in the spine of Kebara does not reach a value high enough as to be diagnosed as pathological scoliosis. Yet, the combination of the lateral wedging of the thoracic vertebral bodies together with the small lateral deviation from the sagittal plane of the spinous processes (T8–T12) indicates a mild asymmetry of the thoracic spine of Kebara 2 in the coronal plane. Based on this conclusion, we aligned the lower

Table 18.2 Thoracic and lumbar vertebral body dimensions for Kebara 2 and modern humans

Vertebra	Kebara 2				Modern humans[a]				Modern humans[b]		Kebara 2				Modern humans[b]			
	Ventral height (M1)	Dorsal height (M2)	Dorsoventral diameter (M4)	Wedging, sagittal plane	Ventral height (M1)	Dorsal height (M2)	Dorsoventral diameter (M4)	Wedging sagittal plane	Ventral height (M1)	Dorsal height (M2)	Right lateral height	Left lateral height	Vertebral body width	Wedging coronal plane	Right lateral height	Left lateral height	Vertebral body width	Wedging coronal plane
T1	15.7	15.9	15.3	0.5	15.9 ± 1.0	16.9 ± 1.2	15.6 ± 1.2	3.9 ± 2.9	15.1 ± 2	17.3 ± 3	14.6	15.1			17.6 ± 3	15.6 ± 3		
T2	17.5	17.7			17.5 ± 1.1	17.8 ± 1.1	16.8 ± 1.3	1.1 ± 2.7	17.6 ± 1	18.1 ± 3		16.9			18.4 ± 3	16.5 ± 2		
T3	17.2	18.1	21.3	2.5	18.3 ± 1.1	18.3 ± 1.2	19.0 ± 1.87	0.2 ± 2.3	18.0 ± 2	18.8 ± 2		16.6			18.0 ± 3	17.3 ± 2		
T4	17.5	18.7	21.3	3.0	18.9 ± 1.0	19.1 ± 1.1	20.9 ± 1.6	0.7 ± 2.5	17.8 ± 2	18.8 ± 3		16.5			18.3 ± 3	18.3 ± 2		
T5	17.8	19.8	21.5	5.5	19.1 ± 1.0	19.7 ± 1.1	22.7 ± 1.6	2.1 ± 2.2	18.1 ± 6	19.0 ± 3	19.0	18.4			19.5 ± 3	18.6 ± 2		
T6	18.3	19.6	22.5	3.3	19.0 ± 1.1	20.4 ± 1.1	24.1 ± 1.7	3.2 ± 2.1	18.0 ± 6	19.7 ± 3	18.1	18.4			19.9 ± 3	18.9 ± 3		
T7	19.7	19.8	23.2	0.1	19.4 ± 1.1	20.9 ± 1.1	25.8 ± 1.82	3.4 ± 2.1	17.9 ± 2	20.7 ± 3	18.9				20.5 ± 3	19.4 ± 2		
T8	19.3	19.9	24.8	1.3	19.6 ± 0.9	21.3 ± 1.1	27.3 ± 2.04	3.3 ± 1.69	18.4 ± 2	21.0 ± 3	19.0	19.1			20.3 ± 3	20.1 ± 3		
T9	20.0	20.2	broken		20.2 ± 1.2	21.7 ± 1.1	28.5 ± 2.0	2.1 ± 1.8	19.6 ± 2	21.4 ± 3					21.3 ± 3	20.7 ± 3		
T10	20.74	21.4	broken		22.1 ± 1.4	22.9 ± 1.2	28.9 ± 2.0	1.8 ± 1.6	20.7 ± 2	22.4 ± 3	19.6				22.2 ± 3	21.5 ± 3		
T11	20.8	24.7	26.8	8.4	22.9 ± 1.2	24.5 ± 1.6	29.3 ± 2.1	3.5 ± 2.4	21.3 ± 2	23.6 ± 4	21.9	22.7	38.2	-1.25	22.7 ± 3	22.7 ± 3	36.7 ± 3.9	-1.3 ± 3.9[d]
T12	21.4	27.6	28.2	12.4	23.7 ± 1.8	26.0 ± 1.3	30.0 ± 2.1	4.4 ± 2.6	22.5 ± 2	24.7 ± 3	22.55	23.65	39.2	-1.6	24.0 ± 3	23.5 ± 2	38.8 ± 3.9	-1.8 ± 5.2[d]
Thoracic vertebrae Total[c]	226.0	243.4			237	249.5			225	245.5								
Lumbar spine[e]																		
L1	24.0	27.8	32	7.0	23.7 ± 2.0	26.1 ± 1.8	28.9 ± 3.5	4.7 ± 2.5					42				40.8 ± 4.1	
L2	23.0	27.6	35.2	7.4	25.1 ± 1.7	26.2 ± 1.7	30.5 ± 3.4	2.1 ± 2.4					44.2				43.3 ± 4.5	
L3	23.3	28.0	36.5	7.4	25.8 ± 1.7	25.7 ± 1.5	32.0 ± 3.2	-0.1 ± 2.5					47.2				45.9 ± 4.3	
L4	26.2	27.6	36.5	2.3	26.0 ± 1.6	24.1 ± 1.6	32.2 ± 2.9	-3.5 ± 2.5					51.1				48.1 ± 4.3	
L5	29.1	23.5	34.4	-9.3	25.9 ± 2.0	21.1 ± 1.7	33.4 ± 2.9	-8.6 ± 3.5					51.9				49.9 ± 4.2	
Lumbar vertebrae total	125.6	134.5		14.8	126.5	123.2	1.4	-5.4										

Vertebral body wedging in the sagittal plane: positive values indicate kyphotic wedging, negative values indicate lordotic wedging

[a] 32 Euromerican and 41 European male individuals (see Gómez-Olivencia et al. 2013 for more information on the sample)

[b] After Masharawi et al. (2008) (n = 210 males and females)

[c] In the case of the sum of samples, the sum of the different means is provided

[d] After Schiess et al. (2014)

[e] After Been (2005)

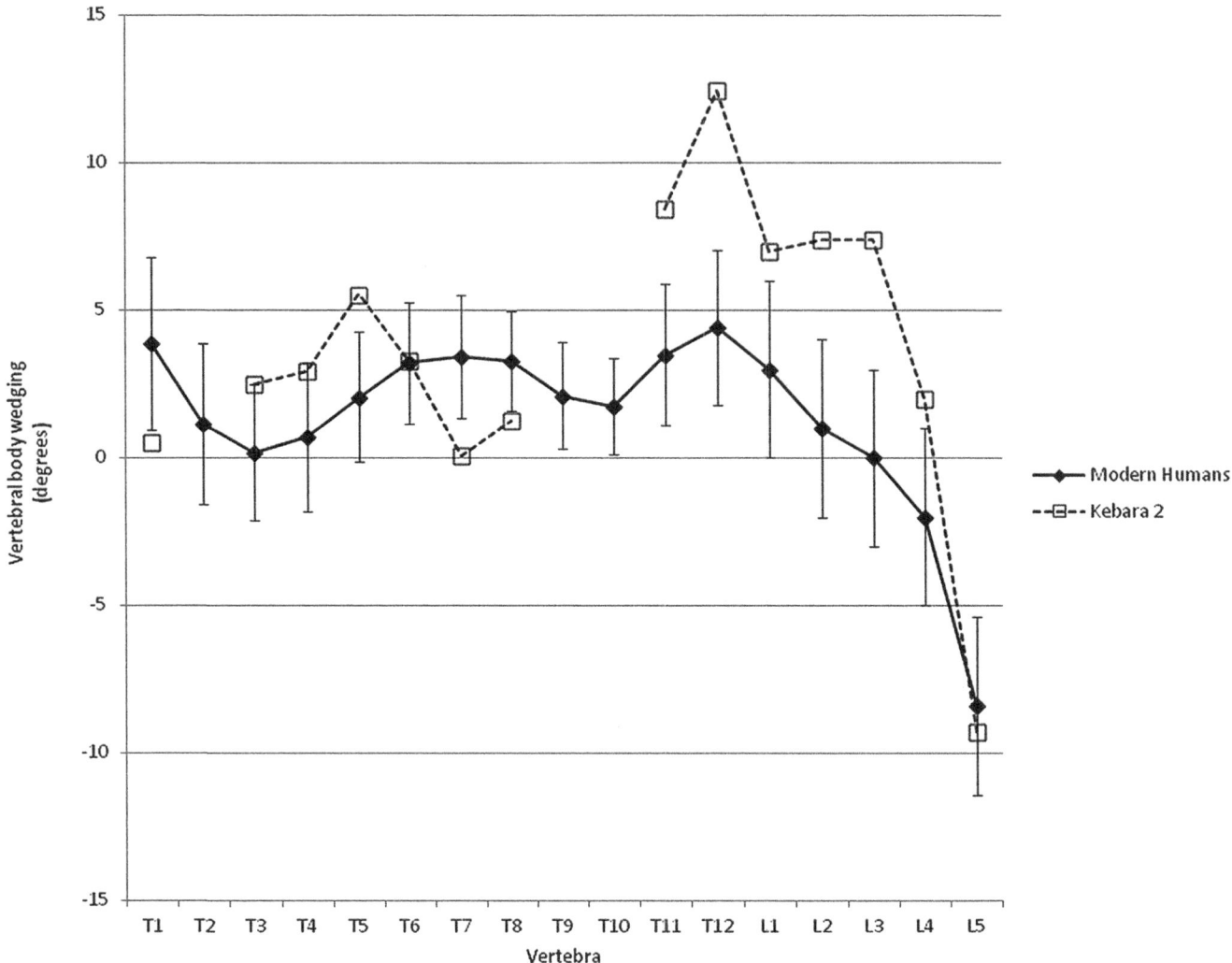

Fig. 18.5 Spinal vertebral body sagittal wedging of modern humans and Kebara 2 (after Table 18.2). Note the extreme kyphotic wedging of the T11-L3 vertebral bodies of Kebara 2 compared to modern humans. Positive values indicate kyphotic wedging; negative values indicate lordotic wedging. Bar equals one standard deviation

thoracic vertebrae (T5–T12) with a small lateral curve (in accordance with Negrini et al. 2012) (Fig. 18.6).[2]

---

[2]Scoliosis is a general term comprising a heterogeneous set of conditions, consisting of changes in the shape and position of the spine, thorax and trunk, and can be defined as a 3D torsional deformity of the spine and trunk. Scoliosis causes a lateral curvature in the frontal and an axial rotation in the horizontal plane (Negrini et al. 2012). Scoliosis can also cause an abnormality in the sagittal plane, but this does not occur in all cases. Scoliotic deformity of the spine is associated with osseous changes in vertebral morphology. Modi et al. (2008) showed lateral vertebral body wedging of five consecutive segments in scoliotic patients. He also showed that the wedging of the apex vertebra is 4.08° ± 2.4° when thoracic scoliosis <30° while the wedging of the apex vertebra is 2.7° ± 5.8° when thoracic scoliosis >30°. Stokes and Aronsson (2001) found that even small scoliotic deformities include vertebral wedging and that the vertebrae generally show larger deformity than the discs in thoracic scoliosis. They also report an average vertebral lateral wedging of 3.7° ± 2.6° with scoliotic

Results

Based on the 3D reconstruction of the spine of Kebara 2, we measured the length of the complete spine from T1 to S1 as 45.6 cm, which is close to the average for modern humans (Nagesh and Kumar 2006, Table 18.3). Nagesh and Kumar (2006) provided three formulae to calculate stature based on spinal length. From these formulae, we estimated the stature of Kebara 2 to have been between 169 and 170 cm (Table 18.3).

This reconstruction also enables us to measure spino-pelvic parameters of Kebara 2 and compare them with those

---

(Footnote 2 continued)
deformity of 20.2° ± 7.3°. Coillard and Rivard (1996) found that in scoliotic vertebrae the spinous process is slightly curved towards the side of convexity. They also report asymmetry in the orientation of the transverse processes.

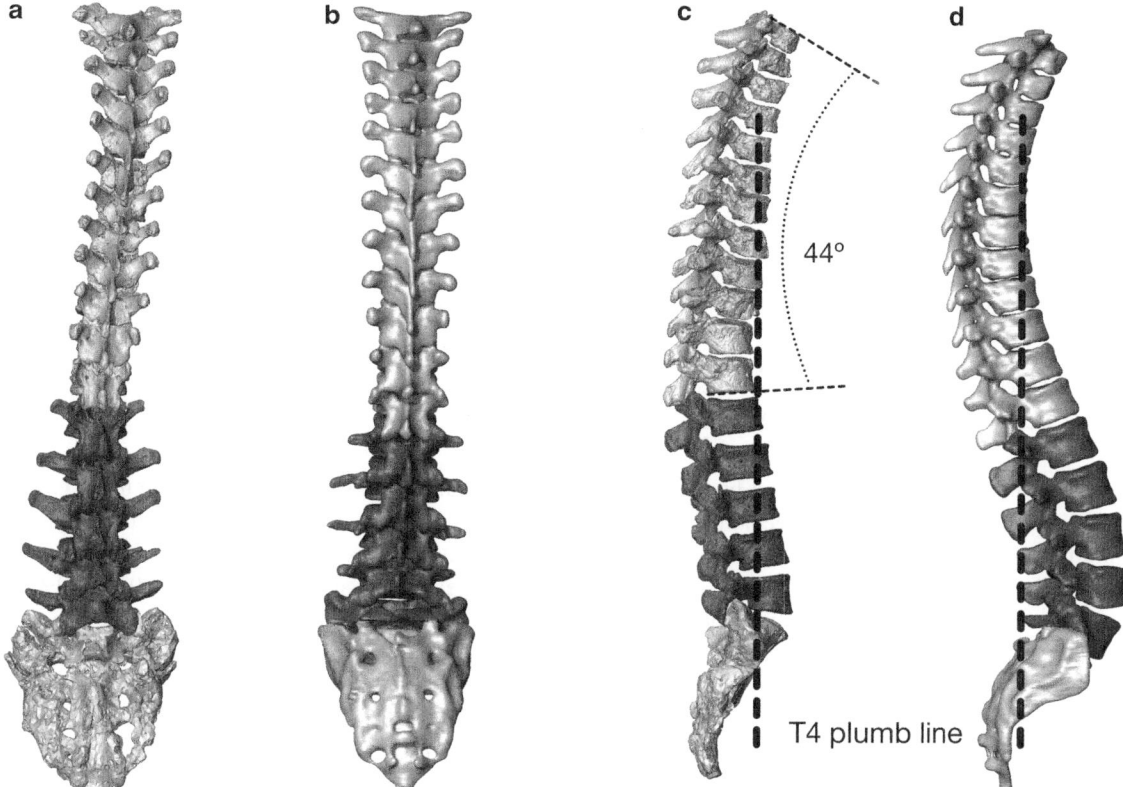

Fig. 18.6 The 3D reconstruction of the complete spine of Kebara 2 and modern human. Sacrum in medium gray, lumbar spine in dark gray and thoracic spine in light gray. **a** Posterior view of the spine Kebara 2, note the slight asymmetry shown in the thoracic spine. **b** Posterior view of the spine of a modern human. **c** Lateral view of the spine of Kebara 2, the thoracic kyphosis (Cobb angle, 44°) and T4 plumb line are shown on the picture. **d** Lateral view of the spine of a modern human, T4 plumb line is shown on the picture

Table 18.3 Spine length and stature for Kebara 2 and modern humans

Research	Population	Sex	Spine length (cm)			Stature (cm)
			Thoraco-lumbar	Thoracic	Lumbar	
Nagesh and Kumar (2006)	South Indian population	Male	44 ± 2	26 ± 1	17 ± 1	166 ± 7
		Female	41 ± 2	25 ± 1	16 ± 1	154 ± 6
Terazawa et al. (1990)	Japanese population	Male			20 ± 1	166
		Female			19 ± 1	154
Current study	Kebara 2 Neanderthal	Male	45.6	27.6	18	Based on thoracolumbar length: 2.419 * 45.6 + 59.989 = **170.3** Based on thoracic length: 3.037 * 27.6 + 85.715 = **169.5** Based on lumbar length: 4.901 * 18 + 80.783 = **169.0**

The three formulae for calculating stature (A, B, and C) are from Nagesh and Kumar (2006)

of modern humans (Table 18.4). These include the apex of the spinal curvatures and the T4 plumb line. In a healthy adult human, all of these variables (i.e., spinal curvatures, apex and plumb line) work in concert to align the vertebral segments in a position that will enable healthy upright posture. In order to effectively balance in an upright posture, the line of gravity of the upper body (e.g., head, arms and trunk) must pass through the pelvis and balance on the supporting legs. Because the line of gravity is hard to measure, as it depends on the contribution of many skeletal and soft tissue components, other reference lines have been established to indicate the relationship between the spine and pelvis in the sagittal plane. One of these is the T4 plumb line (Fig. 18.1), which is a vertical (or plumb) line that passes through the center of T4 vertebral body in the sagittal plane. The anterio-posterior position of this line relative to other

Table 18.4 Spinopelvic variables for Kebara 2 and modern humans

Measurement		Description	Modern human Mean ± SD (range)	Kebara 2 Neanderthal
Thoracic apex	Kuntz et al. (2007)	The vertebra at the apex of the thoracic kyphosis	T7 (T3; T11)	T11
Lumbar apex	Kuntz et al. (2007)	The vertebra at the apex of the lumbar lordosis	L4 (L2; L5)	L5
Spinal balance T4-L4 (mm)	Jackson and Hales (2000)	Horizontal perpendicular distances measured in millimeters between the plumb line from the center of T4 vertebral body and the center of L4 vertebral body	−59.5 ± 21.8 (−124; −10)	+3
Spinal balance T4-S1 (mm)	Jackson and Hales (2000)	Horizontal perpendicular distances measured in millimeters between the plumb line from the center of T4 vertebral body and the posterior angle of S1 vertebral body	−31.3 ± 23 (−101; +13)	+6.5

spinal and pelvic anatomical landmarks can be measured as the horizontal distance (Fig. 18.1) between the plumb line and the landmark. The horizontal distance between the T4 plumb line and the center of L4 vertebral body of Kebara 2 (Fig. 18.6) is 3 mm anterior to L4, while the horizontal distance between the T4 plumb line and the posterior angle of S1 vertebral body is 6.5 mm anterior to S1. The apex of the thoracic kyphosis in Kebara 2 is at T11, while the apex of the lumbar lordosis is at L5.

Discussion

This reconstruction of the sacral, lumbar and thoracic spine of Kebara 2 is the first 3D virtual reconstruction of an adult Neanderthal spine, lacking only the cervical portion. This reconstruction provides a baseline for further 3D reconstructions of the cervical spine, the thorax, and pelvis of Kebara 2. The only previous reconstruction of the spine of Kebara 2 was a plaster reconstruction of a complete Neanderthal skeleton by Sawyer and Maley (2005). They used the thoracic and lumbar vertebrae of Kebara 2 for their model, but in their description there is no mention of any of the remarkable features that are now clear, such as coronal asymmetry or the small lordotic angle. Moreover, it should be noted that Sawyer and Maley's model was based on La Ferrassie 1 (LF1) as the reference individual and several other Neanderthal individuals were added to substitute for the missing elements from LF1. Due to differences in size between the original individuals LF1 and Kebara 2, modifications were done to some of the parts. That was the case of the upper thorax, due to the longer clavicles of LF1 compared to Kebara 2 (Sawyer and Maley 2005).

Our reconstruction demonstrates an upright erect hominin with a somewhat different spinal posture than that of modern humans (Fig. 18.7). When compared to modern humans Kebara 2 shows: a more vertical sacral orientation, a less pronounced lordotic curvature and a thoracic kyphosis within the normal range, but slightly smaller than the average of modern human kyphosis. The recent literature demonstrates that a similar posture was also present in the spines of Shanidar 3 and La Chapelle-aux-Saints 1 (Been et al. 2012, 2013, 2014) and in Pelves 1 and 2 from the Middle Pleistocene site of Sima de los Huesos (Bonmatí et al. 2010; Been et al. 2014). This implies that during hominin evolution upright erect posture might have been achieved through different spinal postures. Australopithecines, for example, have a pelvic incidence that is usually smaller than the average modern human with lumbar lordosis angles that range from slightly below the average of modern humans (e.g., Sts 14, Stw 431) to hyper-lordotic (e.g., Sediba MH2) (Sanders 1998; Whitcome et al. 2007; Been et al. 2012, 2014; Williams et al. 2013). *H. erectus* has pelvic incidence and lumbar lordosis values that are within the range of modern humans but slightly below average (Been et al. 2012, 2014).

The spinal posture of Kebara 2 and the information provided by the lumbar part of other Neanderthal individuals also suggests locomotor and weight-bearing differences between Neanderthals and modern humans. A hypolordotic spine is the posture of choice in static lifting tasks, because a straighter spine can withstand higher compressive loads (Adams et al. 1994; Arjmand and Shirazi-Adl 2005). This preference is probably due to the decreased compression and shear force developed in the hypolordotic lumbar spine and the maximal use of the posterior ligaments and lumbosacral fascia (Gracovetsky et al. 1985; Sanders 1995; Arjmand and Shirazi-Adl 2005). Nonetheless, humans with hypolordotic spines experience certain gait deviations that affect locomotor function and economy. These include: short stride length, slow walking velocity, bent hip bent knee gait, and an anteriorly flexed trunk (Grasso et al. 2000; Sarwahi et al. 2002; Hirose et al. 2004; Jang et al. 2009). These deviations affect locomotor economy in hypolordotic subjects (Fox and Whitcome 2011). If a hypolordotic posture is representative of Neanderthals, Neanderthals might have been better adapted to carry heavy loads and, potentially, to engage in generally more rigorous upper body activities (Pearson 2000; Weaver 2009). On the

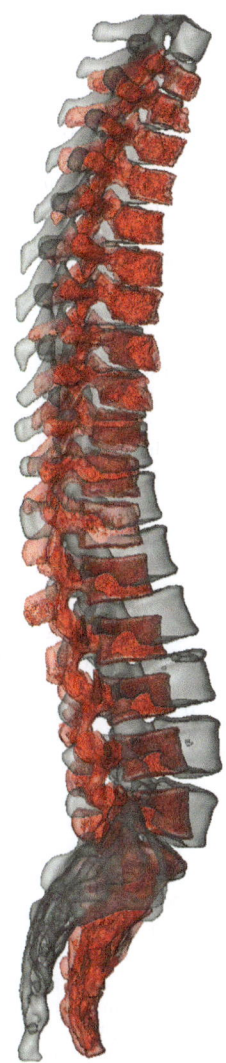

Fig. 18.7 Superimposition of the 3D reconstruction of the complete spine of Kebara 2 (transparent red) and modern human (gray). Note the smaller spinal curvature of Kebara compared to a modern human

other hand, it suggests that Neanderthals potentially had a shorter stride length and slower walking velocity on a flat terrain in comparison with modern humans (Fox and Whitcome 2011; Been et al. 2012).

This reconstruction also revealed mild asymmetry in the thoracic spine of Kebara 2 that would result in slight scoliosis of less than 20° (i.e., not pathological). Given that asymmetries in one spinal area are often associated with asymmetries in another spinal area (Dickson et al. 1984), it is not surprising that asymmetries in the morphology and articulation of the sacrum of Kebara 2 (Duday and Arensburg 1991; Rak 1991) also exist. As noted by Duday and Arensburg (1991) the first sacral vertebra is not completely fused to the rest of the sacrum, and could be regarded as a mild degree of lumbarization. In fact there is another case of homeotic shift of vertebral borders in caudal direction in

Kebara 2: the thoraco-lumbar limit is caudally moved given the presence of lumbar ribs in L1.

Whether or not the asymmetries in the thoracic spine arose as a consequence of (or are related to) changes in the caudal portion of the thoraco-lumbar spine and/or lumbo-sacral borders (as described by Barnes 2012) requires further testing. Because Kebara 2 exhibits an asymmetrical thoracic spine, we anticipate some degree of asymmetry in the size or shape of the ribs of this specimen. The ribs indeed show an endosteal ossification in ribs 5–7 from the right side (Duday and Arensburg 1991). However whether this is also the result of (or it is related to) the asymmetries described in the lumbo-sacral, sacro-iliac and thoracic regions should be further investigated. In any case the described thoracic asymmetry that could also extend to the thoracic cage should be taken into consideration in future reconstruction of the Kebara 2 ribcage.

Based on this reconstruction we can now estimate, for the first time, a few characteristics that stem from it. We calculated the stature of Kebara 2 based on spinal length (Table 18.3), obtaining estimation for the stature of Kebara 2 of 169–170 cm, which conforms to stature estimates of Kebara 2 based on long bones of 166–175.9 cm (Ruff 1991; Vandermeersch 1991; Ruff et al. 2005; Carretero et al. 2012).

We also identified the apices of the spinal curves: T11 is the apex vertebra of the thoracic kyphosis of Kebara 2 and L5 is the apex vertebra for the lumbar lordosis of Kebara 2. The apex vertebrae of Kebara 2 are at the lower end of the normal variation found in the spinal curvatures of modern humans (Table 18.4, Kuntz et al. 2007).

We were also able to measure the position of the T4 plumb line for the first time. The position of the T4 plumb line of Kebara 2 is more ventral than its position in healthy modern humans (Table 18.4, Fig. 18.6). This is in accordance with the findings of Barrey et al. (2007) and Bae et al. (2012) that showed that people with small lumbar lordosis and a vertical sacrum have a more ventral position of the plumb line.

For the reconstruction of the spine of Kebara 2, we employed advanced methods for virtual spinal reconstruction and exploited our personal knowledge of spinal morphology. Yet, in the absence of a living Neanderthal specimen, we acknowledge that the reconstruction presented here is just a proposition regarding how the spine of Kebara 2 was shaped.

Conclusion

This is the first 3D reconstruction of a complete sacral, lumbar and thoracic spine of a Neanderthal. This reconstruction demonstrates that the upright posture of Kebara 2 was slightly different from that of the average modern human. The spine of Kebara 2, when compared to modern humans, exhibits a combination of a vertical sacrum and a small lumbar lordosis together with a nearly average thoracic kyphosis. As a result, the spinopelvic alignment of this specimen was different from modern humans, with a ventral position of the T4 plumb line and a low position of the apex vertebrae in the spinal curves (thoracic kyphosis and lumbar lordosis). This reconstruction provides the basis for a future reconstruction of the Kebara 2 specimen, including the pelvis, the rib cage and the cervical spine.

References

Adams, M. A., McNally, D. S., Chinn, H., & Dolan, P. (1994). Posture and the compressive strength of the lumbar spine. International society of biomechanics award paper. *Clinical Biomechanics, 9*, 5–14.

Adams, M. A., Mannion, A. F., & Dolan, P. (1999). Personal risk factors for first-time low back pain. *Spine, 24*, 2497–2505.

Arensburg, B. (1991). The vertebral column, thoracic cage and hyoid bone. In O. Bar Yosef & B. Vandermeersch (Eds.), *Le squelette Moustérien de Kébara 2* (pp. 113–146). Paris: CNRS.

Arensburg, B., Bar-Yosef, O., Chech, M., Goldberg, P., Laville, H., Meignen, L., et al. (1985). Une sépulture néandertalienne dans la grotte de Kebara (Israel). *Comptes Rendus de l'Académie des Sciences, 300*, 227–230.

Arjmand, N., & Shirazi-Adl, A. (2005). Biomechanics of changes in lumbar posture in static lifting. *Spine, 30*, 2637–2648.

Bae, J. S., Jang, J. S., Lee, S. H., & Kim, J. U. (2012). Radiological analysis of lumbar degenerative kyphosis in relation to pelvic incidence. *The Spine Journal, 12*, 1045–1051.

Barnes, E. (2012). *Atlas of developmental field anomalies of the human skeleton. A paleopathology perspective.* New Jersey: Wiley-Blackwell.

Barrey, C., Jund, J., Noseda, O., & Roussouly, P. (2007). Sagittal balance of the pelvis–spine complex and lumbar degenerative diseases. A comparative study about 85 cases. *European Spine Journal, 16*, 1459–1467.

Been, E. (2005). The anatomy of the lumbar spine of *Homo neanderthalensis* and its phylogenetic and functional implications. PhD Dissertation, Tel Aviv University.

Been, E., Pessah, H., Been, L., Tawil, A., & Peleg, S. (2007). New method for predicting the lumbar lordosis angle in skeletal material. *The Anatomical Record, 290*, 1568–1573.

Been, E., Barash, A., Pessah, H., & Peleg, S. (2010). A new look at the geometry of the lumbar spine. *Spine, 35*, E1014–E1017.

Been, E., Gómez-Olivencia, A., & Kramer, P. A. (2012). Lumbar lordosis of extinct hominins. *American Journal of Physical Anthropology, 147*, 64–77.

Been, E., Pessah, H., Peleg, S., & Kramer, P. (2013). Sacral orientation in hominin evolution. *Advances in Anthropology, 3*, 133–141.

Been, E., Gómez-Olivencia, A., & Kramer, P. A. (2014). Lumbar lordosis in extinct hominins: Implications of the pelvic incidence. *American Journal of Physical Anthropolgy, 154*, 307–314.

Bonmatí, A., Gómez–Olivencia, A., Arsuaga, J.-L., Carretero, J. M., Gracia, A, Martínez, I., et al. (2010). A Middle Pleistocene lower back and pelvis from an aged individual from the Sima de los Huesos site, Spain. *Proceedings of the National Academy of Science USA, 107*, 18386–18391.

Booth, C. K., Bridwell, K. H., Lenke, L. G., Baldus, C. R., & Blanke, K. M. (1999). Complications and predictive factors for the successful treatment of flat back deformity (fixed sagittal imbalance). *Spine, 24*, 1712–1720.

Boulay, C., Tardieu, C., Hecquet, J., Benaim, C., Mouilleseaux, B., Marty, C., et al. (2006). Sagittal alignment of spine and pelvis regulated by pelvic incidence: Standard values and prediction of lordosis. *European Spine Journal, 15*, 415–422.

Carretero, J. M. Rodríguez, L., García-González, R., Arguaga, J.-L., Gómez–Olivencia, A., Lorenzo, C., et al. (2012). Stature estimation from complete long bones in the Middle Pleistocene humans from the Sima de los Huesos, Sierra de Atapuerca (Spain). *Journal of Human Evolution, 62*, 242–255.

Chen, Y. L. (1999). Geometric measurements of the lumbar spine in Chinese men during trunk flexion. *Spine, 24*, 666–669.

Cil, A., Yazic, M., Uzumcugil, A., Kandemir, U., Alanay, A., Alanay, Y., et al. (2005). The evolution of sagittal alignment of the spine during childhood. *Spine, 30*, 93–100.

Cobb, J. R. (1948). Outline for the study of scoliosis. *Instructional course lectures, The American academy of orthopaedic surgeons* (Vol. 5, pp. 261–275). Ann Arbor: JW. Edwards.

Coillard, C., & Rivard, C. H. (1996). Vertebral deformities and scoliosis. *European Spine Journal, 5*, 91–100.

Dickson, R. A., Lawton, J. O., Archer, I. A., & Butt, W. P. (1984). The pathogenesis of idiopathic scoliosis. Biplanar spinal asymmetry. *Bone & Joint Journal, 66*(1), 8–15.

Duday, H., & Arensburg, B. (1991). La pathologie. In O. Bar Yosef & B. Vandermeersch (Eds.), *Le squelette moustérian de Kebara 2* (pp. 179–194). Paris: Editions CNRS.

Farfan, H. F. (1995). Form and function of the musculoskeletal system as revealed by mathematical analysis of the lumbar spine. *Spine, 20*, 1462–1474.

Fox, M., & Whitcome, K. K. (2011). Neanderthal lumbopelvic anatomy and the biomechanical effects of a reduced lumbar lordosis. Ph.D. Dissertation University of Cincinnati.

Gelb, D. E., Lenke, L. G., Bridwell, K. H., Blanke, K. M., & McEnery, K. W. (1995). An analysis of sagittal spinal alignment in 100 asymptomatic middle and older aged volunteers. *Spine, 20*, 1351–1358.

Goh, S., Price, R. I., Leedman, P. J., & Singer, K. P. (1999). The relative influence of vertebral body and intervertebral disk shape on thoracic kyphosis. *Clinical Biomechanics, 14*, 439–448.

Gómez-Olivencia, A., Eaves-Johnson, K. L., Franciscus, R. G., Carretero, J. M., & Arsuaga, J.-L. (2009). Kebara 2: New insights regarding the most complete Neandertal thorax. *Journal of Human Evolution, 57*, 75–90.

Gómez-Olivencia, A., Been, E., Arsuaga, J.-L., & Stock, J. T. (2013). The Neandertal vertebral column. 1—The cervical spine. *Journal of Human Evolution, 64*, 608–630.

Gracovetsky, S., & Iacono, S. (1987). Energy transfer in the spinal cord. *Journal of Biomedical Engineering, 9*, 99–114.

Gracovetsky, S., Farfan, H., & Helleur, C. (1985). The abdominal mechanism. *Spine, 10*, 317–324.

Grasso, R., Zago, M., & Lacquaniti, F. (2000). Interactions between posture and locomotion: Motor patterns in humans walking with bent posture versus erect posture. *Journal of Neurophysiology, 83*, 288–300.

Harrison, D. E., Cailliet, R., Harrison, D. D., Janik, T. J., & Holland, B. (2002). Changes in sagittal lumbar configuration with a new method of extension traction: Nonrandomized clinical controlled trial. *Archives of Physical Medicine and Rehabilitation, 83*, 1585–1591.

Hart, R. A., Badra, M. I., Madala, A., & Yoo, J. U. (2007). Use of pelvic incidence as a guide to reduction of H-type spino-pelvic dissociation injuries. *Journal of Orthopedic Trauma, 21*, 369–374.

Hirose, D., Ishida, K., Nagano, Y., Takahashi, T., & Yamamoto H. (2004). Posture of the trunk in the sagittal plane is associated with gait in community-dwelling elderly population. *Clinical Biomechanics, 19*, 57–63.

Hosman, A. J., Langeloo, D. D., de Kleuver, M., Anderson, P. G., Veth R, P., & Slot, G. H. (2002). Analysis of the sagittal plane after surgical management for Scheuermann's disease: A view on overcorrection and the use of an anterior release. *Spine, 27*, 167–175.

Jackson, R. P., & Hales, C. (2000). Congruent spinopelvic alignment on standing lateral radiographs of adult volunteers. *Spine, 25*, 2808–2815.

Jang, J. S., Lee, S. H., Min, J. H., & Maeng, D. H. (2009). Influence of lumbar lordosis restoration on thoracic curve and sagittal position in lumbar degenerative kyphosis patients. *Spine, 34*, 280–2844.

Kimura, S., Steinbach, G. C., Watenpaugh, D. E., & Hargens, A. R. (2001). Lumbar spine disc height and curvature responses to an axial load generated by a compression device compatible with magnetic resonance imaging. *Spine, 26*, 2596–2600.

Korovessis, P. G., Stamatakis, M. V., & Baikousis, A. G. (1998). Reciprocal angulation of vertebral bodies in the sagittal plane in an asymptomatic Greek population. *Spine, 23*, 700–704.

Kunkel, M. A., Herkommer, M., Reinehr, M., Böckers, T. M., & Wilke, H. J. (2011). Morphometric analysis of the relationships between intervertebral disc and vertebral body heights: An anatomical and radiographic study of the thoracic spine. *Journal of Anatomy, 219*, 375–387.

Kuntz, C., Levin, L. S., Ondra, S. L., Shaffrey, C. I., & Morgan, C. J. (2007). Neutral upright sagittal spinal alignment from the occiput to the pelvis in asymptomatic adults: A review and resynthesis of the literature. *Journal of Neurosurgery Spine, 6*, 104–112.

Legaye, J. (2007). The femoro-sacral posterior angle: An anatomical sagittal pelvic parameter usable with dome-shaped sacrum. *European Spine Journal, 16*, 219–225.

Mac-Thiong, J. M., Roussouly, P., Berthonnaud, E., & Guigui, P. (2010). Sagittal parameters of global spinal balance: Normative values from a prospective cohort of seven hundred nine Caucasian asymptomatic adults. *Spine, 35*, E1193–E1198.

Masharawi, Y., Salame, K., Mirovsky, Y., Peleg, S., Dar, G., Steinberg, N., et al. (2008). Vertebral body shape variation in the thoracic and lumbar spine: Characterization of its asymmetry and wedging. *Clinical Anatomy, 21*, 46–54.

Modi, H. N., Suh, S. W., Song, H. R., Yang, J. H., Kim, H. J., & Modi, C. H. (2008). Differential wedging of vertebral body and intervertebral disc in thoracic and lumbar spine in adolescent idiopathic scoliosis: A cross sectional study in 150 patients. *Scoliosis, 3*, 1–9.

Nagesh, K. R., & Kumar, G. P. (2006). Estimation of stature from vertebral column length in South Indians. *Legal Medicine, 8*, 269–272.

Negrini, S., Aulisa, A. G., Aulisa, L., Circo, A. B., de Mauroy, J. C., Durmala, J., et al. (2012). 2011 SOSORT guidelines: Orthopaedic and rehabilitation treatment of idiopathic scoliosis during growth. *Scoliosis, 7*, 3.

Peleg, S., Dar, G., Steinberg, N., Peled, N., Hershkovitz, I., & Masharawi, Y. (2007). Orientation of the human sacrum: Anthropological perspectives and methodological approaches. *American Journal of Physical Anthropology, 133*, 967–977.

Pearson, O. M. (2000). Postcranial remains and the origins of modern humans. *Evolutionary Anthropology, 9*, 229–247.

Rak, Y. (1991). The pelvis. In O. Bar Yosef & B. Vandermeersch (Eds.), *Le squelette Moustérian de Kebara 2* (pp. 113–146). Paris: Editions CNRS.

Rak, Y., & Arensburg, B. (1987). Kebara 2 Neandertal pelvis: First look at a complete inlet. *American Journal of Physical Anthropology, 73*, 227–231.

Ruff, C. B. (1991). Climate and body shape in hominid evolution. *Journal of Human Evolution, 21*, 81–105.

Ruff, C. B., Niskanen, M., Junno, J. A., & Jamison, P. (2005). Body mass prediction from stature and bi-iliac breadth in two high latitude populations, with application to earlier higher latitude humans. *Journal of Human Evolution, 48*, 381–392.

Sanders, W. J. (1995). Function, allometry, and evolution of the australopithecine lower precaudal spine. Ph.D. Dissertation, New York University.

Sanders, W. J. (1998). Comparative morphometric study of the Australopithecine vertebral series Stw-H8/H41. *Journal of Human Evolution, 34*, 249–302.

Sarwahi, V., Boachie-Adjei, O., Backus, S. I., & Taira, G. (2002). Characterization of gait function in patients with postsurgical sagittal (flatback) deformity—a prospective study of 21 patients. *Spine, 27*, 2328–2337.

Sawyer, G. J., & Maley, B. (2005). Neanderthal reconstructed. *The Anatomical Record, 283*, 23–31.

Scherrer, S. A., Begon, M., Leardini, A., Coillard, C., Rivard, C. H., & Allard, P. (2013). Three-dimensional vertebral wedging in mild and moderate adolescent idiopathic scoliosis. *PLoS ONE, 8*, e71504.

Schiess, R., Boeni, T., Rühli, F., & Haeusler, M. (2014). Revisiting scoliosis in the KNM-WT 15000 *Homo erectus* skeleton. *Jornal of Human Evolution, 67*, 48–59.

Shefi, S., Soudack, M., Konen, E., & Been, E. (2013). Development of the lumbar lordotic curvature in children from age 2 to 20 years. *Spine, 38*, E602–E608.

Simon, P., Espinoza Orías, A. A., Andersson, G. B., An, H. S., & Inoue, N. (2012). In vivo topographic analysis of lumbar facet joint space width distribution in healthy and symptomatic subjects. *Spine, 37*, 1058–1064.

Stokes, I. A., & Aronsson, D. D. (2001). Disc and vertebral wedging in patients with progressive scoliosis. *Journal of Spinal Disorders and Techniques, 14*, 317–322.

Terazawa, K., Alkabane, H., Gotouda, H., Mizukami, K., Nagao, M., & Takatori, T. (1990). Estimating stature from the length of the lumbar part of the spine in Japanese. *Medicine, Science and the Law, 30*, 354–357.

Vandermeersch, B. (1991). La ceinture scapulaire et les membres supérieures. In O. Bar Yosef & B. Vandermeersch (Eds.), *Le squelette Moustérien de Kébara 2* (pp. 157–178). Paris: Editions CNRS.

Vaz, G., Roussouly, P., Berthhonnaud, E., & Dimnet, J. (2002). Sagittal morphology and equilibrium of pelvis and spine. *European Spine Journal, 11*, 80–87.

Vialle, R., Levassor, N., Rillardon, L., Templier, A., Skalli, W., & Guigui, P. (2005). Radiographic analysis of the sagittal alignment and balance of the spine in asymptomatic subjects. *Journal of Bone and Joint Surgery, 87*, 260–267.

Weaver, T. D. (2009). The meaning of Neandertal skeletal morphology. *Proceedings of the National Academy of Science USA, 106*, 16028–16033.

Whitcome, K. K., Shapiro, L. J., & Lieberman, D. E. (2007). Fetal load and the evolution of lumbar lordosis in bipedal hominins. *Nature, 450*, 1075–1078.

Williams, S. A., Ostrofsky, K. R., Frater, N., Churchill, S. E., Schmid, P., & Berger, L. R. (2013). The vertebral column of *Australopithecus sediba*. *Science, 340*, 1232996.

Zhou, S. H., McCarthy, I. D., McGregor, A. H., Coombs, R. R., & Hughes, S. P. (2000). Geometric dimensions of the lower lumbar vertebrae: Analysis of data from digitized CT images. *European Spine Journal, 9*, 242–248.

Chapter 19
Brother or Other: The Place of Neanderthals in Human Evolution

Rachel Caspari, Karen R. Rosenberg, and Milford H. Wolpoff

Abstract Few have provided insights and thoughtful explanations for Neanderthals that equal what have been a central theme in Yoel Rak's publications. One of his deep understandings is that Neanderthals are another way of being human: not inferior, not superior, but different. Looking at what we now understand, Rak has been fundamentally correct in this insight, and where new discoveries have been unexpected, they serve to expand its scope and meaning. Unexpected new information about Neanderthal body form, demography, and even breeding behavior support and flesh out Rak's essential insight about the place of Neanderthals in human evolution. In this paper some of the new discoveries and interpretations of Neanderthals and their evolution are discussed in this context. We examine three aspects of how Neanderthals are another way of being human: body shape (as revealed in the pelvis), population structure (as revealed in their paleodemography), and breeding behavior (as revealed by paleogenetics, in the pattern of ancient gene flow). In these ways Neanderthals are like their ancestors, or more broadly are the plesiomorphic condition.

Keywords Interbreeding • Neanderthal body shape • Neanderthal breeding behavior • Neanderthal pelvic form • Neanderthal population structure

Introduction

Few scholars have provided insights and thoughtful explanations for Neanderthals that equal those of our good friend, Yoel Rak (Fig. 19.1). Our conversations with Rak over many years often centered on his insistence that Neanderthals represent another way of being human: not inferior, not superior, but definitely different. Although we often find ourselves on opposite sides of paleoanthropological debates from Rak, it may surprise him to know that we agree with him on this central tenet. Neanderthals are different, yet human, and in this paper we review some of our work on Neanderthal morphology and culture that reflects the nature of that difference.

Over the last few years, unexpected new information about Neanderthal body form, demography, and breeding behavior support and flesh out Rak's essential insight about the place of Neanderthals in human evolution. In this paper we examine three aspects of Neanderthal biology: body shape (as reflected in the pelvis), population structure (as suggested by paleodemography), and breeding behavior (as suggested by paleogenetics in the pattern of ancient gene flow). We argue, and are sure Rak would agree, that in body form, demography, and population structure, Neanderthals reflect the plesiomorphic condition, and are unlike modern humans (people alive today and their immediate ancestors) in many ways. However, these differences exist within an open genetic system, with ancient contacts between archaic and modern humans that attest to Neanderthal humanity.

Neanderthal Body Form

One area of Neanderthal morphology that has received a great deal of attention is the pelvis. Early discoveries of Neanderthals from the Middle East (Shanidar in Iraq and Tabun in Israel) showed what was thought to be a distinctive morphology in the form of an elongated, thinned superior

R. Caspari (✉)
Department of Sociology Anthropology and Social Work, Central Michigan University, Mount Pleasant, MI 48859, USA
e-mail: caspa1r@cmich.edu

K.R. Rosenberg
Department of Anthropology, University of Delaware, Newark, DE 19716, USA
e-mail: krr@udel.edu

M.H. Wolpoff
Department of Anthropology, University of Michigan, Ann Arbor, MI 48109, USA
e-mail: wolpoff@umich.edu

Assaf Marom and Erella Hovers (eds.), *Human Paleontology and Prehistory*, Vertebrate Paleobiology and Paleoanthropology, DOI: 10.1007/978-3-319-46646-0_19

Fig. 19.1 Exhibit "A"

pubic ramus. Scholars like Stewart (1960) described the trait, which he saw in specimens from Tabun and Shanidar (but not in the specimens from Skhul), as "peculiar" and interpreted it as a derived trait in Neanderthals that indicated contemporaneity of Neanderthals and modern humans in the Levant. The functional significance of the trait was not addressed and it was widely assumed that Neanderthals had a derived condition.

In the 1980s, several scholars considered the functional significance of this trait. A number of hypotheses (Trinkaus 1984; Dean et al. 1986; Rosenberg 1988) were based on the assumption that because in humans, females have a longer pubis than males, the elongated pubis in Neanderthals was indicative of an expanded birth canal, with a range of explanations for why Neanderthals might have had an expanded birth canal relative to modern humans. These hypotheses never satisfactorily explained the genuinely peculiar trait, namely an unusual pattern of sexual dimorphism in (the admittedly small sample of) Neanderthals. Neanderthal males had long pubic bones which were in some cases longer than those of females, in sharp contrast to the pattern we see in all human populations.

With the discovery of the Kebara pelvis (Kebara 2) in 1983, it was possible for the first time to examine a Neanderthal pubic bone in the context of the entire pelvic girdle. In his meticulous description of that specimen, Rak (1990, 1991b; Rak and Arensburg 1987) pointed to some interesting differences between it and the pelvis of more recent humans. This specimen is the only Neanderthal pelvis for which both pelvic inlet breadth and pubis length are known and Rak showed, in contrast to the expectations of earlier scholars, that even though the Kebara specimen has an elongated pubis compared to modern humans, it has an inlet size not much different from that expected based on body size in humans (but a wide overall pelvic girdle). Rak argued then that the long pubic bones were not related to obstetrical constraints, but were a reflection of the anterior position of the Neanderthal pelvic aperture relative to the acetabulum when compared with modern humans. He compared the position of the pelvic inlet within the frame of the pelvic girdle in Kebara 2 with a series of modern humans. Rak noted that as seen from above, the Kebara 2 pelvic inlet was more anteriorly positioned within the frame of the pelvis than the inlet of modern humans. This is associated with a corresponding anterior shift in the position of the sacrum (the inlet's rear), rendering this weight-bearing portion closer to the bi-acetabular line (compared to modern humans). In life the pelvis is tilted forward in most postures (so that the anterior-superior iliac spine and anterior-inferior iliac spine are on a vertical line), positioning the weight bearing surface at the top of the sacrum (this supports the trunk) directly above the acetabulum where weight is transmitted to the lower limbs. In the Kebara 2 specimen, less tilt is required. According to Rak, the front of the inlet is also more anterior because of the forward shift, which explains why the pubic bones are longer in this and certain other archaic specimens (see Fig. 19.2).

Additional features characterizing this pelvis (and some others including some of those attributed to European Neanderthals) are a consequence of the changed orientations required by this differing inlet position. The iliac blades are broader and the acetabula face more laterally, unchanged from the ancestral condition. But these unusual features are not unique to the Western Asian and European specimens. Important remains from this time period in East Asia such as the Jinniushan pelvis (also with an elongated pubis, Rosenberg et al. 2006) have not yet been described in detail. Later East Asians such as the 18 kyr Minatogawa (Okinawa) male also has pubic length elongation compared to acetabulum size, like the Neanderthal pattern. Rak's hypothesized shift in the position of the pelvic aperture relative to the hips has postural implications and this has been reflected in more recent reconstructions of Neanderthals (Sawyer and Maley 2005; Lloyd 2012) (Fig. 19.3). Rak (1991a) also studied pelvic shape in the australopithecine specimen, AL 288-1

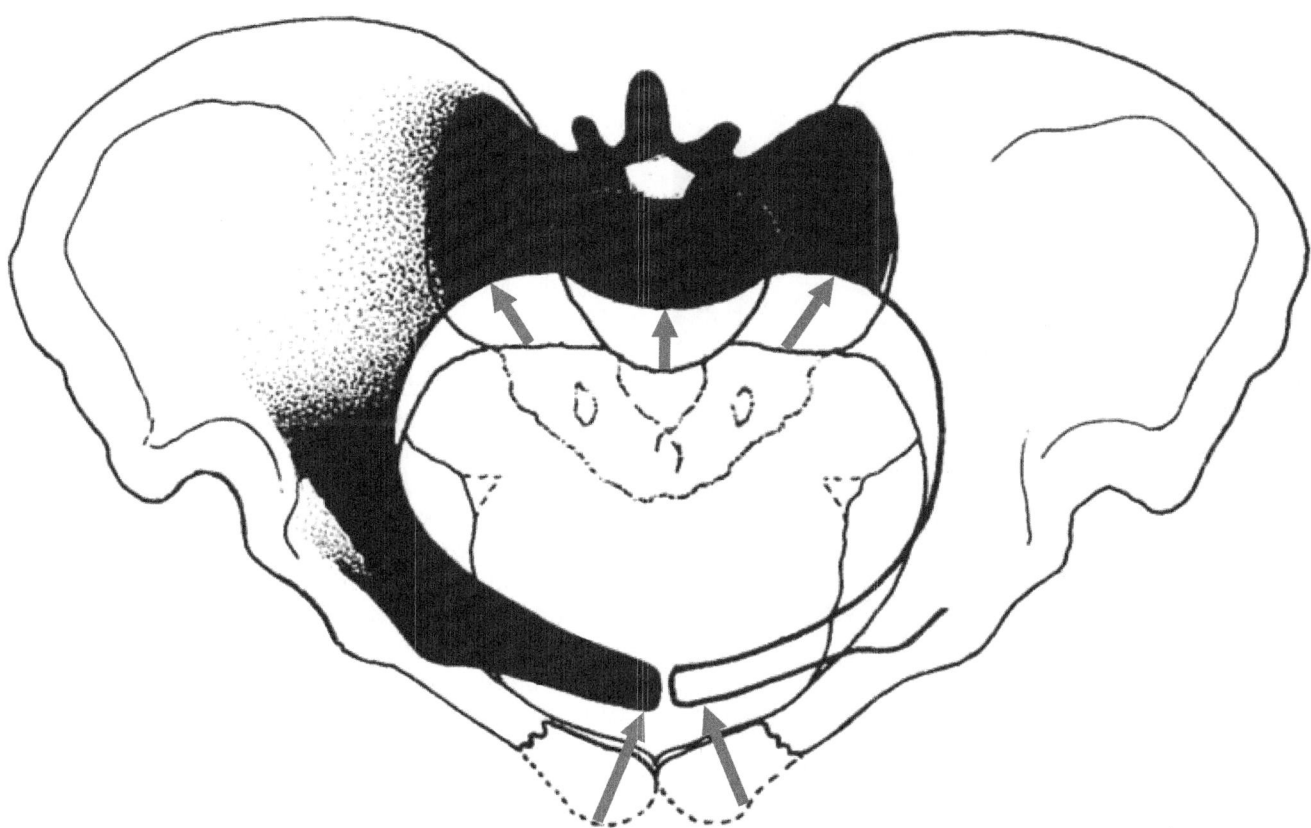

Fig. 19.2 This figure shows the hypothesized changes (*indicated by the arrows*) that would be necessary to go from a Neanderthal pelvis (*shown in white*) to a modern human pelvis (*shown in black or with black stippling*), according to Rak's model. The breadth of the inlet would remain the same in this modeling, but the length of the pubic bone would become shorter as the inlet moves posteriorly within the pelvic girdle. (modified from Rak and Arensburg 1987)

arguing that it, too, was extremely wide relative to body size. He argued "this width, when combined with the horizontal rotation of the pelvis, minimizes the vertical displacement of the center of mass during bipedal walking" (Rak 1991a: 283).

Rak's model of the position of the relationship of the length of the pubis with the position of the pelvic aperture within the pelvic girdle has not been rigorously tested, but his work has led to two robust observations: 1. The elongated pubis in Neanderthals is not a derived trait in that group but a retention of the primitive condition (Rak 1993) and 2. A wide pelvis overall is typical of all hominids.

We are interested here in why Neanderthals (and other hominids) had such a broad pelvis.

"It could be argued that the early *Homo* pelvis from Gona, Ethiopia refutes the climate hypothesis, because it may demonstrate that a wide pelvis was the primitive condition for the genus *Homo*. However, showing that a morphological feature is primitive for a taxonomic group does not explain why this feature persists in some descendant taxa and not others. Even if a wide pelvis was unrelated to climate in early *Homo*, climate adaptation is still the best explanation for why Neanderthals maintained a wide pelvis, early modern humans living closer to

the equator evolved a narrow pelvis, and recent humans who migrated to cold climates regained a wide pelvis." (Weaver 2009: 16032)

"Neanderthals tended to live in cold climates, where wide trunks are advantageous for thermoregulation, so maintaining the primitive pattern of transversely wide outlets would not have interfered with their climatic adaptations" (Weaver and Hublin 2009: 8154).

In his recent book, Churchill (2014) discusses Neanderthal body form in the context of cold adaptation that was important at least some of the time, but also suggests, as others have proposed, that the large Neanderthal thorax might be a retention of the plesiomorphic condition (Gómez–Olivencia et al. 2009). "The Neanderthals, despite their short and stocky build, suffered the legacy of a tropical ancestry, just as do modern Inuit and Eskimo" (Churchill 2014: 129). Yet, there is no question that he considers Neanderthals to be cold adapted (title of Sect. 5.7 is "Neanderthals *were* cold-adapted"). Thorax shape, of course, is reflected in dimensions of the pelvis.

A related question is why do Neanderthals have wide trunks? Ruff (1991, 1994) observed that Neanderthals had relatively wide bodies for their stature compared to modern

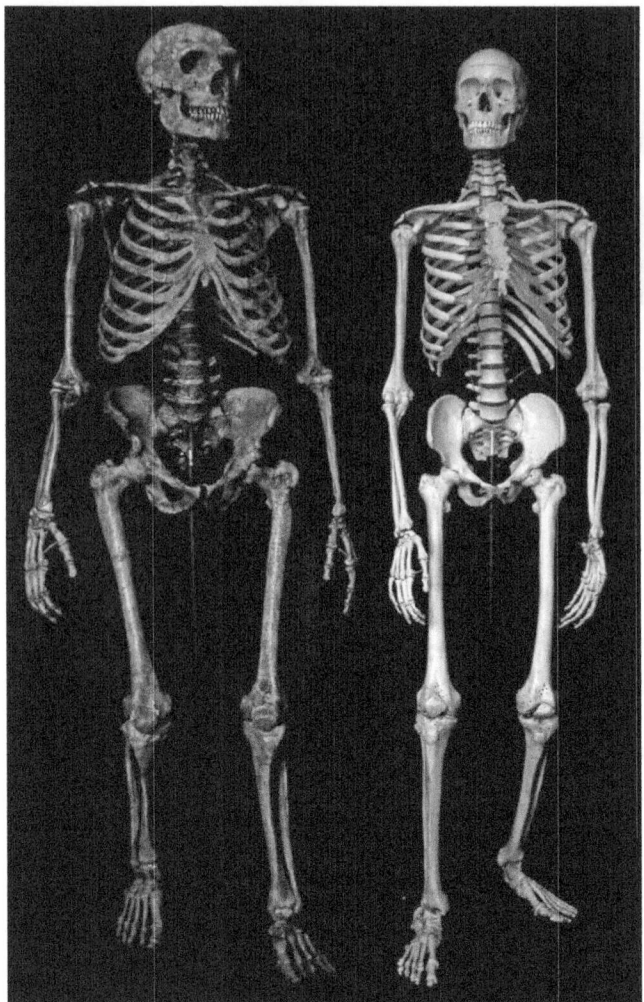

Fig. 19.3 Reconstructed Neanderthal and modern human skeleton. Note the difference in the anterior tilt of the pelvis: less in the Neanderthal, more tilted in the human skeleton shown. (Sawyer and Maley 2005)

humans. Arsuaga et al. (1999) noted that broad trunks were also found at Sima de los Huesos in a sample thought to be one of the ancestors of Neanderthals.

> "Our interpretation is that the pattern of a broad pelvis, a long femoral neck and marked iliac flaring is a shared primitive character already present in *Australopithecus afarensis* (as seen in the A.L. 288-1 specimen), in which case it should be also found in the early *Homo* fossils. However, based on the juvenile and very fragmentary WT 15000 pelvis (West Turkana, Early Pleistocene), the 'adult' bicrestal breadth of *Homo ergaster* has been estimated as narrow. In our opinion, the East African ER 3228 and OH 28 coxal bones are so similar in all the preserved regions to the SH coxal bones, also showing a marked lateral iliac flare, that the inferred narrow bicrestal breadth for WT 15000 might be an error. A narrow pelvis could be a unique modern human condition." (Arsuaga et al. 1999: 257)

Churchill (2014) developed and elaborated the details of Neanderthal body form, not just reconstructing Neanderthal

anatomy but also estimating its physiological needs in a glacial environment. But there is an important comparative context for understanding Neanderthal body form, throughout human evolution and in a wide range of environments. Today, we know much more about the evolution of hominid pelvic shape than we did in 1983 when Kebara 2 was found, because of an expanded fossil record from earlier time periods that includes the Gona specimen (Simpson et al. 2008), the Malapa material attributed to *Australopithecus sediba* (Kibii et al. 2011), the Jinniushan specimen (Rosenberg et al. 2006), and the material from Atapuerca (Arsuaga et al. 1999). In addition, even earlier specimens such as the australopithecines AL 288-1 and Sts 14 have pelvic girdles that are broad relative to stature. This is visually accentuated by the common reconstruction of australopithecine trunks as conically shaped. However, the cone-shaped reconstruction is likely incorrect. The Woranso-Mille skeleton from ca. 3.8 Ma is the earliest hominid skeleton to have sufficient information about trunk shape (preserving both a clavicle and a first rib) to demonstrate a broad, barrel-shaped trunk (Haile-Selassie et al. 2010). Australopithecines reconstructed to have a conical trunk invariably lack the anatomical information provided by the clavicle and first rib, and it now seems clear that a barrel-shaped trunk is the normal trunk shape for later hominids.

It seems to us that Rak was correct in his assertion that a broad pelvis is the primitive condition for hominids, just as Churchill was correct in describing a Neanderthal legacy from a tropical ancestry. In addition to knowing more, we have also unlearned something that we thought was true – namely that the Nariokotome *Homo erectus* specimen which was reconstructed based on very fragmentary pieces of the pelvis of a juvenile male, may not have been as narrow as had been thought (Ruff and Walker 1993; Walker and Ruff 1993). That specimen represented the only narrow-hipped individual in the human fossil record and its narrowness now appears to have been exaggerated in the reconstruction (Ruff 1995; Arsuaga et al. 1999; Simpson et al. 2008).

To further examine the Neanderthal condition in an evolutionary context, we examined pelvic dimensions relative to stature in all human fossils for which these dimensions could be measured or reliably estimated (Table 19.1). It should be noted that these data come from a range of sources and some include estimates (shown in parentheses). An explanation for the source of the data is provided with the table. Taking a broad comparative perspective, we examined relative pelvic breadth in fossil hominids over the last 3.5 million years. We plotted our best estimate of stature (based on whichever long bones were preserved for each specimen) for each fossil against bi-iliac breadth and transverse, anterior-posterior and estimated circumferential dimensions of the pelvic inlet (Fig. 19.4). We compared the

Table 19.1 Fossil pelvic remains and stature estimates. All measurements in mm

Specimen	Max bi-iliac breadth (M2)	Max inlet breadth	Max inlet A-P	Calculated circum. of inlet	Femur length (M1)	Arm bone length (M1) when femur length is unknown	Estimated stature
AL 288-1	247 (Wolpoff, measured on cast of Lovejoy reconstruction)	132 (Tague and Lovejoy 1986)	76	338	281 (max. length) (Wolpoff 1980, measured on cast of Lovejoy reconstruction)		1052
Sts 14	230 Wolpoff reconstruction	104	94	311	(276) (Lovejoy and Heiple 1970)		1034
Sediba (MH 2)	250 (Kiibi et al. 2011)	118	82	319	316 (De Silva pc)		1183
Gona	288 (Simpson et al. 2008)	125	98	353			1350
SH Pelvis 1	335 (Bonmati et al. 2010)	138	121	408	475 (Arsuaga et al. 1999)		1779
Jinniushan	344 (Rosenberg et al. 2006)	149 Rosenberg reconstruction	124	430		260 (ulna) (Rosenberg et al. 2006)	1689 (Rosenberg et al. 2006)
Kebara 2	313 (Rak 1990)	142 (Tague 1992)	118	410		324 (humerus) (Vandermeersch 1991)	1707 (Vandermeersch 1991)
Tabun C1	260 (McCown and Keith 1939); or 270 (Churchill 2014 citing Weaver and Hublin 2009)	144 (Ponce de León et al. 2008)	115	408	416 (McCown and Keith 1939)		1558
Skhul 4	280 (McCown and Keith 1939)				492.5 (McCown and Keith 1939)		1844
La Chapelle-aux-Saints	(295) (measured by Ruff 1994 from Boule's (1911) drawing): or 292 (Churchill 2014)				(433) (Ruff 1994)		1621

Bi-iliac breadth: Bi-iliac breadth of AL 288-1 was measured by Wolpoff on cast reconstruction by Lovejoy. Sts 14 was measured by Wolpoff on the original specimen; the three pieces were propped up into anatomical position with missing part of the pubis on the right side represented in clay. Simpson et al. (2008) reported the same number for Sts 14 and cited Robinson (1972). Arsuaga et al. (1999) initially published a value for SH Pelvis 1 of 340 but this was corrected to 335 in a later paper by Bonmati and colleagues (2010) (Table S2). The value for Tabun C1 is from McCown and Keith (1939). We do not have a great deal of confidence in this measurement and await estimates from Weaver (2009) and/or Ponce de Leon et al. (2008). The Skhul 4 measurement was taken from McCown and Keith (1939) and looks reasonable based on the completion of the specimen shown in drawings in the monograph. Although Qafzeh 9, Skhul 5 and 9 also include fragmentary pelvic elements, we were not confident enough in the estimates provided to include them

We have used Ruff's (1994) estimate of the La Chapelle-aux-Saints bi-iliac breadth, a measurement that was based on Boule's reconstruction of the skeleton which included a sacrum but lacked the pubic portions. Boule reconstructed these from plaster, but the reconstruction was done before elongated pubic bones had been observed in Neanderthals. For this reason we suspect that the pubic bones were longer and therefore the true breadth of the La Chapelle-aux-Saints pelvis was most probably greater than the 295 mm estimated by Ruff from Boule's drawing and given in Churchill (2014), or Trinkaus' (2011) estimate of 292 mm

Femoral Length and Stature Estimation

Whenever possible, when femora were preserved, we used femoral length to estimate stature (Table 2) using the Feldesman et al. regression formula given in Simpson et al. (2008) Supplementary Material because it gives estimates for AL 288-1 and WT 15000, at different ends of the size range, that were close to estimates of other workers. Femur length for Sts 14 was estimated geometrically from the STS 14 proximal femur and unassociated distal remains from Sterkfontein (Lovejoy and Heiple 1970)

The femur length for MH2 was estimated by DeSilva. He estimated tibia length for MH2 based on the tibia of MH4; in comparable parts MH2 is on average 97.9% as big as MH4, DeSilva used 97.9% of the MH4 tibial length as the estimate for MH2. He then examined the ratio between the lengths of the tibia and femur in the Turkana Boy and the Dmanisi skeleton (the only two Plio-Pleistocene skeletons with both tibia & femur preserved). The Dmanisi tibia is 80% length of femur; Nariokotome's is 88%. Applied to MH2 this suggests the femur was between 301 and 331 mm, an average of 316 mm – the value we have used for the MH2 stature estimate

For Atapuerca SH Pelvis 1, we accepted Arsuaga and colleagues' (1999) argument that two partial femora probably belong to the same individual: Femur X, a proximal left fragment that fits in the acetabulum and has "an atypical remodeling" on its head that marches a corresponding "atypical remodeling" on the surface of the acetabulum; and AT 432, a right diaphysis. These were reconstructed to a femur length between 47 and 48 cm

We used the length of forearm bones to reconstruct height in two other specimens. For Jinniusham, we use the estimate of Rosenberg et al. (2006) from the length of the complete ulna. Kebara 2 lacks a femur and we estimate its stature from its virtually complete humerus (Vandermeersch 1991)

Finally, Gona lacks a femur, or any other limb bone. We have taken the average of the maximum and minimum of Simpson et al.'s (2008) estimates for the Gona stature

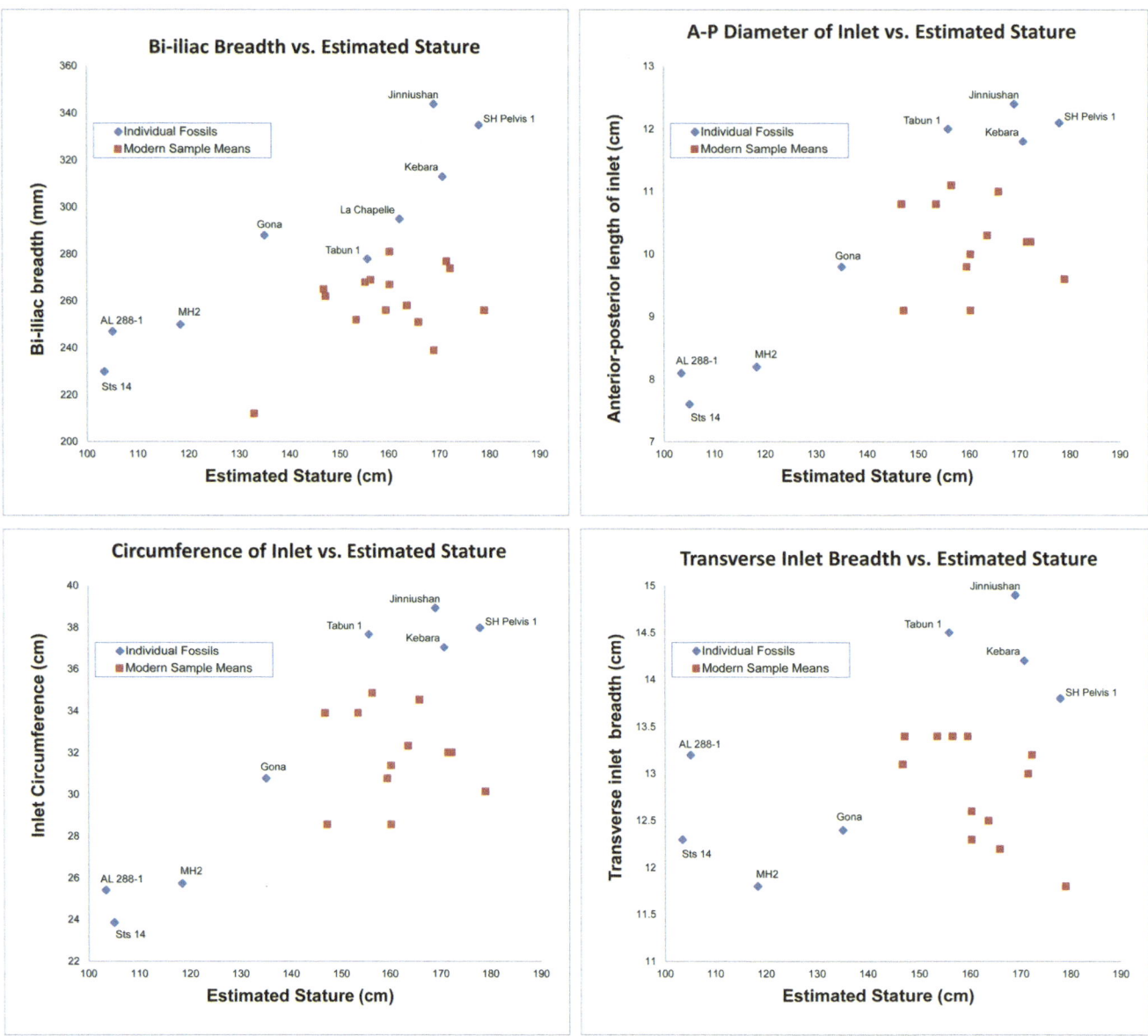

Fig. 19.4 Each graph shows fossil specimens (diamonds) and sex-specific sample means from modern human populations (squares). Stature for each fossil specimen represents our best estimate of stature depending on what skeletal elements were preserved. All fossil data are given in Table 19.1 with explanations about sources of measurements. (from Tague 1989)

fossil data to published mean sex-specific values for recent human populations from a wide geographic and climatic range (Tague 1989).

Across fossil specimens, there is a positive significant correlation between our estimate of stature and each of these dimensions of the pelvic aperture. For example, in Fig. 19.4, each graph shows a dimension of the pelvic girdle plotted against estimated stature. The blue dots represent the individual fossil specimens and the red dots the sex-specific means (or in one case a mixed-sex mean) of modern human populations. The fossil specimens all appear to have a greater bi-iliac breadth relative to stature than the modern

means. The modern human range (given that the red dots represent sample means) is at or below (but certainly never above) the fossils for bi-iliac breadth for a given stature. There is a systematic pattern of difference between the fossil and modern data; the fossil data lie around a line that falls above the modern human means. That is, modern humans are narrower in bi-iliac breadth relative to stature than the fossils.

We also examined direct measurements of the pelvic aperture at the level of the first pelvic plane (in females this dimension would be the inlet of the birth canal). Dimensions of the midplane and outlet are difficult given the poor preservation of many of the fossils. Keep in mind that these are measurements of the pelvic aperture, which in females, but not in males, functions as the birth canal. Here, we examined the relationship between stature and transverse diameter of the inlet, AP diameter of the inlet and circumference of the inlet. Circumference of the inlet was calculated from the transverse and AP dimensions using the formula for the circumference of an ellipse given the lengths of the two axes. Because the neonatal head is malleable and is molded (the shape is changed) during labor, the circumference may in fact be the most important dimension for obstetrical purposes). These comparisons are based on data published by Tague (1989). The pattern of inlet dimensions relative to stature is similar to that of bi-iliac breadth discussed above, although in this case, some of the modern human samples fall between the early and later hominids, i.e., for their stature, they seem to have a transverse inlet dimensions similar to that of earlier humans. For most

dimensions of the pelvic inlet relative to stature, like bi-iliac breadth, modern human sex-specific sample means are the same size or smaller than the fossil humans.

Finally, we examined the relationship between bi-iliac breadth and pelvic inlet circumference to see if the difference between modern humans and earlier humans in bi-iliac breadth could be accounted for by a wider overall pelvis which would be reflected in greater inlet dimensions (Fig. 19.5). There is no clear difference between the fossil specimens and modern samples in this relationship suggesting that we cannot exclude obstetric differences as the source of the difference between modern humans and fossil specimens in dimensions of pelvic breadth.

Table 19.2 gives the correlation coefficients between stature and each dimension of the pelvis for fossil human individuals and modern human sex-specific means. The correlation coefficients which are statistically significant at the 0.05 level are shown in boldface.

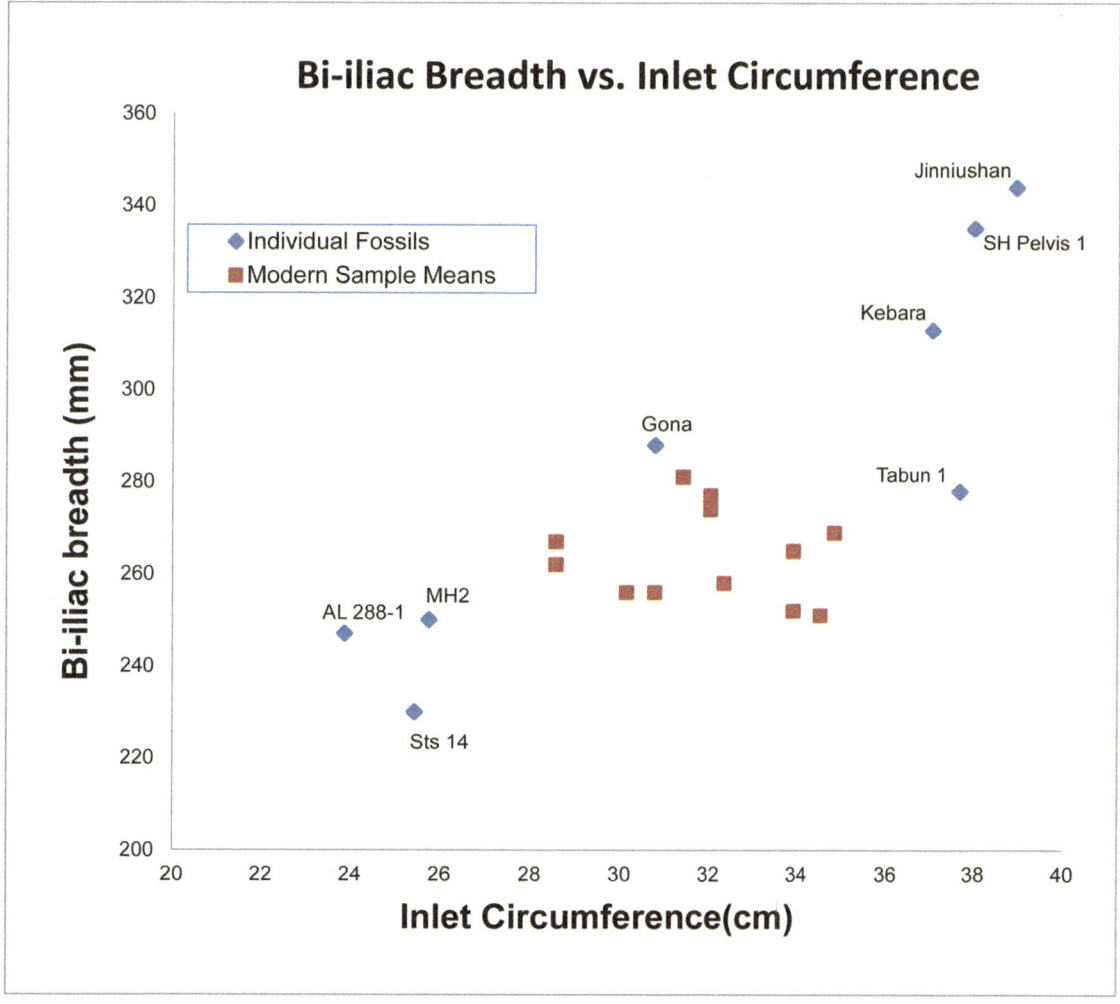

Fig. 19.5 Bi-iliac breadth vs. Inlet circumference. Each graph shows fossil specimens (diamonds) and sex-specific sample means from modern human populations (squares). All fossil data are given in Table 19.1 with explanations about sources of measurements. (from Tague 1989)

Table 19.2 Correlation coefficients across fossil individuals and modern sex-specific sample means. Values that are statistically significant at 0.05 level are shown with yellow highlights

	Stature vs bi-iliac breadth	Stature vs. inlet transverse	Stature vs. inlet ap	Stature vs. circumference
Fossil	**0.935**	**0.733**	**0.971**	**0.922**
Modern	0.086	**−0.588**	−0.129	−0.463

The fossil specimens span a much greater range of stature than the modern humans and it is hence not surprising that the correlation coefficients for stature vs. each of the pelvic dimensions are significant and positive. The correlation across modern population sex-specific means span a small range of stature and in those, only one correlation coefficient is statistically significant (for stature against pelvic inlet transverse diameter) but in a negative direction. That is, across the smaller range of modern populations, those populations that are on average taller, have relatively narrower transverse inlets.

What could account for the differences that we see between fossil specimens and modern samples? One possibility is thermoregulation which has long been regarded as an important determinant of body proportions and shape (Ruff 1991, 1994). However, both the fossil and the modern samples cover a wide and similar range of climates. Another possible explanation might be body size, but the pattern of differences between the two groups persists in specimens of similar body size.

It is possible that there could be locomotor differences between the groups. This hypothesis can be tested by looking for other evidence of locomotor differences between these groups. Finally, there could be obstetric-related differences between the groups – we can test this by looking at dimensions of the birth canal directly. If the differences between the fossil sample and modern humans are obstetrically related, we would expect to find differences in the birth canal as well as bi-iliac breadth.

In summary, although the samples are small and in some ways not comparable (note that the fossils are individuals, while the moderns are mean values for sex-specific population samples), the pattern is consistent. Relative to stature, modern humans have similar to or smaller bi-iliac pelvic breadths and all dimensions of the pelvic inlet than fossil humans. Most interesting, in Neanderthals, which are most similar to modern humans in body size (and brain size), this difference is still apparent. Although as a male, Kebara does not have a birth canal, Tague (1992) observed that the inlet (the top part of the pelvic aperture at the level of the arcuate lines) in Kebara 2 was relatively spacious compared to modern humans, although the lower planes of the pelvic aperture (the midplane and outlet) were more constrained. Because there seems to be a similar relationship between bi-iliac breadth and pelvic inlet dimensions in modern humans and Neanderthals, we suggest that the differences we discuss here may be related to differences in obstetric constraints, such that modern humans have smaller pelvic inlet dimensions relative to stature than Neanderthals. Two hypotheses that could account for this are:

- modern humans give birth to babies which are smaller than Neanderthals (Rosenberg 1988)
- modern humans have some unique obstetrically-related behavior that allows them to give birth to the same sized babies through a slightly smaller passage.

It appears that in overall body shape and specifically in pelvic breadth (transverse dimensions), Neanderthals reflect the primitive condition for hominids. This is in spite of the fact that there were significant changes in the mechanism of birth over this time period (from australopithecines up until Neanderthals) that accompanied increasing body size and encephalization and that probably required accommodations of the pelvis in the anterior-posterior dimension.

Population Structure

Cultural behavior has been frequently cited as an area in which Neanderthals and modern humans differ significantly, with archaeological evidence often interpreted as evidence of Neanderthal cognitive inferiority compared to modern humans (reviewed by Shea 2011, and Henshilwood and Marean 2003; see also Tattersall 2002; Mellars 2005, 2006; Teyssandier 2008). But even the most entrenched precepts in paleoanthropology can change. More recently, there has been widespread recognition that Neanderthals were capable of complex behaviors reflecting symbolic thought (Soressi and d'Errico 2007), including the use of pigments (Roebroeks et al. 2012), jewelry (Zilhão et al. 2010), and feathers and raptor claws as ornaments (Peresani et al. 2011; Morin et al. 2012; Radovčić et al. 2015). All of this is probably evidence of symbolic social signaling, complementing the evidence of complexity of thought demonstrated by the multi-stage production of Levallois tools. Neanderthals were human. Yet, as Rak has suggested, Neanderthals may have been human in a different way. Symbolic associations are less frequent and arguably less sophisticated in Mousterian contexts than in Upper Paleolithic assemblages, perhaps reflecting another way of being human.

Behavioral evolution is exceptionally complex because "both genetic and non-genetic inheritance, and the interactions between them, have important effects on evolutionary outcomes" (Danchin et al. 2011: 475). Demography, as an interface between the biological and the cultural, reflects this complexity, and many authors have considered it to be the cause of the behavioral differences underlying the Middle and Upper Paleolithic (e.g., Shennan 2001; Caspari and Lee 2004, 2006; Hovers and Belfer-Cohen 2006; Zilhão 2007; Powell et al. 2009; Richerson et al. 2009). In this section we review Neanderthal and Upper Paleolithic modern human demography, and suggest ways that demography may account for these different ways of being behaviorally human (Hovers and Belfer-Cohen 2006; Caspari and Wolpoff 2013).

There are a number of difficulties that affect the demographic reconstruction of fossil groups; in particular, small sample sizes, dependence on the age structure of juvenile data, and difficulties determining ages in fragmentary adult skeletal material have long been problematic. There are a few Pleistocene sites such as Atapuerca (Sima de los Huesos) and Krapina whose large samples and presence of juveniles make them amenable to demographic reconstruction. These sites are unusual, however, and most Neanderthal material is fragmentary and isolated and therefore cannot be used in traditional analyses. Methods that employ coarser, categorical age estimates and pooled samples must be used to circumvent many problems of preservation and sampling (Caspari and Lee 2004, 2006). As we show below, demographic profiles of both Krapina and Sima de los Huesos have been individually characterized by high levels of young adult mortality, so high that they have sometimes been interpreted as anomalies. However, when compared to each other, and to the pooled Neanderthal sample, the demographic characteristics of these sites appear to reflect their life history (and can be broadly interpreted as the life history of Neanderthals) and not a consequence of environmental or taphonomic factors unique to these two sites. We believe it is therefore valid to interpret these data as reflecting part of a larger scale archaic demographic pattern, created and maintained by social strategies that appear to have been different from those of modern humans.

Archaic Human Survivorship: The Krapina Death Distribution

Over a century ago the Croatian paleontologist Dragutin Gorjanović-Kramberger excavated and described Neanderthal fossils from a rock shelter near Krapina, 40 km northwest of Zagreb. Although the fossils from the site are fragmentary, they represent the remains of perhaps more than 70 individuals, most found within two stratigraphic layers, with U-series and ESR dates of about 130,000 years BP (Rink et al. 1995). Because of the number and proximity of the fossils, because the sediments at the site accumulated rapidly over a short period of time, and because a number of the bones share unique non-metric features, many workers have treated the remains from Krapina as a single population. As such, Krapina is one of the few archaic European sites amenable to demographic analysis. Its large sample size and preservation of juveniles have made it possible to successfully age adult dentitions using wear-based seriation and reconstruct the demography of the population (Wolpoff 1979).

The dental remains at Krapina are extensive; there are over 190 isolated teeth, some associated as "dental individuals" (Wolpoff 1979), and many teeth are associated with the maxillae and mandibles in the collection. While demographic data have been assessed for other skeletal elements, none are as comprehensive as those based on the dentition and the demography discussed here is based solely on dental remains. Age at death was estimated using several methods: juveniles were aged based on dental eruption, and adult age estimates were based on Miles (1963) method ages (wear-based seriation) that depend on the juvenile estimations. The wear-based dental age estimates were further validated with assessments determined from relative pulp volume using μCT (Wolpoff 1979; Wolpoff and Caspari 2006; Caspari et al. 2009). All the age estimations used in this research are founded on conservative assumptions about eruption times in Neanderthals and other ancient samples; these have been independently validated, holding up well in comparisons with other recent age assessments for the same juvenile specimens (Smith et al. 2010, 2013).

The Krapina Survivorship Distribution

Figure 19.6 depicts the death distribution at Krapina transformed to a survivorship curve compared to the survivorship curve from Libben, a prehistoric Native American sample, and assuming the same survivorship in the earliest interval because infant survivorship cannot be obtained from the fossil sample. Sex determination was not possible for most of the isolated Krapina teeth, and sex differences in survivorship could not be addressed. Krapina is recognized as a sample with high mortality rates. Atapuerca S-H (not shown in Fig. 6) closely resembles the Krapina sample in this and other ways (Bermúdez de Castro 1995; Bocquet-Appel and Arsuaga 1999). To standardize the distributions (to compensate for the missing children in the fossil samples), we assumed 69% survivorship at 5 years of age, the number observed for the

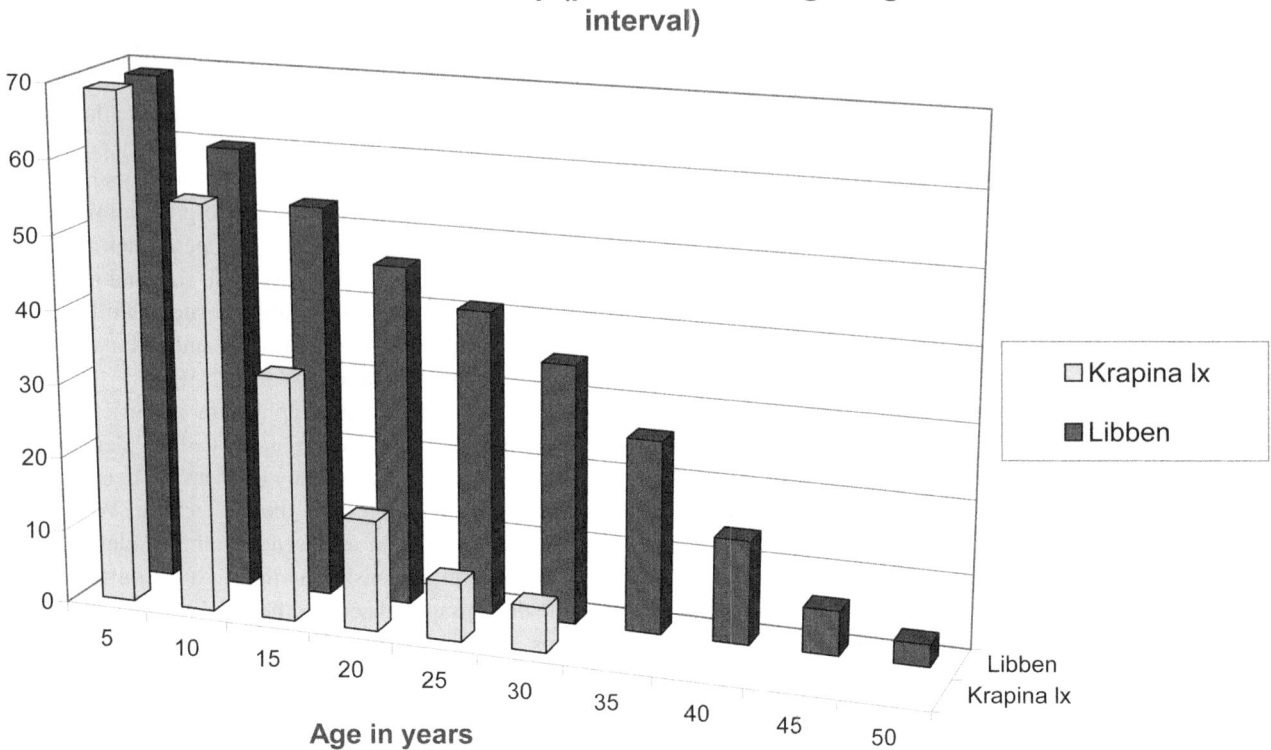

Fig. 19.6 The Krapina and Libben Survivorship Curves, adapted from Wolpoff and Caspari (2006)

Libben sample (Lovejoy et al. 1977; Lovejoy 1985). We also assumed a stationary population with a stable age distribution (Hoppa and Vaupel 2002), a reasonable assumption for comparisons, although it is unlikely to consistently describe individual populations at specific times, which probably fluctuated (Harpending 1997; Meindl et al. 2001). As shown, the Krapina survivorship pattern is unlike Libben or any recent or living human survivorship curve, with very high young adult mortality. Although Krapina is remarkable in having virtually no older adults (above age 30), the general pattern of high young adult mortality is not unusual. At Atapuerca, like Krapina, there are very high levels of juvenile and young adult mortality, with few individuals surviving to age 35. It is possible that these distributions reflect the peculiarities of specific sites – taphonomic or catastrophic occurrences that somehow selected against the preservation of older individuals (Bocquet-Appel and Arsuaga 1999). This, however, is unlikely since the death distribution does not resemble the age structure of living high-mortality hunter-gathers (e.g., the Ache), nor are the very young or old overly represented, two expectations of catastrophes (Wolpoff and Caspari 2006). Moreover, research focusing on the human fossil record more broadly suggests these curves were not exceptional; rather, they reflect an archaic life history pattern characterized by high young adult mortality (Caspari and Lee 2004).

OY Ratios and the Emergence of Modern Survivorship Patterns

High young adult mortality in archaic hominid groups was demonstrated by Caspari and Lee (2004) using Miles Method age estimations that were subsequently verified by assessments based on relative pulp volume using µCT (as cited above). OY ratios, the ratio of older to younger adults were calculated for large aggregates of fossil samples, as a simple approach to deal with longevity independent of understandings of juvenile mortality patterns, precise age estimates, and the problems of small samples of different sizes. It is important that the components of the OY ratio are of relative categories, independent of actual age; third molar eruption indicates adulthood, and we accept the implication that different times of eruption reflect different times that adulthood was attained. For our purposes double that age marks the beginning of older adulthood, the age one could first become a grandparent. Therefore groups with different maturation rates can be compared and variation in the OY ratio provides insights into longevity without assessing variation in actual lifespan, or treating the samples as populations. This approach circumvents some of the problems associated with paleodemography, yet allows the evaluation of evolutionary hypotheses.

Caspari and Lee (2004) reported a large and significant increase in the OY ratio between Neanderthals (and other archaic hominids) and the modern humans of the Upper Paleolithic: the ratio for Upper Paleolithic Europeans is approximately 2.0 (as is the OY ratio of Libben); that for European Neanderthals is 0.4. Thus, for every 10 young Neanderthal adults in the death distribution (between ages 15–30), there are only 4 older adults (over 30). Re-analysis limited to non-burials yielded similar results, and the systematic nature of the changes and the many different site histories sampled made a purely taphonomic explanation unlikely (Caspari and Lee 2004). The major conclusion to emerge from this study was that adult survivorship increased dramatically in the European Upper Paleolithic. A subsequent study comparing Middle Paleolithic modern humans from western Asia with Neanderthal and Upper Paleolithic European samples indicated that the increase in survivorship was an attribute of the Upper Paleolithic rather than an attribute of modern humans as a whole (Caspari and Lee 2006). Despite living in much harsher conditions approaching the glacial maximum, the OY ratio of the Upper Paleolithic Europeans was more than double that of the Middle Paleolithic modern humans. Phylogeny and ecology dismissed as causes of Upper Paleolithic longevity, we concluded that the changes in mortality patterns were likely caused by cultural changes. It can be argued that this shift represents a change from an archaic to a modern life history pattern that had implications for the success of modern humans.

Model Mortality Curves

Model mortality curves based on OY ratios further suggest that the Krapina and Atapuerca distributions may not be anomalies, but reflect the archaic life history pattern. Adult mortality curves were calculated based on OY ratios assuming a constant number of mortalities per generation (Van Arsdale 2009). This is not a constant rate, but rather a uniform decline in the standing population. Thus, for example if OY is 0.4, a cohort of 100 individuals entering young adulthood at age 15 should lose 71.4 of their members by age 30 and the remaining 28.6 as old adults (over 30), or 4.76 individuals a year, assuming a constant number of deaths/year. Therefore, in this archaic scenario, the mortality risk of a 16 year old is very high (4.76%), increasing to 14.27% in a 30 year old and by 37, there are no survivors. This is consistent with the pattern seen in the Krapina and Atapuerca samples. Models based on higher OY ratios yield mortality curves with increases in mortality rate occurring at a later age, and again assuming a constant number of deaths/year the maximum age in the population will increase. With an OY value of 2 (the value of both the Upper Paleolithic sample and Libben) the deaths/year with an incoming cohort of 100 young adults is 2.22; the mortality risk of a 30 year old is only 3.3% and the maximum age of members of the population is 61. Figure 19.7 compares the mortality curves across three scenarios. The significantly higher mortality rates of Neanderthals would have important

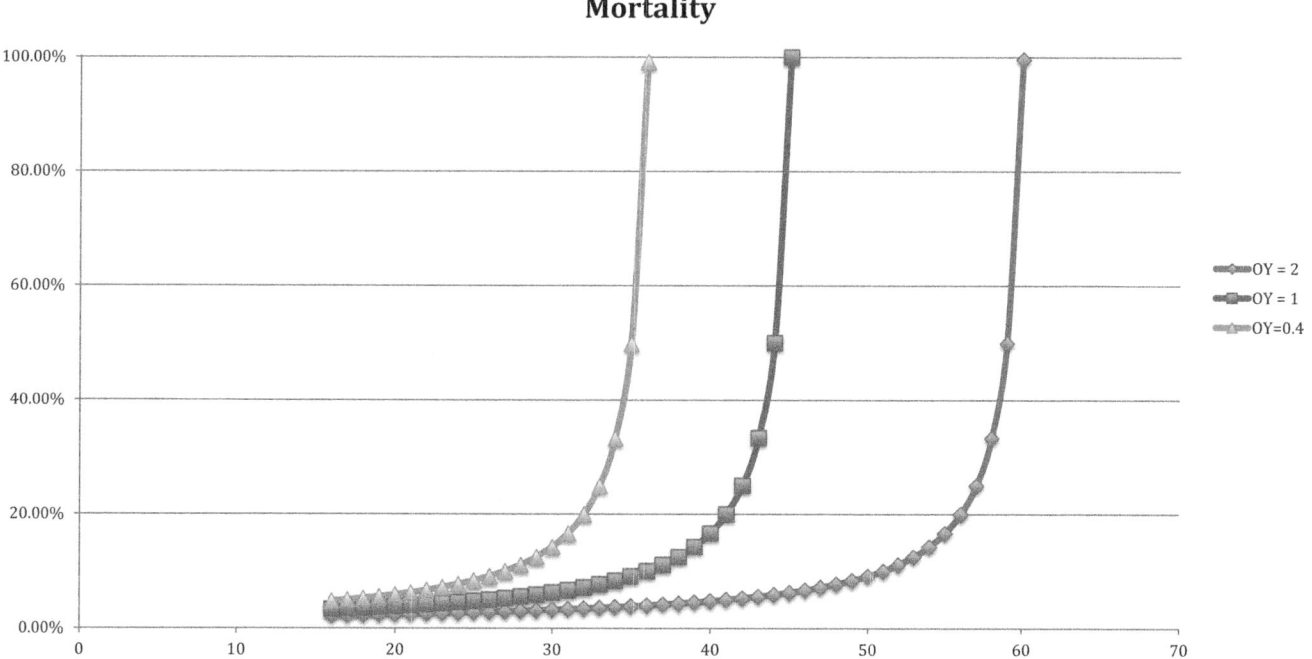

Fig. 19.7 Mortality curves for populations with different OY ratios. Figure courtesy of Adam Van Arsdale

implications for other demographic factors: with OY ratios of 0.4, fertility rates would need to be high, and populations would be less resilient to stochastic fluctuations in population size.

Fertility and Population Stability

Using the Brass Polynomial, a model of declining fertility with age (Gage 1998) Van Arsdale (2009; Caspari et al. 2010) calculated the total fertility rates necessary to replace birth cohorts in populations with different OY ratios. A 50% child mortality rate was assumed. Neanderthals, with OY ratios of 0.4 would need a total fertility rate of over 6.3 births, Upper Paleolithic populations, with OY ratios of 2.0, would require only 5.3 births. This also would affect population stability. Van Arsdale's stochastic models testing the outcomes of variation in fertility rates suggest that Upper Paleolithic populations would avoid population collapse much more easily than Neanderthals (Van Arsdale 2009; Caspari et al. 2010).

Behavioral Implications

By many estimates, Neanderthal population numbers and overall densities were very low (Castellano et al. 2014; Churchill 2014), with relatively few contacts with other groups and little opportunity for specialization. In contrast, the increased survivorship and longevity in the Upper Paleolithic would allow for increases in population growth and expansion that could foster behavioral modernity. It has been argued that with population growth, resources become scarcer (Stiner et al. 1999) and modern behaviors and technologies are the human response. It has also been suggested that modern behaviors appear more often, are more persistent (for reasons discussed below) and disperse more effectively because of population growth and an increased number of interpopulational contacts (Shennan 2001), and the rate of cultural change accelerates. Therefore, the cultural changes associated with the Upper Paleolithic may reflect a ratcheting, positive feedback process.

The same may be the case for increased longevity in the Upper Paleolithic. Longevity itself, in addition to its effects on population growth, may have contributed to the emergence and persistence of modern behaviors, and increased longevity may have persisted through a positive feedback process with these behavioral outcomes. Initially the result of cultural adaptations and/or perhaps climatic factors (d'Errico and Stringer 2011), longevity may have become a prerequisite for the unique and complex behaviors that mark modernity, innovations that in turn promoted both the importance and the survivorship of older adults. Caspari and Lee (2004, 2006) suggested that adult survivorship increased in response to cultural factors promoting the importance of older adults whose experience benefited their kin groups in the harsh conditions of Upper Pleistocene Europe. The experience of older members could also underlie the material expressions associated with the Upper Paleolithic. There are a number of ways in which the demographic changes and intergenerational transfer effects associated with increased adult survivorship could result in the complex behaviors thought to be reflected in Upper Paleolithic archaeology (Lee 2003; Rosenberg 2004).

In the emerging complex adaptations of the Upper Paleolithic, older adult survivorship was likely beneficial to social groups as a whole, promoting intergenerational transfer of a variety of economic and cultural resources (Lee 2003). In humans, as in other social species, there is transfer of resources among individuals, which contributes to the inclusive fitness of a kin group. Intergenerational transfer is particularly important for humans, where it extends over several generations. Grandparents routinely contribute economic and social resources to their descendants, increasing the fertility of their children and the survivorship of their grandchildren. The importance of the economic contributions of older adults to their social groups has been well documented (Kaplan and Robson 2002; Hawkes 2003). In fact, studies of living hunter/gatherers indicate that because of the skill-intensive techniques of resource acquisition, peak production rates occur in individuals over age 30 (Robson and Kaplan 2003). Cultural information is effectively transmitted by older members of society, reinforcing complex social connections. Multiple aspects of cultural knowledge are transmitted, from social identity, to experiences dealing with unusual environmental conditions to technological innovations that promote the survival of social groups. Such knowledge is often embedded in oral traditions in which cross-generational transmission plays an important role.

Survival of periodic subsistence crises is largely dependent on a group memory of past crisis situations and of the strategies appropriate for dealing with the altered environmental conditions. One mechanism utilized by non-literate societies for the preservation of survival knowledge is its incorporation in oral tradition. As a body of reference knowledge, oral traditions potentially operate over two time scales. Secular oral traditions (folktales, songs, and histories) depend on repetition for perpetuation with inherent potential for distortion. In contrast, sanctified oral traditions, such as ritual performances, rely on a correct reproduction of the ritual order to achieve supernatural efficacy. Rituals accordingly assume an invariant character appropriate for the transmission of survival information over extended periods of time (Minc 1986: 39).

Moreover, as has been recently modeled by Strimling and colleagues (2009), repetition – the repeated learning of cultural traits – is a critical factor in cultural learning more generally, and it has the potential to drive cultural evolution:

> Here we show that repeated learning and multiple characteristics of cultural traits make cultural evolution unique … We find that the possibility to predict long-term cultural evolution by some success index, analogous to biological fitness, depends on whether individuals have few or many opportunities to learn. If learning opportunities are few, we find that the existence of a success index may be logically impossible, rendering notions of' 'cultural fitness' meaningless. On the other hand, if individuals can learn many times, we find a success index that works, regardless of whether the transmission pattern is vertical, oblique, or horizontal (Strimling et al. 2009: 13870).

Multigenerational families have more (and more knowledgeable) members to teach and re-teach important lessons. We suggest longevity promoted the intergenerational accumulation and transfer of information that allowed for complex kinship systems and other social networks that are uniquely human.

However, the population growth discussed above is perhaps the most important consequence of increased adult survivorship, the basis of the Upper Paleolithic population expansions reflected in archeological and genetic evidence (Shennan 2001; Templeton 2002; Powell et al. 2009). Not only does increased survivorship create the potential for greater lifetime fertility for individuals who are living longer, but the investment of older individuals in their children's families influences their inclusive fitness both by increasing the fertility of their children and the survivorship of their grandchildren. These selective advantages promote continued population increase. Therefore the increase in survivorship we observe is a significant factor in the evolution of modernity not only through its importance for intergenerational information transfer, but because of its relationship to population expansion.

These demographic changes provide social pressures that we believe led to extensive trade networks, increased mobility, and more complex systems of cooperation and competition between groups, resulting in increased personal ornamentation, material expressions of individual and group identity, and other forms of material information exchange between groups (Wobst 1977). Modern human behavior, then, is a response to demographic pressures.

So where does this leave Neanderthals? While possessing the capacity for symbolic behavior evidenced by recent archaeological discoveries, their symbolic associations are less frequent and less sophisticated than in Upper Paleolithic assemblages. This is likely a reflection of their archaic life history pattern, and a different way of being human.

Neanderthals as a Human Subspecies

We agree with Rak about most things, but phylogenetics is an area where long ago we agreed to disagree! However, recent genetic data have caused us to revisit our ideas about Neanderthal taxonomy. These data have done much to elucidate phylogenetic issues concerning Neanderthals, but introduce a paradox. While demonstrating what we consider widespread gene flow between archaic and modern humans, paleogenetics also indicates Neanderthal differences; there is much greater population structure in the late Pleistocene than we see today. We review these findings in the context of the history of the taxonomic placement of Neanderthals. Modern human population structure is a poor model for the population structure of Neanderthals. Modern humans have no races, but it is very likely that Neanderthals were a human subspecies.

In the second part of the 20th century, the interpretation of Neanderthals as a separate species gained ground and eventually became the majority opinion. This was more a consequence of changes in how species were defined and identified, than a reflection of new Neanderthal discoveries. The species interpretation of Neanderthal variation began with the first Neanderthal to be recognized. The Feldhofer Neanderthal was first described as a species by William King (1864) because of how different from living humans it appeared to be. King wrote "so closely does the fossil cranium resemble that of the chimpanzee, as to lead one to doubt the propriety of generically placing it with Man." In the half century following the Feldhofer discovery, evolution of humans was rejected by many scientists. A number of them found the Neanderthal to be somewhat less different than King described, actually no different in type from other human races, although more primitive. But in those times, recognized races were often named as species because they were interpreted in a polygenic framework (Wolpoff and Caspari 1997, 2013) where races were thought to have had separate origins, sometimes in different primate species. The practice persisted well into the 20th century, largely without comment, until it resulted in absurdity. Thus, in describing the human remains from the Lower Cave at Zhoukoudian, the "Peking Man" sample that many paleoanthropologists formally place in the species *Homo erectus*, Weidenreich named the remains "*Sinanthropus pekinensis*" out of respect for Davidson Black who named the first Zhoukoudian specimens in a scientific paper (Black 1929). In reality, however, Weidenreich believed they were *Homo sapiens*, and wrote that his use of "*Sinanthropus pekinensis*" was a convenience

. . . without any "generic" or "specific" meaning or, in other words, as a "latinization" of Peking Man.. . . it would not be correct to call our fossil "*Homo pekinensis*" or "*Homo erectus pekinensis*"; it would be best to call it "*Homo sapiens erectus pekinensis*." Otherwise it would appear as a proper "species," different from "*Homo sapiens*," which remains doubtful, to say the least (Weidenreich 1943: 246).

But this was not the end. A subsequent series of publications addressed the issue of how fossil species can be recognized and whether or not there should be an attempt to reconcile the definition of fossil species with the biological species concept as defined by Mayr (1942) and others: a group of populations that can actually or potentially interbreed and produce fertile offspring, and which are reproductively isolated from populations in other species. For most of those who did not think such reconciliation was possible, Neanderthals became a distinct species (e.g., Eldredge and Cracraft 1980; Tattersall 1986, 1992).

The final twist in this story came with the discovery of nuclear DNA in human fossils (Green et al. 2010), because this demonstrated the possibility that the interbreeding criterion for biological species could actually be applied to some human fossils (Hawks 2013); in particular, to Neanderthal fossils. Continued discoveries of Neanderthal introgressions into other human populations demonstrate (see below) without question that Neanderthals are a variety of *Homo sapiens*. The demonstration that effective biological barriers to interbreeding can be expected to take a million years or more to become established (Curnoe et al. 2006; Holliday 2006) is compatible with these new data.

The question of some Neanderthal ancestry is informed by the establishment of significant gene flow from Neanderthals (Lohse and Frantz 2014), in many cases bringing adaptive features enhancing climatic selection into other populations (including pigmentation features such as the melanocyte-stimulating hormone receptor gene MC1R (Ding et al. 2014a), red hair and freckles, as well as increased skin thickness with more hair and fewer pores (Vernot and Akey 2014), Neanderthal alleles that affect skin and hair such as keratin filaments (Sankararaman et al. 2014), and specific to certain populations, European lipid catabolism (Khrameeva et al. 2014), the cellular response to ultraviolet-B irradiation in Asians (Ding et al. 2014b). Other cases of gene flow from Neanderthals involve disease adaptation, including HLA class 1 alleles (Abi-Rached et al. 2011), but some genes inherited from Neanderthals may also heighten the risk of diseases such as Type-2 diabetes, liver cirrhosis, lupus and Crohn's disease (Sankararaman et al. 2014).

Several other adaptive genes are shared with Neanderthals (this work is in its infancy), although the direction of gene flow is uncertain and may well be bidirectional. These are cases where the presence of the gene in a common ancestor is unlikely. Human FOXP2 is well-studied, including in Neanderthals (Krause et al. 2007), because its homologues are widespread and because of its importance in speech production (Enard et al. 2002). The initial suggestion was that "[d]ata [may be] consistent with low rates of gene flow between modern humans and Neanderthals" (Coop et al. 2008: 1257). Later, Maricic and colleagues (Maricic et al. 2013) reported on a regulatory variant in a transcription factor affecting the expression of the FOXP2 gene that is found in two Iberian Neanderthals (Sidron cave) and a Croatian Neanderthal (Vindija cave), and is fairly common in some human populations. Theirs is a complex reconstruction in which the Neanderthal FOXP2 and its regulators fall within the human range of variation, while at the same time "this is the only nucleotide variant in that region where the majority of present-day people carry a derived variant that is not present in Neanderthals and Denisovans. Thus, it is possible that this change was positively selected recently during the evolution of fully modern humans (Maricic et al. 2013: 849)".

When gene flow is paired with positive selection, the minimum magnitude of gene flow allowing the genes to be established in other populations is difficult to determine. However, the direct evidence of gene flow negates all arguments that the number of mating events between Neanderthals and other populations were too small to have been important. They could not validly be described as rare, or occasional given that there are many different Neanderthal genes in different human populations, and their effects were anything but too small to be important (Hawks and Throckmorton 2013). With a significant role for selection guiding the dynamics of genetic exchanges, increasing evidence that Neanderthal behavioral capacities fall well within the human range (Villa and Roebroeks 2014) should not be surprising.

This pattern of introgression from Neanderthals demonstrates that past human evolution, like the present, occurs within a network of on-and-off interconnected populations within a single evolving lineage. It has been observed over a time period long enough for the complete replacements of human populations by one or more successive new African species to be evident if they had happened. Our working hypothesis is that Pleistocene human evolution is an example of evolution within a species lineage (Wolpoff et al. 1994; Wildman et al. 2003) described by a geographically diverse widely dispersed network of (intermittently) interconnected populations. If past human variation is within that lineage, it is possible that such variation could be described as subspecies (Wolpoff 2009), even though human subspecies do not exist today (Marks 1995; Templeton 1998, 2013; Caspari 2003, 2010; Wolpoff and Caspari 2013).

The dismissal of human races as an organizing structure for living human biology occurred for many reasons, including political reasons, but there is a firm biological basis for it in the distribution of genetic variation (Templeton 1998), that to some extent is reflected in the distribution of anatomical variation.

- Extant human anatomical variation does not attain the subspecies level; populations are neither different enough, nor separated enough, for a subspecies interpretation of their variation to be valid.
- The ratio of within group to between group variance is very high in humans.
- There is no treeness for human groups (Templeton 1998, 2013).

Thus, the idea that there were once pure human races is dead and buried, and if race cannot reflect unique common descent, and if there is no validity to the precept that human races are constellations of biological characters that show greater differences between each other than variation within one of them, race can only have a social definition (Marks 1995; Caspari 2003; and many others). There simply are no clearly distinct types of humanity (Graves 2001), and there is no racial taxonomy for the living.

Were Neanderthals a Past Human Race?

Outside of anthropology, race is most often used as a synonym for subspecies (Mayr 1969, p. 44; Futuyma 1986: 107–109; Templeton 1998), and for most of its history this has also been true within paleoanthropology (Boule 1923; Dobzhansky 1944; Weidenreich 1946, 1947). Subspecies, however, are not a favored topic in modern biology; they don't exist in the indexes of many recent textbooks, and when they do appear there are some times when subspecies refer to a taxonomically distinct variety of a species, but others when they are used to describe "a species in the making". Subspecies are traditionally defined as geographically circumscribed, genetically differentiated populations. Subspecies are also described as distinct evolutionary lineages within a species. A good example would be the three different subspecies of gorilla, three groups that are physically and geographically distinct (Relethford 2008: 379).

Has this always been the case? Given that there are no human races today – accepting that human geographic variation is not taxonomic – does this mean that there were no races in the human past? Or, is it possible that Neanderthals fit the description of a subspecies as we understand it today? The modern understanding of subspecies comes from the New Synthesis, especially from the works of Mayr (1942) and Dobzhansky (1944). For them, subspecies combined groups of local populations by anatomical similarity and geographic distinctness, in a taxonomic grouping (by descent). Although criticized by Wilson and Brown

(1953), subspecies continue to describe intraspecies variation when it is distinctly geographic; but admittedly, for the most part modern usage is not common because intraspecies variation is not often studied. However, this happens to be a significant problem in human studies where, as discussed above, this variation is almost never regarded as taxonomic.

Dobzhansky (1944) directly addressed the question of whether past hominid samples such as Neanderthals might be subspecies. For him the compelling support for identifying a Neanderthal subspecies came from the newly published Mount Carmel remains (Skhul and Tabun; McCown and Keith 1939), which he interpreted as the result of mixture between two subspecies that were obviously not reproductively isolated, and not as a single population "in the throes of evolutionary change", as McCown and Keith had interpreted the sample. Dobzhansky (1944: 259) noted that "The Mount Carmel population also shows that . . . a morphological gap as great as that between the Neanderthal and the modern types may occur between races, rather than between species."

Jolly (2001) also noted that Neanderthals fit the description of subspecies as allotaxa ("morphologically diagnosable yet not reproductively isolated" populations). Jolly (2001: 1767) proposed "Neanderthals and AfroArabian 'pre-modern' populations may have been analogous to extant baboon (and macaque) allotaxa".

But in our view the most important new evidence for regarding Neanderthals as a past subspecies of *Homo sapiens* is discussed above. Neanderthal genes dispersed under selection into populations with descendants wherein they persist today. Many of these genes led to significant adaptive changes. The fact that so many Neanderthal genes persist as different genes making up different combinations in different individuals is the strongest argument that Neanderthals are *Homo sapiens*. The percentage contribution of Neanderthal genes to gene pools of non-Africans today approximates estimates of the number of Neanderthals that lived at any time as a fraction of the human population at that time.

A good number of the Neanderthal genes that dispersed into other populations were under selection, as we noted above. The fact that many of these gene dispersals are described as introgressions is also important. Introgressions are the transfers of genes that evolved at a much earlier time. Introgressions in Neanderthals suggest that Neanderthal populations were significantly (but not completely) isolated from other human populations, as they may well have been from each other (helping account for significant Neanderthal population structure). The evidence of restricted gene flow with Neanderthals, combined with older observations of a distinct geographic range, and the magnitude of anatomical differences between Neanderthals and their penecontemporaries, suggest that unlike any population today, it is reasonable to interpret Neanderthals as a human subspecies.

Because the human species does not have subspecies today, this supports the notion that Neanderthals are another way of being human. Neanderthals, it would appear, are the best-established demonstration that humans in the past, like many other mammals (Mayr 1963), formed distinct races.

Conclusion

Even as paleogenetics has affirmed Neanderthal humanity, it also brings focus on their difference. As Rak has long recognized, Neanderthals represent another way to be human. The observations on Neanderthal body form, demography, and breeding behavior that we reviewed here reinforce ideas of Neanderthal difference and hopefully provide insight into the nature of that difference. We conclude that in body form, demography and population structure, Neanderthals are unlike modern humans, in some cases reflecting the ancestral condition. We view them therefore as both "brother" and "other," simultaneously expressing some archaic hominid characteristics as well as aspects of the modern condition to which they contributed genetically, anatomically and behaviorally.

Acknowledgements We are very grateful to Erella Hovers and Assaf Marom for putting together this volume in honor of Yoel Rak and for inviting us to contribute. We thank Tom Roček for heroic emergency assistance, Jeremy DeSilva for access to unpublished data and Adam Van Arsdale for the use of his figure (our Fig. 19.7). Finally, we thank Yoel Rak who has been an inspiring model of civility in science and who has shown us how mutual respect, affection, humor and decades-long friendship need not be threatened by differences in scientific perspective.

References

Abi-Rached, L., Jobin, M. J., Kulkarni, S., McWhinnie, A., Dalva, K., Gragert, F., et al. (2011). The shaping of modern human immune systems by multiregional admixture with archaic humans. *Science, 334,* 89–94.

Arsuaga, J. L., Lorenzo, C., Carretero, J. M., Gracia, A., Martínez, I., García, N., et al. (1999). A complete human pelvis from the Middle Pleistocene of Spain. *Nature, 399,* 255–258.

Bermúdez de Castro, J.-M. (1995). The hominids from the Sima de los Huesos of the Sierra de Atapuerca karst: Minimum number of individuals, age at death, and sex. In J-M. Bermúdez, J.-L. Arsuaga, & E. Carbonell (Eds.), *Human evolution in Europe and the Atapuerca evidence. Volume 1.* (pp. 263–281). Valladolid: Sever-Cuesta.

Black, D. (1929). Sinanthropus pekinensis: The recovery of further fossil remains of this early hominid from the Chou Kou Tien deposits. *Science, 6,* 674–676.

Bocquet-Appel, J. P., & Arsuaga, J.-L. (1999). Age distributions of hominid samples at Atapuerca (SH) and Krapina could indicate accumulation by catastrophe. *Journal of Archaeological Science, 26,* 327–338.

Bonmatí, A., Gómez-Olivencia, A., Arsuaga, J. L., Carretero, J. M., Gracia, A., Martínez, I., et al. (2010). Middle Pleistocene lower back and pelvis from an aged human individual from the Sima de

los Huesos site, Spain. *Proceedings of the National Academy of Sciences USA, 107,* 18386–18391.

Boule, M. (1911). *L'Homme fossile de La Chapelle-aux-Saints. Annales de Paléontologie VI.*

Boule, M. (1923). *Les Hommes fossiles. Eléments de paléontologie humaine.* Paris: Masson.

Caspari, R. (2003). From types to populations: A century of race, physical anthropology and the American Anthropological Association. *American Anthropologist, 105,* 63–74.

Caspari, R. (2010). Deconstructing race: Racial thinking, geographic variation, and implications for biological anthropology. In C. S. Larsen (Ed.), *A companion to biological anthropology* (pp. 104–123). Chichester, West Sussex: Wiley-Blackwell.

Caspari, R. (2011). The evolution of grandparents. *Scientific American, 305,* 44–49.

Caspari, R., & Lee, S.-H. (2004). Older age becomes common late in human evolution. *Proceedings of the National Academy of Sciences USA, 101,* 10895–10900.

Caspari, R., & Lee, S.-H. (2006). Is human longevity a consequence of cultural change or modern human biology? *American Journal of Physical Anthropology, 129,* 512–517.

Caspari, R., Lee, S.-H., & Van Arsdale, A. (2010). Implications of reduced mortality risk for late Pleistocene humans. *American Journal of Physical Anthropology, 141,* 79.

Caspari, R., Meganck, J. A., Radovčić, J., Begun, D., Kroll, T., & Goldstein, S. A. (2009). Assessing age at death in adult dentitions: A new approach using three-dimensional microcomputed tomography and its application to fossil samples. *American Journal of Physical Anthropology, 138,* 104.

Caspari, R., & Wolpoff, M. H. (2013). The process of modern human origins: The evolutionary and demographic changes giving rise to modern humans. Chapter 11. In F. H. Smith & J. C. M. Ahern (Eds.), *The origins of modern humans: Biology reconsidered* (pp. 355–390). New York: Wiley.

Castellano, S., Parra, G., Sánchez-Quinto, F. A., Racimo, F., Kuhlwilm, M., Kircher, M., et al. (2014). Patterns of coding variation in the complete exomes of three Neandertals. *Proceedings of the National Academy of Sciences USA, 111,* 6666–6671.

Churchill, S. E. (2014). *Thin on the ground: Neandertal biology, archeology and ecology.* Oxford: Wiley-Blackwell.

Coop, G., Bullaughey, K., Luca, F., & Przeworski, M. (2008). The timing of selection at the human FOXP2 gene. *Molecular Biology and Evolution, 25,* 1257–1259.

Curnoe, D., Thorne, A., & Coate, J. A. (2006). Timing and tempo of primate speciation. *Journal of Evolutionary Biology, 19,* 59–65.

Danchin, É., Charmantier, A., Champagne, F. A., Mesoudi, A., Pujol, B., & Blanchet, S. (2011). Beyond DNA: Integrating inclusive inheritance into an extended theory of evolution. *Nature Reviews Genetics, 12,* 475–486.

Dean, M. C., Stringer, C. B., & Bromage, T. G. (1986). Age at death of the Neanderthal child from Devil's Tower, Gibraltar and the implications for studies of general growth and development in Neanderthals. *American Journal of Physical Anthropology, 70,* 301–309.

d'Errico, F., & Stringer, C. B. (2011). Evolution, revolution or saltation scenario for the emergence of modern cultures? *Philosophical Transactions of the Royal Society B, 366,* 1060–1069.

Ding, Q., Hu, Y., Xu, S., Wang, C. C., Li, H., Zhang, R., et al. (2014a). Neanderthal origin of the haplotypes carrying the functional variant Val92Met in the MC1R in modern humans. *Molecular Biology and Evolution, 31,* 1994–2003.

Ding, Q., Hu, Y., Xu, S., Wang, J., & Jin, L. (2014b). Neanderthal introgression at chromosome 3p21. 31 was under positive natural selection in East Asians. *Molecular Biology and Evolution, 31,* 683–695.

Dobzhansky, T. H. (1944). On species and races of living and fossil man. *American Journal of Physical Anthropology, 2*, 251–265.

Eldredge, N., & Cracraft, J. (1980). *Phylogenetic patterns and the evolutionary process: Method and theory in comparative biology.* New York: Columbia University Press.

Enard, W., Przeworski, M., Fisher, S. E., Lai, C. S., Wiebe, V., Kitano, T., et al. (2002). Molecular evolution of FOXP2, a gene involved in speech and language. *Nature, 418*, 869–872.

Futuyma, D. J. (1986). *Evolutionary biology.* Sunderland: Sinauer.

Gage, T. B. (1998). The comparative demography of primates: With some comments on the evolution of life histories. *Annual Review of Anthropology, 27*, 197–221.

Gómez-Olivencia, A., Eaves-Johnson, K. L., Franciscus, R. G., Carretero, J. M., & Arsuaga, J.-L. (2009). Kebara 2: New insights regarding the most complete Neandertal thorax. *Journal of Human Evolution, 57*, 75–90.

Graves, L. L. (2001). *The emperor's new clothes: Biological theories of race at the millennium.* New Brunswick: Rutgers University Press.

Green, R. E., Krause, J., Briggs, A. W., Maricic, T., Stenzel, U., Kircher, M., et al. (2010). A draft sequence of the Neandertal genome. *Science, 328*, 710–722.

Haile-Selassie, Y., Latimer, B. M., Alene, M., Deino, A. L., Gibert, L., Melillo, S. M., et al. (2010). An early *Australopithecus afarensis* postcranium from Woranso-Mille, Ethiopia. *Proceedings of the National Academy of Sciences USA, 107*, 12121–12126.

Harpending, H. (1997). Living record of past population change. In R. R. Paine (Ed.), *Integrating archaeological demography: Multidisciplinary approaches to prehistoric population* (pp. 89–100). Carbondale: Southern Illinois University.

Hawkes, K. (2003). Grandmothers and the evolution of human longevity. *American Journal of Human Biology, 15*, 380–400.

Hawks, J. (2013). Significance of Neandertal and Denisovan genomes in human evolution. *Annual Review of Anthropology, 42*, 433–449.

Hawks, J., & Throckmorton, Z. (2013). The relevance of archaic genomes to modern human origins. In F. H. Smith & J. C. M. Ahern (Eds.), *The origins of modern humans: Biology reconsidered* (pp. 339–354). New York: Wiley.

Henshilwood, C. S., & Marean, C. W. (2003). The origin of modern human behaviour: Critique of the models and their test implications. *Current Anthropology, 44*, 627–665.

Holliday, T. W. (2006). Neanderthals and modern humans: An example of a mammalian syngameon? In K. Harvati & T. Harrison (Eds.), *Neanderthals revisited: New approaches and perspectives* (pp. 281–298). Dordrecht: Springer.

Hoppa, R. D., & Vaupel, J. W. (2002). *Paleodemography: Age distribution from skeletal samples* (Vol. 31). Cambridge: Cambridge University Press.

Hovers, E., & Belfer-Cohen, A. (2006). "Now you see it, now you don't" – modern human behavior in the Middle Paleolithic. In E. Hovers & S. L. Kuhn (Eds.), *Transitions before the transition: Evolution and stability in the Middle Paleolithic and Middle Stone Age* (pp. 295–304). New York: Springer.

Jolly, C. (2001). A proper study of mankind: Analogies from the papionin monkeys and their implications for human evolution. *Yearbook of Physical Anthropology, 44*, 177–204.

Kaplan, H. S., & Robson, A. (2002). The emergence of humans: The coevolution of intelligence and longevity with intergenerational transfers. *Proceedings of the National Academy of Sciences USA, 99*, 10221–10226.

Khrameeva, E. E., Bozek, K., He, L., Yan, Z., Jiang, X., Wei, Y., et al. (2014). Neanderthal ancestry drives evolution of lipid catabolism in contemporary Europeans. *Nature Communications, 5*, 1–8.

Kibii, J. M., Churchill, S. E., Schmid, P., Carlson, K. J., Reed, N. D., De Ruiter, D. J., et al. (2011). A partial pelvis of *Australopithecus sediba*. *Science, 333*, 1407–1411.

King, W. (1864). On the Neanderthal skull, or reasons for believing it to belong to the Clydian Period and to a species different from that represented by man. *British Association for the Advancement of Science, Notices and Abstracts for, 1863*, 81–82.

Krause, J., Lalueza-Fox, C., Orlando, L., Enard, W., Green, R. E., Burbano, H. A., et al. (2007). The derived FOXP2 variant of modern humans was shared with Neandertals. *Current Biology, 17*, 1908–1912.

Lee, R. D. (2003). Rethinking the evolutionary theory of aging: Transfers, not births, shape senescence in social species. *Proceedings of the National Academy of Sciences USA, 100*, 9637–9642.

Lloyd, J. (2012). Prehistoric autopsy: Our ancient past. *Focus (BBC) Science and Technology*, Issue 246.

Lohse, K., & Frantz, L. A. (2014). Neandertal admixture in Eurasia confirmed by maximum-likelihood analysis of three genomes. *Genetics, 196*, 1241–1251.

Lovejoy, C. O. (1985). Dental wear in the Libben population: Its functional pattern and role in the determination of adult skeletal age at death. *American Journal of Physical Anthropology, 68*, 47–56.

Lovejoy, C. O., & Heiple, K. G. (1970). A reconstruction of the femur of *Australopithecus africanus*. *American Journal of Physical Anthropology, 32*, 33–40.

Lovejoy, C. O., Meindl, R. S., Pryzbeck, T. R., Barton, T. S., Heiple, K. G., & Kotting, D. (1977). Paleodemography of the Libben site, Ottawa County, Ohio. *Science, 198*, 291–293.

Maricic, T., Günther, V., Georgiev, O., Gehre, S., Ćurlin, M., Schreiweis, C., et al. (2013). A recent evolutionary change affects a regulatory element in the human FOXP2 gene. *Molecular Biology and Evolution, 30*, 844–852.

Marks, J. (1995). *Human biodiversity: Genes, race, and history.* New York: Aldine de Gruyter.

Mayr, E. (1942). *Systematics and the origin of species from the viewpoint of a zoologist.* New York: Columbia University Press.

Mayr, E. (1963). *Animal species and evolution.* Cambridge: Belknap Press of Harvard University Press.

Mayr, E. (1969). *Principles of systematic zoology.* New York: McGraw-Hill.

McCown, T. D., & Keith, A. (1939). *The Stone Age of Mount Carmel: The fossil human remains from the Levalloiso-Mousterian* (Vol. 2). Oxford: Clarendon Press.

Meindl, R. S., Mensforth, R. P., & York, H. P. (2001). Mortality, fertility, and growth in the Kentucky Late Archaic: The paleodemography of the Ward site. In O. H. Prufer, S. E. Pedde, & R. S. Meindl (Eds.), *Archaic transitions in Ohio and Kentucky prehistory* (pp. 87–109). Kent, OH: Kent State University Press.

Mellars, P. A. (2005). The impossible coincidence: A single species model for the origins of modern human behaviour in Europe. *Evolutionary Anthropology, 14*, 12–27.

Mellars, P. A. (2006). Archeology and the dispersal of modern humans in Europe: Deconstructing the "Aurignacian". *Evolutionary Anthropology, 15*, 167–182.

Miles, A. E. W. (1963). The dentition in the assessment of individual age in skeletal material. In D. R. Brothwell (Ed.), *Dental anthropology* (pp. 191–209). Oxford: Pergamon Press.

Minc, L. D. (1986). Scarcity and survival: The role of oral tradition in mediating subsistence crises. *Journal of Anthropological Archaeology, 5*, 39–113.

Morin, E., & Laroulandie, V. (2012). Presumed symbolic use of diurnal raptors by Neanderthals. *PLoS ONE, 7*, e32856.

Peresani, M., Fiore, I., Gala, M., Romandini, M., & Tagliacozzo, A. (2011). Late Neandertals and the intentional removal of feathers as evidenced from bird bone taphonomy at Fumane Cave 44 ky B.P., Italy. *Proceedings of the National Academy of Sciences USA, 108*, 3888–3893.

Ponce de León, M. S. P., Golovanova, L., Doronichev, V., Romanova, G., Akazawa, T., Kondo, O., et al. (2008). Neanderthal brain size at birth provides insights into the evolution of human life history. *Proceedings of the National Academy of Sciences USA, 105,* 13764–13768.

Powell, A., Shennan, S., & Thomas, M. G. (2009). Late Pleistocene demography and the appearance of modern human behavior. *Science, 324,* 1298–1301.

Radovčić, D., Sršen, A. O., Radovčić, J., & Frayer, D. W. (2015). Evidence for Neandertal jewelry: Modified white-tailed eagle claws at Krapina. *PLoS ONE, 10,* e0119802.

Rak, Y. (1990). On the differences between two pelvises of Mousterian context from the Qafzeh and Kebara caves, Israel. *American Journal of Physical Anthropology, 81,* 323–332.

Rak, Y. (1991a). Lucy's pelvic anatomy: Its role in bipedal gait. *Journal of Human Evolution, 20,* 283–290.

Rak, Y. (1991b). The pelvis. In O. Bar-Yosef & B. Vandermeersch (Eds.), *Le squelette Moustérien de Kébara 2 (Cahiers de Paléoanthropologie)* (pp. 147–156). Paris: Editions CNRS.

Rak, Y. (1993). Morphological variation in *Homo neanderthalensis* and *Homo sapiens* in the Levant: A biogeographic model. In W. H. Kimbel & L. B. Martin (Eds.), *Species, species concept and primate evolution* (pp. 523–536). New York: Plenum Press.

Rak, Y., & Arensburg, B. (1987). Kebara 2 Neanderthal pelvis: First look at a complete inlet. *American Journal of Physical Anthropology, 73,* 227–231.

Relethford, J. H. (2008). *The human species* (7th ed.). Boston: McGraw-Hill.

Richerson, P. J., Boyd, R., & Bettinger, R. L. (2009). Cultural innovations and demographic change. *Human Biology, 81,* 211–235.

Rink, W. J., Schwarcz, H. P., & Smith, F. H. (1995). ESR ages for Krapina hominids. *Nature, 378,* 24.

Robinson, J. R. (1972). *Early hominid posture and locomotion.* Chicago: University of Chicago Press.

Robson, A., & Kaplan, H. (2003). The evolution of human life expectancy and intelligence in hunter-gatherer economies. *American Economic Review, 93,* 150–169.

Roebroeks, W., Sier, M. J., Nielsen, T. K., De Loecker, D., Parés, J. M., Arps, C. E. S., et al. (2012). Use of red ochre by early Neandertals. *Proceedings of the National Academy of Sciences USA, 109,* 1889–1894.

Rosenberg, K. R. (1988). The functional significance of Neandertal pubic length. *Current Anthropology, 29,* 595–617.

Rosenberg, K. R. (2004). Living longer: Information revolution, population expansion, and modern human origins. *Proceedings of the National Academy of Sciences USA, 101,* 10847–10848.

Rosenberg, K. R., Lü, Z., & Ruff, C. B. (2006). Body size, body proportions, and encephalization in a Middle Pleistocene archaic human from northern China. *Proceedings of the National Academy of Sciences USA, 103,* 3552–3556.

Ruff, C. B. (1991). Climate, body size and body shape in hominid evolution. *Journal of Human Evolution, 21,* 81–105.

Ruff, C. B. (1994). Morphological adaptation to climate in modern and fossil hominids. *Yearbook of Physical Anthropology, 37,* 65–107.

Ruff, C. B. (1995). Biomechanics of the hip and birth in early *Homo. American Journal of Physical Anthropology, 98,* 527–574.

Ruff, C. B., & Walker, A. (1993). Body size and body shape. In A. Walker & R. E. Leakey (Eds.), *The Nariokotome Homo erectus skeleton* (pp. 234–265). Cambridge: Harvard University Press.

Sankararaman, S., Mallick, S., Dannemann, M., Prüfer, K., Kelso, J., Pääbo, S., et al. (2014). The genomic landscape of Neanderthal ancestry in present-day humans. *Nature, 507,* 354–357.

Sawyer, G. W., & Maley, B. (2005). Neanderthal reconstructed. *The Anatomical Record, 283B,* 23–31.

Shea, J. J. (2011). *Homo sapiens* is as *Homo sapiens* was: Behavioral variability versus "behavioral modernity" in Paleolithic archaeology. *Current Anthropology, 52,* 1–35.

Shennan, S. (2001). Demography and cultural innovation: A model and its implications for the emergence of modern human culture. *Cambridge Archaeological Journal, 11,* 5–16.

Simpson, S. W., Quade, J., Levin, N. E., Butler, R., Dupont-Nivet, G., Everett, M., et al. (2008). A female *Homo erectus* pelvis from Gona, Ethiopia. *Science, 322,* 1089–1092.

Smith, T. M., Machanda, Z., Bernard, A. B., Donovan, R. M., Papakyrikos, A. M., Muller, M. N., et al. (2013). First molar eruption, weaning, and life history in living wild chimpanzees. *Proceedings of the National Academy of Sciences USA, 110,* 2787–2791.

Smith, T. M., Tafforeau, P., Reid, D. J., Pouech, J., Lazzari, V., Zermeno, J. P., et al. (2010). Dental evidence for ontogenetic differences between modern humans and Neanderthals. *Proceedings of the National Academy of Sciences USA, 107,* 20923–20928.

Soressi, M., & d'Errico, F. (2007). Pigments, gravures, parures: les comportements symboliques controversés des néandertaliens. In B. Vandermeersch & B. Maureille (Eds), *Les néandertaliens. Biologie et cultures* (pp. 297–309). Paris: Éditions du CTHS.

Stewart, T. D. (1960). Form of the pubic bone in Neanderthal man. *Science, 131,* 1437–1438.

Stiner, M. C., Munro, N. D., Surovell, T. A., Tchernov, E., & Bar-Yosef, O. (1999). Paleolithic population growth pulses evidenced by small animal exploitation. *Science, 283,* 190–194.

Strimling, P., Enquist, M., & Eriksson, K. (2009). Repeated learning makes cultural evolution unique. *Proceedings of the National Academy of Sciences USA, 106,* 13870–13874.

Tague, R. G. (1989). Variation in pelvic size between males and females. *American Journal of Physical Anthropology, 80,* 59–71.

Tague, R. G. (1992). Sexual dimorphism in the human bony pelvis, with a consideration of the Neandertal pelvis from Kebara Cave, Israel. *American Journal of Physical Anthropology, 88,* 1–21.

Tague, R. G., & Lovejoy, C. O. (1986). The obstetric pelvis of AL 288-1 (Lucy). *Journal of Human Evolution, 15,* 237–255.

Tattersall, I. (1986). Species recognition in human paleontology. *Journal of Human Evolution, 15,* 165–175.

Tattersall, I. (1992). Species concepts and species identification in human evolution. *Journal of Human Evolution, 22,* 341–350.

Tattersall, I. (2002). *The monkey in the mirror. Essays on the science of what makes us human.* Oxford: Oxford University Press.

Templeton, A. R. (1998). Human races: A genetic and evolutionary perspective. *American Anthropologist, 100,* 632–650.

Templeton, A. (2002). Out of Africa again and again. *Nature, 416,* 45–51.

Templeton, A. R. (2013). Biological races in humans. *Studies in History and Philosophy of Science Part C: Studies in History and Philosophy of Biological and Biomedical Sciences, 44,* 262–271.

Teyssandier, N. (2008). Revolution or evolution: The emergence of the Upper Paleolithic in Europe. *World Archaeology, 40,* 493–519.

Trinkaus, E. (1984). Neandertal pubic morphology and gestation length. *Current Anthropology, 25,* 509–514.

Trinkaus, E. (2011). The postcranial dimensions of the La Chapelle aux Saints 1 Neandertal. *American Journal of Physical Anthropology, 14,* 461–468.

Van Arsdale, A. P. (2009). Reduced adult mortality and the expansion of the human bio-cultural niche in the Late Pleistocene. *American Journal of Physical Anthropology, 138,* 259.

Vandermeersch, B. (1991). La ceinture scapulaire et les member supérieurs. *Le squelette Moustérien de Kébara 2 (Cahiers de Paléoanthropologie)* (pp. 157–178). Paris: Editions du CNRS.

Vernot, B., & Akey, J. M. (2014). Resurrecting surviving Neandertal lineages from modern human genome. *Science, 343,* 1017–1021.

Villa, P., & Roebroeks, W. (2014). Neandertal demise: An archaeological analysis of the modern human superiority complex. *PLoS ONE, 9*, e96424.

Walker, A., & Ruff, C. B. (1993). Reconstruction of the pelvis. In A. Walker & R. E. Leakey (Eds.), *The Nariokotome Homo erectus skeleton* (pp. 221–233). Cambridge: Harvard University Press.

Weaver, T. D. (2009). The meaning of Neandertal skeletal morphology. *Proceedings of the National Academy of Sciences USA, 106,* 16028–16033.

Weaver, T. D., & Hublin, J.-J. (2009). Neandertal birth canal shape and the evolution of human childbirth. *Proceedings of the National Academy of Sciences USA, 106,* 8151–8156.

Weidenreich, F. (1943). The skull of *Sinanthropus pekinensis*: A comparative study of a primitive hominid skull. *Palaeontologia Sinica*, New Series D, Number 10.

Weidenreich, F. (1946). Generic, specific, and subspecific characters in human evolution. *American Journal of Physical Anthropology, 4,* 413–431.

Weidenreich, F. (1947). Are human races in the taxonomic sense "races" or "species"? *American Journal of Physical Anthropology, 5,* 369–371.

Wildman, D. E., Uddin, M., Liu, G., Grossman, L. I., & Goodman, M. (2003). Implications of natural selection in shaping 99.4% non-synonymous DNA identity between humans and chimpanzees: enlarging genus *Homo*. *Proceedings of the National Academy of Sciences USA, 100,* 7181–7188.

Wilson, E. O., & Brown, W. L. (1953). The subspecies concept and its taxonomic application. *Systematic Zoology, 2,* 97–111.

Wobst, H. M. (1977). Stylistic behavior and information exchange. In C. E. Cleland (Ed.), *For the Director: Research essays in honor of James B. Griffin* (pp. 317–342) (Anthropological Papers Number 61). Ann Arbor: Museum of Anthropology.

Wolpoff, M. H. (1979). The Krapina dental remains. *American Journal of Physical Anthropology, 50,* 67–113.

Wolpoff, M. H. (1980). *Paleoanthropology*. New York: Knopf.

Wolpoff, M. H. (2009). How Neandertals inform human variation. *American Journal of Physical Anthropology, 139,* 91–102.

Wolpoff, M. H., & Caspari, R. (1997). *Race and human evolution*. New York: Simon and Schuster.

Wolpoff, M. H., & Caspari, R. (2006). Does Krapina reflect early Neandertal paleodemography? *Periodicum Biologorum, 108,* 425–432.

Wolpoff, M. H., & Caspari, R. (2013). Paleoanthropology and race. In D. Begun (Ed.), *A companion to paleoanthropology* (pp. 321–338). London: Wiley-Blackwell.

Wolpoff, M. H., Thorne, A. G., Jelínek, J., & Zhang, Y.-Y. (1994). The case for sinking *Homo erectus*. 100 years of *Pithecanthropus* is enough! *Courier Forschungsinstitut Senckenberg, 171,* 341–361.

Zilhão, J. (2007). The emergence of ornaments and art: An archaeological perspective on the origins of "behavioral modernity". *Journal of Archaeological Research, 15,* 1–54.

Zilhão, J., Angelucci, D. E., Badal-García, E., d'Errico, F., Daniel, F., & Dayet, L. (2010). Symbolic use of marine shells and mineral pigments by Iberian Neandertals. *Proceedings of the National Academy of Sciences USA, 107,* 1023–1028.

Index

Note: Page numbers followed by *f*, *t* and n indicate figures, tables and footnotes, respectively

A

Acetabulum, 57*f*, 58–60, 60*f*, 61*f*, 62, 63*f*, 66, 179n, 241, 254, 257n
Acheulian, 191
Acheulo-Yabrudian, 187–191, 191*t*, 193*f*, 197
Adaptation, 3, 6, 7, 12n, 17, 22, 53–56, 62, 64, 66–69, 103, 104, 109, 114, 118, 120, 121, 161, 162, 167, 169, 173, 175, 176, 255, 264, 266
Adaptive zone, 3, 4
Amud, 175, 177*t*, 194*t*, 198, 205*t*, 211*t*, 217, 228*t*
Amud 1 (fossil), 175, 177*t*, 205*t*, 211*t*
Analogy, 103
Apes, 2, 4, 11–14, 21, 22, 22n, 23, 46, 47*f*, 56, 59, 62, 64, 69, 76, 79, 80, 82, 84, 89, 113, 128, 135, 141, 234, 235, 236
Apomorphy, 53, 127, 157
Arago cave, 146
Ardipithecus, 54, 87, 92, 128
Artificial selection, 16, 23
Australopith(ecine)
    gracile, 50, 100–102
    megadont, 95, 100, 101
    robust, 50, 71, 72, 91, 92, 100–102, 105, 109, 120, 134
Australopithecus
    afarensis, 2*f*, 23, 31, 34*t*, 42, 46*f*, 48*f*, 51*t*, 87, 92, 112*f*, 127, 128, 134*f*, 134–140, 141–143*t*, 233, 234, 256
    africanus, 22, 47, 50, 51*t*, 96*t*, 98–100, 102, 104, 112*f*, 112, 113, 120, 134, 134*f*, 135, 135*t*
    anamensis, 87, 92, 119, 128, 134, 135, 138, 141
    garhi, 95, 100
    sediba, 2, 48, 50, 51, 256
Autapomorphy, 225

B

Bayesian analysis, 215, 218
Bending, 110–113, 113*f*, 114, 114*f*, 115*f*, 176
Biogeography, 104
Biomechanics, 109
Bipedality, 53, 54, 56, 62, 64, 66–69
Bite force, 71, 73, 85, 85*f*, 86, 87, 89, 91, 92, 118, 121, 156
Bodo, 145, 146, 151–154, 154*t*, 155*t*, 156, 157
Body size, 7, 60, 61*f*, 62*f*, 63*f*, 114–116, 161, 162, 172, 173, 176, 177, 178*f*, 183, 255, 260
Bovids, 31, 33, 36, 37, 37*f*, 41, 54, 56, 58*f*, 59, 60, 61*f*, 62, 62*f*, 63*f*, 64, 64*t*, 65*f*, 68, 103, 104
Brain
    lateralization, 235
    size, 1, 4, 6, 7, 23, 46, 103, 145, 146, 155*t*, 260

C

Canine
    reduction, 71, 87, 88, 92, 127, 141
Carbon isotopes, 36, 37*f*, 104, 119
Catarrhines, 71–73, 75, 77*t*, 78, 78*t*, 79, 81, 85, 88*f*, 89–92
Chesowanja, 96*t*, 97, 99*t*, 104
Chimpanzee, 2*f*, 3, 22, 46, 47*f*, 47, 49, 49*f*, 50*f*, 51, 53, 104, 118, 127, 128, 128*t*, 134, 135, 135*t*, 136, 136n, 137, 138, 138n, 139, 140, 234, 265
Chin, 205, 206, 211
Cladistic analysis, 103
Climate
    change, 33, 41
Cognition, 4
Cognitive abilities, 18
Cold adaptation theory, 255
Convergence, 55, 103
Cortex
    frontal, 47, 49, 51*t*
    motor, 235
Cranial base, 79, 97, 100, 101, 145, 153, 155*t*, 156, 234
Cross sectional area
    geometry, 85
    second moment, 176, 178
C-3 foods, 104
CT scanning, 147
Cultural evolution, 4, 11, 18, 18n, 19, 26, 265

D

Darwin, Charles Robert, 11, 12, 12n, 13, 14, 14n, 15, 15n, 16, 17, 17n, 18, 18n, 19–23, 25, 26, 26n, 71
Demography, 253, 261, 268
Dental modernity, 216, 225
Dentition, 50, 76*f*, 96, 96*t*, 97–101, 101*t*, 102, 104, 109, 110*f*, 118, 120, 121, 127, 128, 135, 136, 141, 143, 206, 215–217, 261. *See also* megadontia
Diet, 3, 36, 59*t*, 86, 87, 92, 101*t*, 104, 105, 109–111, 113, 114n, 117–121, 127
Dikika, 233, 234
Dispersal, 40, 104, 215, 216, 223, 267
Dmanisi, 6, 161, 162, 163*t*, 164, 164–168*t*, 167–169, 170*t*, 171, 172, 257*t*

Broken Hill section

Broken Hill, 145–147, 147*f*, 148–151, 151*f*, 152–154, 154*t*, 155, 155*t*, 156, 157

© Springer International Publishing AG 2017
Assaf Marom and Erella Hovers (eds.), *Human Paleontology and Prehistory*,
Vertebrate Paleobiology and Paleoanthropology, DOI: 10.1007/978-3-319-46646-0

DNA
   mitochondrial, *see mtDNA*
   nuclear, 217, 235, 266
Dobzhansky, Theodosius, 21, 25, 26, 267
Domestication, 23
Drimolen, 96, 96*t*, 99*t*, 100
Durophagy, 109, 118, 120

E

Early *Homo, see Homo*
Ecosystem, 31, 33, 38, 40–42
El Sidrón, 234
Enamel
   microwear, 104, 109, 119, 120
   thickness, 87, 120
Endocast, 45, 46*f*, 47*f*, 47–50, 49–51*f*, 235
Endocranial volume, 7, 8, 45, 153
Encephalization, 260
Eugenics, 24, 25, 25n, 27
Evolutionary psychology, 1, 4
Evolutionary rates, 1, 2, 4, 6, 7, 11, 14n, 19

F

Facial skeleton, 146, 149, 150, 152, 154, 155*t*, 156
Feeding behavior, 56, 66, 109, 117, 118, 135, 141
Femur, 54*f*, 56, 57*f*, 59, 62, 100, 162, 163, 163*t*, 164, 175, 176, 178*t*,
   182*t*, 183*f*, 257*t*, 257n
Fertility, 264, 265
fMRI, 235
Foramen spinosum, 148
FOXP2 gene, 236, 266

G

Galton, Francis, 16, 17, 21, 23–25
Gape, 71–76, 72*f*, 77*t*, 78*t*, 79–82, 80*t*, 81–85*f*, 84–87, 88*f*, 88–92
Gene flow, 215, 222, 265–267
Generalized Procrustes Analysis, 157, 206
Genetics, 25, 28
Genius, 11, 15, 15n, 16, 23, 24, 192
Genome, 22n, 28
Geology, 26, 27
Geometric morphometric method, 157, 203, 204, 207
Gerenuk, 53–56, 57*f*, 58–60, 62, 62*f*, 63*f*, 64–67, 68*f*. *See also*
   Lithocranius valleri
Giraffids, 36
Gondolin, 96, 96*t*, 99*t*, 100
Gorilla, 22, 23, 49*f*, 53, 76, 77*t*, 79*t*, 82, 101*t*, 104, 116*f*, 117*f*, 134, 267
Gould, Stephen Jay, 11, 17–20, 22, 101, 114

H

Habitual activity, 175
Hadar
   formation, 128, 131
   fossils, 133–136, 138–139
Haeckel, Ernest, 22, 26
Haldane, John B. Sanderson, 25, 26, 28
Handaxe, 5, 190, 191, 191*t*, 192, 193*f*
Hayonim Cave, 198
Heat
   loss, 161, 162, 166–169, 167*t*, 168*t*, 170*t*, 171–173, 171*t*
   transfer, 166, 167
Herbivores, 36, 41, 90, 91

Hominids, 3–7, 22, 23, 67, 233, 234, 255, 256, 259, 260, 263
Hominins, 31, 32*f*, 33, 49, 53, 54, 56, 66, 71, 86–89, 92, 95, 96, 98,
   99*t*, 100, 102–105, 116, 120, 127, 128, 135, 139, 140, 145, 146,
   148–150, 153, 154, 156, 157, 162, 167, 169, 172, 176, 181, 184,
   199, 216, 239, 240
Hominoids, 2, 64, 66, 103, 104, 117, 117*f*, 121, 134, 135
Homo
   early, 6, 51*t*, 115, 216n, 255, 256
   erectus, 51*t*, 54, 115, 145–150, 147*f*, 152–154, 156, 161, 205, 256, 265
   floresiensis, 216n
   heidelbergensis, 145, 146, 157, 204, 205
   neanderthalensis, 115, 145, 146, 161, 162, 163*t*, 164*t*, 167, 169,
      171–173, 217, 218. *See also* Neanderthal
   rhodesiensis, 145, 146
   sapiens, 1, 2, 4–8, 19, 22n, 51*t*, 78*t*, 79*t*, 128*t*, 146, 148, 149, 153,
      205, 215, 216, 265–267
Homoplasy, 102, 103, 105
Hooker, Joseph, 22, 64, 67
Horotely, 2
Human brain, 1, 7, 8, 13, 14, 46
Human races, 11, 13, 17, 23, 265–267
Humerus, 162, 163*t*, 175, 176, 178*t*, 180*t*, 181*f*, 183, 184, 257*t*
Huxley, Thomas Henry, 14n, 19, 22, 26
Hybridization, 205, 212, 225
Hylobates, 2, 4*f*, 77*t*, 78*t*, 80, 81
Hyoid, 233, 234*f*, 236
Hypodigm, 95, 97–99, 99*t*, 100, 102, 112, 115, 145, 146

I

Ilium, 56, 57, 60*f*, 62, 62*f*
Intelligence, 11, 16–19, 23, 25, 26
Inuits, 163, 163*t*, 164, 164–168*t*, 169, 170*t*, 177, 177*t*, 179*t*, 180*t*, 182*t*,
   255
Ischium, 56, 57, 60*f*, 62

J

James, William, 17

K

Kabwe, 145
Kebara, 175, 177*t*, 180*t*, 181, 182*t*, 184, 198, 217, 228*t*, 233, 234, 234*f*,
   236, 239–244, 242*f*, 243*f*, 243n, 244*t*, 245*t*, 246*f*, 247*f*, 247*t*,
   246–250, 248*t*, 249*f*, 254, 256, 257*t*, 257n, 260
Kebara 2 Neanderthal, 177*t*, 179n, 184, 228*t*, 233, 234, 234*f*, 239, 240,
   244*t*, 247*t*, 248, 248*t*, 249, 254, 256, 260
King, William, 265
Koobi Fora formation, 96*t*, 97, 99*t*, 104
Kromdraai, 95, 96, 96*t*, 98, 99*t*

L

Lamarck, Jean-Baptiste, 21, 23n
Laminar technology, 187, 197–199
Language, 11, 12, 12n, 13, 14, 18, 19, 19n, 26, 233, 235, 236
Levallois technology, 187, 197–199
Levant, 184, 187, 197–199, 204, 254
Limb proportions, 1, 167, 172
Lithic technology, 187, 198
Lithocranius valleri, see Gerenuk
Locomotion, 1, 3, 53, 54, 56, 64, 69, 162, 167, 173
Long bone robusticity, 176
Lordosis, 53, 54*f*, 55, 55*f*, 64, 66, 239–242, 240*f*, 243*f*, 244*t*, 248*t*,
   248–250

Lower-Middle Paleolithic transition, 175, 187, 188, 190–192, 197
Lyell, Charles, 21, 26, 26n, 27, 28

M

Macaca fascicularis, 71, 72, 72*f*, 89
Mandible, 72, 73, 73n, 74*f*, 85, 86, 89, 96, 96*t*, 97, 98, 99*t*, 100–102, 109, 110, 110*f*, 111, 112, 113*f*, 114*f*, 114, 115*f*, 115–117, 117*f*, 118, 121, 127–129, 133–136, 138, 140, 141, 145, 146, 156, 203, 204, 204*f*, 205–208, 212, 219*t*, 222, 225, 233, 235*f*, 261
Mandibular
    alveolus, 128–131, 135, 146
    condyle, 72, 73, 73n, 74, 75, 90, 91, 131, 133, 205, 206*f*, 212
    hypertrophy, 109, 121
    notch, 133, 134, 205, 206*f*, 207, 211
    ramus, 101*t*, 110, 131, 135, 205, 206*f*
    robusticity index, 116, 117, 117*f*
    symphysis, 141, 211
Mastication, 88, 101*t*, 109–112, 118–120
Material culture, 1, 6, 7
Mauer, 145, 146, 205*t*, 211*t*
Mayr, Ernst Walter, 266–268
Megadontia, 102, 103, 110, 111, 117, 120, 121
Mendel, Gregor, 18, 24
Mental foramen, 132, 133, 205, 207
Metabolic rate, 166
Metabolism, 166
Metopic suture, 47, 48*f*
Misliya, 187, 190, 191*t*, 191, 192, 194, 194*t*, 195*f*, 197–199
Morphocline, 103, 104
Monophyly, 101*f*, 102–105
Morphological change, 4
Mortality curves, 263, 263*f*
Mount Carmel, 187, 188, 197, 267
Mousterian, 187–190, 194*t*, 197–199, 260
mtDNA, 22n
Mutation, 19, 236

N

Nariokotome, 256, 257n
Natural selection, 5, 11–18, 22, 23, 26, 26n, 103
Naturalism
    birth canal, 254, 258–260
    body shape, 253, 260
    population structure, 253, 260, 265, 267, 268
Neanderthal, 6, 8, 22, 22n, 115, 145, 146, 149–151, 156, 157, 161, 162, 163*t*, 164*t*, 167–169, 171*t*, 171–173, 175–177, 177*t*, 179, 179*t*, 179n, 180, 180*t*, 181*f*, 181, 182*t*, 183*f*, 183, 184, 203–205, 205*t*, 207, 212, 215–218, 220, 222, 222*f*, 223*f*, 223–225, 226–230*t*, 233–235, 235*f*, 236, 239, 240, 244*t*, 247*t*, 248, 248*t*, 249, 253–256, 255*f*, 255n, 256*f*, 257n, 260, 261, 263–268
Near East, 203
Neo-Darwinian synthesis, 17n
Ngandong, 147, 153

O

Obstetric constraints, 260
Ohalo 2 fossil, 175, 177*t*
Olduvai Gorge, 22, 96, 96*t*, 98, 104
Omo, 97, 99*t*, 100, 112, 113*f*, 153
Ontogeny, 127, 128, 135–141, 138*f*, 139*f*
Out of Africa, 6, 215, 216
Oxygen isotopes, 37, 38, 38*t*, 203

P

Paleoecology, 31, 32, 33, 37, 38, 40, 41
Paleolithic
    lower, 88, 187, 190, 191, 197
    middle, 175, 179, 184, 187–189, 192, 194, 197, 198, 233, 263
Paleoneurology, 45, 46
Pan, 2, 2*f*, 77*t*, 79*t*, 81, 89, 90, 115–117*f*, 128, 128*t*, 136. *See also* Chimpanzee
    paniscus, 77*t*, 79*t*, 81, 128
    troglodytes, 2*f*, 77*t*, 79*t*, 89, 90, 128, 128*t*
Parallelism, 103
Paranthropus
    aethiopicus, 31, 34*t*, 96*t*, 97*f*, 98*f*, 99*t*, 100
    boisei, 96*t*, 97*f*, 98*f*, 98, 99*t*, 100, 101*t*, 102, 109, 110*f*, 111, 112*f*, 114*f*, 117*f*, 121. *See also Zinjanthropus boisei*
    crassidens, 98
    robustus, 95, 96*t*, 97*f*, 98*f*, 99*t*, 101*t*, 102, 112
PCA, *see* Principal Component Analysis
Peking Man, 266
Pelvic inlet, 254, 256, 259, 260
Pelvis, 54*f*, 56, 57*f*, 58, 60, 60*f*, 239–241, 242*f*, 243, 247, 248, 250, 253, 254, 255*f*, 255, 256, 256*f*, 257*t*, 257n, 259, 260
Petralona, 145, 146, 150–154, 151*f*, 154*t*, 155*t*, 156, 157
Phylogeny, 3*f*, 76, 103, 216, 263
Pleistocene
    early, 206, 212
    middle, 145, 146, 154, 157, 205, 205*t*, 211, 212, 225, 248
    late, 7, 146, 157, 175, 206, 215, 225, 265
Plesiomorphy, 6, 157, 253, 255
Pliocene, 31, 32*f*, 33, 128
Population genetics, 28
Posture, 11, 13, 14n, 21, 23, 53, 54, 54*f*, 55, 56, 58, 59, 64, 66, 69, 239–241, 244*t*, 247–249, 254
Primates, 6, 14n, 23, 26, 34*t*, 54, 56, 60, 64, 66, 69, 71–73, 92, 109, 114n, 120, 127, 135, 140
Principal Component Analysis, 157, 207, 209*f*, 210*f*, 211, 212
Pubis, 57, 60*f*, 242*f*, 254, 255, 257n

Q

Qafzeh, 175–177, 177*t*, 179*t*, 179n, 180, 180*t*, 181*f*, 181, 182*t*, 183, 183*f*, 184
Qafzeh-Skhul group, 175–177, 177*t*, 179*t*, 180*t*, 181, 181*t*, 182*t*, 183, 183*f*, 184

R

Radius, 162, 163*t*, 175, 176, 178*f*, 178*t*, 180, 180*t*, 181*f*, 183, 184
Reproduction, 16, 17, 19, 25, 264

S

Sacrum, 239–241, 242*f*, 243*f*, 244, 247*f*, 249, 250, 254, 257
Sangiran, 147, 153
Scoliosis, 240, 241, 246n
Segmented model, 166
Sexual dimorphism, 87, 96*t*, 110, 115*f*, 206
Shanidar, 162, 163–168*t*, 169, 170*t*, 171, 177*t*
Shape analysis, 203–206, 209*f*, 212
Shungura Formation, 96*t*, 97, 100
Sima de los Huesos, 145, 146, 148, 153, 157, 205*t*, 216, 233, 234, 248, 256
Skhul, 175, 176, 177*t*, 179*t*, 180*t*, 181*f*, 182*t*, 183*f*, 184, 204, 205, 212, 216, 254, 267
Skin temperature, 161, 162, 166, 167, 172, 173

Speciation, 7, 31, 41, 145, 146
Sphenoid, 145, 146, 148, 149, 151, 156
Spinal
    column, 239, 240f
    curvature, 239–241, 247, 249f
Spino-pelvic morphology, 239, 247
Spinous process, 55f, 66, 241, 244, 246n
Sterkfontein, 96, 98, 257n
Sulcus
    intraparietal, 46, 47f, 50f
    lunate, 46f, 47f, 48, 49f, 50
    lateral calcarine, 48, 49f
Supra-laryngeal space, 233, 234
Swartkrans, 95, 96t, 98, 99t, 134
Swing pattern, 166
Synapomorphy, 102
Systematics, 145

T
Tabun, 217, 224, 253–254, 267
    Tabun C1, 13t, 181, 203–205, 207, 211t, 212, 228t, 257t
    Tabun C2, 203–207, 209f, 211, 212, 220t, 228t
Tachytely, 3, 4, 6, 7
Taung, 22, 45, 47, 48, 48f, 51f, 98, 133, 134, 134f
Taxonomy, 21, 98, 265, 267
Thermoregulation, 161, 162, 168, 169, 171–173, 255, 260
Tibia, 54f, 162, 163t, 175, 176, 178t, 182t, 183f, 257t
Tool making, 5

Tool use, 12, 18
Torsion, 112, 114, 115f, 178t
Trait
    derived, 156, 215, 216, 222, 225, 254, 255
    diagnostic, 222, 225
    primitive, 156, 212, 215, 216, 222, 225

V
Variation
    genetic, 19n, 23, 173, 216, 261, 267
    intraspecific, 135, 136
    phenotypic, 23, 173
Vertebra, 60f, 64, 65f, 66, 67f, 68f, 234, 239–244, 245t, 246n, 248t, 249
Vocal capacity, 12, 234–236
Vocal tract, 233–235

W
Walking cycle, 165, 166
Wallace, Alfred Russel, 13, 17
Weidenreich, Franz, 128, 149, 150, 153, 265–267
West Turkana, 96t, 97, 98, 99t, 100, 162, 167, 256

Z
Zhoukoudian, 147, 149, 150, 153, 265
Zinjanthropus boisei, 95, 96t, 98
Zygomatic arch, 150, 156

The manufacturer's authorised representative in the EU is Springer
Nature Customer Service Centre GmbH, Europaplatz 3, 69115 Heidelberg,
Germany. If you have any concerns regarding our products, please
contact ProductSafety@springernature.com

Printed and bound by CPI Group (UK) Ltd, Croydon, CR0 4YY

06/05/2026

02103609-0001